Lecture Notes in Computer Science

Lecture Notes in Computer Science

Vol. 60: Operating Systems, An Advanced Course. Edited by R. Bayer, R. M. Graham, and G. Seegmüller. X, 593 pages. 1978.

Vol. 61: The Vienna Development Method: The Meta-Language. Edited by D. Bjørner and C. B. Jones. XVIII, 382 pages. 1978.

Vol. 62: Automata, Languages and Programming. Proceedings 1978. Edited by G. Ausiello and C. Böhm. VIII, 508 pages. 1978.

Vol. 63: Natural Language Communication with Computers. Edited by Leonard Bolc. VI, 292 pages. 1978.

Vol. 64: Mathematical Foundations of Computer Science. Proceedings 1978. Edited by J. Winkowski. X, 551 pages. 1978.

Vol. 65: Information Systems Methodology, Proceedings, 1978. Edited by G. Bracchi and P. C. Lockemann. XII, 696 pages. 1978.

Vol. 66: N. D. Jones and S. S. Muchnick, TEMPO: A Unified Treatment of Binding Time and Parameter Passing Concepts in Programming Languages. IX, 118 pages. 1978.

Vol. 67: Theoretical Computer Science, 4th GI Conference, Aachen, March 1979. Edited by K. Weihrauch. VII, 324 pages. 1979.

Vol. 68: D. Harel, First-Order Dynamic Logic. X, 133 pages. 1979.

Vol. 69: Program Construction. International Summer School. Edited by F. L. Bauer and M. Broy. VII, 651 pages. 1979.

Vol. 70: Semantics of Concurrent Computation. Proceedings 1979. Edited by G. Kahn. VI, 368 pages. 1979.

Vol. 71: Automata, Languages and Programming. Proceedings 1979. Edited by H. A. Maurer. IX, 684 pages. 1979.

Vol. 72: Symbolic and Algebraic Computation. Proceedings 1979. Edited by E. W. Ng. XV, 557 pages. 1979.

Vol. 73: Graph-Grammars and Their Application to Computer Science and Biology. Proceedings 1978. Edited by V. Claus, H. Ehrig and G. Rozenberg. VII, 477 pages. 1979.

Vol. 74: Mathematical Foundations of Computer Science. Proceedings 1979. Edited by J. Bečvář. IX, 580 pages. 1979.

Vol. 75: Mathematical Studies of Information Processing. Proceedings 1978. Edited by E. K. Blum, M. Paul and S. Takasu. VIII, 629 pages. 1979.

Vol. 76: Codes for Boundary-Value Problems in Ordinary Differential Equations. Proceedings 1978. Edited by B. Childs et al. VIII, 388 pages. 1979.

Vol. 77: G. V. Bochmann, Architecture of Distributed Computer Systems. VIII, 238 pages. 1979.

Vol. 78: M. Gordon, R. Milner and C. Wadsworth, Edinburgh LCF. VIII, 159 pages. 1979.

Vol. 79: Language Design and Programming Methodology. Proceedings, 1979. Edited by J. Tobias. IX, 255 pages. 1980.

Vol. 80: Pictorial Information Systems. Edited by S. K. Chang and K. S. Fu. IX, 445 pages. 1980.

Vol. 81: Data Base Techniques for Pictorial Applications. Proceedings, 1979. Edited by A. Blaser. XI, 599 pages. 1980.

Vol. 82: J. G. Sanderson, A Relational Theory of Computing. VI, 147 pages. 1980.

Vol. 83: International Symposium Programming. Proceedings, 1980. Edited by B. Robinet. VII, 341 pages. 1980.

Vol. 84: Net Theory and Applications. Proceedings, 1979. Edited by W. Brauer. XIII, 537 Seiten. 1980.

Vol. 85: Automata, Languages and Programming. Proceedings, 1980. Edited by J. de Bakker and J. van Leeuwen. VIII, 671 pages. 1980.

Vol. 86: Abstract Software Specifications. Proceedings, 1979. Edited by D. Bjørner. XIII, 567 pages. 1980

Vol. 87: 5th Conference on Automated Deduction. Proceedings, 1980. Edited by W. Bibel and R. Kowalski. VII, 385 pages. 1980.

Vol. 88: Mathematical Foundations of Computer Science 19 Proceedings, 1980. Edited by P. Dembiński. VIII, 723 pages. 19

Vol. 89: Computer Aided Design - Modelling, Systems Engineeri CAD-Systems. Proceedings, 1980. Edited by J. Encarnacao. 461 pages. 1980.

Vol. 90: D. M. Sandford, Using Sophisticated Models in Re lution Theorem Proving.
XI, 239 pages. 1980

Vol. 91: D. Wood, Grammar and L Forms: An Introduction. IX, pages. 1980.

Vol. 92: R. Milner, A Calculus of Communication Systems. VI, pages. 1980.

Vol. 93: A. Nijholt, Context-Free Grammars: Covers, Normal Fo and Parsing. VII, 253 pages. 1980.

Vol. 94: Semantics-Directed Compiler Generation. Proceedi 1980. Edited by N. D. Jones. V, 489 pages. 1980.

Vol. 95: Ch. D. Marlin, Coroutines. XII, 246 pages. 1980.

Vol. 96: J. L. Peterson, Computer Programs for Spelling Correc VI, 213 pages. 1980.

Vol. 97: S. Osaki and T. Nishio, Reliability Evaluation of Some F Tolerant Computer Architectures. VI, 129 pages. 1980.

Vol. 98: Towards a Formal Description of Ada. Edited by D. Bjø and O. N. Oest. XIV, 630 pages. 1980.

Vol. 99: I. Guessarian, Algebraic Semantics. XI, 158 pages. 1

Vol. 100: Graphtheoretic Concepts in Computer Science. Edite H. Noltemeier. X, 403 pages. 1981.

Vol. 101: A. Thayse, Boolean Calculus of Differences. VII, 144 pa 1981.

Vol. 102: J. H. Davenport, On the Integration of Algebraic Functi 1–197 pages. 1981.

Vol. 103: H. Ledgard, A. Singer, J. Whiteside, Directions in Hu Factors of Interactive Systems. VI, 190 pages. 1981.

Vol. 104: Theoretical Computer Science. Ed. by P. Deussen. 261 pages. 1981.

Vol. 105: B. W. Lampson, M. Paul, H. J. Siegert, Distributed Syste Architecture and Implementation. XIII, 510 pages. 1981.

Vol. 106: The Programming Language Ada. Reference Manua 243 pages. 1981.

Vol. 107: International Colloquium on Formalization of Program Concepts. Proceedings. Edited by J. Diaz and I. Ramos. VII, pages. 1981.

Vol. 108: Graph Theory and Algorithms. Edited by N. Saito T. Nishizeki. VI, 216 pages. 1981.

Vol. 109: Digital Image Processing Systems. Edited by L. Bolc Zenon Kulpa. V, 353 pages. 1981.

Vol. 110: W. Dehning, H. Essig, S. Maass, The Adaptation of V Man-Computer Interfaces to User Requirements in Dialogs. X pages. 1981.

Vol. 111: CONPAR 81. Edited by W. Händler. XI, 508 pages.

Vol. 112: CAAP '81. Proceedings. Edited by G. Astesiano and C. B VI, 364 pages. 1981.

Vol. 113: E.-E. Doberkat, Stochastic Automata: Stability, Nond minism, and Prediction. IX, 135 pages. 1981.

Vol. 114: B. Liskov, CLU, Reference Manual. VIII, 190 pages.

Vol. 115: Automata, Languages and Programming. Edited by S. and O. Kariv. VIII, 552 pages. 1981.

Vol. 116: M. A. Casanova, The Concurrency Control Proble Database Systems. VII, 175 pages. 1981.

Lecture Notes in Computer Science

Edited by G. Goos and J. Hartmanis

170

7th International Conference on Automated Deduction

Napa, California, USA
May 14–16, 1984
Proceedings

Edited by R. E. Shostak

Springer-Verlag
Berlin Heidelberg New York Tokyo 1984

Editor

R. E. Shostak
SRI International
333 Ravenswood Avenue
Menlo Park, CA 94025
U.S.A.

Library of Congress Cataloging in Publication Data
International Conference on Automated Deduction
(7th : 1984 : Napa, Calif.)
Seventh International Conference on Automated
Deduction.
(Lecture notes in computer science ; 170)
1. Automatic theorem proving–Congresses.
2. Logic, Symbolic and mathematical–Congresses.
I. Shostak, Robert. II. Title.
III. 7th International Conference on Automated
Deduction. IV. Series.
QA76.9.A96I58 1984 511.3 84-5441

CR Subject Classification (1982): I.1, J.2.

Printed and bound by R.R. Donnelley & Sons, Harrisonburg, Viriginia.
Printed in the United States of America.

9 8 7 6 5 4 3 2 1

3-540-96022-8 Springer-Verlag Berlin Heidelberg New York Tokyo
0-387-96022-8 Springer-Verlag New York Heidelberg Berlin Tokyo

FOREWORD

The Seventh International Conference on Automated Deduction was held May 14-16, 1984, in Napa, California. The conference is the primary forum for reporting research in all aspects of automated deduction, including the design, implementation, and applications of theorem-proving systems, knowledge representation and retrieval, program verification, logic programming, formal specification, program synthesis, and related areas.

The presented papers include 27 selected by the program committee, an invited keynote address by Jorg Siekmann, and an invited banquet address by Patrick Suppes. Contributions were presented by authors from Canada, France, Spain, the United Kingdom, the United States, and West Germany.

The first conference in this series was held a decade earlier in Argonne, Illinois. Following the Argonne conference were meetings in Oberwolfach, West Germany (1976), Cambridge, Massachusetts (1977), Austin, Texas (1979), Les Arcs, France (1980), and New York, New York (1982).

Program Committee

P. Andrews (CMU)
W.W. Bledsoe (U. Texas) *past chairman*
L. Henschen (Northwestern)
G. Huet (INRIA)
D. Loveland (Duke) *past chairman*
R. Milner (Edinburgh)
R. Overbeek (Argonne)
T. Pietrzykowski (Acadia)
D. Plaisted (U. Illinois)
V. Pratt (Stanford)
R. Shostak (SRI) *chairman*
J. Siekmann (U. Kaiserslautern)
R. Waldinger (SRI)

Local Arrangements

R. Schwartz (SRI)

CONTENTS

7th International Conference on Automated Deduction

UNIVERSAL UNIFICATION

Jörg H. Siekmann
Universität Kaiserslautern
FB Informatik
Postfach 3049
D-6750 Kaiserslautern

Überhaupt hat der Fortschritt das an sich, daß er viel größer ausschaut, als er wirklich ist.
J.N. Nestroy, 1859

ABSTRACT: This article surveys what is presently known about first order unification theory.

CONTENTS

0. INTRODUCTION

Unification theory is concerned with problems of the following kind: Let f and g be function symbols, a and b constants and let x and y be variables and consider two *first order terms* built from these symbols; for example:

$$t_1 = f(x,g(a,b))$$
$$t_2 = f(g(y,b),x).$$

The first question which arises is whether or not there exist terms which can be substituted for the variables x and y such that the two terms thus obtained from t_1 and t_2 become equal: in the example g(a,b) and a are two such terms. We shall write

$$\sigma_1 = \{x \leftarrow g(a,b), y \leftarrow a\}$$

for such a unifying substitution : σ_1 is a *unifier* of t_1 and t_2 since $\sigma_1 t_1 = \sigma_1 t_2$.

In addition to the *decision problem* there is also the problem of finding a *unification algorithm* which generates the unifiers for a given pair t_1 and t_2.

Consider a variation of the above problem, which arises when we assume that f is commutative:

(C) $$f(x,y) = f(y,x).$$

Now σ_1 is still a unifying substitution and moreover $\sigma_2 = \{y \leftarrow a\}$ is also a unifier for t_1 and t_2, since

$$\sigma_2 t_1 = f(x,g(a,b)) =_C f(g(a,b),x) = \sigma_2 t_2.$$

But σ_2 is *more general* than σ_1, since σ_1 is an instance of σ_2 obtained as the *composition* $\lambda \circ \sigma_2$ with $\lambda = \{x \leftarrow g(a,b)\}$; hence a unification algorithm only needs to compute σ_2.

There are pairs of terms which have more than one most general unifier (i.e. they are not an instance of any other unifier) under commutativity, but they always have at most *finitely many*. This is in contrast to the first situation (of free terms), where every pair of terms has at most *one* most general unifying substitution.

The problem becomes entirely different when we assume that the function denoted by f is associative:

(A) $$f(x,f(y,z)) = f(f(x,y),z).$$

In that case σ_1 is still a unifying substitution, but

$$\sigma_3 = \{x \leftarrow f(g(a,b),\ g(a,b)),\ y \leftarrow a\}$$

is also a unifier:

$$\sigma_3 t_1 = f(f(g(a,b),\ g(a,b)),\ g(a,b)) =_A f(g(a,b),\ f(g(a,b),\ g(a,b))) = \sigma_3 t_2 \quad .$$

But $\sigma_4 = \{x \leftarrow f(g(a,b),\ f(g(a,b),\ g(a,b))),\ y \leftarrow a\}$ is again a unifying substitution and it is not difficult to see that there are *infinitely many* unifiers, all of which are most general.

Finally, if we assume that both axioms (A) and (C) hold for f then the situation changes yet again and for any pair of terms there are at most *finitely many* most general unifiers *under (A) and (C).*

The above examples as well as the practical applications of unification theory quoted in the following paragraph share a common problem, which in its most abstract form is as follows:

> *Suppose two terms s and t are given, which by some convention denote a particular structure and let s and t contain some free variables. We say s and t are unifiable iff there are substitutions (i.e. terms replacing the free variables of s and t) such that both terms become equal in a well defined sense.*

If the structure can be axiomatized by some first order theory T, unification of s and t under T amounts to solving the equation s = t in that theory.

However, the mathematical investigation of equation solving in certain theories is a subject as old as mathematics itself and, right from the beginning, very much at the heart of it: It dates back to Babylonian mathematics (about 2000 B.C.).

Universal unification carries this activity on in a more abstract setting: just as universal algebra abstracts from certain properties that pertain to specific algebras and investigates issues that are common to all of them, universal unification addresses problems, which are typical for equation solving as such.

Just as traditional equation solving drew its impetus from its numerous applications (the - for those times - complicated division of legacies in Babylonian times and the application in physics in more modern times), unification theory derives its impetus from its numerous applications in computer science, artificial intelligence and in particular in the field of computational logic.

Central to unification theory are the notion of a *set of most general unifiers* $\mu U\Sigma$ (traditionally: the set of base vectors spanning the solution space) and the *hierarchy of unification problems* based on $\mu U\Sigma$ (see part II for an exact definition of this hierarchy):

(i) a theory T is *unitary* if $\mu U\Sigma$ always exists and has at most one element;

(ii) a theory T is *finitary* if $\mu U\Sigma$ always exists and is finite;

(iii) a theory T is *infinitary* if $\mu U\Sigma$ always exists and there exists a pair of terms such that $\mu U\Sigma$ is infinite for this pair;

(iv) a theory T is of *type zero* otherwise.

We denote a *unification problem* under a theory T by

$$\langle s = t \rangle_T \quad .$$

In many practical applications it is of interest to know for two given terms s and t if there exists a *matcher* (a one-way-unifier) μ such that $\mu(s)$ and t are equal under T. We denote a *matching problem* under a theory T by

$$\langle s \geq t \rangle_T \quad .$$

In other words, in a matching problem we are allowed to substitute into one term only (into s using the above convention) and we say s *matches* t with *matcher* μ.

A unification problem (a matching problem) under a theory T poses two questions:

Q1: *is the equality of two terms under T decidable? If so:*

Q2: *are these two terms unifiable and if so, is it possible to generate and represent all unifiers?*

Q1 is the usual word problem, which has found a convenient computational treatment for equational logics [KB70], [HO80]. These techniques, called *term rewriting systems* are discussed in section II. 2.2. An affirmative answer to Q1 is an important prerequisite for unification theory.

Q2 summarizes the actual interest in unification theory and is the subject of this article.

> *It is reasonable to expect that the relationship between computer science and mathematical logic will be as fruitful in the next century as that between physics and analysis in the last.*
>
> *John McCarthy, 1963*

I. EARLY HISTORY AND APPLICATIONS

There is a wide variety of areas in computer science where unification problems arise.

1. *Databases*

A deductive database [GM78] does not contain every piece of information explicitely. Instead it contains only those facts from which all other information the user may wish to know can be deduced by some inference rule. Such inference rules (deduction rules) heavily rely on unification algorithms.

Also the user of a *relational database* [DA76] may logically AND the properties she wants to retrieve or else she may be interested in the NATURAL JOIN [CO70] of two stored relations. In neither case, she would appreciate if she constantly had to take into account that AND is an associative and commutative, or that NATURAL JOIN obeys an associative axiom, which may distribute over some other operation.

2. *Information retrieval*

A patent office may store all recorded electric circuits [BC66] or all recorded chemical compounds [SU65] as some graph structure, and the problem of checking whether a given circuit or compound already exists is an instance of a test for graph isomosphism [UL76], [UN64], [CR68]. More generally, if the nodes of such graphs are labelled with universally quantified variables ranging over subgraphs, these problems are practical instances of a *graph matching problem.*

3. *Computer vision*

In the field of *computer vision* it has become customary to store the internal representation of certain external scenes as some net structure [CL71], [WN75]. The problem to find a particular object - also represented as some net - in a given scene is also an instance of the *graph matching problem* [RL69]. Here one of the main problems is to specify as to what constitutes a successfull match (since a strict test for endomorphism is too rigid for most applications) although serious investigation of this problem is still pending (see para-unification in section IV).

4. *Natural Language Processing*

The processing of natural language [TL81] by a computer uses *transformation rules* to change the *syntax* of the input sentence into a more appropriate one.

Inference rules are used to manipulate the *semantics* of an input sentence and to disambiguate it.

The world knowledge a natural language understanding system must have is represented by certain (syntactic) *descriptions* and it is paramount to detect if two descriptions describe the same object or fact.

Transformation rules, inference rules and the matching of descriptions are but a few applications of unification theory to this field.

5. *Expert Systems*

An expert system is a computer program to solve problems and answer questions,which up to now only human experts were capable of [SH76]. The power of such a system largely depends on its ability to *represent* and *manipulate* the knowledge of its field of expertise. The techniques for doing so are yet another instance of the application of unification theory within the field of *artificial intelligence.*

6. *Computer Algebra*

In *computer algebra (*or *symbol manipulation)* [SG77] matching algorithms also play an important rôle: for example the integrand in a symbolic integration problem [MO71] may be matched against certain patterns in order to detect the class of integration problems it belongs to and to trigger the appropriate action for a solution (which in turn may involve several quite complicated matching attempts[BL71], [CK71], [FA71], [HN71], [MB68], [MO74].

7. *Programming Language*

An important contribution of artificial intelligence to programming language design is the mechanism of *pattern-directed* invocation of procedures [BF77], [HT72], [HT76], [RD72], [WA77]. Procedures are identified by patterns instead of procedure identifiers as in traditional programming languages. Invocation patterns are usually designed to express goals achieved by executing the procedure. Incoming messages are tried to be matched against the invocation patterns of procedures in a procedural data base, and a procedure is activated after having completed a successful match between message and pattern. So, matching is done (1) for looking up an appropriate procedure that helps to accomplish an intended goal, and (2) transmitting information to the involved procedure.

For these applications it is particularly desirable to have methods for matching objects belonging to high level data structures such as strings, sets, multisets etc.

A little reflection will show that for very rich matching structures, as it has e.g. been proposed in MATCHLESS in PLANNER [HT72], the matching problem is undecidable. This presents a problem for the designer of such languages: on the one hand, very rich and expressive matching structures are desirable, since they form the basis for the invocation and deduction mechanism. On the other hand, drastic restrictions will be necessary if matching algorithms are to be found. The question is just how severe do these restrictions have to be.

The fundamental mode of operation for the programming language *SNOBOL* [FG64] is to detect the occurrence of a substring within a larger string of characters (like e.g. a program or some text) and there are very fast methods known, which require less than linear time [BM77]. If these strings contain the SNOBOL 'don't care'-variables, the occurrence problem is an instance of the stringunification problem mentioned in the following paragraph.

Current attempts to use *first order predicate logic* [KO79] as a programming language [CM81] heavily depend on the availability of fast unification algorithms. In order to gain speed there are attempts at present to have a *hardware realization* of the unification procedure[GS84]

8. *Algebra*

A famous decidability problem, which inspite of many attacks remained open for over twenty-five years, has only recently been solved: *the monoid problem* (also called Löb's Problem in Western Countries, Markov's Problem in Eastern Countries and the Stringunification Problem in Automatic Theorem Proving [HJ64], [HJ66], [HJ67], [LS75], [MA54], [SS61], [PL72]) is the problem to decide whether or not an equation system over a free semigroup possesses a solution. This problem has been shown to be decidable [MA77]. The monoid problem has important practical applications inter alia for Automatic Theorem Proving (stringunification [SI75] and second order monadic unification [HT76], [WN76]) for Formal Language Theory (the crossreference problem for van Wijngaarden Grammars [WI76]), and for pattern directed invocation languages in artificial intelligence as mentioned above.

Another wellknown *matching problem* is *Hilbert's Tenth Problem* [DA73], which is known to be undecidable [MA70]. The problem is to decide whether or not a given polynomial $P[x_1,x_2,...,x_n] = 0$ has an integer solution (a Diophantine solution). Although this problem was posed originally and solved within the framework of traditional equation solving, unification theory has shed a new light upon this problem (see III.1.).

Semigroup theory [HO76], [CP61] is the field traditionally posing the most important unification problems (i.e. those involving associativity). Although scientifically more mature than unification theory is today, interesting semigroup problems have been solved using the techniques of unification theory (see e.g. [SS82], [LA80], [LA79]).

9. *Computational Logic*

All present day *theorem provers* have a procedure to unify first order terms as their essential component: i.e. a procedure that substitutes terms for the universally quantified variables until the two given terms are symbolwise equal or the failure to unify is detected.
This problem was first studied by Herbrand [HE30], who gives an explicit algorithm for computing a most general unifier. But unification algorithms only became of real importance with the advent of automatic theorem provers (ATP) and algorithms to unify two first order terms have independently been discovered by [GO67], [RO65] and [KB70]. Because of their paramount importance in ATP's there has been a race for the fastest such algorithm [RO71], [BA73], [VZ75], [HT76], [MM79] resulting in a linear first order unification algorithm for the free algebra of terms [PW78], [KK82].

Also for almost as long as attempts at proving theorems by machines have been made, a critical problem has been well known [GO67], [CK65], [NE71]: Certain equational axioms, if left without precaution in the data base of an automatic theorem prover, will force the ATP to go astray. In 1967, Robinson [RN67] proposed that substantial progress ("a new plateau") would be achieved by removing these troublesome axioms from the data base and building them into the deductive machinery.

Four approaches to cope with equational axioms have been proposed:

(1) To write the axioms into the data base, and use an additional rule of inference, such as paramodulation [WR73].
(2) To use special "rewrite rules" [KB70], [WR67], [HT80], [HO80].
(3) To design special inference rules incorporating these axioms [SL72].
(4) To develop special unification algorithms incorporating these axioms [PL72].

The last approach (4) still appears to be promising, however it has the drawback that for every new set of axioms a new unification algorithm has to be found. Also recently there has been interesting work on combinations of approach (2) and (4); see section III 2.2.

The work on higher unification by G. Huet [HT72], [HT75], [HT76], has been very influential for first order unification theory also and was fundamental in shaping the field as it is known today.

G. Plotkin has shown in a pioneering paper [PL72] that whenever an automatic theorem prover is to be refutation complete, its unification procedure must generate a set of unifiers satisfying the three conditions completeness, correctness and minimality, which are defined below.

Summarizing unification theory rests upon two main pillars: *Universal Algebra* and *Computational Logic* and we shall now turn to a brief survey of the important notions, which form the theoretical framework of the field.

"... *but we need notions, not notation.*"

A. Tarski, 1943

II. A FORMAL FRAMEWORK

1. *Unification from an Algebraic Point of View*

As usual let $\mathbb{N}$ be the set of natural numbers. A set of 'symbols with arity' is a mapping Ω: M $\rightarrow$ $\mathbb{N}$, where M is some set. For f$\in$M Ωf is the *arity* of f. The *domain of* Ω is used to denote certain n-ary operations and is sometimes called a *signature*. (f,n)$\in\Omega$ is abbreviated to f$\in\Omega$.

A *Universal Algebra A* is a pair (A,Ω), where A is the *carrier* and f$\in\Omega$ denotes a mapping

$$f: A^n \rightarrow A, \text{ where } \Omega f = n \text{ (and if } a_1,\ldots,a_n \in A$$

then we write $f_A(a_1,\ldots,a_n)$ for the *realization* of the denoted mapping). Note that if Ωf = 0 then f is a distinguished constant of the algebra *A*. COD(Ω), the *codomain* of Ω, is its *type*.

If *A* and *B* are algebras, φ: $A \rightarrow B$ is a *homomorphism* if $\varphi f_A(a_1,\ldots,a_n) = f_B(\varphi a_1,\ldots,\varphi a_n)$; a bijective homomorphism is called an *isomorphism*, in symbols $\simeq$.

For a subset $A_o \subseteq A$, $\varphi_o = \varphi|_{A_o}$ is the *restriction* of φ to A_o. An equivalence relation ρ is a *congruence* relation iff $a_1\rho b_1,\ldots,a_n\rho b_n$ implies $f_A(a_1,\ldots,a_n)\ \rho\ f_A(b_1,\ldots,b_n)$.

$A/_\rho = (A/_\rho,\Omega)$ is the *quotient algebra modulo* ρ. $[a]_\rho$ is the congruence class generated by a$\in$A.

For a class of algebras $\mathbf{K}_o$ of fixed type, the algebra *A* = (A,Ω) is *free in* $\mathbf{K}_o$ *on the set X*, in symbols $A_{\mathbf{K}_o}(X)$, iff

(i) $(A,\Omega) \in \mathbf{K}_o$
(ii) $X \subseteq A$
(iii) if $B \in \mathbf{K}_o$ and φ_o : X$\rightarrow$B is any mapping, then there exists a unique homomorphism φ: $A \rightarrow B$ with $\varphi_o = \varphi|_X$.

If $\mathbf{K}$ is the class of *all* algebras of the fixed type, then $A_{\mathbf{K}}(X)$ is *the* (since it exists and is unique up to isomorphism) *absolutely free algebra on X*. The elements of $A_{\mathbf{K}}(X)$ are called *terms* and are given a concrete representation W_Ω^X by:

(i) $x \in X$ is in W_Ω^X

(ii) if $t_1, t_2, \ldots, t_n$ are terms and $\Omega f = n$, $n \geq 0$, then $f(t_1, \ldots, t_n)$ is in W_Ω^X.

We assume that Ω consists of the disjoint sets Φ and Γ such that

$$f \in \Phi \text{ iff } \Omega f \geq 1 \quad \text{and}$$
$$f \in \Gamma \text{ iff } \Omega f = 0 \ .$$

Φ is called the set of function symbols, Γ the set of constants and X the set of variables.
We define operations

$$\hat{f}: (W_\Omega^X)^n \rightarrow W_\Omega^X \quad \text{for } n = \Omega f$$

by $\hat{f}(t_1, \ldots, t_n) = f(t_1, \ldots, t_n)$. Let $\hat{\Omega}$ be the set of these (term building) operations. Let $\emptyset$ denote the empty set.

$F_\Omega^X = (W_\Omega^X, \hat{\Omega})$ is isomorphic to $A_{\boldsymbol{K}}(X)$ and hence is called the absolutely free *term algebra* on X. $F_\Omega^\emptyset$ is the *initial* term algebra (or Herbrand universe). We shall write F_{Ω_0} for $F_\Omega^\emptyset$. Our interest in F_{Ω_0} is motivated by the fact that for every algebra $A = (A, \Omega)$ there exists a unique homomorphism

$$h_A: F_{\Omega_0} \rightarrow A \ .$$

But then instead of investigating A, we can restrict our attention to a quotient of F_{Ω_0} modulo the congruence induced by h_A.

In order to have variables at our disposal in the initial algebra we define $\Omega_X = \Omega \cup X$, that is we treat variables as special constants. Since

$F_\Omega^X \simeq F_{\Omega_X}^\emptyset$ we simply write F_Ω if $X \neq \emptyset$ and $X \subset \Omega$ and F_{Ω_0} if $X = \emptyset$. Because terms are objects in F_Ω we shall write $t \in F_\Omega$ instead of $t \in W_\Omega^X$.

An *equation* is a pair of terms $s, t \in F_\Omega$, in symbols $s = t$. The equation $s = t$ is *valid* in the algebra A (of the same type), in symbols

$$A \models s = t \quad \text{iff}$$

for every homomorphism $\varphi: F_\Omega \rightarrow A$

$$\varphi s = \varphi t \quad \text{in } A.$$

Let $\bar{\sigma}: X \rightarrow F_\Omega$ be a mapping which is equal to the identity mapping almost everywhere. A *substitution* $\sigma: F_\Omega \rightarrow F_\Omega$ is the homomorphic extension of $\bar{\sigma}$ and is represented as a finite set of pairs:

$$\sigma = \{x_1 \leftarrow t_1, \ldots, x_n \leftarrow t_n\} \ .$$

Σ is the set of substitutions on F_Ω. The identity mapping on F_Ω, i.e. the *empty substitution*, is denoted by ε. If t is a term and σ a substitution, define

$V: F_\Omega^n \to 2^X$ by $V(t) = \{$set of variables in t$\}$ and $V(t_1, \ldots, t_n) = \bigcup_{i \le n} V(t_i)$

$|t| \in \mathbb{N}$ denotes the length of t (i.e. the number of symbols in t)

$DOM(\sigma) = \{x \in X: \sigma x \neq x\}$

$COD(\sigma) = \{\sigma x: x \in DOM(\sigma)\}$

$XCOD(\sigma) = V(COD(\sigma))$

$\Sigma_o \subset \Sigma$ is the set of *ground* substitutions, i.e. $\sigma \in \Sigma_o$ iff $COD(\sigma) \subset F_{\Omega_o}$.

An equation s = t is *unifiable* (is *solvable*) in *A* iff there exists a substitution $\xi: F_\Omega \to F_\Omega$ such that $\xi s = \xi t$ is valid in *A*.

For a set of equations T let $=_T$ be the finest congruence containing (s,t) and all pairs (σs, σt), for $\sigma \varepsilon \Sigma$ and $s=t \in T$. $F_\Omega/=_T$ is the quotient algebra modulo $=_T$.

A *unification problem for T*, denoted as

$$\langle s = t \rangle_T \quad ,$$

is given by the equation s = t, $s, t \in F_\Omega$. The problem is to decide whether or not s = t is unifiable in $F_\Omega/=_T$.

We denote the constituent parts of the initial algebra $F_\Omega/=_T$ as

$$(F_\Omega, T)\text{- } algebra.$$

2. *Unification from a Logical Point of View*

2.1 *EQUATIONAL LOGIC*

The *well formed formulas* of our logic are equations defined as pairs (s,t) in $W_\Omega^X \times W_\Omega^X$ and denoted as s = t.

A substitution σ is a finite set of pairs in $W_\Omega^X \times W_\Omega^X$ (i.e. classical work confuses the issue a little by identifying the representation with the mapping that is being represented). The application of $\sigma = \{x_1 \leftarrow t_1, \ldots, x_n \leftarrow t_n\}$ to a term t, σt is obtained by simultaneously replacing each x_i in t by t_i.

Let T be a set of equations. The equation p = q *is derivable from* T, $T \vdash p = q$, if $p = q \in T$ or p = q is obtained from T by a finite sequence of the following operations:

(i) $t = t$ is an axiom
(ii) if $s = t$ then $t = s$
(iii) if $r = s$ and $s = t$ then $r = t$
(iv) if $s_i = t_i$, $1 \le i \le n$ then $f(s_1, \ldots, s_n) = f(t_1, \ldots, t_n)$, $n = \Omega f$
(v) if $s = t$ then $\sigma s = \sigma t$ where $\sigma \in \Sigma$.

For a set of equations T, $T \vdash s = t$ iff $s = t$ is valid in all *models* of T.

Theorem (Birkhoff): $T \models s = t$ iff $T \vdash s = t$
We shall abbreviate $T \models s = t$ (and hence $T \vdash s = t$) by $s =_T t$. An equation $s = t$ is *T-unifiable*, iff there exists a substitution σ such that $\sigma s =_T \sigma t$.

Although this is the traditional view of unification, its apparent simplicity is deceptive: we did not define what we mean by a 'model'. In order to do so we should require the notion of an interpretation of our well formed formulas, which is a 'homomorphism' from W_Ω^X to certain types of algebras, thus bringing us back to section 1.

Since neither $\models$ nor $\vdash$ are particularily convenient for a computational treatment of $=_T$, an alternative method is presented below.

2.2 *COMPUTATIONAL LOGIC*
For simplicity of notation we assume we have a box of symbols, GENSYM, at our disposal, out of which we can take an unlimited number of "new" symbols. More formally: for F_Ω, let $\Omega = \Phi \cup \Gamma \cup X$, where $X = X_o \cup \text{GENSYM}$ with $\Omega x = 0$, $x \in X$.

We shall adopt the computational proviso that whenever GENSYM is referenced by $v \in \text{GENSYM}$ it is subsequently 'updated' by $\text{GENSYM}' = \text{GENSYM} - \{v\}$ and $X_o' = X_o \cup \{v\}$ and $\Omega' = \Phi \cup \Gamma \cup X'$, where $X' = X_o' \cup \text{GENSYM}'$. Since $F_{\Omega'} \simeq F_\Omega$ we shall not always keep track of the '-s and just write F_Ω.

A *renaming substitution* $\rho \in \Sigma_X \subset \Sigma$ is defined by
(i) $\text{COD}(\rho) \subset X$
(ii) $x, y \in \text{DOM}(\rho)$: if $x \neq y$ then $\rho x \neq \rho y$.

For $s, t \in F_\Omega$: $s \sim_\rho t$ if $\exists \rho \in \Sigma_X$ such that $\rho s = \rho t$. If $\rho s = t$ then t is called an *X-variant* of s, if in addition $\text{COD}(\rho) \subset \text{GENSYM}$ then t is called a *new* X-variant of s.

In order to formalize the accessing of a subterm in a term, let $\mathbb{N}^*$ be the set of sequences of positive integers, Λ the empty sequence in $\mathbb{N}^*$

and let $\cdot$ be the concatenation operation on sequences. Members of $\mathbb{N}^*$ are called *positions*, and denoted by $\pi \in \mathbb{N}^*$. They are used as follows: for any $t \in F_\Omega$ let $\Pi(t) \subset \mathbb{N}^*$, the set of *positions in t*, be:

(i) if $\Omega t = 0$ then $\Pi(t) = \{\Lambda\}$
(ii) if $t = f(t_1,\ldots,t_n)$ then $\Pi(t) = \{\Lambda\} \cup \{i\cdot\pi : 1\le i\le n, \pi\in\Pi(t_i)\}$.

For example: $f(g(a,y),b) = \{\Lambda,1,2,1\cdot 1, 1\cdot 2\}$.
The *subterm of* $t = f(t_1,\ldots,t_n)$ *at* π, $t|\pi$, is defined as:

(i) $t|\pi = t$ for $\pi = \Lambda$ or $\pi\notin\Pi(t)$
(ii) $t|i\cdot\pi' = t_i|\pi'$ for $\pi = i\cdot\pi'$.

For example: $f(g(a,y),b)|1.2 = y$.
A *subterm replacement* of *t* by *s at* π, $\hat{\rho}t$, with $\hat{\rho} = [\pi\leftarrow s]$ is defined as:

(i) $\hat{\rho}t = s$ if $\pi = \Lambda$
(ii) $\hat{\rho}t = f(t_1,\ldots,\hat{\delta}t_i,\ldots,t_n)$ if $t = f(t_1,\ldots,t_n)$ and $\pi = i\cdot\pi'$ and $\hat{\delta} = [\pi'\leftarrow s]$
(iii) $\hat{\rho}t = t$ if $\pi\notin\Pi(t)$.

We denote replacements by $\hat{\sigma},\hat{\rho},\hat{\delta}$, etc. and substitutions by σ,ρ,δ etc.

A relation $\rightarrow \subseteq F_\Omega \times F_\Omega$ is *Noetherian* (terminating) if there are no infinite sequences: $s_1 \rightarrow s_2 \rightarrow s_3 \rightarrow \ldots$. As usual $\overset{+}{\rightarrow}$ is the transitive and $\overset{*}{\rightarrow}$ the reflexive and transitive closure of $\rightarrow$. A relation $\rightarrow$ is *confluent* if for every $r,s,t\in F_\Omega$ such that $r \overset{*}{\rightarrow} s$ and $r \overset{*}{\rightarrow} t$ there exists a $u\in F_\Omega$ such that $s \overset{*}{\rightarrow} u$ and $t \overset{*}{\rightarrow} u$. A confluent Noetherian relation is *canonical*.

We define two important relations $\rightarrow_R$ and $\rightarrowtail_R$ on $F_\Omega \times F_\Omega$ as follows:

A *rewrite system* $R = \{l_1 \Rightarrow r_1,\ldots,l_n \Rightarrow r_n\}$ is any set of pairs $l_i,r_i \in F_\Omega$, such that $V(r_i) \subseteq V(l_i)$, $1\le i\le n$.

For two terms s and t we say s is *rewritten* to t, $s \rightarrow_R t$, if there exists $\pi\in\Pi(s)$, $\sigma\in\Sigma$ and $\tilde{l}_i \Rightarrow \tilde{r}_i \in R$ such that $s|\pi = \sigma\tilde{l}_i$ and $t = \hat{\sigma}s$, where $\hat{\sigma} = [\pi\leftarrow\sigma\tilde{r}_i]$ and $\tilde{l}_i,\tilde{r}_i$ are new X-variants of l_i,r_i. Occasionally we keep track of the information by writing $s \xrightarrow[{[\pi,i,\sigma]}]{} t$, $s \xrightarrow[{[\pi,i]}]{} t$, $s \xrightarrow[{[\pi]}]{} t$ etc.

For two terms s and t we say s is *paramodulated* to t, $s \rightarrowtail_R t$, if there exists $\pi\in\Pi(s)$, $l_i \Rightarrow r_i \in R$, $\sigma\in\Sigma$ such that $\sigma(s|\pi) = \sigma\tilde{l}_i$ and σ is most general (see 3. below), $\tilde{l}_i$ is a new X-variant of l_i and $\sigma s \xrightarrow[{[\pi,i]}]{} t$.

For example for $R = \{g(x,0) \Rightarrow 0\}$ we have
$s = f(g(a,y),y) \rightarrowtail_R f(0,0) = t$
with $\pi = 1$ and $\sigma = \{x\leftarrow a, y\leftarrow 0\}$.

But note $s \not\rightarrow_R t$, since we are not allowed to substitute into s.

The notation and definitions of term rewriting systems are essentially those of [HT80]; the importance of term rewriting systems (demodulation) for theorem proving was first noticed in [WR67]. Suppose for an equational theory T there is a rewrite system R_T such that for $s,t \in F_\Omega$:

$$s =_T t \text{ iff } \exists p \in F_\Omega \text{ such that } s \xrightarrow{*}_{R_T} p \text{ and } t \xrightarrow{*}_{R_T} p \quad .$$

In that case we say T is *embedded into* R_T and write

$$T \hookrightarrow R_T \quad .$$

For an equational theory T there are techniques to obtain a system R_T such that $T \hookrightarrow R_T$; moreover for many theories of practical interest it is possible to obtain a rewrite system R_T such that $\longrightarrow_{R_T}$ is canonical [KB70], [HT80], [PS81], [HL80]. Canonical relations $\rightarrow^T$ are an important basis for *computations* in *equational logics*, since they define a unique *normal form* $||t||$ for any $t \in F_\Omega$, given by $t \xrightarrow{*} ||t||$ and $\nexists s \in F_\Omega$ such that $||t|| \rightarrow s$. Hence $s =_T t$ iff $||s|| = ||t||$.

In case R_T is Noetherian (i.e. R defines the Noetherian relation $\rightarrow_{R_T}$), we also say it is a *reduction system*.

3. *Universal Unification*

An equational theory T is *decidable* iff $s =_T t$ is decidable for any $s,t \in F_\Omega$. Let $\mathcal{T}_=$ denote the family of decidable finitely based equational theories.

A *T-unification problem* $\langle s = t\rangle_T$ consists of a pair of terms $s,t \in F_\Omega$ and a theory $T \in \mathcal{T}_=$.

A substitution $\sigma \in \Sigma$ is a *T-unifier* for $\langle s = t\rangle_T$ iff $\sigma s =_T \sigma t$. The subset of Σ which unifies $\langle s = t\rangle_T$ is $U\Sigma_T(s,t)$, the *set of unifiers* (for s and t) *under T*. It is easy to see that $U\Sigma_T$ is recursively enumerable (r.e.) for any s and t: Since F_Ω is r.e. so is Σ, now for any $\delta \in \Sigma$, check if $\delta s =_T \delta t$ (which is decidable since $T \in \mathcal{T}_=$) then $\delta \in U\Sigma_T(s,t)$ otherwise $\delta \notin \Sigma_T(s,t)$.

We shall omit the subscript T and (s,t) if they are clear from the context. The composition of substitutions is defined by the usual composition of mappings: $(\sigma \circ \tau)t = \sigma(\tau t)$. If $W \subseteq X$, then T-equality is extended to substitutions by

$$\sigma =_T \tau \; [\![W]\!] \quad \text{iff } \forall x \in W \quad \sigma x =_T \tau x \; ,$$

σ and τ are T-equal in W.

Let $\leq_T$ be a partial order on terms such that $s \leq_T t$ iff there exists $\delta\varepsilon\Sigma$ satisfying $s =_T \delta t$. This relation is extended to substitutions: We say σ is an instance of τ and τ is more *general* than σ, in symbols

$$\sigma \leq_T \tau \,[\![W]\!] \quad \text{iff} \quad \exists\lambda\in\Sigma \quad \text{with}$$

$$\sigma =_T \lambda \circ \tau \,[\![W]\!] \quad \text{for some } W \subset X \ .$$

If $\sigma \leq_T \tau \,[\![W]\!]$ and $\tau \leq_T \sigma \,[\![W]\!]$ then $\sigma \approx_T \tau \,[\![W]\!]$, σ and τ are *T-equivalent in W*.

Similarly extend $\sim_T$ to substitution by: $\sigma\sim_T\tau\,[\![W]\!]$ iff $\sigma x\sim_T\tau x$, $x\varepsilon W$.

Lemma: (i) For $T=\emptyset$: $\sigma\sim_T\tau\,[\![W]\!]$ iff $\sigma\approx_T\tau\,[\![W]\!]$.

(ii) There exists an equational theory T such that

$$\sigma\approx_T\tau\,[\![W]\!] \text{ but } \sigma\not\sim_T\tau\,[\![W]\!]$$

See [HE83] for a proof.

For $\Sigma_1,\Sigma_2 \subseteq \Sigma$ we define $\Sigma_1 \circ \Sigma_2 = \{\sigma_1 \circ \sigma_2 : \sigma_1 \in \Sigma_1, \sigma_2 \in \Sigma_2\}$.

$\Sigma_1 \subseteq_T \Sigma_2 \,[\![W]\!]$ iff $\forall\sigma_1 \in \Sigma_1 \; \exists \sigma_2 \in \Sigma_2$ s. th. $\sigma_1 =_T \sigma_2 \,[\![W]\!]$,

$\Sigma_1 =_T \Sigma_2 \,[\![W]\!]$ iff $\Sigma_1 \subseteq_T \Sigma_2 \,[\![W]\!]$ and $\Sigma_2 \subseteq_T \Sigma_1 \,[\![W]\!]$.

Universal unification is concerned with three fundamental problems:

PROBLEM ONE *(Decidability Problem)*

For a given equational theory $T \in \mathcal{T}_=$, is it decidable for any s and t whether s and t are unifiable?

That is, we are interested in classes of theories such that "s and t are unifiable under T" is decidable for every T in that class.

A unifier σ for $\langle s = t\rangle_T$ is called *a most general unifier* (mgu) if for any unifier $\delta \in U\Sigma_T(s,t)$: $\delta \leq_T \sigma \,[\![W]\!]$, where $V(s,t) = W$. Since in general a single most general unifier does not exist for $\langle s = t\rangle_T$, we define $\mu U\Sigma_T(s,t)$, the *set of most general unifiers*, as:

(i)	$\mu U\Sigma \subseteq U\Sigma$	(correctness)
(ii)	$\forall\delta\in U\Sigma$ there exists $\sigma\in\mu U\Sigma$ s.th. $\delta\leq_T\sigma\,[\![W]\!]$	(completeness)
(iii)	$\forall\ \sigma_1\leq\sigma_2 \in \mu U\Sigma$: if $\sigma_1\leq_T\sigma_2\,[\![W]\!]$ then $\sigma_1 = \sigma_2$	(minimality)

From condition (ii) it follows in particular that $U\Sigma =_T \Sigma\circ U\Sigma\,[\![W]\!]$, i.e. $U\Sigma$ is a *left ideal* in the semigroup $(\Sigma,\circ)$ and $U\Sigma$ is *generated by* $\mu U\Sigma$. For theoretical reasons (idempotency of substitutions) as well as for many practical applications, it turned out to be useful to have the additional technical requirement:

(o)	For a set of variables Z with $V(s,t)\subseteq Z$:
	$XCOD(\sigma) \cap Z=\emptyset$ for $\sigma\varepsilon\mu U\Sigma_T(s,t)$ (protection of Z)

If conditions (o) - (iii) are fulfilled we say $\mu U\Sigma$ *is a set of most general unifiers away from Z* [FH83], [PL72].

The set $\mu U\Sigma_T$ does not always exist;when it does then it is unique up to the equivalence $\approx_T$, see [FH83]. For that reason it is sufficient to generate one $\mu U\Sigma_T$. In the following we always take $\mu U\Sigma_T$ as same representative of the equivalence class $[\mu U\Sigma_T]_{\approx}$.

PROBLEM TWO *(Existence Problem)*:

For a given equational theory $T \in \mathcal{T}_=$ *, does* $\mu U\Sigma_T(s,t)$ *always exist for every* $s,t \in F_\Omega$?

PROBLEM THREE *(Enumeration Problem)*:

For a given equational theory $T \in \mathcal{T}_=$ *, is*

$\mu U\Sigma_T(s,t)$

recursively enumerable for any $s,t \in F_\Omega$*?*

That is, we are interested in an algorithm which generates all mgu's for a given problem $\langle s = t\rangle_T$. Section III.1 summarizes the major results that have been obtained for special theories T.

The central notion $\mu U\Sigma_T$ induces the following fundamental classes of equational theories based on the cardinality of $\mu U\Sigma_T$:

(i) A theory T is *unitary* if $\forall s,t$ $\mu U\Sigma_T(s,t)$ exists and has at most one element. The class of such theories is $\mathcal{U}_1$ *(type one)*.

(ii) A theory T is *finitary* if it is not unitary and if $\forall s,t$ $\mu U\Sigma_T(s,t)$ exists and is finite.
The class of such theories is $\mathcal{U}_\omega$ *(type* ω*)*.

(iii) A theory T is *infinitary* if $\forall s,t$ $\mu U\Sigma_T(s,t)$ exists and there exists $\langle p = q\rangle_T$ such that $\mu U\Sigma_T(p,q)$ is infinite.
The class of such theories is $\mathcal{U}_\infty$ *(type* ∞*)*.

(iv) A theory T is of *type zero* if it is not in one of the above classes
The class of these theories is $\mathcal{U}_0$.

(v) A theory is *unification-relevant* if it is not of type zero. The class of these theories is $\mathcal{U}$.

Several examples for unitary, finitary and infinitary theories as well as type zero theories are discussed in III.1. A *matching problem* $\langle s \geq t\rangle_T$ consists of a pair of terms and a theory $T \in \mathcal{T}_=$. A substitution $\nu \in \Sigma$ is a T-*matcher* (or one-way-unifier) if $\nu s =_T t$. $M\Sigma_T$ is the set of matchers and a set of *most general matchers* $\mu M\Sigma_T$ is defined similarily to $\mu U\Sigma_T$.

The set $\mu M\Sigma_T$ induces the classes of *matching-relevant theories* similar to the classes based on $\mu U\Sigma_T$: a theory T is *unitary matching* if $\mu M\Sigma_T$ always exists and has at most one element. The class of such theories is $\mathcal{M}_1$. Analogeously we define $\mathcal{M}_\omega$ $\mathcal{M}_\infty$ $\mathcal{M}_0$ and the class $\mathcal{M}$.

A *unification algorithm* $U\mathbf{A}_T$ *(a matching algorithm* $M\mathbf{A}_T$*) for a theory T* is an algorithm which takes two terms s and t as input and generates a set $\Psi_T \subseteq U\Sigma_T (\subseteq M\Sigma_T)$ for $\langle s = t\rangle_T$ (for $\langle s \geq t\rangle_T$). A *minimal* algorithm $\mu U\mathbf{A}_T$ $(\mu M\mathbf{A}_T)$ is an algorithm which generates a $\mu U\Sigma_T$ $(\mu M\Sigma_T)$.

For many practical applications this requirement is not strong enough, since it does not imply that the algorithm terminates for theories $T \in \mathcal{U}_1 \cup \mathcal{U}_\omega$. On the other hand, for $T \in \mathcal{U}_\omega$ it is sometimes too rigid, since an algorithm which generates a finite superset of $\mu U\Sigma_T$ may be far more efficient than the algorithm $\mu U\mathbf{A}_T$ and for that reason preferable. For that reason we define:

An algorithm $U\mathbf{A}_T$ is *type conformal* iff:

(i) $U\mathbf{A}_T$ generates a set Ψ_T with $U\Sigma_T \supseteq \Psi_T \supset \mu U\Sigma_T$ for some $\mu U\Sigma_T$.

(ii) $U\mathbf{A}_T$ terminates and Ψ_T is finite if $T \in \mathcal{U}_1 \cup \mathcal{U}_\omega$ and

(iii) if $T \in \mathcal{U}_\infty$ then $\Psi_T \in [\mu U\Sigma_T]_\approx$.

Similarly: algorithm $M\mathbf{A}_T$ is type conformal iff (i) - (iii) hold with U replaced by M.

"However to generalize, one needs experience ..."

G. Grätzer
Universal Algebra, 1968

III. RESULTS

"... a comparative study necessarily presupposes some previous separate study, comparison being impossible without knowledge."

N. Whitehead
Treatise on Universal Algebra, 1898

1. *Special Theories*

This section is concerned with Problem Two and Three (the existence resp. the enumeration problem) mentioned in II.3: *For a given equational theory T, does there exist an algorithm, which enumerates* $\mu U\Sigma_T(s,t)$ *for any terms s and t?*

The following table summarizes the results that have been obtained for special theories, which consist of combinations of the following equations:

A	(associativity)	$f(f(x,y),z) = f(x,f(y,z))$
C	(commutativity)	$f(x,y) = f(y,x)$
D	(distributivity)	$D_R: f(x,g(y,z)) = g(f(x,y),f(x,z))$ $D_L: f(g(x,y),z) = g(f(x,z),f(y,z))$
H,E	(homomorphism, endomorphism)	$\varphi(x \circ y) = \varphi(x) \cdot \varphi(y)$
I	(idempotence)	$f(x,x) = x$

Abbreviations:

FPA: Finitely Presented Algebras
QG: Quasi-Groups
AG: Abelian-Groups
H10: Hilbert's 10th Problem
Sot: Second order terms
Hot: Higher order terms (i.e. $\geq 3^{rd}$ order)

$S\emptyset_T$: Many sorted first order terms, empty theory T, the sort structure $\langle \mathcal{S}, \leq \rangle$ is a tree.
$S\emptyset_N$: Many sorted first order terms, empty theory T, the sort structure $\langle \mathcal{S}, \leq \rangle$ is not a tree, $\mathcal{S}$ is finite.

The column under $U\mathbf{A}_T$ indicates whether or not a type conformal algorithm has been presented in the literature. The 'type of a theory' and 'type conformal' are defined in section II.3.

Theory T	Type of T	Unification decidable	$\mu U \Sigma_T$ recursive	$_U\mathbf{A}_T$	References
Ø	1	Yes	Yes	Yes	[HE30][RO65][RO71][KB70][G67][PR60] [BA73][HT76][MM79][PW78]
A	∞	Yes	Yes	Yes	[HM67][PL72][SI75][LS75][MA77]
C	ω	Yes	Yes	Yes	[SI82]
I	ω	Yes	Yes	Yes	[RS78][SB82] [SZ82]
A+C	ω	Yes	Yes	Yes	[ST75][LS76][HU79] [FA83][HT78]
A+I	?	Yes	?	No	[SS82] [SZ82]
C+I	ω	Yes	Yes	Yes	[RS78]
A+C+I	ω	Yes	Yes	Yes	[LS76]
D	∞	?	Yes	Yes	[SZ82]
D+A	∞	No	Yes	Yes	[S78][SZ82]
D+C	∞	?	Yes	Yes	[SZ82]
D+A+C	∞	No	Yes	Yes	[SZ82]
D+A+I	?	Yes	?	No	[SZ82]
H,E	1	Yes	Yes	Yes	[VO78]
H+A	∞	Yes	Yes	Yes	[VO78]
H+A+C	ω	Yes	Yes	Yes	[VO78]
E+A+C	∞	?	?	No	[VO78]
QG	ω	Yes	Yes	Yes	[HU80]
AG	ω	Yes	Yes	Yes	[LA79] [LBB84]
H10	?	No	?	No	[MA70][DA73]
FPA	ω	Yes	Yes	Yes	[LA80]
Sot, T = Ø	?	No	-	-	[GO81]
Hot,EØ	0	No	-	-	[HT73] [HT75] [BA78] [LC72]
$SØ_T$	1	Yes	Yes	Yes	[WA84]
$SØ_N$	ω	Yes	Yes	Yes	[WA84]

Except for Hilbert's tenth problem, we have not included the classical work on equation solving in 'concrete' structures such as rings and fields, which is well known. The relationship of universal unification to these classical results is similar to that of universal algebra [GR79] to classical algebra.

Let us comment on a few entries in the above table: The *Robinson Unification Problem*, i.e. unification in the free algebra of terms or *unification under the empty theory* Ø has attrackted most attention so far and was already discussed in section I.9.

Unification under associativity is the famous monoid problem mentioned in I.8. Plotkin gave the first unification algorithm for this theory [PL72] and used it to demonstrate the existence of *infinitary* equational theories. Completeness, correctness and minimality proofs are presented in [SI78], which also discusses the practical implications of these results for theorem proving and programming language design. Makanin showed the decidability of this unification problem [MA77].

Unification under commutativity has a trivial solution, whereas

minimality presents a hard problem. A type conformal algorithm is presented in [SI76]. The main interest in this theory however derives from its *finitary* nature in contrast to the *infinitary* theory of associativity. A nice characterization of this difference is possible in terms of the universal unification algorithm presented below. However a deep theoretical explanation of why two seemingly very similar theories belong to entirely different classes is still an open research problem.

Apart from its practical relevance, *unification under associativity and commutativity* (A+C) poses an important theoretical problem: why is it that the combination of an infinitary theory (A) with a finitary theory (C) results in a *finitary* theory (A+C), whereas the combination of an infinitary theory (D) with the finitary (C) results in an *infinitary* theory (D+C)? Both theories (A+C) and (A+C+I) define common data-structures, namely bags and sets respectively.

Unification under distributivity and associativity provides a point in case that the combination of two infinitary theories is an infinitary theory. Is this always the case? The D+A-Unification Problem is also of theoretical interest with respect to Hilbert's Tenth Problem, which is the problem of Diophantine solvability of a polynomial equation. An axiomatization of Hilbert's Tenth Problem would involve the axioms A and D plus additional axioms for integers, multiplication, etc. Calling the union of these axioms HTP, the famous undecidability result [DA73] shows the undecidability of the *unification problem under HTP*. Now the undecidability of the D+A-Unification Problem demonstrates that *all Hilbert axioms in HTP* can be dropped except for D and A (holding for one function symbol) and the problem *still remains undecidable*. Since A-unification is known to be decidable, the race is open as to whether or not A can be dropped as well and D on its own presents an undecidable unification problem.

More generally: it is an interesting and natural question for an undecidable problem to ask for its "minimal undecidable substructure". Whatever the result may be, the D+A problem already highlights the advantage of the abstract nature of universal unification theory in contrast to the traditional point of view, with its reliance on intuitively given entities (like integers) and structures (like polynomials).

The *undecidability results* for second and higher order logic were the first undecidability results obtained in the framework of unification theory and rely on a coding of known undecidability results (Post's Correspondence Problem and H10) into these problems.

Finally it is important to realize that the results recorded in the above table do not always hold for the whole class of first order terms. The extension of these special results to the whole class of first order terms is but a special case of the *Combination Problem of Theories:*

From the above table we already have

$$
\begin{array}{llll}
D \in \mathcal{U}_\infty, & A \in \mathcal{U}_\infty & \text{und} & D+A \in \mathcal{U}_\infty \\
D \in \mathcal{U}_\infty, & C \in \mathcal{U}_\omega & \text{und} & D+C \in \mathcal{U}_\infty \\
A \in \mathcal{U}_\infty, & C \in \mathcal{U}_\omega & \text{und} & A+C \in \mathcal{U}_\omega \\
C \in \mathcal{U}_\omega, & I \in \mathcal{U}_\omega & \text{und} & C+I \in \mathcal{U}_\omega \\
H \in \mathcal{U}_1, & A \in \mathcal{U}_\infty & \text{und} & H+A \in \mathcal{U}_\infty \\
H \in \mathcal{U}_1, & A+C \in \mathcal{U}_\omega & \text{und} & H+A+C \in \mathcal{U}_\omega \\
D_L \in \mathcal{U}_1, & C \in \mathcal{U}_\omega & \text{und} & D_L+C \in \mathcal{U}_\infty \\
D_L \in \mathcal{U}_1, & D_R \in \mathcal{U}_1 & \text{und} & D_L+D_R = D \in \mathcal{U}_\infty
\end{array}
$$

Using a more informal notation we can write: $\infty + \infty = \infty$, $\infty + \omega = \infty$, $\omega + \omega = \omega$, $1 + \infty = \infty$, $1 + \omega = \omega$, $1 + \omega = \infty$ and even $1+1 = \infty$ for these results.

Here we assume that for example C and A hold for the same function symbol f and the combination of these axioms is denoted as C+A. But what happens if C and A hold for two different function symbols, say C for f and A for g? Even the most trivial extension in this spirit, which is the extension of a known unification result to additional "free" functions(i.e. the empty theory for every function symbol, which is not part of the known unification result) as mentioned above is unsolved.

Summarizing we notice that unification algorithms for different theories are usually based on entirely different techniques.

They provide the experimental laboratory of Universal Unification Theory and it is paramount to obtain a much larger experimental test set than the one recorded above.

"...(theories) are worthy of a comparative study, for the sake of the light thereby thrown on the general theory of symbolic reasoning and on algebraic symbolism in particular"

N. Whitehead
Treatise on Universal Algebra, 1898

2. *The General Theory*

2.1 *A Classification of Equational Theories*

We like to present some important subclasses of equational theories, which turned out to be of practical interest as well as being useful as "basic building blocks" for other equational classes. We shall first present the definitions and then show a few theorems in order to demonstrate the descriptive value of these theories and to give a flavour of the field. Let $\mathcal{T}_=$ be the class of *equational theories*, which are finitely based and have a decidable word problem. At present the most important subclass is

$$\mathcal{T}_{\Rightarrow} := \{T \in \mathcal{T}_= : \text{there exists a term rewriting system s.th. } T \hookrightarrow R\}$$

A theory T is *regular* iff for every $l=r \in T$ $V(l) = V(r)$; we shall write $\mathcal{T}^*$ if $\mathcal{T}$ is a class of regular theories . As an immediate result we have:

$$\mathcal{T}_=^* = \mathcal{T}_{\Rightarrow}^*$$

The fundamental classes for unification theory are $\mathcal{U} := \mathcal{U}_1 \cup \mathcal{U}_\omega \cup \mathcal{U}_\infty$, the class of *unification relevant theories*, and $\mathcal{U}_0$, the class of *type-zero theories*. Similarily we define $\mathcal{M}$, the class of *matching relevant theories*, and $\mathcal{M}_0$. It is not difficult to see that $\mathcal{M}_0$ is a subclass of $\mathcal{U}_0$:

Proposition 1: $\mathcal{M}_0 \subset \mathcal{U}_0$

An important requirement with respect to unification theory is that the matching problem is decidable for T; let $\mathcal{T}_\leq$ denote this class. The class $\mathcal{A} := \mathcal{T}_{\Rightarrow} \cap \mathcal{T}_\leq$ is the class of *admissible* theories. Defining $\mathcal{T}_\downarrow \subset \mathcal{T}_{\Rightarrow}$ as the subclass with a *confluent* rewriting system and $\mathcal{R} \subset \mathcal{T}_{\Rightarrow}$ as the subclass with a *Noetherian* rewriting system and abbreviating $\mathcal{R}_\downarrow = \mathcal{R}\mathcal{T}_\downarrow$ (throughout this section we use the denotational proviso that juxtaposition abbreviates intersection of classes) as the name for the *canonical theories in a generalized sense* (i.e. any canonicalization is allowed). Defining $\mathcal{C} \subset \mathcal{R}_{\downarrow}$ as the class having a (standard) canonicalization and let $\mathcal{M}_{\omega 1} = \mathcal{M}_{\omega} \cup \mathcal{M}_{1}$ we have the classes $\mathcal{A}\,\mathcal{C}\,\mathcal{M}_{\omega 1}$ and $\mathcal{A}\,\mathcal{C}^{*}\mathcal{M}_{\omega 1}$, which turned out to be important for universal unification algorithms: it can be shown that $\mu U\Sigma_{T}$ is recursively enumerable for any T in a subclass of $\mathcal{A}\,\mathcal{C}^{*}\mathcal{M}_{\omega 1}$. Calling this subclass $\mathcal{A}\,\mathcal{C}^{*}\mathcal{M}_{\downarrow}$ we have

Theorem 1 (Szabo): $\mu U\Sigma_{T}$ exists for any $T \in \mathcal{A}\mathcal{C}^{*}\mathcal{M}_{\downarrow}$

This theorem has been extended in [HD82] to a larger class containing the confluent "modulo" and confluent "over" theories

For regular equational theories we have:

Theorem 2: For $T \in \mathcal{A}$: Every complete set of matchers is minimal, i.e. $\mu M\Sigma_{T} = M\Sigma_{T}$

See [SS81]; and [FH83] for a generalization.

Theorem 3: (Huet, Fage) There is an equational theory $T \in \mathcal{T}_{=}$ such that

(i) $\mu U\Sigma_{T}$ (s,t) does not exist for a given pair of terms.

(ii) $\mu M\Sigma_{T}$ (s,t) does not exist for a given pair of terms.

This important result shows that (even for first order theories) a minimal basis is not always altainable [FH83].

Theorem 4: (Huet, Fage) There is a regular canonial theory $T \in \mathcal{C}^{*}$ such that

$\mu U\Sigma_{T}$ (s,t) does not exist for a given pair of terms s and t.

The class of Ω-*free* theories, $\mathcal{F}_{\Omega}$ turned out to be important for its descriptive worth:

$$\mathcal{F}_{\Omega} := \{T \in \mathcal{T}_{=} : \text{if } f(t_1,\ldots,t_n) =_{T} f(s_1,\ldots,s_n) \text{ then } t_i =_{T} s_i;\ 1 \leq i \leq n\}$$

The following results characterize $\mathcal{F}_{\Omega}$ with respect to the basic hierarchy:

Lemma 1: $\mathcal{U}_{\infty} \cap \mathcal{F}_{\Omega} \neq \emptyset$ and $\mathcal{F}_{\Omega} \neq \mathcal{U}_{\infty}$

i.e. there exists an Ω-free infinitary theory, but $\mathcal{F}_{\Omega}$ is different from $\mathcal{U}_{\infty}$.

Lemma 2: $\mathcal{U}_{\omega} \cap \mathcal{F}_{\Omega} \neq \emptyset$

i.e. there exists an Ω-free finitary theory.

Lemma 3: $\mathcal{U}_{1} \cap \mathcal{F}_{\Omega} \neq \emptyset$

i.e. there exists an Ω-free unitary theory.

But: *Problem* $\mathcal{U}_{0} \cap \mathcal{F}_{\Omega} = \emptyset$?

i.e. does $\mu U\Sigma$ exist for every Ω-free theory?

In other words $\mathcal{F}_\Omega$ is somehow 'diagonal' to the basic hierarchy of equational classes. But we have the surprising result, which gives a nice algebraic characterization of the unitary matching theories:

Theorem 5 (Szabo):
$$\mathcal{M}_1 = \mathcal{F}_\Omega$$

i.e. $\mathcal{F}_\Omega$ constitutes exactly the class of unitary matching theories.

Necessary conditions for a theory T to have an effective, minimal and complete unification algorithm is that T is unification relevant and admissible. Therefore let $\mathcal{N} = \mathcal{A}\mathcal{U}$ be the class of *normal* theories and we have by theorem 1

$$\mathcal{A}\,\mathcal{C}^*\mathcal{M}_\downarrow = \mathcal{N}\,\mathcal{C}^*\mathcal{M}_\downarrow \quad \text{resp.} \quad \mathcal{A}\,\tilde{\mathcal{C}}^*\mathcal{M}_\downarrow = \mathcal{N}\,\tilde{\mathcal{C}}^*\mathcal{M}_\downarrow$$

Theorem 4 shows that even the regular theories are not normal theories and here are some results with respect to $\mathcal{T}_=^*$:

Lemma 4:
$$\mathcal{F}_\Omega \subset \mathcal{T}_=^*$$
i.e. the Ω-free theories are regular.

Theorem 6: (Szabo) $\mu M\Sigma_T$ exists for every $T \epsilon \mathcal{T}_=^*$:
$$\mathcal{T}_=^* \subset \mathcal{M}$$

Finally we define the *permutative theories* $\mathcal{P}$ as those that have a finite equivalence class:

$$\forall T \in \mathcal{P} : \forall t \in F_\Omega[t]_{=_T} \text{ is finite.}$$

For this class we have

Proposition 2:
$$\mathcal{P} \subset \mathcal{A}$$

i.e. the permutative theories are admissible.
Also there is the important result:

Theorem 7: (Szabo)
$$\mathcal{P} \subset \mathcal{U}$$

i.e. $\mu U\Sigma_T$ always exists for permutative theories.

Lemma 5:
$$\mathcal{P} = \mathcal{P}^*$$

i.e. permutative theories are always regular.

Proposition 3:
$$\mathcal{P} \subset \mathcal{M}_{\omega 1}^*$$

i.e. permutative theories are regular and finitely matching.

Since $\mathcal{N} = \mathcal{A}\mathcal{U}$ we have by definition:

Corollary: $\mathcal{D} \subsetneq \mathcal{N}$

i.e. the permutative theories are normal theories.

Unification theory has results and hard open problems similar to the wellknown compactness theorems or the Ehrenfeucht Conjecture. These are tied to the important concept of a *local subclass of a class* $\mathcal{T}$:

Let term(T) := {l,r : l=r ∈ T} be the set of terms in T ∈ $\mathcal{T}_=$ and let I(T) be the set of instances of these terms:

$$I(T) := \{\sigma t : t \in term(T),\ \sigma \in \Sigma\} \quad .$$

Similarly we define G(T) as the *finite* set of all generalizations of these terms:

$$G(T) := \{\hat{\sigma} t:\ t \in term(T),\ \ \hat{\sigma} = [\pi \leftarrow x],\ \pi \in \Pi(t),\ x \in X\} \quad .$$

We assume terms equal under renaming to be discarded, i.e. $G(T)/_\sim$. With these two sets we obtain the *characteristic set* of an equational theory T as:

$\chi(T) := I(T) \cup G(T)$ and the finite *local-characteristic set* as:

$$\lambda(T) := term(T) \cup G(T).$$

Let $\mathcal{E}(T)$ be some first order property of T. If the property $\mathcal{E}$ is only considered with respect tc a subset Θ of F_Ω, we shall write $\mathcal{E}(T)|_\Theta$.

Definition 1: For a theory T $\mathcal{E}(T)$ is χ-*reducible* iff there is a property $\overline{\mathcal{E}}$ of T:

$$\overline{\mathcal{E}}(T)|_{\chi(T)} \text{ implies } \mathcal{E}(T)$$

Let $\mathcal{T}_\mathcal{E}$ be the class of theories having property $\mathcal{E}$ then the χ-*subclass*

$\chi\mathcal{T}_\mathcal{E} \subseteq \mathcal{T}_\mathcal{E}$ is the set:

$$\chi\mathcal{T}_\mathcal{E} := \{T \in \mathcal{T}_\mathcal{E} : \mathcal{E}(T) \text{ is } \chi\text{-}reducible\} \quad .$$

A theory T is λ-*reducible* iff there is a property $\overline{\mathcal{E}}$ of T:

$$\overline{\mathcal{E}}(T)|_{\lambda(T)} \text{ implies } \mathcal{E}(T)$$

$\lambda \mathcal{T}_{\mathcal{E}} := \{T \in \mathcal{T}_{\mathcal{E}} : \mathcal{E}(T)$ is λ-*reducible*$\}$ is the λ- *subclass* of $\mathcal{T}_{\mathcal{E}}$

For certain theories it may be possible to reduce $\mathcal{E}(T)$ to a *finite test set* $loc(T) \subset F_{\Omega}$ such that

$$\overline{\mathcal{E}}(T)\,|_{loc(T)} \quad \text{implies} \quad \mathcal{E}(T)$$

and we have in that case

$$loc\, \mathcal{T}_{\mathcal{E}} := \{T \in \mathcal{T}_{=} : \text{a finite test set } loc(T) \text{ exists}\}$$

A typical result, shown in [SZ82] is:

Theorem 8: $\mathcal{A}\mathcal{C}^*\mathcal{M}_{\infty} = \chi\mathcal{A}\mathcal{C}^*\mathcal{M}_{\infty}$

and hence we have $\mathcal{A}\mathcal{C}^*\mathcal{M}_{\omega 1} = \chi\mathcal{A}\mathcal{C}^*\mathcal{M}_{\omega 1}$. This theorem greatly simplifies the test for $T \in \mathcal{A}\mathcal{C}^*\mathcal{M}_{\infty}$ since we only have to show that it holds for matching problems on $\chi(T)$; i.e. for all problems $\langle s \geq t \rangle_T$ with $s,t \in \chi(T)$.

A major research problem of the field is to λ-reduce (or at least to χ-reduce) the property of a theory to be unitary, finitary or infinitary. A first result in this respect is the λ-reducibility of unitary matching theories:

Theorem 9: (Szabo) $\mathcal{M}_1 = \lambda \mathcal{M}_1$

The proof of this theorem demonstrates the intention of the above definition. Setting $\mathcal{E}$ to:

$\mathcal{E}(T)$: iff $T \in \mathcal{F}_{\Omega} = \mathcal{M}_1$ (i.e. the property we wish to show) and

$\overline{\mathcal{E}}(T)$: iff all terms $p,q \in \lambda(T)$ are unifiable with at most one most general unifier (i.e. they are unitary).

It can be shown that $\overline{\mathcal{E}}$ implies $\mathcal{E}$ and hence we only have to test the terms in $\lambda(T)$. In [SZ82] it is shown that this test can be even more simplified.

Theorems of this nature are of considerable practical importance since they allow an immediate classification of a given theory: Usually it is not too hard to find some unification algorithm for a given theory - however it can be very tricky to ensure that it is complete, i.e. that it generates all unifiers. But if we already know that the given theory is unitary or finitary this task is greatly simplified.

The following results are concerned with the reducibility of unitary unification theories.

In 1975 P. Hayes conjectured that Robinson's unification algorithm for free terms may well be the only case with *at most one* most general unifier.

Unfortunately this is not the case: for example let $T_a := \{a = a\}$ for any constant a, then $T_a \in \mathcal{U}_1$.

But the problem turned out to be more complex than anticipated at the time: for example let $T_{aa} := \{f(a,a) = a\}$ for any constant a, then

$$T_{aa} \in \mathcal{U}_1 .$$

We first observe that the unitary unification theories are a proper subset of the unitary matching theories:

Proposition 4: $\mathcal{U}_1 \subsetneq \mathcal{UM}_1 \subseteq \mathcal{M}_1$.

In [SZ82] it is shown that

Theorem 10: $\mathcal{U}_1 = \chi\,\mathcal{U}_1$

i.e. the unitary unification theories are χ-reducible.
But:

Conjecture: $\mathcal{U}_1 = \lambda\,\mathcal{U}_1$.

To illustrate the use of the above theorems let us consider the empty theory T_ε, i.e. the Robinson-unification problem for free terms. In order to show $T_\varepsilon \in \mathcal{U}_1$, in the stone age of unification theory one had to invent a special algorithm and then prove its completeness and correctness [RO65], [KB70].

A more elegant method is contained in [HT76]: factoring F_Ω by $\approx$, it is possible to show that $F_{\Omega/\approx}$ forms a complete semi-lattice under $\leq$. Hence if two terms are unifiable there exists a common instance and hence there exists a l.u.b., which is the most general such instance: thus follows $T_\varepsilon \in \mathcal{U}_1$.

However using the above theorem, this result is immediate: Since the absolutely free algebra of terms is in particular Ω-free: $T_\varepsilon \in \mathcal{M}_1$. Now since $\chi(T_\varepsilon)$ is empty every TEST set is empty. Hence there does not exist a pair in TEST with more than one mgu, thus follows $T_\varepsilon \in \mathcal{U}_1$.

Although the comparative study of theories and classes of theories has uncovered interesting algebraic structures this is without doubt nothing but the tip of an iceberg of yet unknown results.

2.2 *Universal Unification Algorithms*

Experience shows that unification algorithms for different theories are usually based on entirely different methods. For theoretical reasons as well as for heuristic purposes it would be interesting to have a universal unification algorithm for a whole class of theories, however inefficient it might be: A *universal unification algorithm (a universal matching algorithm) for a class of theories* $\mathcal{T}$ is an algorithm which takes as input a pair of terms (s,t) *and a theory* $T \in \mathcal{T}$ and generates a complete set of unifiers (matchers) for $\langle s = t \rangle_T$ (for $\langle s \geq t \rangle_T$). In other words, just as a universal Turing machine takes as its input a specific argument and the description of a special Turing machine, a universal unification algorithm has an input pair consisting of a special unification problem and an (equational) theory T.

To exhibit the essential idea behind the universal algorithms suppose $\langle s = t \rangle_T$ is the unification problem to be solved and let R be the rewrite system for T. Let h be a 'new' binary function symbol (not in Ω) then h(s,t) is a term. Using these conventions we have the following consequence of Birkhoff's theorem, which is the basis for all universal unification algorithms:

There exists $\sigma \in \Sigma$ with $\sigma s =_T \sigma t$

iff

there exist terms p,q and $\delta \in \Sigma$ such that

$$h(s,t) \overset{*}{\rightarrowtail}_R h(p,q) \quad \text{and} \quad \delta p =_{T_\varepsilon} \delta q.$$

Here T_ε is the empty theory, i.e. $=_{T_\varepsilon}$ denotes symbolwise equality.

A first step towards an application of this result is a proper organization of the paramodulation steps $\rightarrowtail$ into a tree, with the additional proviso that we never paramodulate into variables, i.e. if $s \rightarrowtail t$ then $s|\pi \notin X$.

For a given term t the *labeled paramodulation tree* P_t is defined as:

(i) t (the *root*) is a node in P_t

(ii) if r is a node in P_t and $r \rightarrowtail s$, then s (the successor) is a node in P_t

(iii) the edge (r,s), where $r \underset{[\pi,i,\theta]}{\rightarrowtail} s$, is labeled with the triple $[\pi,i,\theta]$.

Using the above result we have: if h(p,q) is a node in $P_{h(s,t)}$ such that p,q are Robinson-unifiable with σ then $\delta = \sigma \circ \theta$ is a correct T-unifier for s and t, where θ is the combination of all the paramodulation substitutions obtained along the path h(s,t) to h(p,q).

And vice versa for every T-unifier τ for s and t there exists a node h(p,q) in $P_{h(s,t)}$ such that p and q are Robinson-unifiable with σ and $\tau \leq_T \sigma \circ \theta$.

Of course the set of unifiers obtained with this tree is far too large to be of any interest and the work of Lankford [LB79] and Hullot [HU80], based on [FA79], is concerned with pruning this tree under the constraint of maintaining completeness. Hullot [HU80] shows the close correspondence between $\rightarrow$ (rewrite) and $\rightarrowtail$ (paramodulation, narrowing) steps and [JKK82] investigate an incremental universal unification algorithm by separating the given theory T into two constituent parts T = R∪E, where only R must be E-canonical.

Since the set of unifiers $U\Sigma_T$ is trivially recursively enumerable for $T \in \mathcal{T}_{=}$, there is the important requirement that a universal unification algorithm generates the *minimal* set $\mu U\Sigma_T$ or is at least type conformal. Since such a result is unattainable in general, there is a strong incentive to find classes of theories, such that a universal unification algorithm is minimal for every theory T within this class. But such a class should be large enough to contain the theories of practical interest. In [SS81] the class $\mathcal{A}\mathcal{C}^*\mathcal{M}_\downarrow$ is proposed and it is shown that the universal unification algorithm based on P_t is correct, minimal and complete for this class. Herold [HE82] gives an extension of this class, which is the widest currently known.

The Next 700 Unification Algorithms

These theoretical results can be applied in practice for the design of an *actual* unification algorithm. So far the design of a special purpose algorithm was more of an art than a science, since for a given theory there was no indication whatsoever of how the algorithm might work. In fact the algorithms recorded in the table of III.1 all operate on entirely different principles.

Using the universal unification algorithm as a starting point this task is now much easier by first isolating the crucial parts in the universal algorithm and then designing a practical and efficient solution.

The universal algorithm has been successfully applied to a special case

[RS78] yielding a minimal algorithm [SZ82], which in addition is much simpler than the one previously known.

A collection of canonical theories [HL80] is a valuable source for this purpose and has already been used to find the first unification algorithms for Abelian group theory and quasi group theory [LA79], [HU80].

IV. OUTLOOK AND OPEN PROBLEMS

The following seven paragraphs give some perspective and sketch some of the likely developments unification theory is to undertake in the near future.

Unification in Sorted Logics
In most practical applications variables do not range over a flat universe of discourse but are *typed*. Unification of two *typed* (or *sorted*) *terms* amounts to solving an equation in the corresponding *heterogeneous* algebra rather than in homogeneous algebras as proposed in section II. The formal framework for doing so is wellknown and has already found a proper place in computer science as a tool for the description of abstract data types.

Depending on the structure of the sorts (usually some form of a lattice) the extension of the known results to sorted domains is not trivial.

Complexity results and Special Purpose Theories
Except for the (SNOBOL) string matching problem and the unification problem in free terms (Robinson) no complexity results are known. Good candidates for the next least-complexity-race may be unification under commutativity or idempotence, since they have fairly simple algorithms, there is a practical demand for efficiency and finally the known techniques of [PW78], [KK82] may be extendable to these cases.

Also there is every incentive to obtain a much larger collection of special purpose unification algorithms.

Combination of Theories
Why is the combination of a finitary theory with an infinitary theory sometimes a finitary theory whereas in other cases it is infinitary? Is it possible to develop a systematic theory of a *combinator of theories*, say $T_1 \oplus T_2$, where T_1 and T_2 are equational theories? A similar problem is known for simplification algorithms.[SH84] [NO80]

What is the algebraic structure of $\oplus$ (i.e. a theory *whose objects are theories*) with respect to unification theory?

Paraunification
For many practical applications the requirement that two terms are unifiable in the strict sense as defined above is too rigid. For example the *matching of descriptions* in artificial intelligence does not demand proper T-equality. Instead there is interest in algorithms which detect whether or not the "essential components" of two descriptions coincide.

Can this problem be expressed within our algebraic framework of unification theory? In [SZ82] *affinity* of two terms s and t is defined such that s and t are *affin*, $s \underset{\Psi}{\sim} t$, if they coincide in their essential components $\Psi \subset \Omega$.

A *paraunification problem* $\langle s \underset{\Psi}{\sim} t\rangle_T$ is the problem to find a substitution δ for s and t such that $\delta s \underset{\Psi}{\sim} \delta t$.

This notion expresses in a more abstract way the classical notion of an *approximation* of a solution.

Subunification
If a term s is a subterm of some $\bar{t} \in [t]_T$ we write $s \subseteq_T t$. The *subunification problem* $\langle s \subseteq t\rangle_T$ is the problem to find a substitution δ for s and t such that $\delta s \subseteq_T \delta t$.

Again there is a practical need for subunification algorithms.

Higher Order Unification
Although the unification of two terms of order ω is outside of the scope of this survey article we like to suggest one interesting aspect related to the work recorded here. The undecidability results for second and higher order unification [HT73], [LC72], [GO81] as well as the enormeous proliferation of unifiers even for small problems [HT76], [HE75] have cast some shadows on earlier hopes for higher order theorem proving [RN67]. But may be T-unification for ω-order logics is not more *but less complex* then free ω-unification? For example the second order monadic unification problem closely resembles the stringunification problem.

Now the stringunification problem is infinitary, it posed a very hard decidability problem and the known stringunification algorithms are almost useless for all practical purposes. However stringunification *under commutativity* (i.e. the A+C-problem) is comparatively simple: it is finitary, decidability is easy to compute and the unification algorithms [ST81], [LS76] are not too far away from practical applicability

Open Problems

Whereas the previous paragraphs listed extensions and interesting fields of investigation we now like to list some specific open problems.

P1: $\mathcal{U}_1 = \lambda\mathcal{U}_1$; i.e. can the test for a unitary theory be further localized to the finite test set $\lambda(T)$?

P2: Characterize the borderline between finitary and infinitary theories, i.e. $\mathcal{U}_\omega$ and $\mathcal{U}_\infty$. This is <u>the</u> major open problem right now.

P3: $\mathcal{M}_\omega = \lambda\mathcal{M}_\omega$? $\mathcal{M}_\omega = \chi\mathcal{M}_\omega$? $\mathcal{M}_\infty = \lambda\mathcal{M}_\infty$?

P4: $T \in \mathcal{M}_1$ decidable? Note: $\mathcal{M}_1 = \lambda\mathcal{M}_1$

P5: $\mathcal{U}_0 \ \mathcal{M}_1 \neq \emptyset$? Does there exist a type-zero theory which is unitary matching?

P6: $T \in \mathcal{N}\mathcal{M}^*_{\omega 1}$ decidable? i.e. in the light of the above results is $T \in \mathcal{D}$ decidable?

P7: Does there exist a minimal (and/or type conformal) universal unification algorithm for the whole class $\mathcal{A}\mathcal{C}^*\mathcal{M}_{\omega 1}$?

P8: Does there exist a type conformal (i.e. terminating) universal matching algorithm for $\mathcal{A}\mathcal{C}^*\mathcal{M}_{\omega 1}$? Note this is a prerequisite for P7.

Does there exist a type conformal universal matching algorithm for $\mathcal{A}\mathcal{C}\mathcal{M}_{\omega 1}$? Since this is probably not the case: show its unsolvability. Where is the exact borderline?

P9: $\mathcal{C} = \mathcal{A}\mathcal{C}$? i.e. are the canonical theories admissible?

(P10): $\mathcal{T}_= = \mathcal{T}_\Rightarrow$? i.e. can every finitely based theory with a decidable word problem be embedded into a rewrite system? This would have strong implications for universal unification algorithms.

P11: (permutative theories). Let $\mathcal{D}_i = \mathcal{U}_i\mathcal{D}$ $i \in \{1,\omega,\omega 1\}$.

Does there exist a type conformal universal unification algorithm for $\mathcal{D}_i$? Is $T \in \mathcal{D}_i$ decidable?

P12: If $T \in \mathcal{M}_1$ is $T \in \mathcal{U}_1$ decidable?

P13: (existence problem). Give an algebraic characterization of classes of theories such that $\mu U\Sigma_T$ exists for T in this class. Is $T \in \mathcal{U}_0$ decidable?

P14: The problematic issue of universal unification algorithms can be reduced to the following question: Given a unifier δ for s and t under T, i.e. $\delta s =_T \delta t$; *is* δ *most general?* Since the question can not be answered in general: for which equational class is it decidable?

P15: In many applications it is useful to have very fast, albeit incomplete unification algorithms. However they should not be "too incomplete".

Because of its theoretical beauty, its fundamental nature as well as its practical significance, unification theory is likely to develop into a major subfield of computer science.

ACKNOWLEDGEMENT

This is an updated version of a paper I wrote two years ago with Peter Szabó, with whom I worked very closely on unification problems for almost five years.

Although I see the disillusionment every day university life can cause not only in a young and idealistic mind, I deeply regret his decision to leave academic life. I like to remember him as one of my best friends and as the most extraordinarily gifted person I had the opportunity to work with.

V. BIBLIOGRAPHY

[BA72] Barrow, Ambler, Burstall: 'Some techniques for recognizing Structures in Pictures', Frontiers of Pattern Recognition, Academic Press Inc., 1972

[BA78] L.D. Baxter:'The Undecidability of the Third Order Dyadic Unification Problem', Information and Control, vol.38, no.2, 1978

[BA73] L.D. Baxter: 'An efficient Unification Algorithm', Rep. CS-73-23, University of Waterloo, Dept. of analysis and Computer Science, 1973

[BC66] H. Bryan, J. Carnog:'Search methods used with transistor patent applications', IEEE Spectrum 3, 2, 1966

[BF77] H.P. Böhm, H.L. Fischer, P. Raulefs: 'CSSA: Language Concepts and Programming Methodology', Proc. of ACM, SIGPLAN/ART Conference, Rochester, 1977

[BL71] F. Blair et al: 'SCRATCHPAD/1: An interactive facility for symbolic mathematics', Proc. of the 2nd Symposium on Symbolic Manipulation, Los Angeles, 1971

[BL77] A. Ballantyne, D. Lankford: 'Decision Procedures for simple equational theories', University of Texas at Austin, ATP-35, ATP-37, ATP-39, 1977

[BM77] R. Boyer, J.S. Moore: 'A Fast String Searching Algorithm', CACM vol. 20, no. 10, 1977

[BO68] D.G. Bobrow (ed):'Symbol Manipulation Languages', Proc. of IFIP, North Holland Publishing Comp., 1968

[BO77] H. Boley:'Directed Recursive Labelnode Hypergraphs: A New Representation Language', Journal of Artificial Intelligence, vol. 9, no. 1, 1977

[CA70] Caviness: 'On Canonical Form and Simplification', JACM, vol. 17, no. 2, 1970

[CD69] CODASYL Systems Committee: 'A survey of Generalized Data Base Management Systems', Techn.Rep. 1969, ACM and IAG

[CD71] CODASYL Systems Committee:'Feature Analysis of Generalized Data Base Management Systems', TR 1971, ACM, BC and IAG

[CK67] Cook: 'Algebraic techniques and the mechanization of number theory', RM-4319-PR, Rand Corp., Santa Monica, Cal., 1965

[CK71] C. Christensen, M. Karr: 'IAM, A System for interactive algebraic Manipulation', Proc. of the 2nd Symposium on Symbolic Manipulation, Los Angeles, 1971

[CL71] M. Clowes: 'On Seeing Things', Techn.Rep. 1969, ACM and IAG

[CM81] W. Clocksin, C. Mellish: 'Programming in PROLOG', Springer 1981

[CO70] E.F. Codd: 'A Relational Model of Data for Large shared Databanks', CACM, 13,6,1972

[CO72] E.F. Codd: 'Relational Completeness of Data Base Sublanguages', in Data Base Systems, Prentice Hall, Courant Comp. Science Symposia Series, vo. 6, 1972

[CP61] A. Clifford, G. Preston: 'The Algebraic Theory of Semigroups', vol.I and vol. II, 1961

[CR68] D.G. Corneil: 'Graph Isomorphism', Ph.D.Dept. of Computer Science, University of Toronto, 1968

[DA71] J.L. Darlington: 'A partial Mechanization of Second Order Logic', Mach.Int. 6, 1971

[DA76] C.J. Date: 'An Introduction to Database Systems', Addison-Wesley Publ. Comp. Inc., 1976

[DA73] M. Davis: 'Hilpert's tenth Problem is unsolvable', Amer.Math.Monthly, vol. 80, 1973

[FA71] R. Fateman: 'The User-Level Semantic Matching Capability in MACSYMA', Proc. of the 2nd Symposium on Symbolic Manipulation, Los Angeles, 1971

[FA79] M. Fay: 'First Order Unification in an Equational Theory', Proc. 4th Workshop on autom. Deduction, Texas, 1979

[FA83] F. Fage: 'Associative Commutative Unification', INRIA report CNRS-LITP4, 1983 (see also this volume)

[FG64] D.J. Farber, R.E. Griswald, I.P. Polonsky: 'SNOBOL as String Manipulation Language', JACM, vol. 11, no.2, 1964

[FH83] F. Fage, G. Huet: 'Complete Sets of Unifiers and Matchers in Equational Theories', Proc. CAAP-83, Springer Lec.Notes Comp. Sci, vol. 159, 1983

[GI73] J.F. Gimpel: 'A Theory of Discrete Patterns and their Implementation in SNOBOL4, CACM 16, 2, 1973

[GM78] H. Gallaire, J. Minker. 'Logic and Databases', Plenum Press, 1978

[GO66] W.E. Gould. 'A matching procedure for ω-order logic', Scientific report no.4, Air Force Cambridge Research Labs., 1966

[GO67] J.R. Guard, F.C. Oglesby, J.H. Benneth, L.G. Settle: 'Semi-Automated Mathematics', JACM 1969. vol. 18, no.1

[GO81] D. Goldfarb: 'The Undecidability of the Second Order Unification Problem', Journal of Theor. Comp.Sci., 13, 1981

[GR79] G. Grätzer. 'Universal Algebra', Springer Verlag, 1979

[GS84] W.K. Giloi, J. Siekmann: 'Entrichtungen von Rechnerarchitekturen für Anwendungen in der Künstlichen Intelligenz', BMFT-Antrag 1984, Univ. Berlin und Kaiserslautern

[GU64] J.R. Guard: 'Automated logic for semi-automated mathematics', Scientific report no.1, Air Force Cambridge Research Labs., AD 602 710, 1964

[HD82] A. Herold: 'Universal Unification and a Class of Equational Theories', Proc. GWAI-82, W. Wahlster (ed) Springer Fachberichte, 1982

[HE82] A. Herold: 'Some Basic Notions of first Order Unification Theory', Univ. Karlsruhe, Interner Report, 1983

[HE30] J. Herbrand. 'Recherches sour la theorie de la demonstration', Travaux de la Soc. des Sciences et des Lettre de Varsovie, no.33, 128, 1930

[HE75] G.P. Huet: 'A Unification Algorithm for typed λ-Calculus', J.Theor. Comp. Sci., 1, 1975

[HJ64] J.I. Hmelevskij: 'The solution of certain systems of word equations', Dokl.Akad. Nauk SSSR, 1964, 749 Soviet Math. Dokl.5, 1964, 724

[HJ66] J.I. Hmelevskij: 'Word equations without coefficients', Dokl. Acad. Nauk. SSSR 171, 1966, 1047 Soviet Math. Dokl. 7, 1966, 1611

[HJ67] J.I. Hmelevskij: 'Solution of word equations in three unknowns', Dokl. Akad. Nauk. SSR 177, 1967, no.5, Soviet Math. Dokl. 8, 1967, no. 6

[HL80] J.M. Hullot: 'A Catalogue of Canonical Term Rewriting Systems, Research Rep. CSL-113, SRI-International, 1980

[HN71] A. Hearn: 'REDUCE2, A System and Language for Algebraic Manipulation', Proc. of the 2nd Symposium on Symbolic Manipulation, Los Angeles, 1971

[HO76] J. Howie: 'Introduction to Semigroup Theory', Acad. Press 1976

[HO80] G. Huet, D.C. Oppen: 'Equations and Rewrite Rules', in 'Formal Languages: Perspectives and Open Problems', Ed. R. Book, Academic Press, 1980

[HR73] S. Heilbrunner: 'Gleichungssysteme für Zeichenreihen', TU München, Abtl. Mathematik, Ber.Nr. 7311,1973

[HT72] C. Hewitt: 'Description and Theoretical analysis of PLANNER a language for proving theorems and manipulating models in a robot', Dept. of Mathematics, Ph.C. Thesis, MIT, 1972

[HT76] C. Hewitt. 'Viewing Control Structures as Patterns of Passing Massages', MIT, AI-Lab., Working paper 92, 1976

[HT72] G. P. Huet. 'Constrained resolution: a complete method for theory', Jenning's Computing Centre rep. 1117, Case Western Reserve Univ., 1972

[HT73] G.P. Huet: 'The undecidability of unification in third order logic', Information and Control 22 (3), 257-267, 1973

[HT75] G. Huet: 'Unification in typed Lambda Calculus', in λ-Calculus and Comp. Sci. Theory, Springer Lecture Notes, No.37, Proc. of the Symp. held in Rome, 1975

[HT76] G. Huet: 'Resolution d'equations dans des langauges d'ordere 1,2,..,ω', These d'Etat, Univ. de Paris, VII, 1976

[HT78] G. Huet: 'An Algorithm to Generate the Basis of Solutions to Homogenous Linear Diophantine Equations', Information Proc. letters 7,3, 1978

[HT80] G. Huet: 'Confluent reductions: Abstract Properties and Applications to Term Rewriting Systems', JACM vo. 27, no.4, 1980

[HU79] J.M. Hullot: 'Associative Commutative Pattern Matching', 5th Int.Joint Conf. on AI, Tokyo 1979

[HU80] J.M. Hullot: 'Canonical Forms and Unification', Proc. of 5th Workshop on Automated Deduction', Springer Lecture Notes, 1980

[JP73] D. Jensen, T. Pietrzykowski: 'Mechanising λ-order type theory through unification', Rep. CS73-16, Dept. of Applied Analysis and Comp. 4, 1972

[JKK82] J. Jouannaud, C. Kirchner, H. Kirchner: 'Incremental Unification in Equational Theories', Université de Nancy, Informatique, 82-R-047, 1982

[KB70] D. E. Knuth, P.B. Bendix: 'Simple word Problems in Universal Algebras', in: Computational Problems in Abstract Algebra, J. Leech (ed), Pergamon Press, Oxford, 1970

[KK82] D. Kapur, M.S. Krishnamoorthy, P. Narendran: 'A new linear Algorithm for Unification', General Electric, Rep. no. 82CRD-100, New York, 1982

[KM72] Karp, Miller, Rosenberg: 'Rapid Identification of repeated Patterns in Strings, Trees and Arrays', ACM Symposium on Th.of Comp. 4, 1972

[KM74] Knuth, Morris, Pratt: 'Fast Pattern Matching in Strings', Stan-CS-74-440, Stanford University, Comp. Sci. Dept., 1974

[KM77] S. Kühner, Ch. Mathis, P. Raulefs, J. Siekmann: 'Unification of Idempotent Functions', Proceedings of 4th IJCAI, MIT, Cambridge, 1977

[KO79] R. Kowalski: 'Logic for Problem Solving', North Holland, 1979

[LA79] D.S. Lankford: 'A Unification Algorithm for Abelian Group Theory', Rep. MTP-1, Louisiana Techn. Univ., 1979

[LA80] D.S. Lankford: 'A new complete FPA-Unification Algorithm', MIT-8, Louisiana Techn. Univ., 1980

[LB79] D.S. Lankford, M. Ballantyne: 'The Refutation Completeness of Blocked Permutative Narrowing and Resolution', 4th Workshop on Autom. Deduction, Texas, 1979

[LBB84] D. S. Lankford, G. Butler, B. Brady: 'Abelian Group Unification Algorithms for elementary terms', to appear in: Contempory Mathematics.

[LC72] C.L. Lucchesi: 'The undecidability of the unification problem for third order languages', Rep. CSRR 2059, Dept. of Applied Analysis and Comp. Science, Univ. of Waterloo, 1972

[LO80] D. Loveland: 'Automated Theorem Proving', North Holland, 1980

[LS75] M. Livesey, J. Siekmann: 'Termination and Decidability Results for Stringunification', Univ. of Essex, Memo CSM-12, 1975

[LS76] M. Livesey, J. Siekmann: 'Unification of Sets and Multisets', Univ. Karlsurhe, Techn.Report, 1976

[LS79] M. Livesey, J. Siekmann, P. Szabo, E. Unvericht: 'Unification Problems for Combinations of Associativity, Commutativity, Distributivity and Indempotence Axioms', Proc. of Conf. on Autom. Deduction, Austin, Texas, 1979

[LS73] G. Levi, F. Sirovich: 'Pattern Matching and Goal directed Computation', Nota Interna B73-12, Univ. of Pisa, 1973

[MA54] A.A. Markov: 'Trudy Mat.Inst.Steklov', no.42, Izdat.Akad. Nauk SSSR, 1954, NR17, 1038,1954

[MA70] Y. Matiyasevich: 'Diophantine Representation of Rec. Enumerable Predicates', Proc. of the Scand. Logic Symp., North Holland, 1978

[MA77] G.S. Makanin: 'The Problem of Solvability of Equations in a Free Semigroup', Soviet Akad, Nauk SSSR, Tom 233, no.2, 1977

[MA77] Maurer: 'Graphs as Strings', Universität Karlsruhe, Techn. Rep., 1977

[MB68] Manove, Bloom, Engelmann: 'Rational Functions in MATHLAB', IFIP Conf. on Symb. Manipulation, Pisa, 1968

[MM79] A. Martelli, U. Montaneri: 'An Efficient Unification Algorithm', University of Pisa, Techn. Report, 1979

[MO71] J. Moses: 'Symbolic Integration: The Stormy Decade', CACM 14, 8, 1971

[MO74] J. Moses: 'MACSYMA - the fifth Year', Project MAC, MIT, Cambridge, 1974

[NE71] A. Nevins: 'A Human oriented logic for ATP', JACM 21, 1974 (first report 1971)

[NI80] N. Nilsson: 'Principles of Artificial Intelligence', Tioga Publ. Comp.,Cal., 1980

[NO80] G. Nelson, D. Oppen: 'Fast Decision Procedures Based on Congruence Closure', JACM, 27, 2, 1980

[PL72] G. Plotkin: 'Building in Equational Theories', Machine Intelligence, vol. 7, 1972

[PR60] D. Prawitz: 'An Improved Proof Procedure', Theoria 26, 1960

[PS81] G. Peterson, M. Stickel: 'Complete Sets of Reductions for Equational Theories with Complete Unification Algorithms', JACM, vol. 28, no.2, 1981

[PW78] M. Paterson, M. Wegman: 'Linear Unification', J. of Comp. and Syst. Science, 1968, 16

[RD72] Rulifson, Derksen, Waldinger: 'QA4: A procedural calculus for intuitive reasoning', Stanford Univ., Nov. 1972

[RL69] J. Rastall: 'Graph-family Matching', University of Edinburgh, MIP-R-62, 1969

[RN67] J.A. Robinson: 'A review on automatic theorem proving', Symp. Appl.Math., vol. 19, 1-18, u967

[RO65] J.A. Robinson: 'A Machine Oriented Logic based on the Resolution Principle', JACM 12, 1965

[RO71] J.A. Robinson: 'Computational Logic: The Unification Computation', Machine Intelligence, vol. 6, 1971

[RS78] P. Raulefs, J. Siekmann: 'Unification of Idempotent Functions', Universität Karlsruhe, Techn. Report, 1978

[RSS79] P. Raulefs, J. Siekmann, P. Szabo, E. Unvericht: 'A short Survey on the State of the Art in Matching and Unification Problems, SIGSAM Bulletin, 13, 1979

[SB82] P. Szabo: 'Undecidability of the D_A-Unification Problem', Proc. of GWAI, 1979

[SB82] J. Siekmann, P. Szabo: 'A Minimal Unification Algorithm for Idempotent Functions', Universität Karlsruhe, 1982

[SH76] E.H. Shortliffe: 'MYCIN: Computer Based Medical Consultations', North Holland Publ. Comp. 1976

[SH75] B.C. Smith, C. Hewitt: 'A Plasma Primer', MIT, AI-Lab., 1975

[SH84] R. Shostak: 'Deciding Combinations of Theories', JACM, vol. 31, no.1, 1984

[SG77] SIGSAM Bulletin: 'ACM special interest group on Symbolic and Algebraic Manipulation, vol. 11, no.3, 1977 (issue no. 43) contains an almost complete bibliography

[SI75] J. Siekmann: 'Stringunification' Essex University, Memo CSM-7, 1975

[SI76] J. Siekmann: 'Unification of Commutative Terms', Uni. Karlsruhe, 1976

[SI78] J. Siekmann: 'Unification and Matching Problems', Ph.D., Essex Univ., Memo CSA-4-78

[SL72] J. R. Slagle: 'ATP with built-in theories including equality, partial ordering and sets', JACM 19, 120-135, 1972

[SL74] J. Slagle: 'ATP for Theories with Simplifiers, Commutativity and Associativity', JACM 21, 1974

[SO82] J. Siekmann, P. Szabo: 'Universal Unification and a Classification of Eqautional Theories', Proc. of Conf. on Autom. Deduction, 1982, New York, Springer Lecture Notes comp. Sci., vol. 87

[SS81] J. Siekmann, P. Szabo: 'Universal Unification and Regular ACFM Theories', Proc. IJCAI-81, Vancouver, 1981

[SS82] J. Siekmann, P. Szabo: 'A Noetherian and Confluent Rewrite System for Indempotent Semigroups', Semigroup Forum, vol. 25, 1982

[SS61] D. Skordew, B. Sendow: 'Z. Math. Logic Grundlagen', Math.7 (1961), 289, MR 31, 57 (Russian) (English translation at Univ. of Essex, Comp. Sci. Dept.)

[ST81] M. Stickel: 'A Unification Algorithm for Assoc. Commutative Functions', JACM, vol. 28, no.3, 1981

[ST74] G.F. Stewart: 'An Algebraic Model for String Patterns', Univ. of Toronto, CSRG-39, 1974

[SU65] E. Sussenguth: 'A graph-theoretical algorithm for matching chemical structures', J. Chem. Doc. 5, 1, 1965

[SU78] P. Szabo, E. Unvericht: 'The Unification Problem for Distributive Terms', Univ. Karlsruhe, 1978

[SZ78] P. Szabo: 'Theory of First Order Unification' (in German, thesis) Univ. Karlsruhe, 1982

[TA68] A. Tarski: 'Equational Logic and Equational Theories of Algebra', Schmidt et al (eds), Contributions to Mathematical Logic, North Holland, 1968

[TE81] H. Tennant. 'Natural Language Processing', Petrocelli Books, 1981

[TY75] W. Taylor: 'Equational Logic', Colloquia Mathematica Societatis Janos Bolya, 1975

[UL76] J.R. Ullman: 'An Algorithm for Subgraph Isomorphism', JACM, vol. 23, no.1, 1976

[UN64] S.H. Unger: ´GIT - Heuristic Program for Testing Pairs of directed Line Graphs for Isomorphism`, CACM, vol.7, no.1, 1964

[VA75] J. van Vaalen: ´An Extension of Unification to Substitutions with an Application to ATP`, Proc. of Fourth IJCAI, Tbilisi, USSR, 1975

[VO78] E. Vogel: ´Unifikation von Morphismen`, Diplomarbeit, Univ. Karlsruhe, 1978

[VZ75] M. Venturini-Zilli: ´Complexity of the Unification Algorithm for First Order Expression`, Calcolo XII, Fasc IV, 1975

[WA77] D.H.D. Warren: ´Implementing PROLOG`, vol. 1 and vol. 2, D.A.I. Research Rep., no. 39, Univ. of Edinburgh, 1977

[WA84] Ch. Walther: ´Unification in Many Sorted Theories`, Univ. Karlsruhe, 1984

[WC76] K. Wong, K. Chandra: ´Bounds for the String Editing Problem`, JACM, vol. 23, no.1, 1976

[WE73] P. Weiner: ´Linear Pattern Matching Algorithms`, IEEE Symp. on SW. and Automata Theory, 14, 19773

[WH98] N. Whitehead: ´Treatise on Universal Algebra`, 1898

[WI76] van Wijngaarden (et all): ´Revised Rep. on the Algorithmic Language ALGOL68`, Springer-Verlag, Berlin, Heidelberg, N.Y., 1976

[WN75] Winston: ´The Psychology of Computer Vision`, McGraw Hill, 1975

[WN76] G. Winterstein: ´Unification in Second Order Logic`, Bericht 3, Univ. Kaiserslautern, 1976

[WR67] L. Wos, G.A. Robinson, D. Carson, L. Shalla: ´The Concept of Demodulation in Theorem Proving`, JACM, vol. 14, no.4, 1967

[WR73] L. Wos, G. Robinson: ´Maximal Models and Refutation Completeness: Semidecision Procedures in Automatic Theorem Proving`, in: Word problems (W.W. Boone, F.B. Cannonito, R.C. Lyndon, eds), North Holland, 1973

A Portable Environment for Research in Automated Reasoning

Ewing L. Lusk

Ross A. Overbeek

Mathematics and Computer Science Division
Argonne National Laboratory
Argonne, Illinois 60439

ABSTRACT

The Interactive Theorem Prover (ITP), an environment that supports research into the theory and application of automated reasoning, is described. ITP is an interactive system providing convenient access to and control of the many inference mechanisms of Logic Machine Architecture (LMA), described elsewhere. LMA itself has been substantially enhanced since the last report on its status, and we describe here some of the enhancements, particularly the addition of a tightly-coupled logic programming component, which provides an integration of the theorem-proving and logic programming approaches to problems represented in the predicate calculus.

1. Introduction

In [2] and [4] the authors defined a layered software architecture called Logic Machine Architecture (LMA), for the construction of inference-based systems. Many of the basic decisions were motivated by the considerations given in [3]. The primary purpose of this paper is to describe ITP, the first major system built with the LMA tools. ITP is itself a tool for research rather than system development. It provides access to and control of all inference mechanisms provided by LMA, allowing its user to experiment easily with a wide variety of inference rules and strategies for their use. In addition, the LMA architecture and its own design make convenient the addition of new capabilities and experimentation with alternatives to current design decisions. The system (written in Pascal) is highly portable, in the public domain, and has been distributed to approximately fifty university and industrial sites. Its purpose is to make unnecessary the large programming investment which used to be required before one could undertake serious experimentation in automated reasoning.

A secondary purpose of this paper is to present the current status of LMA, which has evolved substantially since the publication of its original definition. The primary enhancement has been the addition of a logic programming component, available to the ITP user as an independent Prolog subsystem, or as an inference mechanism integrated into the other inference mechanisms of LMA. For example, in the course of computing a hyperresolvent, a literal recognized as a "Prolog" literal may be "resolved" away by querying the Prolog subsystem. Two-way communication can occur between the theorem-proving and Prolog subsystems. This communication takes place at the literal rather than clause level, permitting close cooperation during an attempt to satisfy a Prolog goal or to compute a resolvent.

The Prolog subsystem is the first step in the creation of an integrated environment in which theorem-proving systems may interact with other "intelligent" systems, such as symbol-manipulation packages like MACSYMA and logic programming systems like Prolog. ITP provides to the researcher in automated reasoning an integrated, uniform environment for exploring

This work was partially supported by the Applied Mathematical Sciences Research Program (KC-04-02) of the Office of Energy Research of the U.S. Department of Energy under Contract W-31-109-Eng-38, and also by National Science Foundation grant MCS82-07496.

these interactions.

In the following sections, we describe the research environment provided by ITP, describe in detail the integrated Prolog component, and discuss the issues raised by combining disparate systems possessing reasoning capabilities. We conclude with a report on the current status of the porting/distribution subproject, and describe our plans for LMA in two widely differing computational environments: microcomputer systems and supercomputers.

2. ITP Facilities

ITP has been designed to provide a rich, powerful, and friendly environment for research in automated reasoning. It offers a large number of features and options to control their use, so that a wide variety of experiments can be conducted. ITP is also easily extendible, so that new features inspired by experiments can be added smoothly to the system.

2.1. Absence of Limits

ITP inherits from LMA the freedom from system-imposed limits on the number of clauses present in the clause space, the number of literals or variables per clause, the length of names, etc. Past experience has painfully taught us that any such limits established in the name of efficiency are soon made obsolete by new types of problems one wants the system to handle.

2.2. Representation Languages

LMA is currently an entirely clause-based system. Thus, it requires that knowledge be represented to it in clauses. However, the user has several options for representing the clauses. One form is that used by LMA as its external (character string) form of data, designed for ease of parsing, not reading. ITP translates a number of different user-oriented languages into and out of this language. Some of the languages supported are:

a) If-then format: If p(x) & q(x) then r1(x) | r2(x)

b) The format used by our earlier system, AURA

c) The Clocksin-Mellish [1] format for Prolog clauses

d) A special-purpose language oriented toward circuit design applications.

Because input and output languages are independent, any of the above languages can be translated into any other by reading in a list of clauses using one language, and writing it out using another. Except for minor incompatibilities due to naming conventions and built-in functions, the languages are interchangeable.

The specific languages supported are not as significant as the ease with which new languages can be added. Because the translation is to the LMA portable format, one need not know anything about the LMA internal data structures to write a new translator for a specific purpose.

2.3. Basic Operation

The basic operation of ITP is quite straightforward. Its power derives from the many options which control its fundamental algorithm.

The clause space of ITP is divided into four lists of clauses: the Axiom List, the Set-of-Support List, the Have-Been-Given List, and the Demodulator List. Each plays a specific role in the fundamental operation, which is repeated many times in the course of one execution of the program. The fundamental operation consists of the following steps:

1. Choose a clause from the Set-of-Support. Call this clause "the given clause".
2. Infer a set of clauses, using one or more of the inference rules listed below, which have the given clause as one parent and the other parent clauses selected from the Axiom List and the Have-Been-Given List.
3. For each generated clause, "process" it (simplify, perform subsumption checks, evaluate it for addition to the clause space, etc.) If the clause is to be kept, add it to the end of the Set-of-Support List.
4. Move the given clause from the Set-of-Support List to the Have-Been-Given List.

The exact way in which each of these steps is carried out (how the given clause is chosen, which inference rules are used, etc.), is determined by user-controlled options, described below.

2.4. Options

ITP has many options to control its operation and to customize its interface to a particular application or user. Options are manipulated through a hierarchical set of self-explanatory menus. A set of options can be saved in a file and restored at the beginning of a later run. Multiple options sets can be used at different times during the same run. In this paper, we will only describe the most significant of the choices available to the user.

2.4.1. Choosing a Given Clause

An option controls how a given clause is chosen from the Set-of-Support. If the first clause is always chosen, a breadth-first search results. If the last clause is always chosen, a depth-first search results. Alternatively, each clause on the Set-of-Support List is evaluated according to a weighting scheme[7], and the "best" clause is selected. Weighting can be used to attempt to focus a search when the user has some idea of how to determine "relevance".

2.4.2. Inference Rules

The inference rules currently supported in ITP are the ones we have found most useful over the years. They are:

1. Hyperresolution
2. Unit-Resulting resolution
3. Paramodulation into given clause
4. Paramodulation from given clause
5. Binary resolution
6. Unit resolution
7. Link-resolution (still experimental)
8. Forward demodulation
9. Backward demodulation
10. Factoring
11. Unit Deletion
12. Negative hyperresolution
13. Linked UR-resolution

Paramodulation has a collection of suboptions controlling the instantiations allowed to occur. The two linked resolution inference rules each have a variety of options controlling their operations.

Some of the inference rules which generate a new clause by performing a sequence of resolution steps have been enhanced. In particular, some literals can be "removed" by evaluating built-in predicates, one of which can be used to query other reasoning systems. For example, if

If p(x) & $LT(0,x) & $ASK(prolog,path(node(x),node(0))) then q(x)

and

p(10)

are clauses, then q(10) can be derived, provided "prolog" can establish that the goal "path(node(10),node(0))" succeeds. The first antecedent literal is removed by the unit clause; the second, by evaluating the special predicate $LT (which evaluates to "true", if the first argument is less than the second); and the third, by evaluating the special predicate $ASK (used to query the prolog system in this example).

The exact inference rule options are not as important as the fact that a rich set is included and it is easy to add new options. New inference rules automatically and consistently interact with both the higher level algorithms and with the database mechanisms in Layer 1 of LMA which provide rapid access to the relevant terms and clauses.

2.4.3. Processing Generated Clauses

New clauses generated by the previous inference rules are (optionally) processed in a number of ways before being added to the clause space.

First they are demodulated. Demodulation is a complete term-rewriting mechanism[8, 9], driven by the unit equality clauses on the Demodulator List. Built-in functions (e.g., arithmetic operations) can also be used to simplify the derived clauses.

After demodulation, a complete subsumption check is performed.

Next, the clause is evaluated, using a weighting mechanism evolved from the one described in [7]. The templates which direct the weighting algorithm are part of the options. The set is used here from the one used to select the given clause.

If the demodulated version of the clause is neither subsumed nor rejected by the weighting mechanism, it is integrated into the clause space. Optionally, "back subsumption" is performed, in which existing clauses in the clause space less general than the new clause are deleted. Back demodulation occurs at this point: If the new clause is a unit equality clause and meets certain other criteria, it is added to the Demodulator List and is immediately applied to all existing clauses. Any new clauses resulting from the process are then completely processed, just as if they had been generated by an inference rule.

If specified by the user, each step of the previous procedure is logged to a file, which in turn is used in the proof examination facility. A proof can be extracted from the log, and each step of the proof can be explained in detail.

2.4.4. Utilities

The logging facility is an example of a utility, a subsystem of the program not strictly necessary for making experimental runs, but one which makes the research process more convenient. Other utilities include the ability to save the status of a run and restore it later, and two interactive subsystems for studying the effects of demodulators and weighting templates in isolation from the rest of the theorem-proving process. It is also possible to run the theorem-proving process itself in interactive mode, adding and deleting clauses, generating new clauses one at at time, and accepting or rejecting them "by hand". This mode is particularly useful when one has in advance some idea of how a deduction should proceed.

2.4.5. Documentation

User-oriented documentation is distributed with the system. One manual[5] is a tutorial on how to use ITP. It contains a description of each command and of the effects of each option. The tutorial contains a sample session and many examples. Another manual[6] is provided for those intending to add more functionality to ITP. Because in LMA terms, ITP is a Layer 3 program, adding functionality requires an understanding of the Layer 2 subroutines. These are listed together with their calling sequences and a discussion of the relevant data types.

Some of the information in [5] is available on-line as well. The help facility is file-driven and easily extended or customized by the user.

3. Integration of Theorem-Proving and Logic Programming

3.1. Tightly-Coupled Logic Programming

LMA and ITP now support a Prolog component, which can be used in two ways. There is an ITP command which initiates a Prolog subsystem. Within this subsystem, the user is interacting with normal Prolog. As throughout LMA, no limits exist on the number of literals (subgoals) per clause or on the number of variables per clause. Because the Prolog subsystem is contained within ITP, all language options are available. We have used as input to the Prolog component clauses that were written with a theorem-proving run in mind, and vice versa.

The most interesting use of the Prolog component occurs when it is tightly bound into the other inference mechanisms of LMA. As an illustration, consider the inference rule hyperresolution. From the theorem-proving point of view, the negative literals of a clause represent subgoals to be achieved in the formation of a hyperresolvent. Some of these subgoals might be best attacked by the use of Prolog. In LMA, the user can identify certain subgoals as appropriate to the Prolog subcomponent. The details are as follows.

A fifth list of clauses is added to ITP, representing the standard Prolog database of facts (positive unit clauses) and procedures (non-unit Horn clauses). The special predicates, $ASK and $TELL, are incorporated into ITP to allow communication with "foreign" systems. For example, if the Axiom List contains:

```
If Status(x,Failed) & $ASK(prolog,ispath(x,y))
 then Status(y,Suspicious)
```

and the Prolog list contains:

```
conn(a,b).
conn(b,c).
ispath(X,Y) :- conn(X,Y).
ispath(X,Y) :- conn(X,Z), ispath(Z,Y).
```

and the Set-of-Support contains:

```
Status(a,Failed)
```

then ITP would deduce

```
Status(b,Suspicious)
Status(c,Suspicious)
```

by hyperresolution in conjunction with the Prolog computations of the ways in which the subgoal ispath(a,y) can succeed.

The $TELL(prolog,<clause>) predicate is used to add facts to the Prolog database during the run. Whenever a positive unit clause of this form is derived, the Prolog component is invoked with the goal $TOLD(lma,<clause>). Normally, the Prolog procedures would include

```
$TOLD(-,X) :- assertz(X).
```

to cause the new clause to be added to the Prolog space. More exotic computations are possible, allowing the Prolog system to screen incoming information.

Similarly, $ASK and $TELL primitives, recognized by the Prolog component, allow it to ask that a subgoal be established by theorem-proving techniques in LMA or to add a clause to the LMA clause space.

To illustrate the utility of such a coupling with a problem that has been previously investigated by others, we consider the problem of dealing with partial orderings. Suppose that a theorem-proving system is investigating a partial ordering in which a1 < a2 < a3 < a4 < a5 < a6 < a7. Most existing systems use transitivity to deduce the 21 separate relationships. Orderings involving more than just a few elements are too costly. The system will spend all of its time deducing the relationships between elements, rather than reasoning about the problem at hand.

One solution is to construct special-purpose packages for storing the known relationships, which can be invoked by the theorem prover to aid in inferences and simplifications. Such an approach can now be implemented under ITP by coding the special package of routines in Prolog. To illustrate, we include a simplified version of such a package:

```
lessthan(X,Y) :- lt(X,Y).
lessthan(X,Y) :- lt(X,Z), lessthan(X,Z).

notlt(X,Y) :- lessthan(X,Y), !, fail.
notlt(X,Y).

$TOLD(TELLER,FACT) :- contradicts(FACT), !, $TELL(TELLER,false).
$TOLD(-,lt(X,Y)) :- notlt(X,Y), !, assertz(lt(X,Y)).
$TOLD(-,-).

contradicts(lt(x,y)) :- lessthan(y,x).
```

These Prolog procedures maintain a set of facts of the form lt(a1,a2). A goal of the form lessthan(a,b) can be established if either lt(a,b) is known, or if transitivity can be used to establish lt(a,b). New facts are integrated into the Prolog space only if they add new information; that is, lt(a1,a2) is added only if lessthan(a1,a2) cannot be established with the existing facts.

By including a literal of the form -$ASK(prolog,lessthan(x,y)) in a nucleus, the clauses in the non-Prolog component of LMA can invoke the services of the Prolog component. Similarly, if a clause of the form $TELL(prolog,lt(a1,a2)) is generated, the Prolog component will receive the new information.

Note that if a contradiction is detected by Prolog (i.e., if it is told that lt(a1,a2), but it can calculate that lessthan(a2,a1)), then it will notify the "teller". This is achieved by execution of the subgoal $TELL(TELLER,false), which informs the teller that the null clause has been deduced.

This short example is intended to illustrate some of the characteristics and potential of interfacing a classical theorem prover to a Prolog system. We have intentionally omitted a number of complexities, and do not wish to imply that the previous example represents a complete solution to the problems involved in reasoning about partial orderings.

3.2. Types of Reasoning

The integration of a classical theorem prover with a Prolog system naturally leads to the question of distribution of effort. What type of computation should be processed by the Prolog system, and what type of computation should be performed in the theorem-proving component? To answer this question, we have partitioned the types of reasoning required to solve problems into four categories:

1. Totally focused reasoning is used to solve computational problems, e.g., adding two numbers or inverting a matrix.
2. Relatively focused reasoning is also well-focused, but may include occasional backtracking.
3. Relatively unfocused reasoning is characterized by some notion of how to identify relevant facts, but no precise idea of which steps will lead to the desired answer.
4. Strategic reasoning is used to form an overall plan of how to attack a given problem.

3.2.1. Totally Focused Reasoning

Finding the eigenvalues of a matrix is a reasoning problem. It is completely algorithmic; the process is well enough understood that no steps are wasted. It is possible to write clauses which allow a theorem prover to solve this problem. It could be done in Prolog, as well; but it is more appropriately done in FORTRAN. This is true not just because of efficiency considerations, but also because numerical subroutine packages for such tasks benefit from the enormous research effort that has gone into their development. Sophisticated reasoning systems must be able to perform totally focused reasoning by accessing the appropriate existing software.

3.2.2. Relatively Focused Reasoning

The next level of reasoning is still algorithmic, but the algorithm may involve a certain amount of trial-and-error, with backtracking to recover from temporary dead ends. Logic programming languages such as Prolog are appropriate for this level, which we call relatively focused.

3.2.3. Relatively Unfocused Reasoning

At this level the path to be followed by the reasoning process becomes difficult to lay out in advance, and there may be substantial exploration of blind alleys. It is not known just which steps will lead to a solution, although there may be heuristics available to guide the reasoning process in the general direction of the goal. At this level, resolution-based and natural deduction theorem provers have demonstrated competence. Much research has been done on the control of unfocused reasoning. It should be pointed out that attainment of complete control (e.g., no clauses generated that do not participate in a proof) may mean that the problem at hand is more appropriate for a different tool.

3.2.4. Strategic Reasoning

Strategic reasoning involves the identification of a sequence of steps leading to the solution of a problem, followed by execution of the steps. It is to this level of reasoning that the AI research in planning and problem solving has been directed. General-purpose strategic reasoning tools are just beginning to emerge.

3.2.5. Event-Driven vs. Goal-Directed Reasoning

Another, almost orthogonal, way to separate types of reasoning is to distinguish event-driven from goal-directed reasoning. Very roughly speaking, event-driven reasoning seeks to ascertain the consequences of a given fact, whereas goal-directed reasoning seeks to determine whether a given goal can be reached. Both types of reasoning can be carried out by both theorem provers and by logic programming systems, although it is sometimes more convenient to express goal-directed reasoning in Prolog and even-driven reasoning in clauses to be processed by a theorem prover. Within LMA one now has a choice, so that the most convenient approach can be taken, and the approaches can be mixed in the same attack on a problem.

3.3. The Integrated Environment

The Prolog subcomponent of the ITP reasoning system is the first of a series of planned additions of subsystems that cooperate with LMA. The next target is a symbolic manipulation system like MACSYMA. We have found that communications with such packages, other than from a user via a terminal, are quite difficult. It seems ironic that workers in the area of machine intelligence so often design systems which are difficult for machines to use. We anticipate better success with MAPLE, a system with many of the features of MACSYMA. As such systems are integrated, communication will continue to be through the $ASK/$TELL mechanism.

The one other system presently supported by $ASK/$TELL is the ITP user himself. For example, messages which are deduced in the course of a run can be sent to the console with the $TELL(user,<list>) term. The elements of <list> can be strings, variables (which, if instantiated, will be replaced by their instantiations) and terms of the form $CHR(<integer>), which can be used to send control characters to the console. This mechanism has been used to disguise ITP as a custom interactive system providing advice on how to run a prototype subsystem of a nuclear plant.

As communication among disparate subsystems increases, it becomes important to define a format for the interchange of problems (and solutions) convenient for the parsing algorithms on each end of the communication link (but not necessarily for the human reader). Communication between systems continually raises the question of exactly how knowledge, problems, and answers should be encoded. The recognition that logic offers a unifying notation certainly has aided in finding solutions. However, the total absence of standards and the surprising lack of interest in program-to-program communication will cause this to remain a problem in the near future.

4. Experience with Porting and Distribution of LMA and ITP

LMA and ITP have been ported from the VAX/UNIX environment in which they were developed to four other environments: VAX/UNIX, IBM/CMS, Perq, and Apollo. Our experience has shown that Pascal is a good language for writing portable software. Given a robust Pascal compiler (support of external compilation, an include mechanism, and generous limits on the number of files and procedures) the port takes a few days at most. Changes are always limited to the I/O routines and to the particular syntax for external compilation.

About fifty copies of LMA and ITP have been distributed, to universities, industrial research groups, and government-sponsored laboratories. The system is being used for research in symbolic logic, and circuit design and validation. A number of groups are using it to explore applications of automated reasoning in nuclear reactor procedure prompting systems, building-wide fire alarm systems, and organic chemical synthesis.

5. Future Plans

LMA was created to support research in automated reasoning systems. We believe that classical theorem-proving systems, logic programming systems, and expert systems represent different approaches towards solving significant reasoning problems. We intend to create an integrated environment which will eventually support the tools and techniques required by each of the three approaches. This will require the addition of the user interface tools characteristic of many of the better expert systems environments. In addition, we hope to develop techniques for rapidly interfacing LMA-based systems with a wide variety of software systems. The increasing awareness that intelligent systems must communicate with one another, as well as with humans, will certainly aid in these efforts.

Within a year or two, there will almost certainly be available machines for less than $3000 that are capable of supporting the entire LMA environment. These will include at least two megabytes of memory, virtual memory, and a hard disk. Such systems should allow anyone wishing to perform experiments in automated reasoning access to the required environment.

In addition to porting LMA to micro-based systems, we are investigating ports to supercomputers. Our intent is to eventually move LMA to one or more supercomputers, recoding sections to take advantage of parallelism. At this time we are investigating the possibility of a port to the HEP, a machine supporting tightly-coupled multiprocessing. The exact point at which we make such a port depends critically on the evolution of the software environments for such machines, which frequently support only a dialect of FORTRAN.

6. Summary

ITP has been serving three goals. It provides a test bed for the LMA subroutine package, because it makes use of all of the LMA tools. It has been used to teach and develop ideas in automated reasoning. Since its distribution, a number of groups have used it to experiment with their own applications to determine the potential utility of automated reasoning in their projects. It is also in limited use as a production tool, particularly in the areas of circuit design and validation, and for research in formal logic.

The development of the integrated logic programming component represents a first step towards the creation of a system built around a wide variety of communicating systems, each with specialized capabilities. It is our hope that the development and widespread distribution of such a system will accelerate progress in building ever more powerful reasoning systems.

References

1. W. F. Clocksin and C. S. Mellish, *Programming in Prolog*, Springer-Verlag, New York (1981).
2. E. Lusk, William McCune, and R. Overbeek, "Logic Machine Architecture: inference mechanisms," pp. 85-108 in *Proceedings of the Sixth Conference on Automated Deduction, Springer-Verlag Lecture Notes in Computer Science, Vol. 138*, ed. D. W. Loveland,Springer-Verlag, New York ().
3. E. Lusk and R. Overbeek, "Data structures and control architecture for the implementation of theorem-proving programs," in *Proceedings of the Fifth Conference on Automated Deduction, Springer-Verlag Lecture Notes in Computer Science, Vol. 87*, ed. Robert Kowalski and Wolfgang Bibel, ().
4. E. Lusk, William McCune, and R. Overbeek, "Logic machine architecture: kernel functions," pp. 70-84 in *Proceedings of the Sixth Conference on Automated Deduction, Springer-Verlag Lecture Notes in Computer Science, Vol. 138*, ed. D. W. Loveland,Springer-Verlag, New York (1982).

5. Ewing L. Lusk and Ross A. Overbeek, "An LMA-based theorem prover," ANL-82-75, Argonne National Laboratory (December, 1982).

6. Ewing L. Lusk and Ross A. Overbeek, "Logic Machine Architecture inference mechanisms - layer 2 user reference manual," ANL-82-84, Argonne National Laboratory (December, 1982).

7. J. McCharen, R. Overbeek, and L. Wos, "Complexity and related enhancements for automated theorem-proving programs," *Computers and Mathematics with Applications* **2** pp. 1-16 (1976).

8. S. Winker and L. Wos, "Procedure implementation through demodulation and related tricks," pp. 109-131 in *Proceedings of the Sixth Conference on Automated Deduction, Springer-Verlag Lecture Notes in Computer Science, Vol. 138*, ed. D. W. Loveland, Springer-Verlag, New York (1982).

9. L. Wos, G. Robinson, D. Carson, and L. Shalla, "The concept of demodulation in theorem proving," *Journal of the ACM* **14** pp. 698-704 (1967).

A NATURAL PROOF SYSTEM BASED ON REWRITING TECHNIQUES

Deepak Kapur and Balakrishnan Krishnamurthy

Computer Science Branch
General Electric Research and Development Center
Schenectady, New York, 12345

ABSTRACT

Theorem proving procedures for the propositional calculus have traditionally relied on syntactic manipulations of the formula to derive a proof. In particular, clausal theorem provers sometimes lose some of the obvious semantics present in the theorem, in the process of converting the theorem into an unnatural normal form. Most existing propositional theorem provers do not incorporate substitution of equals for equals as an inference rule. In this paper we develop a "natural" proof system for the propositional calculus, with the goal that most succinct mathematical proofs should be encodable as short formal proofs within the proof system.

The main distinctive features of NPS are:

1. The substitution principle for the equivalence connective is incorporated as an inference rule.

2. A limited version of the powerful ideas of extension, originally suggested by Tseitin, are exploited. Extension allows the introduction of auxiliary variables to stand for intermediate sub-formulas in the course of a proof.

3. Formulas are standardized by converting them into a normal form, while at the same time preserving the explicit semantics inherent in the formula.

4. A generalization of the semantic tree approach is used to perform case analysis on literals as well as sub-formulas.

5. Additional enhancements such as a generalization of resolution are suggested.

We show that from a complexity theoretic viewpoint NPS is at least as powerful as the resolution procedure. We further demonstrate formulas on which NPS fares better than resolution. Finally, since proofs in NPS usually resemble manual proofs, we feel that NPS is easily amenable to an interactive theorem prover.

1. INTRODUCTION

With increased interest in theorem proving procedures, a collection of proof systems have been proposed in the literature to handle a variety of logical theories. Nevertheless, theorem proving in the propositional calculus lies at the heart of most of those procedures. Even though theoretical evidence has all but foreclosed the possibility of a uniform efficient theorem prover even for the propositional calculus, the hope remains that we can develop practical theorem provers that can cope with a large collection of "naturally occurring" theorems. Perhaps a natural and desirable class of theorems that such proof systems must be capable of handling are those for which there exist succinct mathematical proofs. It would then appear that such a proof system must incorporate the techniques that we employ when proving theorems manually. This paper deals with the development of a "natural" proof system.

We distinguish a proof system from a proof procedure. A proof system is a non-deterministic procedure that defines the notion of a proof by providing the inference rules that may be used. Any sequence of formulas that are derived using the given set of inference rules forms a valid proof of the final formula in the sequence. Non-determinism arises from the fact that the proof system does not specify any order in which the rules are to be applied. The complexity of a proof is then defined as the length of the proof.

In contrast, a proof procedure is a deterministic procedure that not only lays out the inference rules, but also incorporates specific heuristics that precisely determine the order in which the applicability of the rules are checked and subsequently applied. Consequently, given a theorem, a (complete) proof procedure generates a unique proof for the theorem. The complexity of the proof is then defined as the time/space complexity of finding the proof.

Observe that there exists polynomially bounded *proof procedures* for theorem proving in the propositional calculus if and only if P = NP. On the other hand, there exists polynomially bounded *proof systems* for theorem proving in the propositional calculus if and only if NP is closed under complementation (see [4]). It is commonly believed that neither P = NP nor NP is closed under complementation. Thus, we should expect neither a polynomially bounded proof procedure nor such a proof system for the propositional calculus. Nevertheless, we could ask if most mathematical proofs that we normally encounter, can be encoded within a given proof system without a significant increase in the length of the encoded proof. For, if certain mathematical proofs have no efficient encoding within a given proof system, then there is no hope of efficiently finding a proof of the theorem using any proof procedure based on that proof system.

In this paper we argue that most existing proof systems have certain weaknesses that prevent succinct encodings of certain types of mathematical arguments. We also point out that many of these popular proof systems require the formula to be presented in a normal form that is unnatural. This makes it difficult to translate intuitive heuristics that we normally use while proving theorems manually, into a proof procedure based on those proof systems. We propose a new method for theorem proving in the propositional calculus, that we call as "Natural Proof System," wherein we synthesize various attractive features from a variety of proof systems.

In the remainder of this section we develop the necessary terminology and notation. In Section 2 we comment on existing proof systems and their strengths and weaknesses. We deliberately go into some detail to point out the attractive features of these proof systems and the difficulties encountered by them. This is intended to motivate the techniques employed in the proposed proof system. In Section 3 we present an initial version of the proposed theorem proving technique, followed by a number of enhancements to the technique in Section 4. Finally, we conclude in Section 5 with remarks on transforming the proof system to a proof procedure and its implementation.

We call a propositional variable simply as a variable and distinguish it from a literal, which is a variable together with a parity- positive or negative. By a formula we mean a well formed propositional formula using any of the 16 binary connectives. The connectives of primary interest are: AND ($\wedge$), OR ($\vee$), NOT ($\sim$), IMPLIES ($\Rightarrow$), EQUIVALENCE ($\equiv$) and EXCLUSIVE-OR ($+$). The constants TRUE and FALSE will be represented by 1 and 0, respectively.

2. CURRENT PROOF SYSTEMS

2.1 Resolution

The elegance of the resolution proof system, originally suggested by Robinson ([12]), is, in part, due to its simplicity. It is based on a single rule of inference, called the *resolution principle.* This makes the conversion from a proof system to a proof procedure relatively easy, since the heuristics need not be complicated and messy. The only non-determinism that needs to be resolved is the decision of where to apply the rule of inference. Furthermore, this technique can be extended to the predicate calculus efficiently, through the process of unification.

In spite of its simplicity, the resolution proof system compares favorably in its power with most other existing proof system, i.e., the ability to encode succinct proofs. It has been shown [4] that resolution can polynomially simulate most other "reasonable" proof systems. (A proof system Δ_1 can polynomially simulate Δ_2 if for every proof in Δ_2 there is a proof of the same theorem in Δ_1 with at most a polynomial increase in the length of the proof.)

However, there are two main drawbacks of resolution. The requirement that the theorem be represented in conjunctive normal form (CNF) is both restrictive and unnatural. It prevents us from incorporating intuitive heuristics based on the structure of the original formula. The second and more serious drawback is its inability to handle the logical equivalence connective well. Consider the following set of equations:

$$a + b + c = 1$$
$$a + d + e = 0$$
$$b + d + f = 0$$
$$c + e + f = 0$$

where, *a, b, c, d, e and f* are variables. Observe that every variable occurs twice in this set of equations. Hence, the sum of the left-hand sides of the equations, which is 0, is not equal to the sum of the right-hand sides. Consequently, if we encode each equation in CNF then the conjunction of the collection of clauses representing the above equations would be unsatisfiable. However, a resolution proof of this statement seems to be unnecessarily lengthy. The reason is that there is no short representation for a chain of Boolean sums in CNF. Using this fact, Tseitin [13] has demonstrated that theorems of the type illustrated in the above example, have no polynomially bounded proofs in certain restricted forms of resolution. While a similar result for unrestricted resolution has not been theoretically proved, there is ample evidence to believe that unrestricted resolution will fare no better.

2.2 Natural Deduction Systems (Gentzen Systems)

Natural deduction systems incorporate the *deduction theorem* as a rule of inference. One of the earliest natural deduction system was proposed by Gentzen [7]. Gentzen systems do not require the theorem to be represented in any normal form. They operate on each of the binary connectives in a natural way. Gentzen systems develop a tree in which the root is the theorem and the leaves are axioms. This makes the proof structure more comprehensible, and this could prove a useful feature in an interactive theorem prover. A unification based extension of Gentzen systems to the predicate calculus has been recently suggested by Abdali and Musser [1].

On the negative side, the variety of inference rules needed to handle all of the binary connectives makes it more difficult to design a deterministic proof procedure based on Gentzen systems. A more serious difficulty is the fact that a proof tree should really be viewed as a dag (directed acyclic graph) to avoid repeated proofs of the same sub-formulas. Finally, Gentzen systems are also and-or based, and hence can not handle the logical equivalence connective well.

2.3 Semantic Trees

Semantic trees were originally suggested in [8] (also by Davis and Putnam, see [3]). They represent a proof of the theorem based on a case analysis on the variables occurring in the theorem. A tree is developed in which the root is labelled with the theorem, the leaves are labelled with the constant 1, and the sons of a node labelled F are labelled F_1 and F_2 where, F_1 and F_2 are obtained by choosing a variable x in F and evaluating F with $x = 0$ and $x = 1$. An extension of this technique has been suggested by Monien and Speckenmyer [11] where, instead of a variable, a clause is used for splitting the original problem. Yet another generalization, suggested by Bibel [2], calls for splitting over a set of variables instead of a single variable. The simplicity of this technique is appealing. Further, if the tree is viewed as a dag, this proof system is not known to be any less powerful than resolution.

However, a deterministic proof procedure based on this proof system is sensitive to the heuristics used in choosing the variables for instantiation. In addition, efficient extensions of this technique to the predicate calculus have not been investigated. Finally, this too is an and-or based proof system and our earlier comment on the logical equivalence connective prevails.

2.4 Rewrite Rules Technique

Recently, Hsiang [5,6] has reported a proof system based on *rewrite rules.* The negation of the theorem is specified as a conjunction of a set of equations. Each equation is a Boolean-sum of products equated to a constant. These equations are manipulated using the usual rules of equality to obtain a contradiction. While this is a more general version than that reported in [5,6], it is very much in the spirit of Hsiang's technique. The most attractive feature of this proof system is the ability to substitute equals for equals (Hsiang does so in a limited way though)—a principle, fundamental to any manual theorem proving process. The substitution principle is precisely what is needed to efficiently handle the exclusive-or connective.

While the normal form used in this technique is tailored to support a chain of exclusive-or's, it has it's own weaknesses. For example, it does not support an efficient representation for a chain of disjunctions. Further, if the normal form is derived from a CNF representation of the formula, as the author suggests, the advantages of using the exclusive-or connective could well be completely lost. In fact, if the equations in Tseitin's examples are converted from CNF to the normal form used here, the form of the equations changes so drastically that this proof system no longer admits short proofs for the same theorems when presented in this manner!

Another drawback of this proof system is that it relies on susbstitution alone for proving the theorem. Equations are expanded by unifying one term in one equation with another term in another equation, so that the terms may be substituted for each other. This expansion process often increases the length of the equations (not necessarily the number of terms in the equations). It appears that it would be preferable to replace the expansion process with some other technique.

2.5 Extension

Tseitin [13] introduced a technique called *extension* which allows the introduction of auxiliary variables to stand for arbitrary functions of existing variables. He demonstrated that by repeated applications of the principle of extension, one can efficiently manipulate complex sub-formulas by merely operating on the auxiliary variables that stand for the sub-formulas. Cook and Reckhow [4] have shown that extension is so powerful that it overshadows the inference rules of the proof system. That is, whenever two reasonable proof systems - with different inference rules—are enriched with the power of extension, the two resulting proof systems are equally powerful. Krishnamurthy [9] has shown that a variety of mathematical arguments can be succinctly encoded using the principle of extension. Thus, this simple feature of extension can significantly reduce proof lengths.

However, the difficulty is in a deterministic implementation of the principle of extension. How can a proof procedure recognize the need for auxiliary variables and determine how they should be defined? In the proof system suggested in the sequel we provide a technique for a deterministic implementation of a limited form of extension.

3. A NATURAL PROOF SYSTEM

In this section we present the main ideas of the proposed proof system, which we shall call NPS (Natural Proof System). It is a natural deduction system in the sense that (i) the proof preserves the structure of the theorem, and (ii) the deduction theorem is assumed. Theorems are proved by assuming the negation and arriving at a contradiction. Thus, to prove $\Phi \vdash F$ where, Φ is a set of formulas and F is a formula which is a logical consequence of Φ, we assume $(\Phi \wedge \sim F)$ and derive a contradiction.

3.1 Equational Normal Form

The observations about normal forms made in the previous section indicate that while normal forms are desirable to avoid handling every binary connective and to give some structure to the formulas that are manipulated within the proof system, they should preserve the structure of the original formula. In particular, the distributive properties of the Boolean operators should not be used in deriving the normal form. With this in mind we develop a normal form purely for standardization of the formula, called *equational normal form* (ENF). For the sake of convenience, in an interactive implementation of a proof procedure, we might include the remaining connectives as well. However, for clarity, we limit ourselves here to $\vee$, $+$, and $\sim$.

ENF is a representation for not just a formula, but for a statement of the form "$F = X$" where, F is a formula and X is either a literal or a Boolean constant. ENF is a set of equations of two types: *d-type* (disjunction) and *s-type* (summation). A *d-type* equation is of the form $A_1 \vee A_2 \vee \cdots \vee A_n = B$ where, $A_1, A_2, \cdots, A_n$ are literals and B is either a literal or the constant 1. An *s-type* equation is of the form $A_1 + A_2 + \cdots + A_n = B$ where, $A_1, A_2, \cdots, A_n$ are *positive* literals and B is a Boolean constant.

We sketch below a procedure for converting "F = X" into ENF and follow it up with an example. We begin with the expression tree (also know as the parse tree) for F. Recall that the leaves of the expression tree are labelled with variables and the internal nodes are labelled with Boolean operators. Using the usual rules of the Boolean operators, other than the distributive laws, we reduce all operators to $\sim$, $\vee$ and $+$. We also flatten the parse tree by using the associative properties of $\vee$ and $+$. Finally, convert the tree into a dag by identifying common sub-expressions. Now associate a new variable name with each $\vee$-node and with each $+$-node of the dag, X being assigned to the root of the dag. Write an equation for each such node in the obvious way - a *d-type* equation for a node labelled with $\vee$ and an *s-type* equation for a node labelled with $+$. The collection of these equations is the ENF.

Example 1: Consider the formula

$$[((a \wedge \sim b \wedge \sim c) \wedge d) \vee ((c + a) \Rightarrow (b \equiv \sim c))] \wedge [(a \wedge \sim b) \Rightarrow (c + a)]$$

The corresponding equation tree and the modified dag are shown in Figure 1.

The ENF equivalent is shown below:

$$\sim a \vee b \vee c \vee \sim d = Z_1$$

$$c + a = Z_2$$

$$b + c = Z_3$$

$$\sim Z_1 \vee \sim Z_2 \vee Z_3 = Z_4$$

$$Z_2 \vee \sim a \vee b = Z_5$$

$$\sim Z_4 \vee \sim Z_5 = \sim X$$

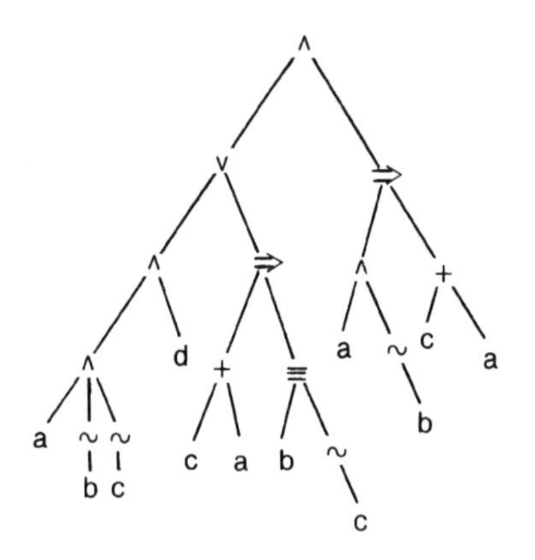

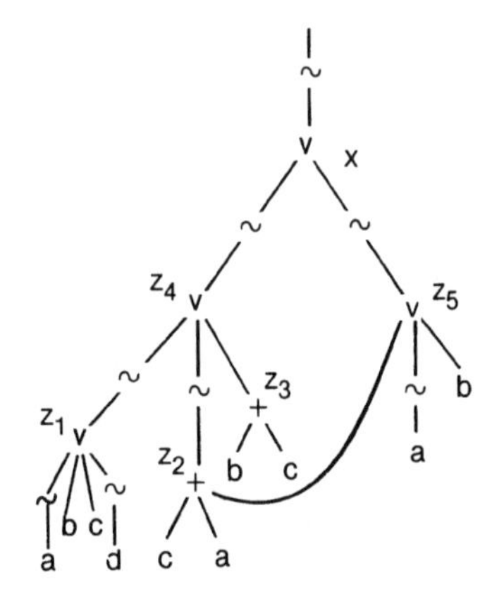

Figure 1

3.2 Simplification Rules for Equations

A set of equations can be simplified using the following rules. Recall that the order of the literals appearing on the left-hand sides of the equations is immaterial, since + and $\vee$ are associative and commutative. Thus, the literals on the left-hand sides of the equations can be rearranged in any manner.

1. *Substitution:* We can substitute equals for equals. Thus from the two equations, $A + X = 0$ and $A + Y = 0$ we can obtain, by substituting the value of A from the first equation into the second equation, the equation $X + Y = 0$. Here X and Y can be an arbitrarily long chain of Boolean sums. Observe that $X + Y = 0$ will automatically be in the normal form. We can require that equations of the form $A = B$ are eliminated by replacing all occurrences of A with B (or, vice versa). We must also point out that in the substitution process ~ 0 should be treated as a 1 and ~ 1 should be treated as a 0.

2. *Reduction of equations:* The following rule can be used to replace equations by simpler equations:

   R1: $A \vee X = 0 \;\rightarrow\; A = 0\,;\; X = 0\,;$

   R2: $\sim A \vee X = A \;\rightarrow\; A = 1\,;\; X = 1\,;$

   Note that X can be a chain of disjunction.

3. *Reduction of formulas:* The following rules can be used to simplify the left-hand sides of equations to simpler forms:

   R3: $X \vee X \;\rightarrow\; X\,;$

   R4: $1 \vee X \;\rightarrow\; 1\,;$

   R5: $0 \vee X \;\rightarrow\; X\,;$

   R6: $\sim A + X \;\rightarrow\; 1 + A + X\,;$

   R7: $X + X \;\rightarrow\; 0\,;$

   R8: $0 + X \;\rightarrow\; X\,;$

3.3 The Proof System

The main core of the proof system described herein is based on a *generalization of the semantic tree approach.* We begin with a formula that we wish to prove to be a theorem or that it is inconsistent. To prove it to be a theorem, we transform the formula into a set of equations asserting that the formula has a value 0. Similarly, to prove it to be inconsistent, we transform the formula into a set of equations asserting that the value of the formula is 1. We then proceed to derive a contradiction. To this end, we construct a proof tree by repeated applications of the simplification process described above and the *splitting* process described below.

We first simplify the set of equations using the rules given in Section 3.2. At any step, if the simplification results in an inconsistent equation $0 = 1$ or $x = \sim x$ in the set, then the set of equations is contradictory and we are done. Otherwise, we choose a sub-expression of the left-hand side of one of the equations. The sub-expression may (and, often will) simply be a literal. We then invoke the splitting rule and create two sub-problems: one in which the sub-expression is equated to 0 and another in which it is equated to 1. These two sub-problems are represented by two sets of equations obtained by adding each of the two equations mentioned above to the original set of equations. The original set of equations is contradictory if and only if each of these new sets of equations is contradictory. We then recursively apply this process. We illustrate the technique on an example in Figure 2.

Example: *Theorem:* $b + c + (b \vee c) + (\sim b \vee \sim c)$

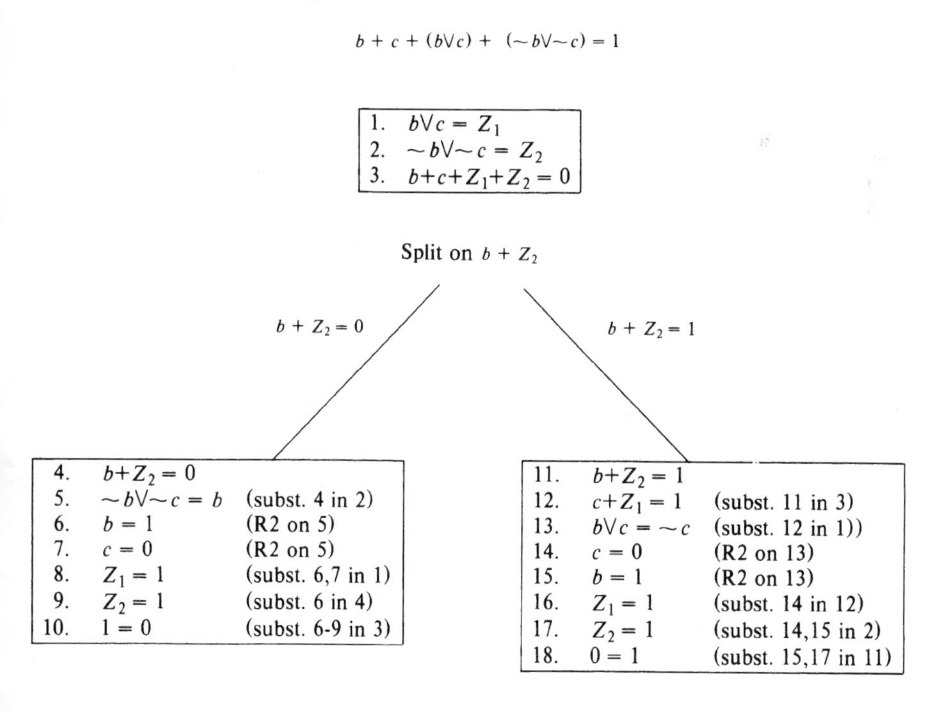

Figure 2

3.4 Completeness

The soundness of this proof system is evident, for, each of the simplification rules can be verified and the soundness of the splitting rule is inherited from the semantic tree approach. We show below that this proof system is also complete.

Theorem 1: NPS is complete for the propositional calculus.

Proof: Given a theorem T in the propositional calculus, we write the assertion "T = 0" in ENF. We need to show that by repeated applications of the simplification process and the splitting process we will necessarily arrive at a contradiction. If we restricted ourselves to the splitting process alone, we will eventually eliminate all the variables. If this does not result in a contradiction along every path in the tree, then by the soundness of the proof system, we would have produced a satisfying truth assignment for the negation of the theorem. This would contradict the hypothesis that T is a theorem. Thus, all that remains to be shown is that the simplification process terminates. In particular, we need to show that the substitution process cannot be applied indefinitely. The number of new variables introduced in transforming T into ENF is bounded by the number of internal nodes in the expression tree for the formula T. This bounds the total number of variables and consequently, the number of possible equations. Hence, the simplification process must terminate. Q.E.D.

4. ENHANCEMENTS

Equations whose left-hand sides are disjunction of literals can be simplified only by splitting on a formula. We would like to avoid the splitting of a problem into subproblems as much as possible in order to avoid an exponential growth in the complexity of the proof. In this section, we discuss two enhancements to the proof system to handle disjunctions— *generalized resolution* and *compaction*. Generalized resolution is a generalization of standard resolution [12]; it can be used to simulate resolution proofs. Even in cases when resolution is not applicable, it generates information which might be useful in subsequently establishing a contradiction. Compaction allows us to recognize a set of equations whose left-hand sides are disjunctions and which have a succinct representation using the boolean sum notation. Later in the section, we discuss ways to identify similar nodes in the proof tree which can be used to avoid repeated independent derivations of identical sets of equations.

4.1 Generalized Resolution

Even though simplification is a powerful tool, it is not sufficient to efficiently handle certain types of arguments. For example, consider the following set of d-equations:

$$\{ X \vee Z = 1;\ \sim X \vee Y = 1;\ \sim Y \vee Z = 1;\ \sim Z_1 \vee \sim Z_2 = \sim Z \}$$

It is easy to deduce from the above equations that Z = 1 thus making $Z_1 = 1$ and $Z_2 = 1$. However, we cannot make any substitutions, as the left-hand sides are disjunctions. On the other hand, if we used the semantic tree approach, we could branch on X and in each of the two sub-trees we would be able to derive $Z_1 = 1$ and $Z_2 = 1$. But then all subsequent arguments that use $Z_1 = 1$ and $Z_2 = 1$ would have to be duplicated in the two subtrees. Instead, we can derive the necessary conclusion without branching, using resolution. If we viewed each of the left-hand sides as a clause and the equations as asserting the conjunction of those clauses, then through two applications of the resolution principle, we can derive Z = 1. This suggests that a generalization of resolution applicable to a set of equations would be a useful tool.

Theorem 2: Let E be a set of equations including $A \vee X = Z$ and $\sim A \vee Y = Z'$. Let Z_1 and Z_2 be variables not occurring in E. We can add $Z \vee Z' = 1$, $Z_1 \vee Z_2 = 1$, $X + Z + 1 = Z_1$, and $Y + Z' + 1 = Z_2$ without affecting the satisfiability of E.

Proof: Since either A or $\sim A$ is always true, at least one of Z and Z' must be true. Thus, we can conclude $Z \vee Z' = 1$. For the remaining 3 equations, consider two cases based on the value of A. If A is 0 then $X = Z$ and $Z_1 = 1$. If A is 1 then $Y = Z'$ and $Z_2 = 1$. In either case $Z_1 \vee Z_2 = 1$. Q.E.D.

The simplification rule corresponding to generalized resolution is:

$$GR:\ A \vee X = Z; \sim A \vee Y = Z' \rightarrow A \vee X = Z; \sim A \vee Y = Z';$$
$$Z \vee Z' = 1;\ Z_1 \vee Z_2 = 1;$$
$$X + Z + 1 = Z_1;\ Y + Z' + 1 = Z_2.$$

When Z = Z' = 1, new variables Z_1 and Z_2 are not introduced because $X = Z_1$ and $Y = Z_2$ in that case; instead, of 4 additional equations, only one new equation, $X \vee Y = 1$ is added.

For example, consider the following set of equations:

1. $P \vee Q = 1$
2. $Q \vee R = 1$
3. $R \vee W = 1$
4. $\sim P \vee \sim R = 1$
5. $\sim Q \vee \sim W = 1$
6. $\sim Q \vee \sim R = 1$

The derivation is as follows:

7. $\sim P \vee Q = 1$ *GR* (2, 4)
8. $Q = 1$ *GR* (1, 7)

When 1 is substituted for Q in the above equations, equations 1, 2 and 7 are discarded; equations 5 and 6 simplify to:

5. $\sim W = 1$ *subs.* (8)
6. $\sim R = 1$ *subs.* (8)

Using them in 3 gives an inconsistent equation 0 = 1.

Theorem 3: NPS with GR can simulate the resolution proof system without an increase in complexity.

Proof: Consider a formula F in CNF which has a proof in the resolution proof system. From F, we get a set of equations such that there is an equation corresponding to each clause and nothing else. The equation corresponding to a clause C is $C = 1$. For every step in a proof of F in the resolution system in which a variable x is resolved, we take the equations corresponding to the two clauses involved in the resolution and apply the generalized resolution rule on them using x. Following the resolution proof tree which gives the empty clause, we would obtain 1 = 0, an inconsistent equation. Q.E.D.

4.2 Compaction

Consider the following set of equations:

$$\{ \sim A \vee B \vee C = Z;\ A \vee \sim B \vee C = Z;\ A \vee B \vee \sim C = Z;\ \sim A \vee \sim B \vee \sim C = Z \ \}$$

If these equations have to be transformed into a Boolean sum of products as required by Hsiang's technique, it will take a considerable effort to get to the equivalent form using +, which is:

$$A + B + C = 0; \quad Z = 1$$

Note that if Z = 0, then we immediately get an inconsistent equation from the original 4 equations. Using the above two equations, it is possible to substitute for any of the variables the remaining part of the equation; such a substitution was not possible from the original set of equations. Further, the new equations provide more insight into the semantics of the original set of equations than the original set does. It is often useful to recognize formulas with such structures and transform them to a more succinct form.

Given 3 variables a, b, c, 8 clauses can be expressed using them; so the possible number of subsets of such clauses is 2^8, which is 256. It turns out that there are only 5 relevant cases as other cases either simplify easily using the rules of inference discussed so far or can be obtained using symmetry (i.e., by renaming the variables) from the 5 cases. Without any loss of generality, we can assume that the clause $(a \vee b \vee c)$ is present in all 5 cases, as any clause can be transformed to this clause by symmetry. We can construct the following five subsets of equations and the corresponding reductions on these subsets are also given:

1. $a \vee b \vee c = z$, which is kept as it is.

2. $a \vee b \vee c = z$ and $\sim a \vee \sim b \vee \sim c = z$.
   This is a hard case; most hard tautologies, including Ramsey formulas [10], can be expressed using this case. Further, the satisfiability problem remains NP-complete even if all clauses in a formula are of these forms. These two equations are kept as they are; in addition, we can derive $z = 1$, as otherwise we get an inconsistent equation.

3. $a \vee b \vee c = z$ and $\sim a \vee \sim b \vee c = z$.

   This case is equivalent to the following set of equations:

   $$\rightarrow \; c \vee z_1 = 1; \; a + b = z_1; z = 1.$$

4. $a \vee b \vee c = z; \; \sim a \vee \sim b \vee c = z; \; \sim a \vee b \vee \sim c = z.$
   $$\rightarrow \; a + b + c = z_1; \; a \vee \sim b \vee \sim c = z_1; \; z = 1.$$

5. And, finally,

   $$a \vee b \vee c = z; \; \sim a \vee \sim b \vee c = z; \; \sim a \vee b \vee \sim c = z; \; a \vee \sim b \vee \sim c = z$$
   $$\rightarrow \; a + b + c = 1; z = 1.$$

The utility of the above rules can be easily demonstrated on the CNF representation of formulas used by Tseitin [13]; for an instance, see example 3.6 in Bibel [2]. Using the last rule (case 5), we get formulas identical to those discussed in subsection 2.1 (except that different variable names are used in example 3.6). Adding those four equations immediately gives an inconsistent equation.

4.3 Identification

A proof tree in any proof system can in general have nodes labelled with identical formulas. Two different paths from the root—the formula being proved or disproved—may lead to identical subproblems (formulas) after splitting and simplification. We can avoid repeated independent derivations of sub-trees corresponding to identical nodes in a proof tree by maintaining a hash table for the formulas associated with those nodes of the proof tree that have been completely explored. Prior to exploring a new node, we would use the hash table to ensure that the formula associated with that node has not been encountered earlier.

In the proposed proof system, a node in a proof tree has a set of d-type and s-type equations associated with it. Before hashing the set, we canonicalize the set so as to obtain a unique representation for a set of equations using associative and commutative properties of $\wedge$, $\vee$, and $+$. One way to achieve this is by using an ordering on variable names and constants 0 and 1, parity, sorting arguments of $\vee$ and $+$ in ascending or descending order and deciding whether d-type equations follow s-type equations or vice versa. This would lead to a check for equality of two sets of equations using hashing.

Notice that the above scheme only checks whether two sets of equations (or two formulas) are identical. Since the inconsistency of a set of equations (formula) does not depend upon the particular names used for variables in the equations, two sets of equations that are identical up to variable renaming (in fact, up to literal renaming) characterize the same Boolean function. If a node has a label E which is identical up to variable renaming to the label E' of a known dead node, then E is also a dead node. (A dead node is one whose proof tree has been completely developed.) A variation of the above canonicalization and hashing scheme can be used to check for this symmetry property of formulas (see [9]). This substantially reduces the complexity of proofs of a class of hard tautologies. For handling symmetry, the canonicalization of a set of equations (formula) also involves standardizing the variable names used in the set of equations. The hashing is done on the standardized canonicalized set of equations. Plaisted [personal communication] has implemented the above two schemes on a resolution based proof procedure and reported a substantial improvement in the performance of the theorem prover as a result of these schemes.

As should be evident, what is really needed is a way to identify whether a node labelled with a set E of equations contains as its subset (up to variable renaming), E', the set of equations associated with a known dead node. This is a hard problem and we do not know how to implement this kind of identification check yet (in resolution and semantic tree based proof systems, such a check would amount to doing subsumption check with symmetry). Even if we sacrifice symmetry, the subsumption check has been found hard to implement; incorporating symmetry makes it harder because standradization of variable names does not quite work.

5. CONCLUSION

We have proposed a proof system in which formulas at any intermediate stage are closely related to the structure of the original formula being proved or disproved. It extensively uses the powerful technique of extension for introducing new variables to stand for formulas in a proof, substitution and simplification derived from expressing formulas in Boolean sum notation and generalized semantic tree approach in which split could be done on a variable in the original formula or the one introduced via extension.

A crucial step involved in transforming a nondeterministic proof system to a deterministic proof procedure is the development of a set of heuristics that determine the order in which various rules of inference in a proof system are selected for possible application. Obviously, many proof procedures can be developed from a proof system by choosing different sets of heuristics. The challenge here is to come up with a proof procedure that makes the right choice in selecting the inference rules in most cases, and results in a minimal proof that could be obtained in a proof system. The overhead involved in implementing the heuristics and selecting it is usually justifiable only if it makes the right choices in most cases. One must then establish a completeness result for the proof procedure with respect to a set of heuristics, namely, that if a proof exists, the proof procedure finds it.

A proof procedure is considered *good* for a set of formulas if the complexity of a proof found by the proof procedure is at most a polynomial function of the complexity of a minimal proof in its proof system.

We believe that the heuristics are often dependent and derived from the application area. Although there may exist a small subset of general purpose heuristics based on the syntactic structure of formulas, the rules of inference and the semantics of various logical connectives, powerful heuristics often make use of the knowledge of the domain of the formulas. We believe that a "natural" proof procedure ought to provide facilities for a user to incrementally introduce heuristics and specify how the built-in and the user specified heuristics interact. This is a hard issue and falls outside the scope of this paper; however, we believe that the proposed proof system can incorporate some useful heuristics based on extension such as the user suggesting an arbitrary formula for a case analysis or a lemma to prove the original formula, etc. We plan to implement a proof procedure based on NPS using various heuristics and analyze their performance on different sets of formulas.

We believe that standard techniques suggested in Kowalski and Hayes [8] and Abdali and Musser [1] can be used to lift the proposed proof system to the first-order predicate calculus. However, an efficient generalization of the proposed proof system to the first-order predicate calculus such as that achieved by unification in the resolution procedure, requires further investigation.

6. REFERENCES

[1] S.K. Abdali and D.R. Musser, "A Proof Method based on Sequents and Unification," Unpublished Manuscript (1982), G.E. R and D Center, Schenectady, NY.

[2] W. Bibel, "Tautology Testing with a Generalized Matrix Reduction Method," *Theoretical Computer Science*, 8 (1979), pp.31-44.

[3] C-L Chang and R.C. Lee, *Symbolic Logic and Mechanical Theorem Proving*. Academic Press (1973), New York.

[4] S. A. Cook and R. A. Reckhow, "The Relative Efficiency of Propositional Proof Systems," *J. of Symbolic Logic*, 44 (1979), pp. 36-50.

[5] J. Hsiang, *Topics in Automated Theorem Proving and Program Generation*. Ph.D. Thesis, Department of Computer Science, University of Illinois, Urbana, Illinois.

[6] J. Hsiang and N. Dershowitz, "Rewrite Methods for Clausal and Non-clausal Theorem Proving," *Proc. 10th EATCS Intl. Collq. on Automata, Languages, and Programming*, (1983), Spain.

[7] S.C. Kleene, *An Introduction to Metamathematics*. (1952), Van Nostrand, New York.

[8] R. Kowalski and P. Hayes, "Semantic Trees in Automatic Theorem Proving," in *Machine Intelligence* 4, Meltzer and Michie, eds., Edinburgh Univ. Press, Edinburgh.

[9] B. Krishnamurthy, "Short Proofs for Tricky Formulas," Unpublished Manuscript, (1982), G.E. R and D Center, Schenectady, NY.

[10] B. Krishnamurthy and R. N. Moll, "Examples of Hard Tautologies in the Propositional Calculus," *Proc. of the Thirteenth ACM Symp. on Th. of Computing*, (1981), pp. 28-37.

[11] B. Monien and E. Speckenmeyer, "3-Satisfiability is Testable in $O(1.62^r)$ Steps," Bericht Nr. 3/1979 (1979), GH Paderborn, Fachbereich Mathematik-Informatik.

[12] J.A. Robinson, "A Machine Oriented Logic Based on the Resolution Principle," *JACM*, 12 (1965), pp. 23-41.

[13] G. S. Tseitin, "On the Complexity of Derivations in Propositional Calculus," *Structures in Constructive Mathematics and Mathematical Logic*, Part II, A. O. Sliosenko, ed., (1968), pp. 115-125.

EKL—A Mathematically Oriented Proof Checker

Jussi Ketonen
Stanford University

Abstract

EKL is an interactive theorem-proving system currently under development at the Stanford Artificial Intelligence Laboratory.

A version of EKL transportable to all TOPS-20 systems has been used for simple program verification tasks by students taking CS206, a LISP programmming course at Stanford .

The EKL project began in 1981 and has grown into a large and robust theorem-proving system within a relatively short span of time. It currently runs at SAIL (a KL10-based system at the Stanford Computer Science Department), comprising about 10000 lines of code written in MACLISP.

We describe some of the features of the language of EKL, the underlying rewriting system, and the algorithms used for high order unification. A simple example is given to show the actual operation of EKL.

Research supported by NSF grant MCS-82-06565 and ARPA contract N00039-82-C-0250.

1. Introduction

An interactive theorem-prover is to be judged on the basis of its ability to imitate actual mathematical practice. It can be viewed as a testing ground for the study of representation of facts and modes of reasoning in mathematics. A proof-checker may be used to give immediate feedback on the correctness of our ideas of formal representation. For example, one can judge a mechanically presented proof for its clarity and compactness. We may also compare machine-checked proofs with their informal counterparts.

We see mathematics as a discipline that thrives on highly abbreviated symbolic manipulations. Their logical complexity tends to be low—the use of logic in such contexts is a relatively straightforward matter. Most of the burden of theorem-proving is taken by rewriting processes.

We regard the ongoing development of EKL as an experimental science. In this sense, neither the existing control structures nor the underlying logic is sacred. However, we do not want to stray too far afield into "exotic" logics by violating basic proof-theoretic properties of first order logic. In particular, we want to pay close attention to such obvious requirements as consistency and sound semantics. Our primary criterion is that of *expressibility* and the ability to talk about the intrinsic properties of the concepts in question. Such accidents of formalisation as completeness, decidability, or the complexity of a decision procedure are of only secondary importance; we recognize the fact that as of now we know of very few decidable theories that correspond to natural fragments of mathematics. Since even the simplest known decision procedure for a non-trivial part of logic has an exponential worst-case performance, it is more useful for us to tailor our algorithms to naturally occurring inputs rather than use arbitrary syntactic constraints. Indeed, we expect such trivial syntactic criteria as the number of quantifiers in a formula to appeal only to the most technically oriented logician.

Our intention is to go through major proofs from several areas of mathematics and computer science and see what difficulties arise as a result of trying to formalize them. This kind of "ongoing dialogue" ([Wos 1982]) will in turn be used to design future versions of EKL.

The main distinguishing characteristic of EKL is its flexibility and adaptability to different mathematical environments. The emphasis has been on creating a system and a language which would allow the expression and verification of mathematical facts in a direct, readable and natural way. The development of EKL has so far been heavily weighted in favor of expressibility and user-friendliness as opposed to sophistication in automatic procedures. Our goal is to provide a good environment for formal manipulation. Ultimately we would like to see EKL as an editor with capabilities for symbolic computing and verification.

EKL has strong display features. The recent E/MACLISP interface developed by Richard Gabriel and Martin Frost [Gabriel and Frost 1984] has made it possible to run EKL through the SAIL E editor. Further enhancements by Joseph Weening have made the "EEKL" mode of operation vastly superior to standard terminal interaction.

EKL is written in MACLISP. In fact, the top level of EKL is LISP. Commands given to EKL are simply S-expressions to be evaluated. The user is given a set of programs to manipulate and access proofs; they can be accessed only through their names, using the given EKL routines. Thus EKL becomes automatically programmable—the programming language is LISP itself. This is a way to implement simple proof strategies and to "customize" EKL for particular user requirements. Our approach should be contrasted with that of LCF ([Gordon, Milner and Wadsworth 1979]) — the first programmable proof-checker ever built. LCF can be programmed using a special-purpose meta-language, whereas in EKL much of the burden of control is taken by the rewriting machinery used within the LISP environment.

Complex formulas need not be given to EKL in a LISP format: All EKL commands accept formulas presented as atoms which are then exploded and parsed into an internal form for further processing.

Our current library of EKL proofs consists of facts about LISP functions (APPEND, REVERSE, MAPCAR, NTH, SIZE, MEMBER, LENGTH, SAMEFRINGE, and various permutation functions on lists), elementary arithmetic, and combinatorial set theory. For example, we have been able to produce an elegant and eminently readable proof of Ramsey's theorem in under 50 lines. The shortest computer-checked proof previously known is due to [Bulnes 1979] who proved it in about 400 lines. In addition, we are going through theorems in Landau's Grundlagen in order to make a comparative study of EKL and AUTOMATH proofs [Jutting 1976] of the same theorems.

Simple facts about LISP functions can usually be presented as one-line applications of the induction schema. This represents both the strength of the rewriting system and its lack of specialisation in any mathematical subfield. We do not have any built-in induction heuristics (or indeed, any heuristics at all) as in [Boyer and Moore 1979A], since this would tend to violate the generality of EKL as a system of formal manipulation. One has to specify the instances of induction used unless the high order unification mechanism of EKL can be exploited for the same purposes.

One of the most recent proofs on LISP programs involved SAMEFRINGE, a function defined by a complicated recursion involving the size of S-expressions. Its existence (i.e., the termination of the corresponding program) was proved using a high order inductive schema for natural numbers. A simpler example—the definition of APPEND—will be given later.

EKL is specifically designed to manipulate proofs and to handle several proofs at the same time. A proof in EKL consists of lines. Each line in a proof is a result of a command. A state in EKL consists of the currently active proof and the currently active context. A context is simply a of list declarations for atoms giving their syntactic meaning. Declarations are generated using the DECL function or automatically through the EKL default declaration mechanism. Associated with each line is its context and dependencies. If a line contains a formula, then its context is the set of all declarations needed to make sense of that formula. The current context is the cumulative subtotal of all the context manipulation that has happened thus far.

In a typical command several lines may be used. We first combine the contexts of the cited lines. If an incompatibility turns up, the command is aborted. This context is combined with the previous active context. All the incompatible declarations from the previous context are thrown out. The resulting context is used for parsing of terms and type computations in the command.

It follows that one can use conflicting declarations in different parts of the same proof provided that we do not try to refer to those lines within the same command. The language used is ultimately local to the line in question. This locality allows us to build libraries of already-proven theorems. We can change notation and basic definitions as easily as a mathematician can in switching from one book to another.

There are about ten primitive commands for EKL. They have the effect of introducing new axioms or dependencies: AXIOM, ASSUME, DEFAX, new definitions: DEFINE, specializing universal variables: UE, rewriting: RW,TRW, and manipulating dependencies: CASES; proof by cases and CI; conditional introduction (a form of the deduction theorem). Most of these commands use the EKL rewriting mechanism.

In addition, there are commands for manipulating and editing the proof itself: LINK for goal structuring, CHANGE and DELETE for redoing or deleting a line and all the lines that depend on it, and, finally, COPY and TRANSFER for copying and transferring lines within proofs.

2. The Language of EKL

The language of EKL consists of a finite-order predicate logic with typed λ-calculus and a facility for talking about metatheoretic objects. Each EKL atom has a type (a syntactic entity), a sort, and a syntype which either VARIABLE, CONSTANT, BINDOP, DEFINED or SPECIAL. Of these syntypes, SPECIAL and DEFINED are not user declarable.

The syntype SPECIAL refers to symbols in the standard context of EKL that are heavily overloaded in that they operate on all type levels; strictly speaking, these symbols don't have a type and are regarded as absolute constants. DEFINED atoms are introduced thru the DEFINE command — in proof-theoretic terms, they can be viewed as "eigenvariables" resulting from the elimination of existential quantifiers.

The central notion in the logic of EKL is that of types. Informally, they are represented as a classes of objects in a set-theoretic universe. The main purpose of types in EKL is to restrict the class of acceptable formulas in the language in order to prevent logical contradictions from occurring. Our motivation is quite different from the use of typing in traditional programming languages: The type structures encountered in mathematical reasoning are much simpler than the ones found in programs. The intent is to gain maximum expressibility in our language while preserving consistency — we want to prevent the expression of inconsistencies like $\lambda x. \neg x(x)$. At the same time, we want to give the user the means of rigidly (syntactically) excluding formulas. The user may want to specify the number and types of arguments to a function. For example, expressions of type **ground** → **ground** can be applied to terms of type **ground** resulting in an object of type **ground**. Of particular interest is the notion of *list types*, which allows us to talk in a natural way about parameterized formulas and functions taking an arbitrary number of variables. A term of type **ground*** → **ground** can be applied to any number of terms of type **ground** resulting in an expression of type **ground**. Thus objects of this type could be regarded as having variable arity. Formulas and terms are treated in a completely uniform manner; formulas are simply terms of type **truthval**. All deduction rules manipulate terms of type **truthval**. EKL checks terms for correct typing.

More formally, the EKL type structure is an algebra with an arbitrary set of atoms including a special atom **empty** (representing the null tuple () of **empty** type), together with type constructors "$\otimes$" (product), "$\vee$" (disjunction), "$\rightarrow$" (application), "$*$" (list types) and relation "$\leq$" (for a type being a subtype of another type). In addition, the user can introduce variable types. For example, "?Foo" denotes a variable type with the name Foo.

Bindops are operators that bind variables. They must have the BINDOP syntype. For example, the set-theoretic comprehension operator $\{x|P(x)\}$ can be construed as a bindop of type $\langle$**ground**$\rangle \otimes$ **truthval** → **ground**. This means that the bound variable has to be of type **ground**, the matrix of type **truthval** and that an entity of type **ground** always results.

An applicative operator of variable arity can be declared left or right associative or simply associative. A bindop of variable arity can be declared right associative. Internally the resulting expressions will be automatically "flattened" by the EKL rewriter. Thus $2+3+(1+2)$ becomes $2+3+1+2$ if $+$ has been declared right associative and $\forall x\, y.\forall u\, v.\pi$ becomes $\forall x\, y\, u\, v.\pi$. Note that declarations of this type have implied semantic content.

Tupling is the most fundamental operation in EKL. We regard all functions as applying to ntuples. The term ntuple(x, y, z) is written most of the time as (x, y, z). Our operators are unary: $f(x, y, z)$ is thought of applying f to the triple (x, y, z). In this sense, we do not have 1-tuples: (x) means x. We allow empty tuples; $f()$ refers to a function of no arguments. We associate to the right: $f(x, y, (u, v, w))$ is the same as $f(x, y, u, v, w)$. In particular, the empty tuple is deleted when it appears at the end: $(x, y, ())$ is rewritten to (x, y).

Perhaps the semantically trickiest part is in the introduction of metatheoretic operators $\uparrow$ (for naming EKL objects) and $\downarrow$ (roughly corresponding to evaluation) as a proper part of our language. One has to be careful about the interaction of bound variables with metatheoretic evaluation. For instance, we cannot universally generalize the variable x in in the valid formula $x == \downarrow\uparrow x$. Our approach avoids the separation of metatheory into another domain; the notion of "reflection" in the sense of [Weyhrauch 1978] is absent. Metatheory and the use of semantically attached functions are simply a part of the rewriting process. In order to guarantee the soundness of the system, attachment of functions cannot be completely controlled by the user. EKL has a list of standard attached functions like PLUS, TIMES, APPEND,... etc. Should the user declare a function with an internal name found in this list, the rewriter will do the appropriate computations on quoted entitities to replace the corresponding expression with the (quoted) result if the type information matches what is expected.

More details of the EKL language, including formal semantics and the proofs of consistency and soundness, can be found in [Ketonen and Weening 1983B] .

Even though we have implemented a metatheoretic facility, there has been comparatively little use for it in day-to-day theorem-proving activities. It seems that most of the concepts regarded as "metatheoretic" can more naturally be expressed in terms of high-order predicate logic. The need for metatheory has in many cases been an artifact of restriction into first order expressions; for example, facts about simple schemata can be formulated in terms of second-order quantifiers: the induction schema

$$P(0) \wedge \forall n.P(n) \supset P(n') \supset \forall n.P(n)$$

can be equally well expressed as the second-order sentence

$$\forall P.P(0) \wedge \forall n.P(n) \supset P(n') \supset \forall n.P(n).$$

Our belief is bolstered by the nearly uniform denial by practicing mathematicians that they ever use what logicians call metatheory—it appears to be too crude and simplistic to capture even remotely the processes occurring in mathematics. The intrinsic structure of facts is obscured at the expense of emphasizing the particular choice of a language and syntactic forms for representation. As a disguised form of programming it may not present any apparent increase in either correctness or clarity.

3. The Use of Definitions

Definitions play a key role in mathematics. They seem to be one of the principal ways of controlling complexity of formal discourse.

Defined symbols in EKL have a special syntype—in rough correspondence with the notion of an eigenvariable in proof theory. They can be introduced in two different ways: through the use of the DEFINE command, which checks the validity of the proposed definition, or the DEFAX command, which allows an axiom to be regarded as a definition of some symbol occurring in it.

Definitions are heavily used in EKL by both the rewriter and the unifier.

4. Sophisticated Unification

EKL is based on high-order logic since we felt strongly that many important mathematical facts admit a more natural representation in this context. There has been much recent work on the topic of high-order unification (for example, [Huet 1975], [Miller, Cohen and Andrews 1982] and [Jensen and Pietrzykowski 1976]), which has shown it to be a feasible alternative to first-order methods.

The most critical use of unification in EKL takes place within rewriting where one may expect all the high-order unifiable variables to occur only on one side, for instance the left-hand side, of the match. One can show that in this case the Huet algorithm [Huet 1975] actually converges—it converges even when we allow first-order unifiable variables to occur on both sides. Given a reasonable definition of the size of a term, we can easily prove that the value of the pair ⟨size(**lhs**), size(**rhs**)⟩ decreases lexicographically during the course of unifying **lhs** against **rhs**. In fact, one can show that the cardinality of the set of unifiers generated by this process is exponentially bounded in the size of **lhs**. The algorithm terminates very rapidly in practice —we have yet to worry about exponential blow-ups. We use two-sided unification in the sense explained above in rewriting situations. Thus one can avoid the problem of "free variables" mentioned by [Boyer and Moore 1979A]. This also allows us to do some existential verification in the process of rewriting.

Our implementation is optimized in many ways. For example, explicit substitutions are never made. For efficiency and in order to deal with implicit lambda eliminations, we have used the rather complex data structures for substitution lists suggested by [Lusk, McCune and Overbeek 1982].

The second modification to the Huet algorithm involves the way EKL treats tuples of variables: For example, a variable occurring at the end of a list may match to the constant (), which need not appear explicitly on the other side.

Finally, the unifier may in the process of matching (implicitly) expand definitions of atoms occurring on either side. This has turned out to be a powerful tool, though it is the current bottleneck in the unification process.

As an example of unification we present the EKL verification of the correctness of a definition of APPEND based on a high-order function existence axiom. This is an actual EKL run through the SAIL E editor.

```
;retrieve the basic lisp axioms

(GET-PROOFS LISPAX PRF PRF JK)

(PROOF APPEND)

;declare a new operator taking one or more arguments

(DECL NEWAPPEND (TYPE: |GROUND⊗(GROUND*)→GROUND|) (SYNTYPE: CONSTANT)
      (INFIXNAME: **) (BINDINGPOWER: 840))

;We will be using listinduction.
```

```
;Note the distinct uses of ``.'' in LISTINDUCTIONDEF:
;first, as a delimiter for quantified variables, and then as the infix
;operator-name for CONS.
;Expressions surrounded by bars are regarded as terms by EKL.

(SHOW LISTINDUCTION LISTINDUCTIONDEF)

;labels: LISTINDUCTION
29. (AXIOM |∀PHI.PHI(NIL)∧(∀X U.PHI(U)⊃PHI(X.U))⊃(∀U.PHI(U))|)

;labels: LISTINDUCTIONDEF
33. (AXIOM
      |∀DF NILCASE DEF.(∃FUN.(∀PARS X U.FUN(NIL,PARS)=NILCASE(PARS)∧
                                         FUN(X.U,PARS)=
                                         DEF(X,U,FUN(U,DF(X,PARS)),PARS)))|)

;Note that there are 6 unifiable variables occurring in this line:
;DF of type GROUND⊗(GROUND*)→(GROUND*),
;DEF of type GROUND⊗GROUND⊗GROUND⊗(GROUND*)→GROUND,
;NILCASE of type (GROUND*)→(GROUND*),
;PARS of type GROUND*, and finally X,U of type GROUND.
;
;The variables PARS,X,U occur inside an existential quantifier.
;In the actual unification process they are replaced by functionally
;interpreted higher order variables.

(DEFINE NEWAPPEND |∀V X U.NIL**V=V∧(X.U)**V=X.(U**V)|
        (USE LISTINDUCTIONDEF))

;EKL accepts this definition because it is able to rewrite the formula
;∃NEWAPPEND. ∀V X U.NIL**V=V∧(X.U)**V=X.(U**V) into TRUE
;by matching it against LISTINDUCTIONDEF with the additional unifiable
;variable NEWAPPEND coming from the other side.
;
;After translation from internal forms, the unifier found
;by EKL can be expressed as the following set of pairs:
;(DEF,|λX Y Z X1.X.Z|),(DF,|λX Y.Y|),(U,|U|),(X,|X|),(NILCASE,|λX.X|)
;(NEWAPPEND,|λX Y.FUN(X,Y)|) and (PARS,|V|)
;
;We can now go on and prove facts about NEWAPPEND.
;For example, we can show that U**V is a list
;for any two lists U,V by induction on U.
;This is done by instantiating the universal variable PHI
;in the list induction schema to the term λU.∀V.LISTP (U**V),
;opening the definition of NEWAPPEND,
;and letting the rewriter do the rest through
;the universal elimination (UE) command.
;It should be noted that the variables X,Y,Z have the sort SEXP
;and the variables U,V have the sort LISTP.
;Sorts are unary predicates representing ``semantic'' restrictions
;on atoms. For example, since the variable U is declared to have
;the sort LISTP, the formula LISTP U is true.

(UE (PHI |λU.∀V.LISTP (U**V)|) LISTINDUCTION (OPEN NEWAPPEND))
∀U V.LISTP U ** V
```

```
;This is a useful fact to add to the context of generally known facts
;about NEWAPPEND. The label SIMPINFO as a name to a line has
;special significance to EKL: all rewriting commands will
;use SIMPINFO lines automatically in the default mode.

(LABEL SIMPINFO)

;GENERAL PRINCIPLE: Many meta theoretic objects (axiom schemas etc.)
;can be replaced by constructions involving higher types.
```

Our approach to verifying the existence of LISP-like functions is quite different from the one chosen by [Boyer and Moore 1979A]. We do not use special-purpose definitional mechanisms. Any definition in EKL arises from axioms which often contain variables of higher type. Indeed, we have no desire to represent programs directly in the formalism of EKL: Our approach is purely extensional.

The axiom LISTINDUCTIONDEF given above is sufficient for simple primitive recursive functions with parameters. Many other LISP functions can be defined by primitive recursion on a higher type. Consider, for example, the function *flat* with the property

$$flat(x.y, z) = flat(x, flat(y, z)).$$

While *flat* is not primitive recursive, the function $\lambda y.\, flat(x, y)$ is.

5. Rewriting

Almost all the primitive commands of EKL use rewriting—even the decision procedures are viewed as a part of the rewriting process.

Rewriting immediately poses the problem of control. Indiscriminate use of equalities may easily lead to infinite loops or worse consequences: Unintended replacements. One cannot expect problems of termination to be of great relevance in our context. Thus we need a language for rewriting—how to control the process through simple instructions to EKL.

Many formal manipulation programs use the paradigm of "reasoning experts" operating on a "knowledge base", following the traditional separation of programs from data. Our first attempt at controlling rewriting was based on this approach, following a suggestion made by McCarthy. Regular expressions of strategies were used to tell the rewriter what to do. The results from this experiment were very valuable. While we were able to produce very compact proofs, often the expressions employed were incomprehensible to the casual user. One may argue that the fault lay in the design of the rewriting language. We are inclined to believe otherwise. In our opinion the point of departure from familiar mathematical practice occurs with the very attempt to separate programs from data. Mathematical statements often contain implicit procedural information.

Let us look at a simple example: What is the intended meaning of the fact

$$P(x) \supset A = B\,?$$

One can immediately enumerate several possibilities:

(1) Replace $P(x) \supset A = B$ by **true**, whenever it appears.

(2) Replace $A = B$ by **true** if one can prove $P(x)$ in the current situation.

(3) Replace $P(x)$ by **false** if one can prove $\neg A = B$.

(4) Replace A by B whenever one can prove $P(x)$.

(5) Replace B by A whenever one can prove $P(x)$.

(6) Replace A by B whenever one can prove $P(x)$, but not in terms resulting from this substitution.

Some of the interpretations listed above subsume others: For example, (2) is more general than (1). Lines (4) and (5) are completely contradictory in intent. It is obvious that one can go on listing many more possibilities in this vein. With quantified statements the situation can get even more involved.

EKL terms are complicated data structures tagged by information about the applicability of various rewriting procedures. We view the interpretation of facts as a mapping of the form:

$$\langle \textbf{fact} \rangle \otimes \langle \textbf{mode of use} \rangle \rightarrow \langle \textbf{rewriting procedure} \rangle.$$

A rewriting procedure can be expressed in terms of a tuple consisting of left-hand side, right-hand side, list of unifiable variables, and conditions.

The conditions can either be procedural in nature or consist of formulas to be verified before the result of the procedure can be accepted. The user can impose procedural conditions by listing arbitrary LISP S-expressions that have to evaluate to T in order for the application to be accepted. A rewriting procedure returns either the right-hand side together with a substitution list or else reports failure.

The verification of the conditions of rewriting is considered separate from the process of rewriting. We have a small decision procedure that performs this task quite quickly.

Currently there are three possible modes of use for a fact: Given a non-failing application of a rewrite, the mode **default** accepts the result only if it is simpler, the mode **always** accepts the result always, and the mode **exact** accepts the resulting term but does not allow applications of this rewrite in it.

In addition to actions described above the rewriter does standard simplifications: For example, logical simplifications, λ-eliminations, and removal of unnecessary quantifiers are done automatically. Applications of associative operators are "flattened." Existential statements are replaced by **true** if they can be verified through the use of unification. Equalities are treated in a similar fashion.

Metatheoretic simplifications are done automatically. The rewriter will replace expressions of the form $\downarrow\uparrow t$ by t if t is an EKL expression of the right type and has no free variables that are captured by the current binding environment. In addition, computations involving absolute constants are performed. If an EKL symbol F is attached to the LISP function FOO, then the rewriter may replace $F(\uparrow X_1 \ldots \uparrow X_n)$ with $\uparrow Y$ (assuming the types are consistent), where Y is the result of applying FOO to the S-expressions representing $X_1 \ldots X_n$.

One can ask the rewriter to apply the EKL decision procedure DERIVE. The appropriate term will be replaced by **true** when the procedure succeeds. DERIVE invokes a program which tests the validity of deductions in a fragment of predicate calculus designed to capture the notion of "trivial" inferences ([Ketonen and Weyhrauch 1983C]).

Rewriting procedures can be induced by conditional branches. For example, in the formula

$$\text{if } P \text{ then } A \text{ else } B$$

we use $\neg P$ in DEFAULT mode when rewriting B. In addition, $\neg P$ is added to the set of facts known to EKL.

A more complicated example of the use of conditional branches occurs in the treatment of conjunctions and disjunctions. In the expression

$$P \wedge Q$$

we may use P when rewriting Q and vice versa. Similarly, in

$$P \vee Q$$

$\neg P$ is used when rewriting Q. Conditional branches of this kind are more difficult to handle. The facts used need to be recomputed every time there is a change. A typical effect of this sort of conditional reduction is the removal of redundant expressions in conjunctions.

6. An Example of Rewriting in EKL

Continuing the example presented before, we will prove the associativity of the NEWAPPEND function. This is done by proving by induction on the list variable U the statement

$$(U**V)**W = U**(V**W).$$

Here U, V, W are variables of the sort LISTP. The LISTINDUCTION axiom is invoked through the universal elimination (UE) command by instantiating the variable PHI to the term $\lambda U.(U**V)**W = U**(V**W)$.

Turning on the debugging switch REWRITEMESSAGES forces EKL to give detailed information about its actions.

```
(SETQ REWRITEMESSAGES T)

(UE (PHI |λU.((U**V)**W)=(U**(V**W))|) LISTINDUCTION (OPEN NEWAPPEND))

;the term NIL ** V is replaced by:
;V
;
;the term NIL ** (V ** W) is replaced by:
;V ** W
;
;the term V ** W=V ** W is replaced by:
;TRUE
;
;the term X.U ** V is replaced by:
;X.(U ** V)
;
```

```
;the term X.(U ** V) ** W is replaced by:
;X.((U ** V) ** W)
;
;the term (U ** V) ** W is replaced by:
;U ** (V ** W)
;
;the term X.U ** (V ** W) is replaced by:
;X.(U ** (V ** W))
;
;the term X.(U ** (V ** W))=X.(U ** (V ** W)) is replaced by:
;TRUE
;
;the term (U ** V) ** W=U ** (V ** W)⊃TRUE is replaced by:
;TRUE
;
;the term ∀X U.TRUE is replaced by:
;TRUE
;
;the term TRUE∧TRUE is replaced by:
;TRUE
;
;the term TRUE⊃(∀U.(U ** V) ** W=U ** (V ** W)) is replaced by:
;∀U.(U ** V) ** W=U ** (V ** W)

∀U.(U ** V) ** W=U ** (V ** W)
```

References

Bledsoe, W. W., *Non-resolution Theorem-proving*, Artificial Intelligence 3, 1–36, 1977.

Boyer, R.S., Moore, J. S., *A computational logic*, Academic Press, New York, 1979.

Boyer, R.S., Moore, J. S., *Metafunctions: Proving them correct and using them efficiently as new proof procedures*, SRI International, Technical Report CSL-108, 1979.

Bulnes, J., *Goal: a goal oriented command language for interactive proof construction*, Stanford AI Memo AIM-328, 1979.

de Bruijn, N.G., *AUTOMATH—A Language for Mathematics*, Technological University Eindhoven, Netherlands, 1968.

Gabriel, R.P., Frost, M.E., *A Programming Environment for a Timeshared System*, to be presented at the 1984 Software Engineering Symposium on Practical Software Development Environments, 1984.

Gordon, M.J.C., Milner, R., Wadsworth, C., *Edinburgh LCF*, Springer-Verlag, New York, 1979.

Huet, G.P., *A unification Algorithm for Typed λ-Calculus*, Theoretical Computer Science 1, 27–57, 1975.

Jensen, D.C., Pietzrykowski, T., *Mechanizing ω-Order Type Theory Through Unification*, Theoretical Computer Science 3, 123–171, 1976.

Jutting, L.S., *A translation of Landau's "Grundlagen" in AUTOMATH*, Eindhoven University of Technology, Dept. of Math., 1976.

Ketonen, J., Weening, J., *EKL—An interactive proof checker*, Users' Reference Manual, 40 pp., Stanford University, 1983.

Ketonen, J., Weening, J., *The Language of an Interactive Proof Checker*, 34 pp., Stanford University, CS Report STAN-CS-83-992, 1983.

Ketonen, J., Weyhrauch, R., *A semidecision procedure for predicate calculus*, 16 pp., to appear in the Journal of Theoretical Computer Science 1984.

Kreisel, G., *Neglected Possibilities of Processing Assertions and Proofs Mechanically: Choice of Problems and Data*, in University-Level Computer-Assisted Instruction at Stanford: 1968–1980, edited by Patrick Suppes, IMSSS, Stanford, 1981.

Lusk, E.L, McCune, W.W, Overbeek, R.A., *Logic Machine Architecture: Kernel functions*, in 6th Conference on Automated Deduction, New York, edited by D.W.Loveland, Lecture Notes in Computer Science, No.138, Springer-Verlag, 70–84, 1982.

McCarthy, J., *Computer programs for checking mathematical proofs*, Proc. Symp. Pure Math., Vol. 5, American Mathematical Society, 219–227, 1962.

McCarthy, J., Talcott, C., *LISP: programming and proving*, to appear, available as Stanford CS206 Course Notes, Fall 1980.

Miller, D.A., Cohen, E., Andrews, P.B., *A look at TPS*, in 6th Conference on Automated Deduction, New York, edited by D.W.Loveland, Lecture Notes in Computer Science, No.138, Springer-Verlag, 70–84, 1982.

Shostak, R.E., Schwartz, R., Melliar-Smith, P.M., *STP: A Mechanized Logic for Specification and Verification*, in 6th Conference on Automated Deduction, New York, edited by D.W.Loveland, Lecture Notes in Computer Science, No.138, Springer-Verlag, 32–49, 1982.

Weyhrauch, R., *Prolegomena to a theory of mechanized formal reasoning*, Stanford AI Memo AIM-315, 1978.

Wos, L., *Solving open questions with an automated theorem-proving program*, in 6th Conference on Automated Deduction, New York, edited by D.W.Loveland, Lecture Notes in Computer Science, No.138, Springer-Verlag, 1–31, 1982.

A LINEAR CHARACTERIZATION OF NP-COMPLETE PROBLEMS

Silvio Ursic

Madison, Wisconsin

ABSTRACT

We present a linear characterization for the solution sets of propositional calculus formulas in conjunctive normal form. We obtain recursive definitions for the linear characterization similar to the basic recurrence relation used to define binomial coefficients. As a consequence, we are able to use standard combinatorial and linear algebra techniques to describe properties of the linear characterization.

1. INTRODUCTION

This paper develops polyhedral combinatorics methods for the problem of finding satisfying truth assignments to propositional calculus formulas in conjunctive normal form. In particular, we wish to detect formulas with no solution. The methods to be developed have applicability in the detection of tautologies in the predicate calculus, in analogy to the manner that ground resolution is lifted to resolution. Clauses of a formula in conjunctive normal form become faces of a polytope. An unsatisfiable formula becomes an unsatisfiable set of linear inequalities. In this context, the fact that two clashing clauses can be resolved, producing a third clause, can be interpreted as being a manifestation of the fact that the polytopes in question are highly degenerated. A supporting hyperplane to the polytopes in question can be written in more than one way as a positive linear combination of faces.

Two very influential papers are responsible for setting background goals which led to the "rise of the conjunctive normal form". They are (Robinson 65) and (Cook 71). Robinson's proof that ground resolution will detect unsatisfiable formulas is simple and elegant. A distinct proof of this fact can be found in (Quine 55) as ammended by (Bing 56). The two proofs complement each other by addressing two distinct aspects of the problem. Robinson's ground resolution and Quine's consensus can ultimately be traced to (Blake 37) and (Blake 46). To this respect, see the note (Brown 68). Blake's syllogistic expansion is, exactly, what we today know as consensus, or resolution, of two clauses. His proof of its validity, theorem (10.3), which he characterizes as having "fundamental importance" is precisely the proof of the validity of consensus as given by (Quine 55). It is particularly fascinating to, after so many years, read the review to (Blake 46). The reviewer characterizes the value of concrete applications of Blake's expansion process (what we today know as resolution) as "virtually nil". Almost fifty years after Blake introduced it, the computing time of iterated resolution to detect tautologies in the propositional calculus continues to be unanalized. A not very restrictive particularization of it, regular resolution, was shown to have exponential computing time,(Tseitin 70), (Galil 77).

The paper (Buchi 62) contributed to our understanding of conjunctive normal forms from another direction. He showed it is possible to simulate the behavior of a Turing machine, in a very simple way, with a predicate calculus formula. If the Turing machine to be simulated is known to halt after a finite number of steps, Buchi predicate calculus formulas can be expanded to propositional calculus formulas. The expanded formula has the very important property of being describable with a number of symbols proportional to the Turing machine computing time. This proportionality was noted and used in a fundamental way in (Cook 71). Much followed from Cook work. The book (Garey & Johnson 79) and the ongoing column on NP-complete problems by Johnson in the Journal of Algorithms will introduce the reader to the field. The reductions from Turing machines to conjunctive normal forms in (Buchi 62) are important for another reason. They show that the predicate calculus and, in particular, propositional calculus for computations known to require a finite number of steps, can be considered a programming language. We can specify algorithms with them in a very natural way.

The methods to be pursued in this paper can be collectively labeled as "Methods in Polyhedral Combinatorics". Polyhedral combinatorics, as a topic of research, predates the advent of NP-complete problems. Integer programming was recognized as a powerful unifying technique for a variety of combinatorial problems long before (Cook 71) and (Karp 72) found unambiguous reasons for it. The survey by (Balinsky 65) is a good overview of the state of affairs in the pre NP-completeness era. More recent surveys of the field can be found in (Padberg 80) and (Pulleyblank 83).

Reasons for the tenacity and permanence of integer programming as a technique to solve combinatorial problems are two fold. It provides methods for the practical solution, via the simplex method, of hard combinatorial problems. A success history is exemplified by (Grotschel 82). It also provides a constructive framework for the study of NP-complete problems. The discovery of the ellipsoid algorithm (Khachian 79) and its variants (Ursic 82), has created new reasons to pursue this line of attack. They are exemplified by (Karp & Papadimitriou 80), (Grotschel, Lovasz & Schrijver 81), (Papadimitriou 81) and (Lenstra 82).

This paper presents a new family of polytopes, the Binomial Polytopes. Section 2, motivation, presents them informally and provides reasons for their study. Section 3, the binomial matrices, recursively defines some families of matrices and develops some of their properties and identities among them. Section 4, the binomial polytopes, presents some characterizations of their faces. Section 5, applications, presents some examples of the potential uses of the binomial polytopes. Section 6, conclusion, presents a final comment.

2. MOTIVATION

Consider a boolean formula in N variables, in conjunctive normal form and with p literals per clause (in short: $CNF_{N,p}$). Construct a matrix as follows: label each row with a distinct truth assignment to a $CNF_{N,p}$; label each column with a distinct clause with p literals in N variables; set an element of the matrix to zero if the row label makes the column label true; if the row label makes the column label false, set it to one. Each column of a matrix so defined is the incidence vector, sometimes called the characteristic function, defining the set of truth assignments for which a specific clause is false.

We wish to study the convex hull of the set of points whose coordinates are given by the rows of the matrices defined above. The following example, with N = 3 and p = 2, illustrates the idea.

$$\begin{array}{cccccccccccc} xy & x\bar{y} & \bar{x}y & \bar{x}\bar{y} & xz & \bar{x}z & x\bar{z} & \bar{x}\bar{z} & yz & y\bar{z} & \bar{y}z & \bar{y}\bar{z} \end{array}$$

$$\begin{bmatrix} 1&0&0&0&1&0&0&0&1&0&0&0 \\ 1&0&0&0&0&0&1&0&0&1&0&0 \\ 0&1&0&0&1&0&0&0&0&0&1&0 \\ 0&1&0&0&0&0&1&0&0&0&0&1 \\ 0&0&1&0&0&1&0&0&1&0&0&0 \\ 0&0&1&0&0&0&0&1&0&1&0&0 \\ 0&0&0&1&0&1&0&0&0&0&1&0 \\ 0&0&0&1&0&0&0&1&0&0&0&1 \end{bmatrix} \quad \begin{array}{l} F_x F_y F_z \\ F_x F_y T_z \\ F_x T_y F_z \\ F_x T_y T_z \\ T_x F_y F_z \\ T_x F_y T_z \\ T_x T_y F_z \\ T_x T_y T_z \end{array} \qquad (2.1)$$

We use as variables x, y and z. We label rows and columns with truth assignments and clauses. To indicate clauses we omit the logic operator and use an overbar for negation. To indicate truth assignments we use T for true and F for false with the subscript indicating the variable to which they are assigned.

With matrix (2.1) we obtain a polytope with eight vertices in a space of dimension twelve. This polytope is in a subspace of dimension six and has sixteen faces. Its description is given by three groups of linear inequalities and equations.

$$xy \geq 0,\ \bar{x}y \geq 0,\ x\bar{y} \geq 0,\ \bar{x}\bar{y} \geq 0,$$

$$xz \geq 0,\ \bar{x}z \geq 0,\ x\bar{z} \geq 0,\ \bar{x}\bar{z} \geq 0,$$

$$yz \geq 0,\ \bar{y}z \geq 0,\ y\bar{z} \geq 0,\ \bar{y}\bar{z} \geq 0. \tag{2.2}$$

$$\bar{x}\bar{y} - \bar{x}z + yz \geq 0,$$

$$x\bar{y} - xz + yz \geq 0,$$

$$\bar{x}y - \bar{x}z + \bar{y}z \geq 0,$$

$$\bar{x}\bar{y} - \bar{x}\bar{z} + y\bar{z} \geq 0. \tag{2.3}$$

$$xy + \bar{x}y + x\bar{y} + \bar{x}\bar{y} = 1,$$

$$xz + \bar{x}z + x\bar{z} + \bar{x}\bar{z} = 1,$$

$$yz + \bar{y}z + y\bar{z} + \bar{y}\bar{z} = 1,$$

$$xy + x\bar{y} + \bar{x}z + \bar{x}\bar{z} = 1,$$

$$xy + \bar{x}y + \bar{y}z + \bar{y}\bar{z} = 1,$$

$$xz + \bar{x}z + y\bar{z} + \bar{y}\bar{z} = 1. \tag{2.4}$$

Inequalities (2.2) correspond to the clauses of the $CNF_{3,2}$. This correspondence is the key property we wish to have. The entries in matrix (2.1) were chosen so that this would happen. Inequalities (2.3) do not correspond to clauses. Much of what follows was originated by these inequalities that do not correspond to clauses. The equations (2.4) are a consequence of the fact that the columns of matrix (2.1) are linearly dependent. They describe the linear subspace which contains our polytope. The form assumed by equations and inequalities (2.2), (2.3) and (2.4) depends on how we choose a basis for the column space of matrix (2.1).

Notice that each of the faces corresponding to clauses, namely (2.2), touches all the vertices labeled with truth assignments that make the corresponding clause true. Equivalently, each face does not touch vertices labeled with truth assignments that make the clause false. Hence it is possible to test a $CNF_{3,2}$ for satisfiability by converting to an equality each face corresponding to a clause present in the $CNF_{3,2}$. Each conversion eliminates a few vertices from consideration. If they are all eliminated, we obtain a system of equations and inequalities with no feasible solution.

Notice that for each N and p the polytope in question is fixed. Its vertices are labeled with all the 2^N possible truth assignments and its faces do not depend on a specific $CNF_{N,p}$. The main purpose of this paper is to provide a description as accurate as possible for the faces of this family of polytopes. The task was started in (Ursic 75) where the family was defined and some of its properties were given.

3. THE BINOMIAL MATRICES

We will recursively define some families of matrices. The recursive definitions closely follow the basic recurrence relation with which binomial coefficients can be defined, namely:

$$\binom{0}{0} = 1;\ \text{for } i \neq 0,\ \binom{0}{i} = 0;\ \binom{N}{i} = \binom{N-1}{i-1} + \binom{N-1}{i}. \tag{3.1}$$

The building blocks for the recursive definitions of the binomial matrices are given by the following two by two matrix, with labeled rows and columns.

$$G_x = \overset{\begin{matrix}\phi & x\end{matrix}}{\begin{bmatrix}1 & 0\\ 1 & 1\end{bmatrix}}\begin{matrix}\phi\\ x\end{matrix} = \overset{\begin{matrix}\phi & x\end{matrix}}{\begin{bmatrix}a_x & b_x\\ c_x & d_x\end{bmatrix}}\begin{matrix}\phi\\ x\end{matrix} \tag{3.2}$$

The subscript x of G_x represents a boolean variable and is used to form row and column labels. The label ϕ represents the empty set. It will be used to represent either the empty label or the empty matrix. In small examples we will use as labels and subscripts the variables x, y, z and w. In recursive definitions we will use $x_1, x_2, \ldots, x_N$ and write, for example, G_i instead of G_{x_i}.

The four entries in matrix G_x can be considered one by one labeled matrices. From (3.2) we have:

$$a_x = \overset{\phi}{[1]}\phi;\ b_x = \overset{x}{[0]}\phi;\ c_x = \overset{\phi}{[1]}x;\ d_x = \overset{x}{[1]}x. \tag{3.3}$$

We will use + and . to indicate standard matrix sums and products. We will use $\oplus$ ($\oplus^T$) to indicate the row-wise (column-wise) join, or concatenation, of two matrices, with the same number of identically labeled rows (columns). All row and column labels are retained in the results. We can now define some additional matrices:

$$E_i = a_i \oplus^T c_i = \overset{\phi}{\begin{bmatrix}1\\ 1\end{bmatrix}}\begin{matrix}\phi\\ x_i\end{matrix};\ V_i = b_i \oplus^T d_i = \overset{x_i}{\begin{bmatrix}0\\ 1\end{bmatrix}}\begin{matrix}\phi\\ x_i\end{matrix}. \tag{3.4}$$

Hence we have:

$$G_i = E_i \oplus V_i = (a_i \oplus b_i) \oplus^T (c_i \oplus d_i) = (a_i \oplus^T c_i) \oplus (b_i \oplus^T d_i).$$

We will use $\otimes$ to indicate standard tensor product of two matrices. Labels of the result are sets of row and column labels of the factors. See, for example, (Pease 65). So, we have:

$$G_x \otimes G_y = \overset{\begin{matrix}\phi & x & y & xy\end{matrix}}{\begin{bmatrix}1 & 0 & 0 & 0\\ 1 & 1 & 0 & 0\\ 1 & 0 & 1 & 0\\ 1 & 1 & 1 & 1\end{bmatrix}}\begin{matrix}\phi\\ x\\ y\\ xy\end{matrix} \tag{3.5}$$

Our main family of matrices are in many respects the matrix counterparts of binomial coefficients. They are recursively defined as follows:

$$r_{0,0,0} = \overset{\phi}{[1]}\phi;\ \text{for } i, j \neq 0,\ r_{0,i,j} = \phi; \tag{3.6}$$

$$r_{N,i,j} = (r_{N-1,i,j} \otimes a_N \oplus r_{N-1,i,j-1} \otimes b_N) \oplus^T (r_{N-1,i-1,j} \otimes c_N \oplus r_{N-1,i-1,j-1} \otimes d_N)$$

Definition (3.6) closely mirrors definition (3.1), applied twice to the indices i and j of $r_{N,i,j}$. The resulting matrices are defined for $N \geq 0$ and are non-void in the range $0 \leq i, j \leq N$.

The next theorem is our main tool for the study of some of the combinatorial properties of boolean formulas in conjunctive normal form. It is the exact counterpart of the identity

$$\sum_{0 \leq i \leq N} \binom{N}{i} = 2^N \tag{3.7}$$

The matrices G_i assume the role of the constant two and the matrices $r_{N,i,j}$ are "doubly binomial", either by the row index i or by the column index j.

Theorem B (The binomial identity). The following is a matrix identity:

$$\bigotimes_{1 \leq i \leq N} G_i = \bigoplus^T_{0 \leq i \leq N} \Big(\bigoplus_{0 \leq j \leq N} r_{N,i,j} \Big). \tag{3.8}$$

Proof. We start by giving a detailed proof of the induction step in a proof by induction on N of (3.7). This level of detail is hardly necessary. It illustrates, however, the complete parallelism between binomial coefficients and binomial matrices.

$2^N = 2^{N-1} \,.\, 2$ (separate a "two")

$= \Big(\sum_{0 \leq i \leq N-1} \binom{N-1}{i} \Big) \,.\, (1 + 1)$ (use induction hypothesis and definition of constant "two")

$= \sum_{0 \leq i \leq N-1} \binom{N-1}{i} + \sum_{0 \leq i \leq N-1} \binom{N-1}{i}$ (perform product)

$= \sum_{0 \leq i \leq N} \binom{N-1}{i} + \sum_{-1 \leq i \leq N-1} \binom{N-1}{i}$ (extend sums using $\binom{N-1}{N} = 0$ and also $\binom{N-1}{-1} = 0$)

$= \sum_{0 \leq i \leq N} \binom{N-1}{i} + \sum_{0 \leq i \leq N} \binom{N-1}{i-1}$ (shift indices in second sum)

$= \sum_{0 \leq i \leq N} \Big(\binom{N-1}{i} + \binom{N-1}{i-1} \Big)$ (merge sums)

$= \sum_{0 \leq i \leq N} \binom{N}{i}$ (use defining recurrence relation)

We will repeat the procedure in a proof by induction on N of (3.8). The identity is true for $N = 1$. In this case it reduces to

$G_1 = (a_1 \oplus b_1) \oplus^T (c_1 \oplus d_1)$,

which is true by definition. The induction step is taken care as follows:

$\bigotimes_{1 \leq i \leq N} G_i = \Big(\bigotimes_{1 \leq i \leq N-1} G_i \Big) \otimes G_N$ (separate G_N from the product)

$= \Big(\bigoplus^T_{0 \leq i \leq N-1} \Big(\bigoplus_{0 \leq j \leq N-1} r_{N-1,i,j} \Big) \Big) \times ((a_N \oplus b_N) \oplus^T (c_N \oplus d_N))$

(use induction hypothesis and definition of the matrix G_N)

$$= \bigoplus^{T}_{0 \le i \le N-1} \Big(\bigoplus_{0 \le j \le N-1} ((r_{N-1,i,j} \otimes a_N \oplus r_{N-1,i,j} \otimes b_N) \oplus^{T} (r_{N-1,i,j} \otimes c_N \oplus r_{N-1,i,j} \otimes d_N))$$

(perform tensor product)

$$= \bigoplus^{T}_{0 \le i \le N} \Big(\bigoplus_{0 \le j \le N} ((r_{N-1,i,j} \otimes a_N \oplus r_{N-1,i,j-1} \otimes b_N) \oplus^{T} (r_{N-1,i-1,j} \otimes c_N \oplus r_{N-1,i-1,j-1} \otimes d_N))$$

(extend sums using the fact that the matrices $r_{N,i,j} = \phi$ outside the index range $0 \le i,\ j \le N$, and then shift indices)

$$= \bigoplus^{T}_{0 \le i \le N} \Big(\bigoplus_{0 \le j \le N} r_{N,i,j} \Big) .$$

(use defining recurrence relation)

We conclude that (3.8) is indeed an identity. □

Matrix identity (3.8) essentially permutes rows and columns of the same matrix in two different ways. The following example, with N = 3, illustrates its effect. We have:

$$\begin{bmatrix} r_{3,0,0} & r_{3,0,1} & r_{3,0,2} & r_{3,0,3} \\ r_{3,1,0} & r_{3,1,1} & r_{3,1,2} & r_{3,1,3} \\ r_{3,2,0} & r_{3,2,1} & r_{3,2,2} & r_{3,2,3} \\ r_{3,3,0} & r_{3,3,1} & r_{3,3,2} & r_{3,3,3} \end{bmatrix} = \begin{array}{c} \begin{array}{cccccccc} \phi & x_1 & x_2 & x_3 & x_1x_2 & x_1x_3 & x_2x_3 & x_1x_2x_3 \end{array} \\ \left[\begin{array}{c|ccc|ccc|c} 1 & 0 & 0 & 0 & 0 & 0 & 0 & 0 \\ \hline 1 & 1 & 0 & 0 & 0 & 0 & 0 & 0 \\ 1 & 0 & 1 & 0 & 0 & 0 & 0 & 0 \\ 1 & 0 & 0 & 1 & 0 & 0 & 0 & 0 \\ \hline 1 & 1 & 1 & 0 & 1 & 0 & 0 & 0 \\ 1 & 1 & 0 & 1 & 0 & 1 & 0 & 0 \\ 1 & 0 & 1 & 1 & 0 & 0 & 1 & 0 \\ \hline 1 & 1 & 1 & 1 & 1 & 1 & 1 & 1 \end{array}\right] \end{array} \begin{array}{l} \phi \\ x_1 \\ x_2 \\ x_3 \\ x_1x_2 \\ x_1x_3 \\ x_2x_3 \\ x_1x_2x_3 \end{array} ;$$

$$G_1 \otimes G_2 \otimes G_3 = \begin{array}{c} \begin{array}{cccccccc} \phi & x_1 & x_2 & x_1x_2 & x_3 & x_1x_3 & x_2x_3 & x_1x_2x_3 \end{array} \\ \left[\begin{array}{cc|cc|cc|cc} 1 & 0 & 0 & 0 & 0 & 0 & 0 & 0 \\ 1 & 1 & 0 & 0 & 0 & 0 & 0 & 0 \\ \hline 1 & 0 & 1 & 0 & 0 & 0 & 0 & 0 \\ 1 & 1 & 1 & 1 & 0 & 0 & 0 & 0 \\ \hline 1 & 0 & 0 & 0 & 1 & 0 & 0 & 0 \\ 1 & 1 & 0 & 0 & 1 & 1 & 0 & 0 \\ \hline 1 & 0 & 1 & 0 & 1 & 0 & 1 & 0 \\ 1 & 1 & 1 & 1 & 1 & 1 & 1 & 1 \end{array}\right] \end{array} \begin{array}{l} \phi \\ x_1 \\ x_2 \\ x_1x_2 \\ x_3 \\ x_1x_3 \\ x_2x_3 \\ x_1x_2x_3 \end{array} .$$

It should be emphasized that identity (3.8) is an identity among the entries of the matrices on the right and left hand side having the same row and column labels.

The next theorem presents essential information about the entries and rank of the matrices $r_{N,i,j}$.

Theorem P (Properties of the binomial matrices). The following is true:

(A) Matrix $r_{N,i,j}$ has $\binom{N}{i}$ rows and $\binom{N}{j}$ columns;

(B) Matrix $r_{N,i,j}$ is a matrix of zeros and ones with exactly $\binom{i}{j}$ ones in each row a and $\binom{N-j}{N-i}$ ones in each column;

(C) For $N \geq p \geq q \geq s \geq 0$ we have

$$r_{N,p,q} \cdot r_{N,q,s} = \binom{p-s}{q-s} \otimes r_{N,p,s} \; ; \tag{3.9}$$

(D) For $i \geq j$, matrix $r_{N,i,j}$ has full rank. If $\binom{N}{i} \leq \binom{N}{j}$ then its rows are linearly independent. If $\binom{N}{i} \geq \binom{N}{j}$ then its columns are linearly independent.

Proof. All four properties are proven by induction on N using the defining recursion, namely (3.6). The proof of properties (A) and (B) is uncomplicated. Identity (3.9) is obviously true for $N = 0$. We also have

$r_{N,p,q} \cdot r_{N,q,s} =$

$((r_{N-1,p,q} \otimes a_N \oplus r_{N-1,p,q-1} \otimes b_N) \oplus^T (r_{N-1,p-1,q} \otimes c_N \oplus r_{N-1,p-1,q-1} \otimes d_N))$.

$((r_{N-1,q,s} \otimes a_N \oplus r_{N-1,q,s-1} \otimes b_N) \oplus^T (r_{N-1,q-1,s} \otimes c_N \oplus r_{N-1,q-1,s-1} \otimes d_N)) =$

(use defining recurrence (3.6))

$((r_{N-1,p,q} \cdot r_{N-1,q,s}) \otimes a_N \oplus$

(a matrix of zeros) $\otimes\, b_N) \oplus^T$

$((r_{N-1,p-1,q} \cdot r_{N-1,q,s} \oplus r_{N-1,p-1,q-1} \cdot r_{N-1,q-1,s}) \otimes c_N +$

$(r_{N-1,p-1,q-1} \cdot r_{N-1,q-1,s-1}) \otimes d_N) =$

(perform product taking into account the fact that the entry of b_N is zero)

$(r_{N-1,p,s} \otimes \binom{p-s}{p-q} \otimes a_N \oplus$

(a matrix of zeros) $\otimes\, b_N) \oplus^T$

$((r_{N-1,p-1,s} \otimes \binom{p-s-1}{p-q-1} \oplus r_{N-1,p-1,s} \otimes \binom{p-s-1}{p-q}) \otimes c_N +$

$r_{N-1,p-1,s-1} \otimes \binom{p-s}{p-q} \otimes d_N) =$

(use induction hypothesis)

$\binom{p-s}{q-s} \otimes r_{N,p,s}$ (use defining recurrences (3.1) and (3.6))

We therefore established property (C). Property (D) follows, also by induction, from the fact that the entry in b_1, defined in (3.2), is zero. As a consequence, the matrices $r_{N,i,j}$, as given by recursion (3.6), are block triangular. The two blocks on the diagonal, matrices $r_{N-1,i,j}$ and $r_{N-1,i-1,j-1}$, have either row or column full rank (by induction) and we are done with one special case to be considered. We can have $\binom{N-1}{i} < \binom{N-1}{j}$ and $\binom{N-1}{i-1} > \binom{N-1}{j-1}$ or also, concievably, $\binom{N-1}{i} > \binom{N-1}{j}$ and $\binom{N-1}{i-1} < \binom{N-1}{j-1}$. This system of inequalities can be solved in i and j. We obtain the feasible solution $N = i + j$. Hence $\binom{N}{i} = \binom{N}{j}$ and as a consequence we have $\binom{N-1}{i} > \binom{N-1}{j}$. Full rank for the square matrix $r_{N,i,j}$ follows from the following linear combination of columns of $r_{N,i,j}$.

The entries in matrix $r_{N,i,j}$ are given by

$$\begin{bmatrix} r_{N-1,i,j} & 0 \\ r_{N-1,i-1,j} & r_{N-1,i-1,j-1} \end{bmatrix} . \tag{3.10}$$

Multiply the matrices $r_{N-1,i,j}$ and $r_{N-1,i-1,j}$ in (3.10) by $r_{N-1,j,j-1}$ and then use property (C) and subtract them from the other two matrices in (3.10). We obtain the matrix

$$\begin{bmatrix} r_{N-1,i,j} & r_{N-1,i,j-1} \otimes (P - \binom{j}{1} + 1) \\ r_{N-1,i-1,j} & 0 \end{bmatrix}$$

which is block triangular and, by induction, is of full rank. □

Theorem P establishes basic information about the matrices generated by recursion (3.6). The next theorem expands these properties in a form useful to the task outlined in section 2.

<u>Theorem V (Generalized Vandermonde)</u>. Let a_i, $1 \le i \le k \le N$, be distinct integers such that $\binom{N}{i} \le \binom{N}{a_i}$. Then the matrix

$$\bigoplus^{T}_{1 \le i \le k} \; \bigoplus_{1 \le j \le k} r_{N,a_i,j} \tag{3.11}$$

has linearly independent columns.

<u>Proof.</u> Vandermonde matrix can be written as:

$$\bigoplus^{T}_{1 \le i \le k} \; \bigoplus_{1 \le j \le k} (a_i)^j = \begin{bmatrix} a_1 & (a_1)^2 & (a_1)^3 & \dots & (a_1)^k \\ a_2 & (a_2)^2 & (a_2)^3 & \dots & (a_2)^k \\ & & \dots & & \\ a_k & (a_k)^2 & (a_k)^3 & \dots & (a_k)^k \end{bmatrix} . \tag{3.12}$$

We know that its determinant is

$$\prod_{1 \le j \le k} a_j \prod_{1 \le i < j \le k} (a_i - a_j)$$

and hence is a full rank matrix. Similarly to (3.12), we can define a "binomial Vandermonde matrix", with entries being given by binomial coefficients. We obtain the matrix

$$\bigoplus^{T}_{1 \le i \le k} \; \bigoplus_{1 \le j \le k} \binom{a_i}{j} \tag{3.13}$$

The entries on (3.13) can be considered fractions whose numerators are polynomials in a_i, and hence can be obtained as linear combinations of the columns of (3.12). The denominators are factorials and can be factored out of the matrix. The determinant of (3.13) ends up being the determinant of (3.12) divided by a product of factorials.

Matrix (3.11) is very similar to matrix (3.13). We will prove that (3.11) has linearly independent columns by reducing it to block triangular form, with each matrix in the diagonal having full column rank. The technique is similar to the one used to compute the determinant of (3.12). See, for example, (Knuth 68).

For $j = 2, 3, \ldots, k$, in this order, multiply column j by $r_{n-1,j,j-1}$, multiply column j-1 by $a_k - j+1$ and subtract column j from column j-1. By column of index j we mean here the submatrix given by

$$\bigoplus^T_{1 \le i \le k} r_{N,a_i,j} .$$

Use the fact that $r_{N,a_i,j} \cdot r_{N,j,j-1} = r_{N,j,j-1} \otimes \binom{a_i - j + 1}{1}$, that is property (C) of theorem P.

For $j = 1, \ldots, N-1$ the submatrices $r_{N,a_k,j}$ will be zeroed. We factor $(a_k - a_j)$ from the columns of the result and continue by induction on a matrix of the same kind of order k-1. Theorem V follows from the fact that matrix $r_{N,a_k,k}$ has linearly independent columns, by property (D) of theorem P. □

The matrices $r_{N,i,j}$ behave, in matrix identities, like binomial coefficients. Theorem B should be considered just a sample of what is possible. Consider, for example, the binomial coefficient identity (Vandermonde convolution):

$$\binom{N + M}{k} = \sum_i \binom{N}{i} \binom{M}{k - i} . \tag{3.14}$$

The corresponding matrix identities for binomial matrices are:

$$r_{N+M,k,1} = \bigoplus^T_i (r_{N,i,1} \otimes r_{M,k-i,1})$$

or, applying (3.14) to both row and column indices

$$r_{P+Q+R+S,k,1} = \bigoplus^T_i \bigoplus_j (r_{P,i,j} \otimes r_{Q,i,1-j} \otimes r_{R,k-i,j} \otimes r_{S,k-i,1-j}).$$

It is convenient to define two additional families of matrices. They are:

$$R_{N,j} = \bigoplus^T_{0 \le i \le N} r_{N,i,j} ; \tag{3.15}$$

$$B_{N,p} = \bigoplus_{0 \le j \le p} R_{N,j} = \bigoplus_{0 \le j \le p} \left(\bigoplus^T_{0 \le i \le N} r_{N,i,j} \right). \tag{3.16}$$

From (3.4), (3.6), (3.15) and (3.16) we obtain:

$$R_{0,0} = 1; \text{ for } i \ne 0,\ R_{0,i} = \phi;\ R_{N,i} = R_{N-1,i-1} \otimes V_N \oplus R_{N-1,i} \otimes E_N; \tag{3.17}$$

$$\text{for } i \ge 0,\ B_{0,i} = 1; \text{ for } i < 0,\ B_{0,i} = \phi;\ B_{N,i} = B_{N-1,i-1} \otimes V_N \oplus B_{N-1,i} \otimes E_N. \tag{3.18}$$

From definition (3.16) and identity (3.8) we derive the following identity, which provides an alternate description of $B_{N,N}$:

$$B_{N,N} = \bigotimes_{1 \le i \le N} G_i \tag{3.19}$$

In identities, matrices $B_{N,p}$ behave similarly to the sum

$$\sum_{0 \le i \le p} \binom{N}{i} = \binom{N}{0} + \binom{N}{1} + \binom{N}{2} + \ldots + \binom{N}{p}. \tag{3.20}$$

(Knuth 68) on page 64 mentions that there seems to be no simple formula for (3.20) meaning it does not seem possible to obtain identities involving (3.20) without the sum sign. It is nevertheless possible to derive meaningful identities involving (3,20). (Riordan 68), pp 128-130, reports some results in this direction. The matrix identity involving $B_{N,p}$ needed for our purposes is the analogue of

$$\sum_{0 \le i \le p} \binom{N+M}{i} = \sum_{0 \le i \le N} \binom{N}{i} \sum_{0 \le j \le p-i} \binom{M}{j} = \sum_{0 \le i+j \le p} \binom{N}{i} \binom{M}{j} \qquad (3.21)$$

namely

$$B_{N+M,p} = \bigoplus_{0 \le i \le N} (R_{N,i} \otimes B_{M,p-i}) = \bigoplus_{0 \le i+j \le p} (R_{N,i} \otimes R_{M,j}). \qquad (3.22)$$

Identity (3.22) can be proven by induction on N. The necessary index manipulations closely follow the corresponding ones in a proof by induction of (3.21). This concludes our study of "Binomial Matrixology". It would be interesting, and useful, to precisely characterize which binomial coefficient identities can be converted into binomial matrices identities.

4. THE BINOMIAL POLYTOPES

We define the binomial polytopes as being the convex hull of the 2^N points whose coordinates are given by the rows of the matrices $B_{N,p}$. We will call the polytope associated with matrix $B_{N,p}$ by the name of the matrix that generates it, namely polytope $B_{N,p}$. Definition (3.16) and identity (3.19) inform us of two facts: matrix $B_{N,N}$ has full rank and hence the associated polytope is a simplex; matrices $B_{N,p}$ are formed with a subset of the columns of $B_{N,N}$ and hence the polytopes $B_{N,p}$ are a projection of polytope $B_{N,N}$.

Polytope $B_{N,N}$ is a simplex of dimension $2^N - 1$ in a space of dimension 2^N. To obtain a simplex whose dimension matches the dimension of the space we have to eliminate one of the columns of $B_{N,N}$. The column of $B_{N,N}$ with the empty label, matrix $R_{N,0}$, performs very nicely the role of the homogenization column. We will keep this column in the binomial matrices with the understanding that to obtain the description of our convex hulls we must add the equality $\phi = 1$ to the system of inequalities describing faces.

The inverse of matrix $B_{N,N}$ is easy to compute. From (3.19) we obtain:

$$B_{N,N}^{-1} = \bigotimes_{1 \le i \le N} G_i^{-1} = \bigotimes_{1 \le i \le N} \begin{array}{c} \begin{array}{cc} \phi & x_i \end{array} \\ \begin{bmatrix} 1 & 0 \\ -1 & 1 \end{bmatrix} \end{array} \begin{array}{c} \\ \phi \\ x_i \end{array} . \qquad (4.1)$$

Identity (4.1) tells us what the faces of $B_{N,N}$ are. Each column of $B_{N,N}^{-1}$ describes a face of $B_{N,N}$. The polytopes $B_{N,p}$ are projections of $B_{N,N}$. Hence it is possible to obtain a face of $B_{N,p}$ by Fourier elimination on the known faces of $B_{N,N}$.

We will indicate by $f_{N,p}$ a supporting hyperplane to $B_{N,p}$. As $B_{N,p}$ is a projection of $B_{N,N}$ we can write:

$$f_{N,p} = B_{N,N}^{-1} \cdot h_{N,p} \quad \text{with } h_{N,p} \ge 0 \ . \qquad (4.2)$$

The vector $h_{N,p}$ contains the 2^N Fourier multipliers defining the positive linear combination of faces of $B_{N,N}$ which generate $f_{N,p}$ and has its components labeled with the column labels of $B_{N,N}^{-1}$. The vector $f_{N,p}$ will have its components labeled with the row labels of $B_{N,N}^{-1}$. We also require that all the components of $f_{N,p}$ whose labels are not present in $B_{N,p}$ to be zero. This last condition reflects the fact that $B_{N,p}$ is a projection of $B_{N,N}$. As a consequence we can write

$$h_{N,p} = B_{N,p} \cdot f_{N,p}$$

disregarding all the coordinates of $f_{N,p}$ not present in $B_{N,p}$, which have zero entries.

We say that $f_{N,p}$ defines a face, or is a face, of $B_{N,p}$ if the collection of vertices of $B_{N,p}$ touched by it defines a submatrix of $B_{N,p}$ of maximum rank. Otherwise $f_{N,p}$ defines, or is, a facet of $B_{N,p}$. It is convenient to explicitly list the subscripts of the matrices used to generate $B_{N,p}$. We write

$$B_{3,3} = G_x \otimes G_y \otimes G_z = B_{3,3}(x, y, z).$$

We proceed analogously for faces. We write $f_{4,3}(x, y, z, w)$ for a face of $B_{4,3}(x, y, z, w)$. We are now ready to start to describe faces of the polytopes $B_{N,p}$.

Theorem R (Representative faces). Let $s_{N,j}$ be column vectors with $\binom{N}{j}$ components labeled with all the labels in the columns of $r_{N,i,j}$ and with all the entries being equal to one. Let $a_1, \ldots, a_p$ be the constants used in theorem V with the additional condition that $a_{i+1} = a_i + 1$ for i odd. We will refer to this condition as the "no odd gap condition".

Compute the constants $b_0, b_1, \ldots, b_p$ by solving the linear system

$$\sum_{0 \le k \le p} b_k \binom{a_j}{k} = 0, \quad \text{for } j = 1, \ldots, p. \tag{4.3}$$

Then

$$f_{N,p} = f_{N,(a_1, a_2, \ldots, a_p)} = \bigotimes_{0 \le k \le p} s_{N,k} \cdot b_k \tag{4.4}$$

is a face of $B_{N,p}$.

Proof. We have

$$B_{N,p} \cdot f_{N,p}$$

$$= \left(\bigoplus^T_{1 \le i \le N} \bigoplus_{0 \le j \le p} r_{N,i,j} \right) \cdot \left(\bigoplus^T_{0 \le j \le p} s_{N,j} \cdot b_j \right)$$

(use definitions (3.16) and (4.4))

$$= \bigoplus^T_{0 \le i \le N} s_i \cdot \sum_{0 \le j \le p} b_j \binom{i}{j}$$

(use identity $r_{N,i,j} \cdot s_{N,j} = s_{N,i} \cdot \binom{i}{j}$, a consequence of property (B) of theorem P)

But, for $i = a_1, a_2, \ldots, a_p$, we have

$$\sum_{0 \le j \le p} b_j \binom{i}{j} = 0. \tag{4.5}$$

The constants b_j were chosen, by condition (4.3), so that this would happen. Hence, by theorem V, hyperplane $f_{N,p}$, as defined in (4.4), touches a subset of the vertices of $B_{N,p}$ having maximum rank.

To verify that all the other vertices of $B_{N,p}$, not touched by $f_{N,p}$, are located in only one of the two halfspaces defined by $f_{N,p}$, we must check the sign of the sum in (4.5) for values of i outside the range $a_1, \ldots, a_p$. To this respect, add to the linear system (4.3) the linear equation

$$\sum_{0 \le j \le p} b_j \binom{i}{j} = c_i$$

in order to compute the value of c_i. The quantity c_i will be a function of $a_1, a_2, \ldots, a_p$ and i. We use the comments following (3.13) and conclude that c_i is the ratio of the two Vandermonde determinants, given by the constants $a_1, a_2, \ldots, a_p$ and $a_1, a_2, \ldots, a_p$, i. We have

$$c_i = (a_1 - i)(a_2 - i) \ldots (a_p - i) \tag{4.6}$$

We conclude that $c_i \geq 0$, for $i = 0, \ldots, N$, whenever the sequence $a_1, \ldots, a_p$ satisfies the "no odd gap" condition. □

Some examples are in order. The explicit solution of the linear system (4.3) is simplified by the following facts. We obtain the binomial Vandermonde matrix (3.13) from Vandermonde matrix (3.12) using the Stirling numbers of the second kind and the inverse to the Vandermonde matrix is known explicitly. Hence the Stirling numbers of the first kind and this known inverse permit us to write solutions to (4.3) without having to use Gaussian elimination on (4.3). For details consult, for example, (Knuth 68). The linear system (4.3) is homogeneous. We have used this extra degree of freedom to obtain the constants $b_0, \ldots, b_p$ in a denominator free form. We obtain

for $p = 2$,

$$b_0 = a_1a_2,$$
$$b_1 = 1 - a_1 - a_2,$$
$$b_2 = 2,$$

for $p = 3$,

$$b_0 = a_1a_2a_3,$$
$$b_1 = -1 + a_1 + a_2 + a_3 - a_1a_2 - a_1a_3 - a_2a_3,$$
$$b_2 = -6 + 2a_1 + 2a_2 + 2a_3,$$
$$b_3 = -6.$$

Theorem R generates, for example:

$$f_{3,(1,\ 2)}(x, y, z) = 1 - x - y - z + xy + xz + yz \geq 0, \quad f_{1,1}(x) = 1 - x \geq 0,$$

$$f_{3,(1,\ 2,\ 3)}(x, y, z) = 1 - x - y - z + xy + xz + yz - xyz \geq 0,$$

$$f_{5,(2,\ 3)} = 3 - 2x - 2y - 2z - 2w - 2t + xy + xz + xw + xt + yz + yw + yt + zw + zt + wt \geq 0.$$

It should be noted that xy is not "x times y". The equation $f_{3,(1,\ 2)}(x, y, z)$ is a linear equation and represents a hyperplane defining one of the faces of a polytope and xy is the <u>name</u> of a variable which happens to be a set whose members are the two labels x and y.

A face $f_{N,p}$ of a binomial polytope $B_{N,p}$ is defined by the collection of vertices it touches. Conversely, a face $f_{N,p}$ of $B_{N,p}$ can be considered as defining a function which selects some of the vertices of $B_{N,p}$. The polytope $B_{N,p}$ has 2^N vertices labeled, by our construction, with all the 2^N subsets of N symbols. We can associate a truth assignment with each label (and vertex) as follows. Assign true to x_i if x_i does not appear in the label, otherwise assign false to x_i. For example, the vertex of $B_{4,2}(x, y, z, w)$ labeled xy is associated with the truth assignment: false for x; false for y; true for z; true for w. Hence a face $f_{N,p}$ of $B_{N,p}$ defines or is, a boolean function. With this interpretation, the faces considered in theorem R are some of the boolean symmetric functions with defining parameters $a_1, a_2, \ldots, a_p$.

Symmetric functions have been appearing regularly in connection with methods to obtain solutions to boolean functions. Our $s_{N,i}$ are exactly the functions defined by (Whitehead 01) with his investigation on the "Algebra of Symbolic Logic". His motivation to consider them comes from efforts to decompose, or factor, boolean functions into "prime factors" whose further decomposition seems impossible. The decomposition leads to interesting methods of expressing all the solutions to a boolean function. It is quite remarkable that a similar concept arises here. It is not possible to obtain a face of a polytope as a linear combination of faces with positive multipliers. Symmetric functions continued to be approached, for similar reasons, in (Lowenheim 08) and (Skolem 34). A result in (Shannon 38) increased considerably their usefulness. He presented a very simple circuit to implement them. The many interesting properties of boolean symmetric functions have continued to direct attention on them from an algebraic and combinatorial point of view, as in (Semon 61), (Arnold & Harrison 62), (Cunkle 63).

Propositional calculus formulas exhibit two well known symmetries (Slepian 53). We can permute, or interchange, variables and we can negate them, without altering in any essential way the properties of the formula. It is reasonable to suppose that the binomial polytopes will inherit these two symmetries. This is indeed so. In addition, a third symmetry has been found present. All three symmetries can be found by permuting in all the 4! = 24 ways the 4 vertices of a $B_{2,2}$, and hence can be considered a manifestation, or consequence, of a common underlying symmetry.

So far our presentation has followed the traditional approach of marking sets with zeros and ones. As a consequence, matrix $B_{N,N}$ is triangular and quite sparse, two good characteristics for computational purposes. We found that to describe the symmetries of the binomial polytopes it is more convenient to use -1 and +1, instead of 0 and 1. In this -1+1 setting the generating matrix G_x and the matrix $B_{N,N}$ become:

$$G'_x = \overset{\quad\phi \quad\; x}{\begin{bmatrix} 1 & 1 \\ 1 & -1 \end{bmatrix}} \begin{matrix} \phi \\ x \end{matrix} \; ; \quad B'_{N,N} = \bigotimes_{1 \le i \le N} G'_x . \tag{4.7}$$

The transformation from zero-one variables to minus-one-plus-one variables is given by the matrix T_x, computed as follows:

$$T_x = G_x^{-1} \cdot G'_x = \overset{\quad\phi \quad\; x}{\begin{bmatrix} 1 & 1 \\ 0 & -2 \end{bmatrix}} \begin{matrix} \phi \\ x \end{matrix} \; ; \quad T_x^{-1} = \overset{\quad\phi \quad\; x}{\begin{bmatrix} 1 & 1/2 \\ 0 & -1/2 \end{bmatrix}} \begin{matrix} \phi \\ x \end{matrix} . \tag{4.8}$$

With (4.8) we define our linear transformations:

$$T_N = \bigotimes_{1 \le i \le N} T_i \; ; \quad T_N^{-1} = \bigotimes_{1 \le i \le N} T_i^{-1} .$$

Hence if $f_{N,p}$ is a face of $B_{N,p}$, the corresponding face of $B'_{N,p}$, obtained by changing the entry in b_x of (3.3) to +1 and the entry in d_x to -1, is

$$T_N^{-1} \cdot f_{N,p} = f'_{N,p} . \tag{4.9}$$

In this -1+1 form, the binomial polytopes are projections of a regular simplex. This is a consequence of the fact that matrix $B'_{N,N}$ is an orthogonal matrix. Triangularity has been transformed to orthogonality.

Before being ready to present the symmmetries we need to define a few additional matrices.

<u>Negate x.</u>

$$P_{x,\bar{x}} = \overset{\quad\phi \quad\; x}{\begin{bmatrix} 0 & 1 \\ 1 & 0 \end{bmatrix}} \begin{matrix} \phi \\ x \end{matrix} \; ; \quad Q'_{x,\bar{x}} = \overset{\quad\phi \quad\; x}{\begin{bmatrix} 1 & 0 \\ 0 & -1 \end{bmatrix}} \begin{matrix} \phi \\ x \end{matrix} \; ; \quad Q_{x,\bar{x}} = \overset{\quad\phi \quad\; x}{\begin{bmatrix} 1 & 1 \\ 0 & -1 \end{bmatrix}} \begin{matrix} \phi \\ x \end{matrix} \; ;$$

$$(E_{N,i})' = Q'_{x_i,\bar{x}_i} \otimes \bigotimes_{\substack{1 \le j \le N \\ j \ne i}} I_j . \tag{4.10}$$

The matrices P and Q have been chosen so that $P_{x,\bar{x}} \cdot G_x \cdot Q_{x,\bar{x}} = G_x$. We also have $P_{x,\bar{x}} \cdot G'_x \cdot Q'_{x,\bar{x}} = G'_x$. The matrices P are row permutation matrices. The matrices Q undo, with a linear combination of columns, the permutation of rows performed by P. The matrices Q and Q' are related by $Q = T_x \cdot Q' \cdot T_x^{-1}$. The matrices I_j are identities matrices

$$I_x = \begin{bmatrix} 1 & 0 \\ 0 & 1 \end{bmatrix} \begin{matrix} \phi \\ x \end{matrix} \quad (\text{columns } \phi, x).$$

The identity matrices I_x are used in (4.10) as a padding for all the variables which are not negated. The other symmetries are

Permute x and y.

$$P_{x,y} = \begin{bmatrix} 1 & 0 & 0 & 0 \\ 0 & 0 & 1 & 0 \\ 0 & 1 & 0 & 0 \\ 0 & 0 & 0 & 1 \end{bmatrix} ; \quad Q'_{x,y} = \begin{bmatrix} 1 & 0 & 0 & 0 \\ 0 & 0 & 1 & 0 \\ 0 & 1 & 0 & 0 \\ 0 & 0 & 0 & 1 \end{bmatrix} ; \quad Q_{x,y} = \begin{bmatrix} 1 & 0 & 0 & 0 \\ 0 & 0 & 1 & 0 \\ 0 & 1 & 0 & 0 \\ 0 & 0 & 0 & 1 \end{bmatrix} ;$$

(rows and columns indexed by ϕ, x, y, xy)

$$(E_{N,i,j})' = Q'_{x_i,\, x_j} \otimes \bigotimes_{\substack{1 \le k \le N \\ k \ne i \\ k \ne j}} I_k \, . \tag{4.11}$$

Permute y with xy.

$$P_{y,xy} = \begin{bmatrix} 1 & 0 & 0 & 0 \\ 0 & 0 & 0 & 1 \\ 0 & 0 & 1 & 0 \\ 0 & 1 & 0 & 0 \end{bmatrix} ; \quad Q'_{y,xy} = \begin{bmatrix} 1 & 0 & 0 & 0 \\ 0 & 1 & 0 & 0 \\ 0 & 0 & 0 & 1 \\ 0 & 0 & 1 & 0 \end{bmatrix} ; \quad Q_{y,xy} = \begin{bmatrix} 1 & 0 & 0 & 0 \\ 0 & 1 & 1 & 1 \\ 0 & 0 & 1 & 0 \\ 0 & 0 & -2 & -1 \end{bmatrix} ;$$

(rows and columns indexed by ϕ, x, y, xy)

$$(E^*_{N,i}) = \prod_{\substack{1 \le j \le N \\ j \ne i}} \Big(Q'_{x_i, x_i x_j} \otimes \bigotimes_{\substack{1 \le k \le N \\ k \ne i \\ k \ne j}} I_k \Big). \tag{4.12}$$

It is interesting to note that $Q'_{x,\bar{x}}$ and $Q'_{y,xy}$ are both simpler than $Q_{x,\bar{x}}$ and $Q_{y,xy}$. Using (-1, +1) instead of (0, 1) leads to polytopes with more apparent symmetries. Note also that $Q'_{x,y} = Q_{x,y}$. The permutation of variables can be expressed in the same way in both settings. We also have:

$$Q_{x,\bar{x}} = (Q_{x,\bar{x}})^{-1}; \; Q_{x,y} = (Q_{x,y})^{-1} \; ; \; Q_{y,xy} = (Q_{y,xy})^{-1}.$$

One undoes the effect of a transformation by applying it again. Before presenting the symmetries in theorem S, one additional comment is in order. In this (-1, +1) form, the binomial matrices have been known for a long time, since the work of (Sylvester 1867). The fact that these matrices have such a simple inverse, namely formula (4.1), is the basis of all the fast transform algorithms, following (Good 58). (Andrews & Kane 70) exposition makes clear the generality of the technique. These matrices have generated a vast field of research. See, for example, the book (Beauchamp 75).

Theorem S (Symmetries). If $f'_{N,p}$ is a face of $B'_{N,p}$ then so are:

(Negation or N-symmetry) for any p

$$f'_{N,p}(x_1, \ldots, \bar{x}_i, \ldots, x_N) = (E_{N,i})' \cdot f'_{N,p}(x_1, \ldots, x_i, \ldots, x_N);$$

(Permutation of variables or P-symmetry) for any p

$f'_{N,p}(x_1, \ldots, x_j, \ldots, x_i, \ldots, x_N) = (E_{N,i,j})' \cdot f'_{N,p}(x_1, \ldots, x_i, \ldots, x_j, \ldots, x_N)$;

(Permutation of variables with unions of variables or A-symmetry) <u>for p even</u>

$f'_{N,p}(x_1, \ldots, x_i^*, \ldots, x_N) = (E^*_{N,i})' \cdot f'_{N,p}(x_1, \ldots, x_i, \ldots, x_N)$.

<u>Proof.</u> Consider equation (4.2) in a (-1, +1) setting, namely:

$$h'_{N,p} = B'_{N,p} \cdot f'_{N,p} \tag{4.13}$$

Consider also two matrices P and Q, with P a permutation matrix and E'a full rank matrix, so that

$$P \cdot B'_{N,p} \cdot E' = B'_{N,p}.$$

From (4.13) we have

$$P \cdot h'_{N,p} = P \cdot B'_{N,p} \cdot E' \cdot (E')^{-1} \cdot f'_{N,p}$$

$$= B'_{N,p} \cdot (E')^{-1} \cdot f'_{N,p} \cdot$$

If $f'_{N,p}$ is a face of $B'_{N,p}$ we have $h'_{N,p} \geq 0$ and hence $P \cdot h'_{N,p} \geq 0$. This and the fact that E' is full rank guarantees that $f'_{N,p}$ is a face of $B'_{N,p}$.

We therefore have to verify that the submatrices of (4.10), (4.11) and (4.12) corresponding to the column labels present in $B'_{N,p}$ are of full rank. We will check each one of the three cases separately.

<u>N-symmetry.</u> The effect of matrix (4.10) consists of changing the signs of all the terms of the equation of a face having in its label the variable x_i. The corresponding matrix is an identity matrix with some of its elements changed from +1 to -1.

<u>P-symmetry</u>. The effect of matrix (4.11) consists in the interchange of variable x_i with variable x_j whenever they occur. The interchange does not alter the number of variables in each label. Hence the matrix in question is a full rank permutation matrix. It should be noticed that all the faces of the binomial polytopes found so far are symmetric functions, so this symmetry does not accomplish much.

<u>A-symmetry.</u> The effect of matrix (4.12) can be described as follows.

(1) Add x_i to all the labels of <u>odd</u> length not having x_i. If the label contains x_i, do nothing to it.

(2) Remove x_i from all the labels of <u>even</u> length having x_i. If the label does not contain x_i, do nothing to it.

If p is even,labels with no more than p variables will stay that way. Hence (1) and (2) describe a permutation matrix of full rank. □

If we interchange odd with even and even with odd in the description of the effect of the A-symmetry on a label we obtain another interesting symmetry, which <u>does not</u> associate faces to faces. This modified A-symmetry is a generalization of a symmetry originally presented in (Goto & Takahasi 62). To this respect, see also (Winder 68). The symmetry was developed to classify threshold functions on a cube (a cube in our notation is a $B_{N,1}$) in order to simplify tables of threshold functions. In our case the symmetry appears as a consequence of the recursive definition given to the binomial matrices in (3.6). The reader is invited to verify the interesting action of the A-symmetry in the zero-one setting, using the matrices in (4.12).

The next result expands considerably the linear characterization. It has been known for quite some time that polytopes associated with hard combinatorial problems have relatively few vertices in relation to their very large number of faces. In other words, they are strongly degenerated, in the sense of (Charnes 52). At the end of (Gomory 64) we find an interesting discussion of some of the earlier work on the subject. A small polytope associated with the traveling salesman problem is characterized by H. W. Kuhn as a "miserable polyhedron". How can a polyhedron with so few vertices have so many faces? In the next page J. Edmonds remarks that as the vertices have a simple description, so might its faces.

Theorem C (Composition of supporting hyperplanes). Let $g_{M,q}$ be a supporting hyperplane of $B_{M,q}$ and let $f_{N,p}$ be a supporting hyperplane of $B_{N,p}$. Then

$$g_{M,q} \otimes f_{N,p} = f_{N+M,p+q} \tag{4.14}$$

is a supporting hyperplane of $B_{N+M,p+q}$.

Proof. We have

$$(B_{N,N} \cdot f_{N,p}) \otimes (B_{M,M} \cdot g_{M,q}) = (B_{N,N} \otimes B_{M,M}) \cdot (f_{N,p} \otimes g_{M,q})$$

$$= B_{N+M,p+q} \cdot f_{N+M,p+q} \ .$$

As $B_{N,N} \cdot f_{N,p} \geq 0$ and $B_{M,M} \cdot g_{M,q} \geq 0$, it follows that $B_{N+M,p+q} \cdot f_{N+M,p+q} \geq 0$. □

In (Ursic 83) it was shown that in a few special cases theorem C transforms faces into faces. Of particular usefulness is the case, covered in (Ursic 83), of $q = 0$. It shows that a face of a $B_{N,p}$ is also a face of $B_{N+M,p}$. As we add variables to build larger binomial polytopes, all the faces of the smaller ones are preserved. A binomial polytope with N + 1 variables is inherently more complicated of all the ones with N variables.

The next theorem formalizes the link between a $B_{N,p}$ and a $CNF_{N,p}$, outlined in section 2. It also shows how to use binomial polytopes to test boolean formulas for satisfiability and how to use the binomial polytopes to solve discrete optimization problems.

Theorem L (The linear characterization). Each clause of a propositional calculus formula in conjunctive normal form in N variables and with P literals per clause has a corresponding face of $B_{N,p}$ whose vertices are labeled with all the truth assignments that make the clause true.

Proof. Let $x_1, x_2, \ldots, x_p$ be the literals in our clause. The corresponding face of $B_{N,p}$ is

$$f_{p,(1,\ 2,\ \ldots\ ,\ p)}(x_1, x_2, \ldots ,x_p). \tag{4.15}$$

It is straightforward to verify that this face touches the required vertices. We use theorem V and the association of truth assignments with labels outlined in connection with symmetric functions. □

As a final example we repeat the face description of $B_{3,2}$ outlined in (2.2), (2.3) and (2.4). We know that $B_{3,2}$ is a projection of $B_{3,3}$. We also have $(B_{3,3})^{-1}$. It is therefore a simple matter to obtain its Gale diagram, which is

$$(1,\ 1,\ 1,\ 1,\ -1,\ -1,\ -1,\ -1). \tag{4.16}$$

Consult (Stoer & Witzgall 70) for an inspired introduction to Gale diagams. Hence $B_{3,2}$ has 16 combinatorially isomorphic faces. They are:

faces of type $f_{2,(1,2)}$,

$$\begin{array}{lll}
1 - x - y + xy \geq 0, & 1 - x - z + xz \geq 0, & 1 - y - z + yz \geq 0, \\
x - xy \geq 0, & x - xz \geq 0, & y - yz \geq 0, \\
y - xy \geq 0, & z - xz \geq 0, & z - yz \geq 0, \\
xy \geq 0, & xz \geq 0, & yz \geq 0;
\end{array}$$

faces of type $f_{3,(1,2)}$,

$$\begin{array}{r}
1 - x - y - z + xy + xz + yz \geq 0, \\
x - xy - xz + yz \geq 0, \\
y - xy + xz - yz \geq 0, \\
z + xy - xz - yz \geq 0.
\end{array}$$

Such a small example is somewhat misleading. Binomial polytopes have more faces than one might think. We need at least $N = 7$ or $N = 8$ before interesting things start to happen.

5. APPLICATIONS

The binomial polytopes have considerably more faces than vertices. As a consequence, each vertex is touched by more faces than the corresponding space dimension. As a further consequence, a supporting hyperplane can be expressed in more than one way as a positive linear combination of faces. Degeneracy in linear programs is usually considered a nuisance. In our case, it can be used to our advantage.

An example will illustrate the idea. Consider the two clauses ab and $\bar{b}c$ of a $CNF_{3,2}$. They produce, by resolution, the clause ac. The corresponding faces of a $B_{3,2}$ are

$$f_{2,(1,2)}(a, b) = 1 - a - b + ab \geq 0,$$

$$f_{2,(1,2)}(\bar{b}, c) = b - bc \geq 0,$$

$$f_{2,(1,2)}(a, c) = 1 - a - c + ac \geq 0.$$

Consider the supporting hyperplane

$$f_{2,(1,2)}(a, b) + f_{1,(1, 2)}(\bar{b}, c) = 1 - a + ab - bc. \tag{5.1}$$

We also have

$$\begin{aligned} f_{2,(1,2)}(a, c) + f_{3,(1,2)}(a, b, \bar{c}) &= (1 - a - c + ac) + (c - ac - bc + ab) \\ &= 1 - a + b - bc. \end{aligned}$$

Notice that by rewriting (5.1), corresponding to the clauses ab and $\bar{b}c$, we obtained the face corresponding to clause ac and a fourth face that does not correspond to clauses (a symmetric function). It should come as no surprise that this is so. In fact, as ac is a consequence of ab aad $\bar{b}c$ we can conclude that

$$f_{2,(1,2)}(a, b) + f_{2,(1,2)}(\bar{b}, c) - f_{2,(1,2)}(a, c) \tag{5.2}$$

is also a supporting hyperplane to $B_{3,2}(a, b, c)$.

The linear equation (5.2) is not a positive linear combination of faces of $B_{3,2}$.

This hyperplane is defined by the truth assignments which make ab and $\bar{b}c$ false, but do not falsify ac. The implied clause ac does not use all the information present in ab and $\bar{b}c$. The alternate linear combination of faces uses this additional information in the form of the face $f_{3,(1,2)}(a, b, \bar{c})$.

The fact that the consensus, or resolution, of two clashing clauses produces a third clause, can be considered a proof that the corresponding binomial polytopes are indeed degenerate. The fact that iterated resolution has been shown to detect unsatisfiable formulas can also be considered a proof that the iterated use of the degeneracy of the binomial polytopes will reach the same end.

From this point of view, using degeneracy to find altenate descriptions of supporting hyperplanes can be considered a generalization of the resolution principle. A very powerful method exists to perform this search, namely the simplex method.

6. CONCLUSION

A few facts emerged from our discussion of boolean formulas.

(A) The fundamental role assumed by symmetric and partially symmetric functions.

(B) The existence of other relevant symmetries besides the negation and permutation symmetries.

(C) The role of linear algebra and of the binomial recursion as useful tools for their study.

We have presented a family of polytopes, the binomial polytopes. Their connection with NP-complete problems has been explored. The vertices of the binomial polytopes can be generated with a recurrence relation identical to the one with which we define binomial coefficients. As a consequence the vast body of knowledge accumulated in classical combinatorial analysis and linear algebra becomes available to the study of the combinatorial properties of propositional calculus formulas in conjunctive normal form. Discrete optimization problems should benefit from it. The binomial matrices can be described with sections of Sylvester "ornamental tile-work". They are indeed interesting.

----------o----------

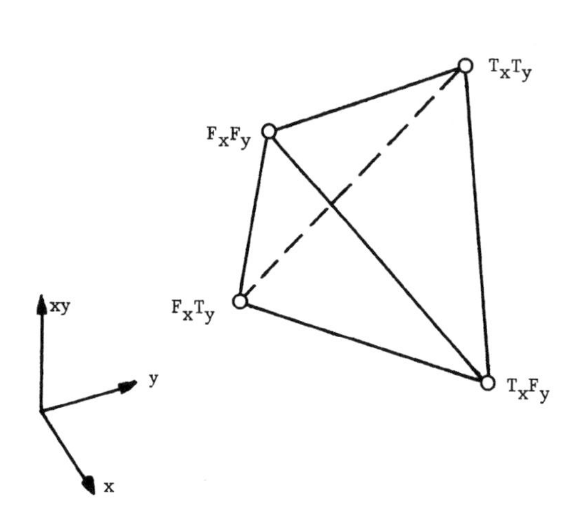

REFERENCES

H. C. Andrews, J. Kane, Kronecker matrices, computer implementation, and generalized spectra, Journal of the ACM 17 (1970) 260-268.

R. F. Arnold, M. A. Harrison, Algebraic properties of symmetric and partially symmetric boolean functions, IEEE Transactions on Electronic Computers 12 (1963) 244-251.

M. L. Balinski, Integer programming: methods, uses, computation, Management Science 12 (1965) 253-313.

K. G. Beauchamp, Walsh functions and their applications, Academic Press, London, 1975.

K. Bing, On simplifying truth-functional formulas, Journal of Symbolic Logic 21, (1956) 253-254.

A. Blake, Canonical expressions in boolean algebra, Ph. D. thesis, Dept. of Mathematics, University of Chicago, August 1937. Review in: J. Symb. Logic 3 (1938) 93.

A. Blake, A boolean derivation of the Moore-Osgood theorem, Journal of Symbolic Logic 11 (1946) 65-70. Review in: J. Symb. Logic 12 (1947) 89-90.

F. M. Brown, The origin of the method of iterated consensus, IEEE Transactions on Electronic Computers 17 (1968) 802.

J. R. Buchi, Turing-machines and the Entscheidungsproblem, Math. Annalen 148 (1962) 201-213.

A. Charnes, Optimality and degeneracy in linear programming, Econometrica 20 (1962) 160-170.

S. A. Cook, The complexity of theorem-proving procedures, Third ACM Symposium on Theory of Computing, (1971) 151-158.

C. H. Cunkle, Symmetric boolean functions, American Math. Montly 70 (1963) 833-836.

Z. Galil, On the complexity of regular resolution and the Davis-Putnam procedure, Theoretical Computer Science 4 (1977) 23-46.

M. R. Garey, D. S. Johnson, Computers and intractability, a guide to the theory of NP-completeness, W. H. Freeman and company, San Francisco, 1979.

I. J. Good, The interaction algorithm and practical Fourier analysis, J. Royal Stat. Soc. Ser. B (1958) 361-372, ibid. (1960) 372-375.

E. Goto, H. Takahasi, Some theorems useful in threshold logic for enumerating boolean functions, Proc. IFIP Congress 62, North-Holland, Amsterdam 1963, pp 747-752.

M. Grötschel, Approaches to hard combinatorial optimization problems, in: B. Korte, ed., Modern Appl. Math., Optimization and Operations Research (1982) 437-515.

M. Grötschel, L. Lovasz, A. Schrijver, The ellipsoid method and its consequences in combinatorial optimization, Combinatoritca 1 (1981) 169-197.

R. M. Karp, C. H. Papadimitriou, On linear characterizations of combinatorial optimization problems, SIAM J. Computing 11 (1982) 620-632.

R. E. Gomory, The traveling salesman problem, in: Proc. IBM Scientific Computing Symposium on Combinatorial Problems (1964) 93-121.

R. M. Karp, Reducibility among combinatorial problems, in: R. E. Miller, J. W. Thacher, eds., Complexity of Copmputer Computations (1972) 85-103.

L.G. Khachian, A polynomial algorithm in linear programming, Soviet Mathematics Doklady 20 (1979) 191-194.

D. E. Knuth, The art of computer programming, volume 1, Addison-Wesley 1968.

H. W. Lenstra, Integer programming with a fixed number of variables, Report 81-03, University of Amsterdam, 1981.

L. Löwenheim, Uber das Auflosungsproblem im logischen Klassenkalkul, Sitzungsberichte der Berliner Mathematischen Gesellschaft 7 (1908) 89-94.

M. W. Padberg, ed., Combinatorial optimization, in; Math. Programming Study 12, North-Holland 1980, pp 1-221.

C. H. Papadimitriou, On the complexity of integer programming, Journal of the ACM 28 (1981) 765-768.

W. R. Pulleyblank, Polyhedral combinatorics, in: Mathematical Programming, the State of the Art, Springer-Verlag 1983, pp 312-345.

J. Riordan, Combinatorial Identities, John Wiley and Sons 1968.

W. V. Quine, A way to simplify truth functions, Am. Math. Montly 62 (1955) 627-631.

J. A. Robinsion, A machine-oriented logic based on the resolution principle, Journal of the ACM 12 (1965) 23-41.

W. Semon, E-algebras in switching theory, Transactions AIEE 80 (1961) 265-269.

C. E. Shannon, A symbolic analysis of relay and switching circuits, Transactions AIEE 57 (1938) 713-723.

T. Skolem, Uber die symmetrisch allgemeinen Losungen im Klassenkalkul, Fundamenta Mathematicae 18 (1932) 61-76.

D. Slepian, On the number of symmetry types of boolean functions of n variables, Canadian J, of Math. 5 (1953) 185-193.

J. Stoer, C. Witzgall, Convexity and Optimization in Finite Dimensions I, Springer-Verlag 1970.

J. J. Sylvester, Thoughts on inverse orthogonal matrices, simultaneous sign-successions, and tesselated pavements in two or more colours, with applications to Newton's rule, ornamental tile-work, and the theory of numbers, Philisophical Magazine 34 (1867) 461-475. (Comment: part II, which the author says will follow, seems to have never been published)

G. S. Tseitin, On the complexity of derivations in propositional calculus, Studies in constructive mathematics and mathematical logic, ed. A. O. Slisenko, Vol. 8 (1968) 115-125.

S. Ursic, A discrete optimization algorithm nased on binomial polytopes, Ph. D. Thesis, University of Wisconsin, Madison, 1975.

S. Ursic, Binomial polytopes and NP-complete problems, in J, F, Traub, ed., Algorithms and Complexity, New directions and Recent Results, Academic Press 1976.

S. Ursic, The ellipsoid algorithm for linear inequalities in exact arithmetic, IEEE Foundations of Computer Science 23 (1982) 321-326.

S. Ursic, A linear characterization of NP-complete problems, Proceedings of the twenty-first annual Allerton conference on communication, control, and computing, (1983) 100-109.

A. N. Whitehead, Memoir on the algebra of symbolic logic, American Journal of Math. 23 (1901) 139-165. Ibid. Part II 23 (1901) 297-316.

R. O. Winder, Symmetry types in threshold logic, IEEE Transactions on Computers 17 (1968) 75-78.

M. C. Pease III, Methods of matrix algebra, Academic Press 1965.

A Satisfiability Tester for Non-Clausal Propositional Calculus

Allen Van Gelder
Computer Science Department
Stanford University
Stanford, CA 94305, USA

Abstract

An algorithm for satisfiability testing in the propositional calculus with a worst case running time that grows at a rate less than $2^{(.25+\epsilon)L}$ is described, where L can be either the length of the input expression or the number of occurrences of literals (i.e., leaves) in it. This represents a new upper bound on the complexity of non-clausal satisfiability testing. The performance is achieved by using lemmas concerning assignments and pruning that preserve satisfiability, together with choosing a "good" variable upon which to recur. For expressions in clause form, it is shown that the Davis-Putnam procedure satisfies the same upper bound.

1. Introduction

An algorithm for satisfiability testing in the propositional calculus with a worst case running time that grows at a rate less than $2^{(.25+\epsilon)L}$ for any positive ϵ is described, where L can be either the length of the input expression or the number of occurrences of literals (i.e., leaves) in it. This represents a new upper bound on the complexity of non-clausal satisfiability testing.

The input to the algorithm is a Boolean expression with connectives **and**, **or**, and **not**. The connectives **iff** and **xor** are not used. The expression need not be in clause form or any normal form. The output is either "satisfiable" followed by a satisfying assignment, or "unsatisfiable". In the latter case, an enumeration proof is available, but it is very long and not very illuminating.

The algorithm works by a variation of Quine's method [Qui50], but for inputs in CNF, is almost equivalent to the Davis-Putnam method [DaP60]. The performance is achieved by using lemmas concerning assignments and pruning that preserve satisfiability, together with choosing a "good" variable upon which to recur.

To determine satisfiability for a given Boolean expression, the algorithm first makes simplifications and assignments that shorten the expression while preserving its satisfiability, then chooses a variable and recursively tests satisfiability after each assignment (true, false) to that variable throughout the expression.

A program implementing this algorithm was able to test expressions with up to 60 leaves within 34 CPU seconds on the Vax 11/780. An expression with 106 leaves and 32 variables took about 5 minutes. The tests were limited, but suggest a growth rate of $2^{\frac{L}{12}}$. The program is written in Franz Lisp.

Previous work on efficient satisfiability testing (see [Gol79] for survey) has been concerned primarily with expressions in clause form. Transformation of an arbitrary expression into clause form can be done in polynomial time only by introducing new variables [Tse68]. This can almost double the number of variables. In addition, the method given there (not necessarily optimal) may increase the length of the expression by a factor of seven. Without introduction of new variables, an exponential blowup is possible. A linear expansion of problem size, while not critical in complexity theory [Coo71], may have a substantial impact in practice. In fact, no non-trivial upper bounds have been shown for non-clausal expressions. Exponential lower bounds for several proof procedures have been shown [Tse68, Gal75, Gal77], but the bound is so low (below $2^{.001L}$) as to be of theoretical significance only. In addition, no non-trivial lower bound is known for *extended* resolution [Tse68, Gal77].

2. Succinct Representation of Boolean Expressions

Boolean expressions are modeled as n-ary trees with two types of nodes, *operation* and *leaf*. Each operation node is either an **and** or an **or**, and may have an any number of children. Each leaf is an occurrence of a literal. A literal is a "polarized" variable, i.e., either a variable (positive polarity) or a complemented variable (negative polarity). Literals can also be polarized. In addition, in interim expressions, an operation node may hold a truth value by containing **and** (for **true**) or **or** (for **false**), and no children.

The input expression may also contain **not** at interior nodes, but the program merely pushes these **nots** down to the leaves, using DeMorgan's rules. This one-time transformation affects neither the number of variables nor the number of leaves, and hence has a negligible effect upon running time. After the **nots** have been pushed down to the leaves, we call the resulting tree an AND-OR tree.

To analyze running time it is convenient to have a uniform and compact representation of Boolean expression trees, which motivates the following definitions.

Definition Let two children of a node both be leaves. If they contain a literal and the negation of that literal, they are *inconsistent*. If they both represent the same literal, one is *redundant*. These definitions do not apply to children that are not leaves.

Definition A *succinct AND-OR tree* is one in which:

(1) The tree contains a truth-value node (empty **and** or **or**) only if that is the only node of the tree.

(2) No **and** has an **and** for a child; no **or** has an **or** for a child.

(3) Every operation node has at least two children (unless (1) applies).

(4) No operation node has *inconsistent* or *redundant* children.

Definition A Boolean expression is *succinct* if it is in the form of a succinct AND-OR tree.

It is clear that any AND-OR tree may be transformed into a logically equivalent succinct AND-OR tree with no increase in the number of leaves. Consequently we shall use the succinct AND-OR tree as the starting point for analysis, and use the number of leaves as the measure of problem size.

3. Notation

We shall frequently denote the set of variables in a Boolean expression by $v_1, \ldots, v_n$ and the literals by $x_1, \ldots, x_n$. Thus $\bar{v}_i$ is v_i negated; x_i means either v_i or $\bar{v}_i$, and $\bar{x}_i$ is the negation of x_i. Here we say v_i is *associated* with x_i and $\bar{x}_i$, and *vice versa*. In the context of a particular discussion, either $x_i = v_i$ or $x_i = \bar{v}_i$ (and $\bar{x}_i = v_i$), but for purposes of analysis we usually do not care which. In some cases we use w, y and z for literals and u for variables, as well.

We denote the number of leaves in an expression by L, and take this as the basic measure of the length of the expression. If we say a *literal* x occurs k times in an expression, we mean that there are k leaves with value x and we imply nothing about leaves with value $\bar{x}$. If we say a *variable* v occurs k times, we mean that there are k leaves with values of either v or $\bar{v}$. Similarly, if we say a literal x occurs in a sub-tree, we mean x and not $\bar{x}$; if we say a variable v occurs in a sub-tree, we mean either v or $\bar{v}$.

We shall use some standardized symbols in diagrams of trees, as illustrated in Fig. 3.1. Circles denote operation nodes; boxes denote leaves; triangles denote subtrees with at least one operation node; a triangle within a box denotes a subtree that may be a leaf. For edges, a single line denotes one edge, a dashed line

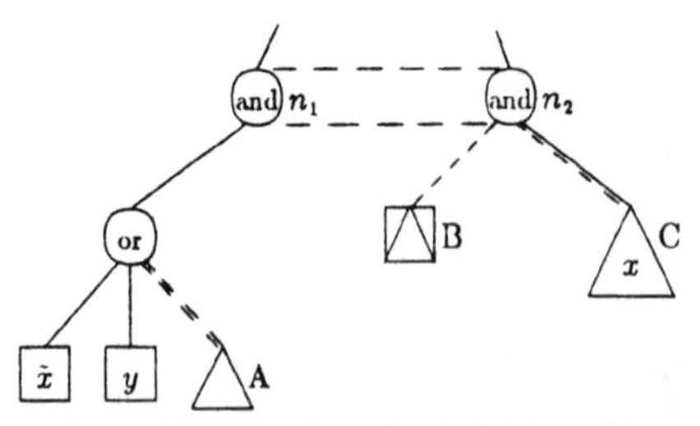

Figure 3.1. Illustration of Symbols in Tree Diagrams.

an optional edge (i.e., the subtree below it is optional). A double dashed line means zero or more edges, and a combination solid and dashed line means one or more edges. The symbol under either type of double line actually represents a set of such symbols. Therefore, in the illustration, A is a possibly null forest, B is completely optional, and C is a non-null forest in which every top level tree has some leaf containing x. Note that names of items are outside the symbols, and contents inside. Finally, the two horizontal dashed lines connecting n_1 and n_2 indicate that they may be the same node; otherwise nodes are taken to be distinct.

4. Overview of the Algorithm

The top level recursive part of the algorithm is described here. It begins after an initialization phase in which the user's input is put into the form of a succinct expression and the variables in the expression are identified. In general, the input to each recursive instance is a succinct expression, a list of the variables in that expression, and a (variable, assignment) pair called the *pending assignment*. For the top level instance, the input expression is the original expression (but in succinct form) and the pending assignment is nil. The output of each recursive instance is "satisfiable" or "unsatisfiable", together with supporting documentation. One recursive instance of the algorithm does the following steps.

1. Make the pending assignment and simplify the resulting expression, producing the ***base expression.*** The base expression becomes the initial value of the ***current expression.***

2. Repeat until no progress is made: Find dominances and prune; find satisfiability preserving assignments and substitute them into the current expression; simplify, producing an updated current expression.

3. Determine satisfiability of the current expression: do the first one that applies of (3a), (3b), and (3c).

3a. If the current expression has no more than a predetermined number of variables (5 in our implementation), then determine its satisfiability by the method of truth tables.

3b. If the root of the current expression is an **or**, then for each subtree of the root: recursively call this algorithm with inputs consisting of the subtree and the null assignment pair. The current expression is satisfiable if any subtree is satisfiable.

3c. Otherwise, choose a variable to "branch" on, say v. Recursively call this algorithm with inputs consisting of the current expression and the assignment (v, **true**). If it returns "unsatisfiable", then try it with the current expression and (v, **false**). If it again returns "unsatisfiable", then the current expression is indeed unsatisfiable.

4. Return "satisfiable" or "unsatisfiable" as determined in (3), together with supporting documentation. The supporting documentation consists of that returned by recursive calls, with this instance's assignments and reasons prepended.

5. Asymptotic Analysis of Worst Case Running Time

An algorithm like this is usually called "divide and conquer", but actually a better name is "chip and conquer", because instead of truly dividing the problem up, it only chips off a piece of constant size as it produces each subproblem. When all subproblems have equal size, the analysis is well known [AHU74]. This section describes the analysis when the subproblems have varying sizes.

Let $T(L)$ be a worst case upper bound on running time for expressions with L leaves, i.e., with length L. Let the cost per recursive step of splitting up the base problem into subproblems, including the cost of applying satisfiability preserving assignments and of simplifying to succinct form, be bounded by $P(L)$. It is easy to implement the algorithm so that $P(L)$ is quadratic. Most analysis and simplification steps can be made linear without much difficulty, but a few present problems, so we take $P(L)$ to be quadratic for asymptotic purposes. We shall show later that, depending upon the structure of the base expression, one of several cases occurs. Each case can be characterized by a small set of positive integers. Say that $A = \{a_1, \ldots, a_{c_A}\}$ represents one case, that $B = \{b_1, \ldots, b_{c_B}\}$ represents another case, etc. The meaning of the set A is as follows: For case A the base problem is split into c_A subproblems. The lengths of the subproblems are at most $L-a_1, L-a_2, \ldots, L-a_{c_A}$. In the worst case, all subproblems are of maximum possible length and all must be solved. Running through the cases, we have:

$$T_A(L) \leq P(L)+T(L-a_1)+\ldots+T(L-a_{c_A}) \tag{5.1}$$

$$T_B(L) \leq P(L)+T(L-b_1)+\ldots+T(L-b_{c_B})$$

$$\ldots$$

and therefore

$$T(L) \leq \max(T_A(L), T_B(L), \ldots) \tag{5.2}$$

Let us advance the inductive hypothesis that there exist K and γ independent of L, whose values will be detemined subsequently, such that for all $m < L$:

$$T(m) < K\gamma^m \tag{5.3}$$

then

$$T_A(L) \leq P(L)+\sum_{j=1}^{c_A} K\gamma^{L-a_j} = P(L)+K\gamma^L \sum_{j=1}^{c_A} \gamma^{-a_j} \tag{5.4}$$

Now consider the equation:

$$f_A(\gamma) \equiv \sum_{j=1}^{c_A} (\gamma^{-1})^{a_j} = 1 \tag{5.5}$$

It is clear that (for $c_A > 1$) this equation has precisely one real root in the range $0 < \gamma^{-1} < 1$, and no other positive roots. Call this root γ_A. Similarly we define $\gamma_B, \ldots$, for all the other cases. Finally, define γ^* to be the maximum γ over all cases. The set of possible cases is a property of the *algorithm*, not the *problem*, as will be clarified later when the specifics of the algorithm are discussed. Consequently, γ^* also depends only on the algorithm. Because $P(L)$ is a polynomial, for any given $\varepsilon > 0$, we can choose a K_ε such that, for all L:

$$P(L) + K_\varepsilon(\gamma^*+\varepsilon)^L f_A(\gamma^*+\varepsilon) < K_\varepsilon(\gamma^*+\varepsilon)^L \tag{5.6}$$

A base case for Eq. 5.3 (enlarging K_ε if necessary) is clearly true, so for all L:

$$T(L) < K_\varepsilon(\gamma^*+\varepsilon)^L \tag{5.7}$$

■

For example, one naive algorithm is to substitute for variables that occur with only one polarity, if any, then pick a variable to "branch" on arbitrarily. There is only one case, and $A = \{2, 2\}$ because there are two subproblems, each at least 2 shorter than the base problem. Eq. 5.5 becomes $2\gamma^{-2} = 1$, and $\gamma^* = \sqrt{2}$.

In order to determine γ^* for the algorithm of this paper, it is necessary to delineate the cases and subproblems that arise. This is done in the following sections.

6. Satisfiability Preserving Assignments and Dominance Pruning

If we adopt the convention that **false** < **true** (cf. [End72] Sec. 1.5), then we may define functions over expressions as follows:

$$eval(E, A) = \text{truth value of } E \text{ with assignment } A \tag{6.1}$$

$$sat(E) = \max_A(eval(E, A)) \tag{6.2}$$

Assignment here means the assignment of a truth value to every variable in E. It is clear that $sat(E)$ is **true** precisely when E is satisfiable. We observe without proof that

$$eval(E_1 \vee E_2, A) = \max(eval(E_1, A), eval(E_2, A)) \tag{6.3}$$

$$eval(E_1 \wedge E_2, A) = \min(eval(E_1, A), eval(E_2, A))$$

$$sat(E_1 \vee E_2) = \max(sat(E_1), sat(E_2))$$

$$sat(E_1 \wedge E_2) = \min(sat(E_1), sat(E_2))$$

Now max and min are monotonically non-decreasing in each argument. Therefore, in an AND-OR tree, *eval* at each node is a monotonically non-decreasing function of *eval* at its children. The same is true of *sat*.

In the following discussion we shall abuse notation somewhat in order to avoid excessive verbiage by writing statements like:

$$eval(E, x = \mathbf{true}) \geq eval(E, x = \mathbf{false})$$

when E has other literals besides x. What we mean by this is that the inequality holds for any pair of assignments that only differ on x (and of course $\tilde{x}$).

The following lemma generalizes the pure literal rule of the Davis-Putnam procedure to non-clausal expressions.

Lemma 6.1 (Triviality Lemma) Let E be a succinct AND-OR tree in which the literal x occurs but $\tilde{x}$ does not occur. Then E is satisfiable if and only if it is satisfiable with an assignment that includes x = **true**.

Proof *If* is immediate. *Only if*: Proceed by induction from a leaf containing x to the root.

$$eval(x, x = \mathbf{true}) \geq eval(x, x = \mathbf{false})$$

For any subexpression S with subtrees S_i, suppose for all subtrees that:

$$eval(S_i, x = \mathbf{true}) \geq eval(S_i, x = \mathbf{false})$$

Then by Eq. 6.3 the same holds for S. That is, whenever E can be satisfied by an assignment that includes $x = \mathbf{false}$, then the same assignment with $x = \mathbf{true}$ also works. ■

Definition For any leaf node n, let $family(n)$ be the set of leaves in (the fringe of) the subtree whose root is the parent of n.

Definition For any literal x, let $families(x)$ be the union of $family(n)$ over all leaves n that contain the literal x.

We note that any two subtrees either are disjoint or nested. Consequently, $families(x)$ can be represented as a disjoint union of $family(n_i)$ over some subset of leaves $\{n_i\}$ that contain x. Moreover, this subset is unique in succinct expressions. This leads to the following

Definition For any literal x in a succinct expression E, the *defining subset* of x is the set of leaf nodes $\{n_i\}$ containing x such that the disjoint union of $family(n_i)$ is $families(x)$.

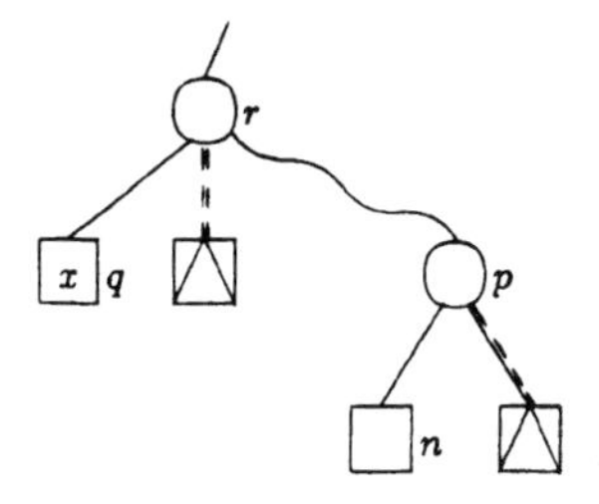

Figure 6.1. Illustration of Dominance.

Definition (Dominance) Let E be a succinct AND-OR tree (see Fig. 6.1) with root node r and other operation node p, such that some child q of r is a leaf containing literal x, and some child n of p is a leaf containing x or $\tilde{x}$. Then we say q is a *dominant* node; q *dominates* n; and n is *dominated* by q.

Lemma 6.2 (Dominance Lemma) Let E, r, q, p, n, x be as in the preceding definition of dominance. (See Fig. 6.1.) Let E_1 be E with node n removed. Let E_2 be E with subtree p removed. Then for all assignments A:

(a) If r and p have the same operation and n contains x,

$$eval(E, A) = eval(E_1, A).$$

(b) If r and p have the same operation and n contains $\tilde{x}$,

$$eval(E, A) = eval(E_2, A).$$

(c) If r and p have different operations and n contains x,

$$eval(E, A) = eval(E_2, A).$$

(c) If r and p have different operations and n contains $\tilde{x}$,

$$eval(E, A) = eval(E_1, A).$$

Proof (a) Suppose r and p contain **and**. Any assignment with x = **false** makes both E and E_1 **false**, so assume x = **true**. Then *eval* (p) is the same in both E and E_1, and the expressions are identical elsewhere. If r and p contain **or**, then the same argument applies with the roles of **true** and **false** interchanged.

(b) Suppose r and p contain **and**. Any assignment with x = **false** makes both E and E_2 **false**, so assume x = **true**. Then *parent* (p) (a descendant of r) contains an **or** and *eval* (p) = **false**. Therefore, *eval* (*parent* (p)) is the same in both E and E_2, and the expressions are identical elsewhere. If r and p contain **or**, then *parent* (p) contains **and**, and the same argument applies with the roles of **true** and **false** interchanged.

(c) The argument is similar to (b). (d) The argument is similar to (a). ∎

The unit clause rule of the Davis-Putnam procedure is a special case of the above Dominance lemma.

The next lemma allows us to identify situations in which two variables may be collapsed into one. For expressions in clause form this lemma is not needed: whenever its conditions apply, the Dominance lemma also applies and is more powerful. Consider this simple example:

> Let E be a succinct AND-OR tree in which the only occurrences of $x, \tilde{x}, y, \tilde{y}$ are as follows: some subtree contains $x \wedge y \wedge \ \ldots$ and some other subtree contains $\tilde{x} \vee \tilde{y} \vee \ \ldots$.

Here x and y could be represented by a single literal with $z = x \wedge y$, $\tilde{z} = \tilde{x} \vee \tilde{y}$. Looking more closely, we see that we can simply assign $y =$ **true**, identify the new literal z with x, and preserve satisfiability.

Definition A node is said to be *under* an **and** or **or**, if the parent of that node contains **and** or **or**, respectively.

Definition A literal x in a succinct expression E is said to be *symmetric* (in E) if the following conditions hold:

(a) Let $\{n_i, i=1, \ldots, k\}$ be the defining subset of x. Then *parent*(n_i) contains **and** for $i=1, \ldots, k$.

(b) Let $\{m_i, i=1, \ldots, p\}$ be the defining subset of $\tilde{x}$. Then *parent*(m_i) contains **or** for $i=1, \ldots, p$.

Definition Let variable v be associated with literals x and $\tilde{x}$ in succinct expression E; i.e., either $x = v$ or $\tilde{x} = v$. Then v is said to be symmetric (in E) if either x or $\tilde{x}$ is symmetric in E.

Definition A variable is said to be *mixed* if it is not symmetric.

The qualification "in E" will be omitted where the meaning is clear without it. The motiviation for the term "symmetric" lies in the effect of assignments to x on the main subexpressions containing x and $\tilde{x}$, i.e., those rooted at the parents of their defining subsets. If $x =$ **true**, none of these subexpressions necessarily become resolved (i.e., evaluate to **true** or **false**); if $x =$ **false**, the defining parents of x resolve to **false**, while the defining parents of $\tilde{x}$ resolve to **true**. Thus the resolving effect of each assignment is symmetric between x and $\tilde{x}$.

Lemma 6.3 (Symmetric Collapsibility Lemma) Let x, y be symmetric literals in succinct expression E, and let $families(x) = families(y)$ and $families(\tilde{x}) = families(\tilde{y})$. Then E is satisfiable if and only if it is satisfiable with an assignment containing $y = \mathbf{true}$.

Proof *If* is immediate. *Only if*: Let $\{n_i, i = 1, \ldots, k\}$ be the defining subset of x, and let $\{m_j, j = 1, \ldots, p\}$ be the defining subset of $\tilde{x}$. Let S_i be the subexpression rooted at $parent(n_i)$, and let R_j be the subexpression rooted at $parent(m_j)$. Then S_i all contain **and** at their roots, and R_j all contain **or** at their roots. Suppose some assignment including $y = \mathbf{false}$ satisfies E. Since x and y have the same *families* and the same is true for $\tilde{x}$ and $\tilde{y}$, it follows for all applicable i, j that:

$$eval(S_i, x = \mathbf{false} \wedge y = \mathbf{true}) = \mathbf{false} = eval(S_i, y = \mathbf{false}),$$

$$eval(R_j, x = \mathbf{false} \wedge y = \mathbf{true}) = \mathbf{true} = eval(R_j, y = \mathbf{false}).$$

But $x, \tilde{x}, y, \tilde{y}$ do not effect $E - \cup S_i - \cup R_j$, so:

$$eval(E, x = \mathbf{false} \wedge y = \mathbf{true}) = eval(E, y = \mathbf{false}).$$

Consequently, E is also satisfiable by an assignment containing $y = \mathbf{true}$. ∎

The next lemma is important because, unlike the preceding two, it can apply to expressions in conjunctive normal form.

Lemma 6.4 (Mixed Collapsibility Lemma) Let x, y be mixed literals in succinct expression E, and let $families(y) \subseteq families(x)$ and $families(\tilde{y}) \subseteq families(\tilde{x})$.

(a) If all nodes in the defining subsets of y and $\tilde{y}$ are under **ors** (see Fig. 6.2), then E is satisfiable if and only if it is satisfiable with an assignment in which $y = \tilde{x}$. In this case, all those **ors** may be replaced by the constant **true**.

(b) If all nodes in the defining subsets of y and $\tilde{y}$ are under **ands**, then E is satisfiable if and only if it is satisfiable with an assignment in which $y = x$. In this case, all occurrences of y and $\tilde{y}$ may be pruned from E.

Proof *If* is immediate. *Only if*: Let $\{n_i, i = 1, \ldots, k\}$ be the defining subset of y, and let $\{m_j, j = 1, \ldots, p\}$ be the defining subset of $\tilde{y}$. Let S_i be the subexpression rooted at $parent(n_i)$, and let R_j be the subexpression rooted at $parent(m_j)$.

Case (a): S_i and R_j all contain **or** at their roots. It follows for all applicable i, j that:

$$eval(S_i, y = \tilde{x}) = \mathbf{true} \geq eval(S_i, y = x),$$

$$eval(R_j, y = \tilde{x}) = \mathbf{true} \geq eval(R_j, y = x).$$

But $y, \tilde{y}$ do not effect $E - \cup S_i - \cup R_j$, so:

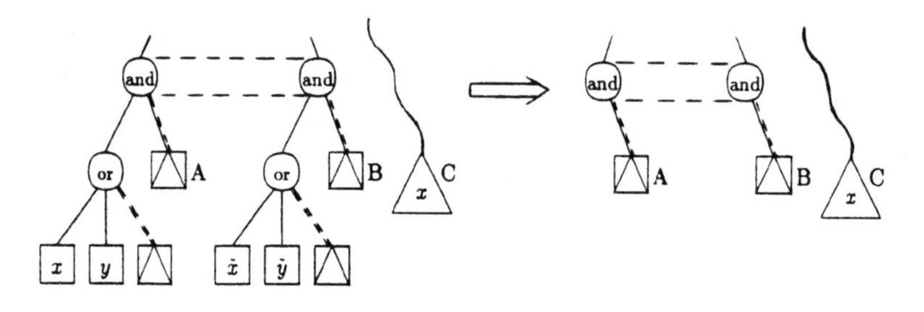

Figure 6.2. Illustration of Mixed Collapsibility.

$eval(E, y = \tilde{x}) \geq eval(E, y = x)$.

Consequently, if E is satisfiable, then it is satisfiable by an assignment in which $y = \tilde{x}$.

Case (b): S_i and R_j all contain **and** at their roots. It follows for all applicable i, j that:

$$eval(S_i, y = x) \geq \textbf{false} = eval(S_i, y = \tilde{x}),$$

$$eval(R_j, y = x) \geq \textbf{false} = eval(R_j, y = \tilde{x}).$$

But $y, \tilde{y}$ do not effect $E - \bigcup S_i - \bigcup R_j$, so:

$$eval(E, y = x) \geq eval(E, y = \tilde{x}).$$

Consequently, if E is satisfiable, then it is satisfiable by an assignment in which $y = x$. With this substitution, the nodes that used to contain y and $\tilde{y}$ become redundant, so may be pruned. ■

By checking for satisfiability preserving rules, we give ourselves the opportunity to reduce the problem size while creating just *one* subproblem. Such rules do not contribute to exponential growth. However, the most important feature of the satisfiability preserving rules embodied in these lemmas is that they enable the exponential growth rate to be *reduced* by attacking some of the worst cases. This idea is developed in the next section.

7. Branching Rules of the Algorithm

In this section we consider a succinct expression in which none of the lemmas of the previous section apply, and investigate how to choose a variable to "branch" on in such a way that the exponential growth rate is low. The basic approach is to try to make the reduction in problem size high. The analysis involves a fairly tedious enumeration of cases. However, the end result is that a worst case growth rate of less than $2^{(.25+\epsilon)L}$ can be achieved.

We note that whenever the operation at the root of the expression is an **or**, then it suffices to solve each subtree of the root independently. Although this possibility must be programmed, it cannot contribute to exponential growth, so for this analysis we assume that the operator at the root is an **and**.

The "branch" variable is of course the variable that is assigned **true** in one subproblem and **false** in the other. In general, each assignment reduces its subproblem's size by some constant, and both subproblems need to be solved to get the solution to the base problem. We choose the branch variable by means of a prioritized list of rules, A, B, C, and D, where A has highest priority. The rule of highest priority whose requirements are met is the one to be applied. With each rule we associate a case of the same name, i.e., case A, case B, etc. Within cases there may be subcases that depend on further particulars of the expression. We will use the fact that case B only arises when case A does not, and so on, without mentioning it again. For each case (or subcase) we seek the list of positive integers that characterizes the worst case reductions in subproblem size for that case. The worst case is that combination that results in the highest value of γ in the solution of Eq. 5.5. In our analysis, all cases produce two subproblems, so are characterized by by two such positive integers, one corresponding to $v = \textbf{true}$ and the other to $v = \textbf{false}$. In general these two integers are not equal. A case would involve more than two subproblems were the analysis to carry two recursive levels down, in order to derive a tighter bound.

Now we summarize the rules used by the algorithm, and what "branch" variable is chosen in each rule. In this summary, x and $\tilde{x}$ will always represent the literals associated with variable v. Recall the definitions of *mixed* and *symmetric* preceding Lemma 6.3.

Rule A Some variable occurs more than 3 times. For each variable v, let $score(v)$ equal the product of the number of occurrences of x and the number of occurrences of $\tilde{x}$. Choose a variable v with maximum *score* as the "branch" variable. If more than one such variable exists, prefer mixed to symmetric.

Rule B Some variable occurs 3 times. Choose one such variable v as the "branch" variable. If more than one such variable exists, prefer mixed to symmetric.

The remaining cases concern expressions in which every variable occurs exactly twice.

Rule C Some variable is mixed. Choose one such variable v as the "branch" variable.

Rule D All variables are symmetric. Choose any variable v as the "branch" variable.

In the following theorem, we analyze the worst case reductions in size that result from using the preceding prioritized list of rules.

Before proceding to the theorem, we present two lemmas. The first sometimes allows us to select the worst case among alternatives without solving Eq. 5.5 for each alternative. The second establishes a simple lower bound on the reduction in length.

Lemma 7.1 Let integers a, b, c, d be such that $a+b = c+d$ and $0 < a < c \leq d < b$. Let $\gamma(m, n)$ be the positive real solution of

$$f_{mn}(\gamma) \equiv \gamma^{-m} + \gamma^{-n} = 1.$$

Then $\gamma(a, b) > \gamma(c, d)$. In other words, between two cases that achieve the same total reduction in subproblem lengths, the more extreme case is the worst.

Proof We restrict our attention to the region of interest, i.e., everything positive. Here f_{mn} is decreasing in γ. Let $e = (m-n)/2$. Then

$$f_{mn}(\gamma) = \gamma^{\frac{m+n}{2}} (\gamma^{e} + \gamma^{-e})$$

For $m+n$ and γ held constant this expression has a unique minimum at $e = 0$. Therefore (with $m+n$ held constant), as e increases in absolute value, γ must also increase to maintain the value of f_{mn} at 1. ∎

Lemma 7.2 If variable v occurs k times in a succinct AND-OR tree with no dominant nodes, and it is chosen as the branch variable, then the sum of the reductions in the two subproblems is at least $3k$.

Proof Each node n in which v or $\bar{v}$ occurs is in the appropriate defining subset. There must be some other leaf in $family(n)$; it contains neither v nor $\bar{v}$ or n would be dominant. One assignment to v eliminates n and the other eliminates $family(n)$, for a total of at least 3 among both assignments. ∎

Theorem 7.3 Let the algorithm outlined in section 4 choose the "branch" variable according to rules A, B, C, and D stated above. Then the resulting cases can be characterized by the following sets of integers.

$A = \{4, 8\}$
$B = \{3, 6\}$
$C = \{4, 4\}$
$D = \{2, 8\}$

Proof In this proof, x and $\bar{x}$ will always represent the literals associated with variable v.

Case A Some variable occurs more than 3 times. Rule A is applied to choose variable v. Each assignment to v reduces the expression length by at least 4. By lemma 7.2, the sum of the reductions for both assignments is at least 12. By lemma 7.1, the worst possibility is that one assignment reduces by 4 and the other by 8. Therefore $A = \{4, 8\}$.

Case B Some variable occurs exactly 3 times. Rule B is applied to choose one such variable v. Without loss of generality, assume x occurs twice and $\bar{x}$ once. There are subcases depending on what operations x and $\bar{x}$ are under. Fig. 7.1 illustrates a typical subcase. In subcases B3 and B6 v is symmetric; in the others it is mixed.

Subcase B1 Both x are under **and**, and $\bar{x}$ is under an **and**. Each assignment to x reduces by at least 4 because either $families(x)$ or $families(\bar{x})$ disappears, as well as the leaves containing x and $\bar{x}$. But by lemma 7.2 the sum of reductions is at least 9. Therefore, $B1 = \{4, 5\}$.

Subcase B2 One instance of x, say n_1, is under an **and**, the other, n_2, is under an **or**, and $\bar{x}$ is under an **and**. Each assignment to x reduces by at least 4 because either $family(n_1)$ or $family(n_2)$ disappears, as well as the leaves containing x and $\bar{x}$. But by lemma 7.2 the sum of reductions is at least 9. Therefore, $B2 = \{4, 5\}$.

Subcase B3 Both x are under **and**, and $\bar{x}$ is under an **or**. Consider the assignment $x =$ **true**. Both instances of x disappear, as well as $\bar{x}$, for a total reduction of at least 3 leaves. Now consider $x =$ **false**. All of $families(x)$ disappears, plus $families(\bar{x})$, for a total reduction of at least 6 nodes. That is, $B3 = \{3, 6\}$.

B4 Both x are under **or**, and $\bar{x}$ is under an **or**. This is similar to B1.

B5 One instance of x is under an **and**, the other is under an **or**, and $\tilde{x}$ is under an **or**. This is similar to B2.

B6 Both x are under **or**, and $\tilde{x}$ is under an **and**. This is similar to B3 with **true** and **false** interchanged.

Therefore, applying lemma 7.1, we conclude that $B = \{3, 6\}$.

The remaining cases concern expressions in which every variable occurs exactly twice, and hence every *literal* occurs exactly once with each polarity.

Case C Some variable is mixed. Apply Rule C to choose one such variable v as the "branch" variable. The parents of x and $\tilde{x}$ must both contain the same operation.

Subcase C1 Both x and $\tilde{x}$ are under an **or**. Consider the assignment $x =$ **true**. Then $families(x)$ and $\tilde{x}$ disappear for a total of at least 3 nodes. But since every variable occurs exactly twice, if an odd number of nodes disappear, some remaining variable occurs once and can be eliminated using the triviality lemma. Therefore, the reduction is at least 4 nodes. A similar argument applies to the assignment $x =$ **false**.

Subcase C2 Both x and $\tilde{x}$ are under an **and**. This is similar to C1 with **true** and **false** interchanged.

Therefore, we conclude that $C = \{4, 4\}$.

Case D All variables are symmetric. Apply Rule D to choose any variable v as the "branch" variable. Without loss of generality, assume that all literals are named so that x_i is under an **and** and $\tilde{x}_i$ is under an **or**.

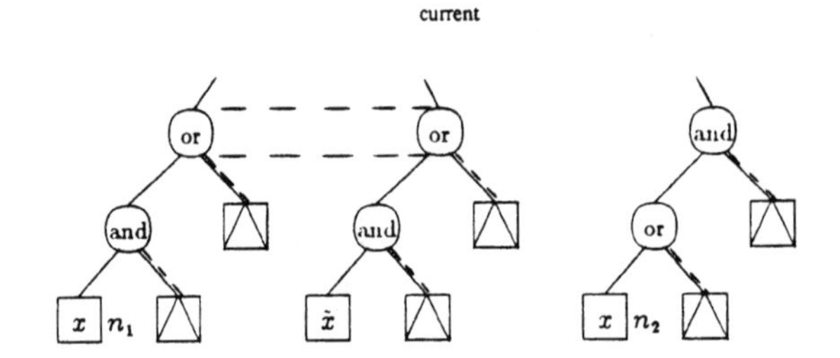

after $x =$ **true** after $x =$ **false**

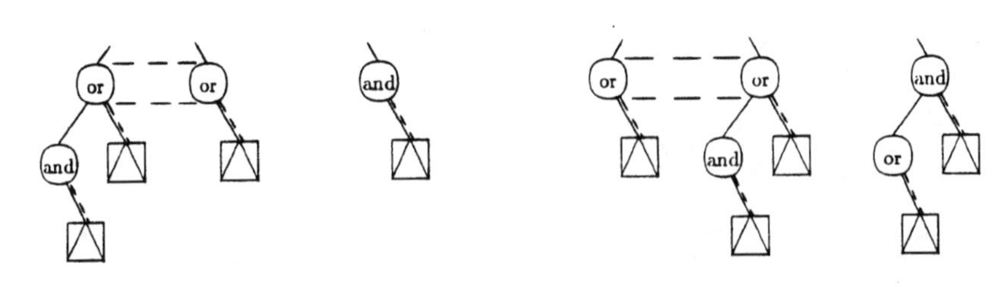

Figure 7.1. Case B2

Subcase D1 At least one of $families(x)$ and $families(\tilde{x})$ contains more than two leaf nodes. First let $families(x)$ contain 3 or more nodes. With the assignment $x =$ **true**, we can only count on x and $\tilde{x}$ disappearing for a reduction of 2 nodes. Consider the assignment $x =$ false. Now both $families(x)$ and $families(\tilde{x})$ disappear, but in addition there is a domino effect. If $families(x)$ contains 5 nodes, then 7 disappear plus an unpaired one that can be eliminated by the triviality lemma, for a total of 8; so consider smaller cases. Let $families(\tilde{x})$ = $\{\tilde{x}, \tilde{w}\}$. (When it is larger, a reduction of 8 is easy to show using a similar argument.) Now examination of all possibilities in which $families(x)$ has 3 or 4 nodes, (keeping in mind the restriction to all symmetric variables, no dominance, and no collapsibility) reveals that there must be at least two *additional* literals (not complements $\tilde{w}$ or each other) that occur in $families(x)$. (See Fig. 7.2.) When one literal associated with a variable disappears, the other can be eliminated by the triviality lemma, so in all, at least 4 variables, hence 8 literals disappear. Thus $D1$ = $\{2, 8\}$.

Subcase D2 Both $families(x)$ and $families(\tilde{x})$ contain exactly two leaf nodes. As in D1, the assignment $v =$ **true** produces a reduction of 2 nodes. Consider the assignment $v =$ **false**. Let w be the other literal in $families(x)$, and let $\tilde{y}$ be the other literal in $families(\tilde{x})$. (See Fig. 7.3.) (w and y cannot be the same literal or they would be collapsible into x.) Both w and $\tilde{y}$ disappear, so $\tilde{w}$ and y become trivial, and may be assigned **true**. But $\tilde{w}$ is under an **or**, so assigning **true** to it causes $families(\tilde{w})$ to disappear. Also, y disappears. If $families(\tilde{w})$ contains 3 or more nodes, this gives an immediate reduction of 7 or more, but exactly 7 would allow elimination of the unpaired node also, for a total reduction of 8. If $families(\tilde{w})$ contains 2 nodes, let $\tilde{z}$ be the other node, as shown in the diagram. It disappears, so z becomes trivial and disappears. This again brings the reduction to 8. Thus D2 achieves the same reductions as D1.

Therefore, we conclude that D = $\{2, 8\}$, concluding the proof. ■

We should emphasize that this theorem provides an upper bound only, which is not necessarily "tight". A tighter bound might be possible by exploring the worst cases above more carefully, including looking at the next level of recursion.

We now turn to the evaluation of γ^*, the exponential growth rate, as defined following Eq. 5.5.

Corollary 7.4 The algorithm's running time is $O(2^{(.25+\epsilon)L})$.

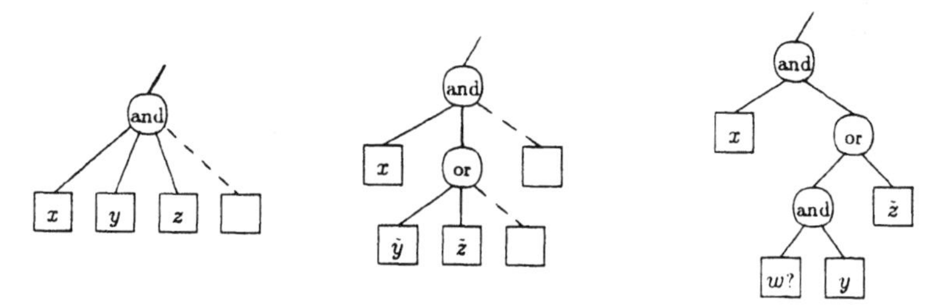

Figure 7.2. Possibilities for $families(x)$ in Subcase D1.

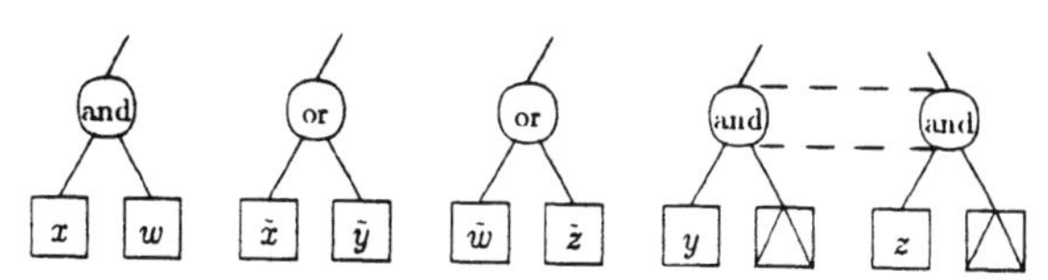

Figure 7.3. Illustration of Subcase D2.

Proof By lemma 7.1, $\gamma_A < \gamma_B$, and obviously $\gamma_C = 2^{.25}$, so it remains to compare B and D to C. The defining equations are:

$$\gamma_B^{-6} + \gamma_B^{-3} = 1 \tag{7.1}$$

$$\gamma_D^{-8} + \gamma_D^{-2} = 1 \tag{7.2}$$

The solutions to five decimal places are $\gamma_B = 2^{.23141}$ and $\gamma_D = 2^{.23248}$. So $\gamma^* = \gamma_C$. ■

Corollary 7.5 The Davis-Putnam procedure has a worst case upper bound of $O(2^{(.25+\varepsilon)L})$ when the input is presented in clause form.

Proof Lemma 6.4 was not used to prove theorem 7.3. Lemma 6.1 applied to expressions in clause form is equivalent to the pure literal rule. Lemmas 6.2 and 6.3 applied to expressions in clause form are equivalent to the unit clause rule. Without lemma 6.4, the algorithm here becomes equivalent to the Davis-Putnam procedure. ■

8. Conclusion and Acknowledgement

The investigation of worst cases in a naive enumeration algorithm has led to the discovery of additional rules that improve performance. The rule embodied in lemma 6.4 can be added to resolution and Davis-Putnam procedures, producing proof procedures that have not been shown to be exponential. The connection between this rule and the extension rule [Tse68, Gal77] should be examined.

This research was supported in part by AFOSR grant 80-0212. I wish to thank Tom Spencer for interesting test cases, Joe Weening for implementation advice, and Vaughn Pratt for helpful discussions.

9. References

AHU74.
Aho, A.V., Hopcroft, J.E., and Ullman, J.D., *The design and analysis of computer algorithms*, Addison-Wesley, Reading, MA (1974).

Coo71.
Cook, S.A., "The complexity of theorem-proving procedures," in *Proc. Third Annual ACM Symposium on Theory of Computing*, (1971).

DaP60.
Davis, M. and Putnam, H., "A computing procedure for quantification theory," *JACM* 7(2?) pp. 201-215 (1960).

End72.
Enderton, H.B., *A Mathematical Introduction to Logic*, Academic Press, New York, NY (1972).

Gal75.
Galil, Z., "The complexity of resolution procedures for theorem proving in the propositonal calculus," TR 75-239, Department of Computer Science, Cornell University (1975).

Gal77.
Galil, Z., "On the complexity of regular resolution and the Davis-Putnam procedure," *Theoretical Computer Science* 4 pp. 23-46 (1977).

Gol79.
Goldberg, A.T., "On the complexity of the satisfiability problem," NSO-16, Courant Institute of Mathematical Sciences, New York University (1979).

Qui50.
Quine, W.V., *Methods of logic*, Henry Holt (1950).

Tse68.
Tseitin, G.S., "On the complexity of derivations in the propositional calculus," pp. 115-125 in *Studies in Constructive Mathematics and Mathematical Logic, Part II*, ed. Slisenko, A.O., (1968).

A DECISION METHOD FOR LINEAR TEMPORAL LOGIC

Ana R. CAVALLI
LITP, Université Paris VII
2, Place Jussieu
75251 Paris Cedex 05
France

Luis FARIÑAS DEL CERRO
LSI, Université Paul Sabatier
118, route de Narbonne
31062 Toulouse Cedex
France

Abstract

In this paper we define a new decision method for propositional temporal logic of programs. Temporal logic appears to be an appropriate tool to prove some properties of programs such as invariance or eventually because in this logic we define operators that enable us to represent properties that are valid during all the development of the program or over some parts of the program.

The decision method that we define is an extension of classical resolution to temporal operators.

We give an example of a mutual exclusion problem and we show, using resolution method, that it verifies a liveness property.

Introduction

In this paper we define a new decision method for propositional linear temporal logic of programs such as defines in [BM],[GPSS], [MZ].

Temporal logic appears to be an appropriate tool to prove some properties of programs like invariance, eventuality and precedence because in this logic we define operators that enable us to represent formally properties that are valid during all the development of the program, over some parts of the program or properties stating that a certain event always precedes another.

The decision method that we define is an extension of classical resolution [RJ], to temporal operators ; to obtain it we will follow the same way as proposed in [FC]. First, we define a conjunctive normal form for the formulas of temporal logic, section 1 ; after that we define transformation rules to obtain comparable formulas, section 2, and finally we define, in section 3, resolution rules for modal formulas in a way analogous of classical resolution and we prove that the rule system is complete, section 4.

In section 5, we give an simple example of a mutual exclusion problem and we show, using resolution method, that it verifies a liveness property.

We define formulas as words in a finite alphabet. A modal formula is either a literal (propositional variable or negation of a propositional variable) or has the form: (A & B), (A v B), $\sim$A, $\Box$A ($\Box$ A means necessary A), o A (o A means next A), where A and B are modal formulas. A modal system is a set of modal formulas.

A w-structure is a pair $< W, \leq >$, where W is a sequence w_0, w_1, w_2, ... the states, and $w_i \leq w_j$ iff $i \leq j$. And an assignement V for $< W, \leq >$ assigns to each propositional variable p a subset V(p) of W.

Given an assignment V for $(W, \leq)$, we define $V(A, w_i) \in \{T, F\}$, where A is a formula and $w_i \in W$, according to the costumary inductive definition ; in particular :

$V(o\, A, w_i) = T$ iff $V(A, w_{i+1}) = T$, and

$V(\Box A, w_i) = T$ iff $V(A, w_j) = T$ for every $w_j \in W$ such that $w_i \leq w_j$

We say that A is valid (unsatisfiable) in W if $V(A, w_i) = T(F)$ for every assignment V for W and every $w_i \in W$.

We consider the following abbreviations :

$\Diamond A = \sim \Box \sim A$ ($\Diamond$ A means possibly A)

$\Diamond_n A = A \vee o\, A \vee \ldots \vee o^{n-1} A$ notes the string $\overbrace{o \ldots o}^{n-1 \text{ times}} A$.

$\Box_n A = A \,\&\, o\, A \,\&\ldots\&\, o^{n-1} A$

The axiomatic system for propositional linear temporal logic is obtained by adding to the usual formalisation of propositional calculus the following axioms and inference rule :

A_1. $\Box(A \to B) \to (\Box A \to \Box B)$
A_2. $o(A \to B) \to (oA \to oB)$
A_3. $o \sim A \leftrightarrow \sim oA$
A_4. $\Box A \to A \ \& \ o \Box A$
A_5. $\Box(A \to oA) \to (A \to \Box A)$
IR. If A then $\Box A$.

1. Conjunctive Normal Form (C.N.F.)

1.1. Let F be a formula, we shall say that F is in C.N.F. if it is of the form :

$$F = C_1 \ \& \ldots \& \ C_m$$

where $m \geqslant 1$ and each C_i (clause) is a disjunction (perhaps with only one disjunct) of the general form :

$$C_i = L_1 \vee \ldots \vee L_n \vee \Box D_1 \vee \ldots \vee \Box D_{n'} \vee \Diamond A_1 \vee \ldots \vee \Diamond A_{n''}$$

where each L_i is a literal preceded by a string of zero or more operators o ; each D_i is a disjunction that possesses the general form of the clauses, and each A_i is a conjunction, where each conjunct possesses the general form of the clauses.

We note by E(E') that E' is a subformula of E.

The *degree* d(A) of a formula A is defined in the following way :

- if A is a literal, $d(A) = 0$
- if $d(A) = n$ and $d(B) = m$, $d(A.B) = \max(m,n)$ provided that . is & or v.
- if $d(A) = n$, $d(\sim A) = n$, $d(oA) = n$
- if $d(A) = n$, $d(\Delta A) = n+1$ where $\Delta = \Box$ or $\Diamond$.

The *modal degree* m(S) of a set of clauses S is defined as follows:

$$m(S) = (n_1, n_2, \ldots)$$

where n_i is the number of clauses in S whose degree is i.

Let S and S' be two sets of clauses, whose modal degrees are $m(S) = (n_1, n_2, \ldots)$ and $m(S') = (n'_1, n'_2, \ldots)$. We say that $m(S) \leqslant m(S')$ if $\exists \, i \in N$, such that $n_i \leqslant n'_i$ and $\forall j > i \ \ n_j = n'_j$.

- We note by *p*(S) the number of o operators in S.

A set of clauses S, where $m(S) = (0, 0, \ldots)$ and $p(S) = 0$ is a set of classical clauses or without modal operators.

1.2. *There is an effective procedure for constructing, for any given temporal formula F, an equivalent formula F' in conjunctive normal form.*

Proof : (by induction on n, degree of the formula). If $(d(F) = 0$, then F' is obtained by classical methods or using the following theorems and axioms:

$$o\ (A_1\ \&\ldots\&\ A_m) \leftrightarrow o\ A_1\ \&\ldots\&\ o\ A_m$$

$$o\ (A_1\ v\ldots v\ A_m) \leftrightarrow o\ A_1\ v\ldots v\ o\ A_m$$

$$\sim o\ A \leftrightarrow o \sim A$$

To complete the induction we consider $d(F) = n+1$. Then, F will have one of the following forms:

a) If F has the form $\Box(A_1\ v\ldots v\ A_m)$ where each A_i, $1 \leqslant i \leqslant m$ has degree $\leqslant n$, then F is in C.N.F. (by induction hypothesis). And if F has the form $\Box(A_1\ \&\ldots\&\ A_m)$ then we apply the following theorem:

$$\Box(A_1\ \&\ldots\&\ A_m) \leftrightarrow \Box A_1\ \&\ldots\&\ \Box A_m$$

b) If F has the form $\Diamond A$, where A has degree $\leqslant$ n. Then in this case F is in C.N.F. (by induction hypothesis).

c) If F has the form $o\ (A_1\ v\ldots v\ A_m)$ (or $o\ (A_1\ \&\ldots\&\ A_m)$) where each A_i has degree $\leqslant$ n, then we use the same theorems and axioms as in case $d(F) = 0$.

And if a A_i is not a literal preceded by a string of zero or more operators o, then we use the two reduction theorems:

$$o\ \Delta\ A \leftrightarrow \Delta\ o\ A \quad \text{where } \Delta \text{ is } \Box \text{ or } \Diamond .$$

as many times as necessary over every subformula A of A_i.

Examples: The following formulas are in conjunctive normal form:

- $o^2 p$
- $o^2(p)\ v \sim q$
- $o^2(p)\ v\ q\ v\ \Box(p\ v \sim q)\ v\ \Diamond\ (o^3 r)$
- $(o^2(p)\ v\ q\ v\ \Box\ (p\ v\ q)\ v\ \Diamond\ (o^3 r))\ \&\ (\Box\ q\ v\ \ (q\ \&\ o\ r))\ \&\ \Box\ (p\ v\ r)$

2. Erasing and transforming

2.1. Erasing Procedure

Let S be the set (conjunctions) of unit clauses

$S = \{\Box A_1,\dots,\Box A_m, B_1^0,\dots,B_{n_0}^0, oB_1^1,\dots,oB_{n_1}^1,\dots,o^nB_1^n,\dots,o^nB_{n_n}^n,$

$\Diamond P_1,\dots,\Diamond P_\ell\}$ where B_i^j is a literal.

We define the following sets:

$e(S,\Box A_i) = \{\Box A_1,\dots,\Box A_{i-1},\Box A_{i+1},\dots,\Box A_m, B_1^0,\dots,o^nB_{n_n}^n,\Diamond P_1,\dots,$

$\Diamond P_\ell, A_i,\dots,o^{p(S)}A_i\}$ where $i = 1,\dots,m$

$e(S,\Diamond P_i) = \{\Box A_1,\dots,\Box A_m,\Diamond P_1,\dots,\Diamond P_{i-1}, P_i,\Diamond P_{i+1},\dots,\Diamond P_\ell\}$
where $i = 1,\dots,\ell$

Examples: - $e(\{\Box p,\Diamond q, o\ r\},\Box p) = \{\Diamond q, o\ r, p, o\ p\}$

- $e(\{\Box p,\Diamond q, o\ r\},\Diamond q) = \{\Box p, q\}$

2.2. Transformation procedure

Let $C = \Diamond C'$ be a unit clause, we define the following transformations:

$t^i(C) = \underset{i}{\Diamond} C' \vee \Diamond(\underset{i}{\Box}\sim C' \ \&\ o^iC')$ where $i = 1,\dots,p(S)$

And t^0 is the identity transformation. If $S = \{C_1,\dots,C_n\}$ be a set of clauses, then a transformation T for S (noted T(S)) is the set $\{t_1(C_1),\dots,t_n(C_n)\}$ where each t_i is one of the transformations.

We must note that when a transformation t_i $i > 0$, is applied, it is possible that the resulting formula is not in normal form. In this case we must obtain the corresponding formula in normal form.

It may be remarked that $m(S) > m(e(T(S),C))\ \forall\ C \in T(S)$ and for every transformation T, where C is of the form $\Box C'$ or $\Diamond C'$.

In what follows, we suppose that $S = \{\Box A_1,\Box A_2, B_0, o\ B_1,\dots, o^nB_n, \Diamond P_1,\Diamond P_2\}$, since the proof is the same as for the general case.

Now we give a decision method for linear temporal logic in Carnap style [CR] [LE] and we will give an extended resolution procedure for this logic based on this decision method.

2.3. *A set of unit clauses S, with $m(S) \neq (0, 0, \dots)$, is unsatisfiable if there is a transformation T such that either $\exists (\Box A) \in T(S)$, for which $e(T(S), \Box A)$ is unsatisfiable or $\forall (\Diamond P) \in T(S)$, $e(T(S), \Diamond P)$ is unsatisfiable.*

Proof: a) Let $S = \{\Box A_1, \Box A_2, B_0, o\, B_1, \dots, o^n B^n, \Diamond P_1, \Diamond P_2\}$ be a set of unit clauses. Suppose that, for every transformation T, $\forall (\Box A) \in T(S)$ and $\exists (\Diamond P) \in T(S)$ (for example $\Diamond P_1$), $e(T(S), \Box A)$ and $e(T(S), \Diamond P)$ are satisfiable. We will prove that S is satisfiable. We distinguish the following two cases:

Case 1. $\exists i \;\; 0 \leqslant i \leqslant p$ (where $p = p(S)$) such that $S_i = e(S, \Box A_1) \cup \{o^i p_1\}$ is satisfiable. Then we have:

1.1. $S_i^! = e(S, \Box A_1) \cup \{o^i P_1, \Box A_1\}$ satisfiable. Then S is satisfiable.

1.2. $S_i^!$ unsatisfiable. Then $S_i'' = \{o^i P_1, o^i \Diamond P_2, \Box A_1, \Box A_2\}$ is unsatisfiable. To prove it we suppose, on the contrary, that S_i'' is satisfiable. Then there is a model $M = \{w_0, w_1, \dots\}$ of S_i (i.e. an assignment V for W satisfying S_i) and a model $M'' = \{w_0^!, w_1^!, \dots\}$ of S_i''. Then $M' = \{w_0, w_1, \dots, w_p, w'_{p+1}, w'_{p+2}, \dots\}$ is a model of S'_i, contradicting the hypotheses. By consequence S''_i is unsatisfiable contradicting that $e(S, \Diamond P_1)$ is satisfiable. Therefore S'_i is satisfiable.

Case 2. $\forall i \;\; 0 \leqslant i \leqslant p$, S_i'' is unsatisfiable. Since $\Diamond P_1 \in e(S, \Box A_1)$ there is a <u>r</u> such that $e(S, \Box A_1) \cup \{o^{p+r+1} p_1\}$ is satisfiable. Let <u>k</u> be the smallest natural number that verifies it.

Then $k \leqslant p$

Proof: By hypothesis $e(S, \Box A_1) \cup \{o^{p+k} P_1\}$is unsatisfiable and $e(S, \Box A_1)$ is satisfiable. Therefore the unsatisfiability must come from a $o^t A_1$ where $0 \leqslant t \leqslant p$, and A_1 possesses the general form $B_0^! \vee o\, B'_1 \vee \dots \vee o^{n'} B_{n'}^! \vee \Box A_1^! \vee \dots \vee \Box A_{m'}^! \vee \Diamond P_1^! \vee \dots \vee \Diamond P_{m''}^!$. In this case the unsatisfiability must come from a $B_j^!$ $(0 \leqslant j \leqslant n')$, because if not then $e(S, \Box A_1)$ will be unsatisfiable. Since $n' \leqslant p, p+k = t+j \leqslant 2p$ $(0 \leqslant t \leqslant p$ and $0 \leqslant j \leqslant p)$ and by consequence $k \leqslant p$.

We suppose $k > 0$. Let S' be set of clauses $\{t^0(\Box A_1), t^0(\Box A_2), t^0(B_0), \dots, t^0(o^n B_n), t^k(\Diamond P_1), t^0(\Diamond P_2)\}$.

By hypothesis $e(S,\Box A_1)$ is satisfiable and $e(S,\Box A_1) \cup \{o^i P_1\}$ $(0 \leqslant i \leqslant k-1)$ is unsatisfiable. Then $e(S', \Box A_1)$ is satisfiable and $e(S', \Box A_1) \cup \{o^i P_1\}$ $(0 \leqslant i \leqslant k-1)$ is unsatisfiable.

And consequentely the set $\{\Box A_2, A_1, \ldots, o^p A_1, \ \Diamond(\Box_k \sim P_1 \ \& \ o^k P_1), \Diamond P_2\}$ is satisfiable. Let $M = \{w_i : i = 0,1,\ldots\}$ be a model of it. Since $e(S, \Diamond P_1)$ is satisfiable, $\{\Box A_1, \Box A_2, \Diamond P_2, o^{p+k+1} P_1\}$ is satisfiable, and $\{\Box A_1, \Box A_2, \Diamond P_2, o^{p+1} \sim P_1, \ldots, o^{p+k} \sim P_1, o^{p+k+1} P_1\}$ is satisfiable too, using the definition of k. Let $M' = \{w_i' : 0,1,\ldots\}$ be a model of it. Then $M'' = \{w_0,\ldots,w_p,w'_{p+1},w'_{p+2},\ldots\}$ is a model of S. Then S is satisfiable.

For the case $k = 0$ the construction of the model is the same, although the transformation t^k with $k > 0$ is not used.

b) Suppose S satisfiable, and let $M = \{w_i : i = 0,1,\ldots\}$ be a model of S. $\forall$ $(\Box A) \in S$, M is a model of $e(S, \Box A)$. And in M there is a w_i and a w_j such that P_1 is true in w_i and P_2 is true in w_j. We suppose, for example, that $w_i < w_j$ then $M' = \{w_h : h = i, i+1, \ldots\}$ is a model of $e(S, \Diamond P_1)$.

Example: Let $S = \{\Box p, \ \Diamond \sim p\}$ be an unsatisfiable set of clauses. Then $e(S, \Diamond \sim p) = \{\Box p, \sim p\}$ is unsatisfiable, and $e(\{\Box p, \sim p\}, \Box p)$ $= \{p, \sim p\}$ is classicaly unsatisfiable.

Let S be a set of unit clauses, which modal degree is $(0,0,\ldots)$ $(m(S) = (0,0,\ldots))$. We note by $\theta^i(S)$ the set of classical clauses obtained, from the clauses in S governed by o^i operators, by erasing the o^i operators.

2.4. *Let S be a set of unit clauses, with* $m(S) = (0,0,\ldots)$. *S is unsatisfiable if there is a i* $(0 \leqslant i \leqslant p(S))$, *such that* $\theta^i(S)$ *is unsatisfiable.*

Proof. Let $S = \{B_1^0, \ldots, B_{n_0}^0, \ o \ B_1^1, \ldots, \ o \ B_{n_1}^1, \ldots, \ o^n B_{n_n}^n\}$ be a set of clauses where each B_i^j is a literal.

a) Suppose that $\theta^i(S)$ is satisfiable for every i. Then for every i there is a w_i that is a model for $\theta^i(S)$. Then $M = \{w_0, w_1, \ldots, w_n, w_n \ldots\}$ is a model of S contradicting the hypothesis.

b) Trivial.

3. Resolution rules.

3.1. Let C_1 and C_2 two unit clauses. We define the operations:

- $\Sigma_i(C_1,C_2)$ $i = 1,\ldots,4$
- $\Gamma(C_1)$

And the properties:

- (C_1,C_2) is resolvable (i.e. C_1 and C_2 are resolvable).
- (C_1) is resolvable

recursively as follows:

3.1.1. Classical operations.

a) . $\Sigma_1(p, \sim p) = \emptyset$ ($\emptyset$ is the empty symbol)

And $(p, \sim p)$ is resolvable.

b) . $\Sigma_1((D_1 \vee D_2),F) = \Sigma_i(D_1,F) \vee D_2$

And if (D_1,F) is resolvable, then $((D_1 \vee D_2),F)$ is resolvable.

3.1.2. Temporal operations.

a) $\Sigma_2(\Box E, \Delta F) = \Delta\, \Sigma_i(E,F)$ provided that Δ is $\Box$, $\Diamond$, o

And if (E,F) is resolvable, then $(\Box E, \Delta F)$ is resolvable

b) $\Sigma_2(\Box E,F) = \Sigma_i(E,F)$

And if (E,F) is resolvable, then $(\Box E,F)$ is resolvable.

c) $\Sigma_2(\Diamond E, \Diamond F) = \Sigma_i(E, \Diamond F) \vee \Sigma_i(\Diamond E,F)$

And if $(E, \Diamond F)$ or $(\Diamond E,F)$ is resolvable, then $(\Diamond E, \Diamond F)$ is resolvable.

d) $\Sigma_2(o\, E, o\, F) = o\, \Sigma_i(E,F)$

And if (E,F) is resolvable, then (oE, oF) is resolvable

e) $\Gamma(E(\Diamond(D \;\&\; D' \;\&\; F)) = E(\Diamond(\Sigma_i(D,D') \;\&\; F))$

And if (D,D') is resolvable, then $(E \Diamond ((D \;\&\; D') \;\&\; F))$ is resolvable.

3.1.3. Transformations operations.

a) $\Sigma_3(\Box E,F) = \Sigma_i(\Box\Box E,F)$

And if $(\Box\Box E,F)$ is resolvable, then $(\Box E,F)$ is resolvable.

b) $\Sigma_3(\Diamond E,F) = \Diamond_n E \vee \Sigma_i(\Diamond(\Box_n \sim E \;\&\; o^n E),F)$

And if $\Diamond_n E$ or $(\Diamond(\Box_n \sim E \;\&\; o^n E),F)$ is resolvable, then $(\Diamond E,F)$ is resolvable.

3.2. If C_1 and C_2 are unit clauses and C_1 and C_2 are resolvable (or C_1 is resolvable), then a clause is called resolvent of C_1 and C_2 $((C_1))$ if it is the result of substituting:

$\emptyset$ for every occurrence of $(\emptyset \;\&\; E)$

E for every occurrence of $(\emptyset \vee E)$

$\emptyset$ for every occurence of $\Delta\,\emptyset$, where Δ is $\Box$, $\Diamond$ or o

in $\Sigma_i(C_1,C_2)$ $(\Gamma(C_1))$, as times as necessary.

We note by $R(C_1,C_2)$ (or $R(C_1)$) a resolvent of C_1 and $C_2((C_1))$.

3.3. Let $C_1 \vee C$ and $C_2 \vee C'$ be two clauses. The resolution rule :

$$1)\quad \frac{C_1 \vee C \qquad C_2 \vee C'}{R(C_1,C_2) \vee C \vee C'}$$

is applied if C_1 and C_2 are resolvables.

And the rule:

$$2)\quad \frac{C_1 \vee C}{R(C_1) \vee C}$$

is applied if C_1 is resolvable.

3.4. Let $E(D \vee D \vee F)$ be a clause. The following rule will be applied

$$3)\quad \frac{E(D \vee D \vee F)}{E(D \vee F)}$$

3.5. Let S be a set of clauses. A deduction of C from S is a finite sequence $C_1,\ldots,C_n$ such that :

- C_n is C

- C_i $(1 \leqslant i \leqslant n)$ is :

  a clause of S, or

  a clause obtained from C_j, $j < i$ using the inference rules 2) or 3) or a clause obtained from C_j and C_k, $j,k < i$, using the inference rule 1).

3.6. A deduction of the empty clause is called a refutation.

3.7. We give an elementary example to the aim to explain the method. Given the two unit clauses $\Box \Diamond$ p and $\Diamond \Box$ $(\sim p \vee q)$, a set of operations and properties corresponding to this set will be :

1. $\Sigma_2(\Box \Diamond p), \Diamond \Box (\sim p \vee q)) = \Diamond \Sigma_2(\Diamond p, \Box (\sim p \vee q))$. And if $(\Diamond p, \Box (\sim p \vee q))$ is resolvable, then $(\Box \Diamond p, \Diamond \Box (\sim p \vee q))$ is resolvable.

2. $\Sigma_2(\Diamond p, \Box(\sim p \vee q)) = \Diamond \Sigma_1(p, \sim p \vee q)$. And if $(p, \sim p \vee q)$ is resolvable then $(\Diamond p, \Box(\sim p \vee q))$ is resolvable.

3. $\Sigma_1(p, \sim p \vee q) = (\Sigma_1(p, \sim p) \vee q)$. Anf if $(p, \sim p)$ is resolvable, then $(p, \sim p \vee q)$ is resolvable.

4. $(\Sigma_1(p, \sim p) \vee q) = (\emptyset \vee q)$. And $(p, \sim p)$ is resolvable. Therefore $(p, \sim p \vee q)$ and $(\Diamond p, \Box (\sim p \vee q)$ are resolvables.

Consequently $(\Box \Diamond p, \Diamond \Box (\sim p \vee q))$ is resolvable, and the inference rule 1) can be applied as follows:

$$\frac{\Box \Diamond p \qquad \Diamond \Box \ (\sim p \vee q)}{\Diamond \Diamond q}$$

because $\Diamond \Diamond q$ is the result of substituting q by $(\emptyset \vee q)$ in $\Diamond \Diamond$ $(\emptyset \vee q)$.

Therefore in classical logic only one operation, $\Sigma_1(p, \sim p)$, is sufficient to apply the resolution rule, while in temporal logic more than one operation can be necessary to make it.

4. *A set of clauses S is unsatisfiable iff S is refutable.*

a/ The proof is obtained by induction on m(S). If m(S) = (0,0,...) then by 2.4. there is a refutation R' of $\theta^i(S)$ that use only the classical operations .To obtain a refutation R of S must put the o operators in its proper place and using, i times, the operation d) in 3.1.2..

Assume that the theorem holds for every set of clauses S" such that $m(S) > m(S")$. Then the proof is obtained by induction on the number of symbols of disjunction that appear between the disjunct of the clauses in S (noted $\nu(S)$).

If $\nu(S) = 0$ and let S' be the set of clauses obtained by 2.3. then by the induction hypothesis there is a refutation R' of S'. In order to obtain a refutation R of S, from R' and S' we distinguish the following two cases:

<u>Case 1</u>. $T = \{t_i^0 : i = 1,\dots,n\}$. Then

1.a. $S' = e(S, \Box A_1)$. The refutation R from R' is obtained by replacing every $o^i A_1$, $i = 1,\dots p$; by $\Box A_1$ and every operation of type $\Sigma_2(\Box B, o^i A_1)$ by $\Sigma_2(\Box B, \Box A_1)$ and $\Sigma_2(o^i B, o^i A_1)$ by $\Sigma_2(o^i B, \Box A_1)$, using the operation a) of 3.1.3., if necessary. For the case $i = 0$, A_1 is replaced by $\Box A_1$ and the operation $\Sigma_i(C, A_1)$ by $\Sigma_2(C, \Box A_1)$ followed by $\Sigma_i(C, A_1)$.

2.a. $S' = \{S_1', S_2'\}$ where $S_j' = e(S, \Diamond P_1)$ and $S_2' = e(S, \Diamond P_2)$. Then there are refutations R'_1 and R'_2 of S'_1 and S'_2 respectively.

Now we have the following two cases:

2.a.1. Any refutation R_1' possesses $\Diamond P_2$ and any R_2' possesses $\Diamond P_1$. In this case there are refutations R_1 and R_2, where $\Diamond$ has been put in its proper place, such that they are identicals until the step where $\Diamond P_1$ and $\Diamond P_2$ appears (Fig. 1.). Then the operation Σ_3 $(\Sigma_3(\Diamond P_1, \Diamond P_2) = \Sigma_i(P_1, \Diamond P_2) \vee \Sigma_i(\Diamond P_1, P_2))$ is used and the refutation R of S is obtained by putting together R_1 and R_2 as in Fig.2.

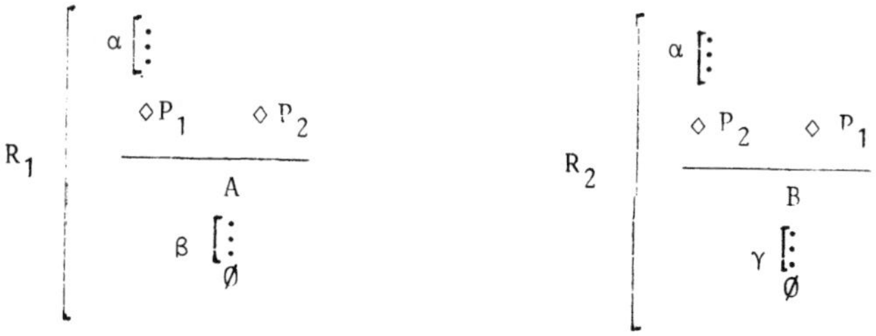

$$R_1 \left[\begin{array}{c} \alpha \left[\vdots \right. \\ \dfrac{\Diamond P_1 \qquad \Diamond P_2}{A} \\ \beta \left[\vdots \right. \\ \emptyset \end{array} \right. \qquad\qquad R_2 \left[\begin{array}{c} \alpha \left[\vdots \right. \\ \dfrac{\Diamond P_2 \qquad \Diamond P_1}{B} \\ \gamma \left[\vdots \right. \\ \emptyset \end{array} \right.$$

Fig. 1.

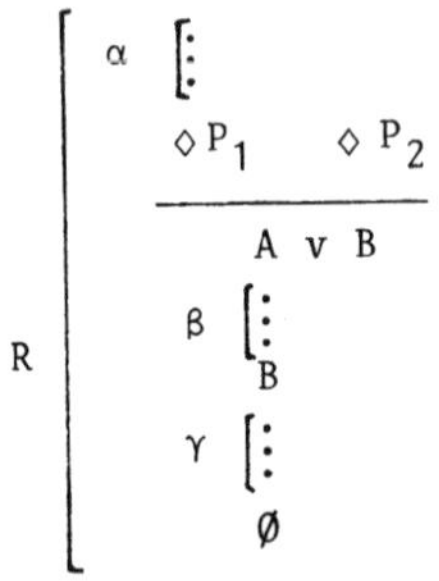

$$R \left[\begin{array}{l} \alpha \left[\vdots \right. \\ \dfrac{\Diamond P_1 \quad \Diamond P_2}{A \vee B} \\ \beta \left[\vdots \right. \\ \quad B \\ \gamma \left[\vdots \right. \\ \quad \emptyset \end{array} \right.$$

Fig. 2.

2.a.2. There is a refutation (R_1 for example) that does not possess $\Diamond P_2$. In this case R_1 can be extended to be a refutation of S, by putting $\Diamond$ in its proper place, and using the transformation a) in 3.1.3. in order to apply the operation (Σ_2) of $\Box$ against the $\Diamond$ that has been added and the operation e) in 3.1.2. if necessary.

Case 2. $T = \{t_1^0,\dots,t_i^k,\dots,t_n^0\}$. The the proof is obtained by induction on k, the number of v, obtained by the transformation t_i^k.

The induction step ($\nu(S) = n+1$) is proved as usual [CHL]. Let $C = E_1 \vee C_1$ be a clause in S. Construct separately refutations R_2 from $(S - \{C\}) \cup E_1$ and R_1 from $(S - \{C\}) \cup C_1$ respectively and put R_2 on the top of R_1 after adding C_1 to all clauses in R_2.

b) If S is refutable then S must be unsatisfiable because the rule preserves the satisfiability.

5. Example.

We illustrate the method by solving a simple mutual exclusion problem for processes P_1 and P_2. Each process is always in one of the two regions code:

$Begin_i$	the process P_i begins the critical section
End_i	the process P_i ends the critical section

and it moves from $Begin_i$ to End_i in the following way:

when the process P_i is in $Begin_i$, it performs "critical section", proceeding in parallel with the other process P_j, that wait until P_i $Ends_i$, to reach itself the "critical section", proceeding in alternatively.

Listed below are the temporal formulas whose conjunctions specify this mutual exclusion system:

1) Star state

$Begin_i$

2) Absence of starvation

$\Box(\sim Begin_i \vee o\ End_i)$

$\Box(\sim End_i \vee o\ Begin_i)$

3) Mutual exclusion

$\Box(\sim Begin_i \vee \sim End_j)$

$\Box(\sim Begin_i \vee \sim Begin_j)$

$\Box(\sim End_i \vee \sim End_j)$, where $i \neq j$

4) Fairness

$\Box \Diamond(Begin_1 \vee End_1)$

$\Box \Diamond(Begin_2 \vee End_2)$

5) Each process P_i is always in exactly one the two code regions

$\Box(Begin_i \vee End_i)$

$\Box(\sim Begin_i \vee \sim End_i)$

In order to show that the specifications verify some properties we can ask: if we begin with process P_1, can P_2 reach the critical section ? i.e. $\Diamond\ o\ Begin_2$?

We assert that process P_2 never reaches the critical section and we add these assertions to the specification: $\Box\, o \sim Begin_2$, and we prove that this set is refutable.

The refutation will be:

$$\frac{\Box\, o \sim Begin_2 \qquad \Box\,(Begin_2 \ v \ End_2)}{}$$ using clause 5. and operations 3.1.3., 3.1.2.a and 3.1.1.

$$\frac{\Box\, o \sim End_2 \qquad \Box\,(\sim End_2 \ v \sim Begin_1)}{}$$ using clause 3. and 3.1.3., 3.1.2.a and 3.1.1.

$$\frac{\Box\, o \sim Begin_1 \qquad \Box\,(Begin_1 \ v \ End_1)}{}$$ using clause 5.

$$\frac{\Box\, o \ End_1 \qquad \Box\,(\sim Begin_1 \ v \ o \ End_1)}{}$$ using clause 2.

$$\frac{\Box \sim Begin_1 \qquad Begin_1}{}$$ using that we begin with process P_1

$$\emptyset$$

6. Some Conclusions

We think that the procedure proposed possesses some advantages with respect to other semantic or syntactical decision methods [BM] [CE]. In particular for the expression of refinement we can follow the ideas developped in classical logic [CHL]. In this way an implementation of a linear refinement of this method has been realized in Prolog [FL].

Bibliography

[BM] BEN-ARI M. - Complexity of proofs and models in programming logics. Ph D., Tel-Aviv University, May 1981.

[CR] CARNAP R. - Modalities and quantification. JSL Vol. 11, 1946, pp. 33-64.

[CHL] CHANG C., LEE R. - Symbolical logic and mechanical theorem proving. Academic Press, New-York, 1973.

[FC] FARIÑAS DEL CERRO L. - A simple deduction method for modal logic. Information Processing Letters, Vol. 14, n°2, 1982

[FL] FARIÑAS DEL CERRO L., LAUTH E. - Raisonnement temporel : une méthode de déduction. Rapport Université Paul Sabatier, Toulouse, 1983.

[GPSS] GABBAY D., PNUELI A., SHEALAH S., STAVI J. - Temporal Analysis of Fairness. Seventh ACM Symposium on Principles of Programming Languages. Las Vegas, NV, Janvier 1980.

[LE] LEMMON E. - An introduction to modal logic. Amer. Phil. Quaterly Monograph Series, 1977.

[MZ] MANNA Z. - Verification of sequential programs: Temporal axiomatization. Report NoSTAN-CS-81-877, Stanford University, 1981.

[RJ] ROBINSON J. - A machine oriented logic based on the resolution principle. J. ACM, 12, 1965, pp. 23-41.

[CE] CLARKE E., EMERSON E. - Design and synthesis of synchronization skeletons using branching time temporal logic, 1981.

A PROGRESS REPORT ON NEW DECISION ALGORITHMS FOR FINITELY PRESENTED ABELIAN GROUPS

D. Lankford, G. Butler, and A. Ballantyne

Louisiana Tech University
Mathematics and Statistics Department
Ruston, Louisiana 71272

ABSTRACT

We report on the current state of our development of new decision algorithms for finitely presented Abelian groups (FPAG) based upon commutative-associative (C-A) term rewriting system methods. We show that the uniform word problem is solvable by a completion algorithm which generates Church-Rosser, Noetherian, C-A term rewriting systems. The raw result is theoretical, and few would contemplate implementing it directly because of the incredible amount of trash which would be generated. Much of this trash can be obviated by a different approach which achieves the same end. First, the uniform identity problem is solved by a modified C-A completion algorithm which generates Church-Rosser bases, and then the desired complete set can be computed directly from the Church-Rosser bases. Computer generated examples of the first part of this two part procedure are given. The second part is still under development. Our computer experiences suggest that Church-Rosser bases may often contain large numbers of rules, even for simple presentations. So we were naturally interested in finding better Church-Rosser basis algorithms. With some minor changes the method of Smith [1966] can be used to generate Church-Rosser bases. The Smith basis algorithm appears promising because computer experiments suggest that the number of rules grows slowly. However, small examples with small coefficients can give rise to bases with very large coefficients which exceed machine capacity so we are not entirely satisfied with this approach.

This work was supported in part by NSF Grant MCS-8209143 and a Louisiana Tech University research grant.

INTRODUCTION

There are at least three different previous methods of solution of the FPAG uniform word problem--the direct sum (or basis) method, which is discussed in detail by Smith [1966], the elementary formula method, see Szmielew [1954], and the linear Diophantine equation method, see Cardoza [1975]. We have not studied the method of Reidemeister [1932] thoroughly enough to classify it. Only one of the above methods has been previously implemented, a variation of the basis method by Smith [1966] who modified the solution of Jacobson [1953] for improved computational efficiency. It should be routine to implement the linear Diophantine equation method, but writing an explicit algorithm for the elementary formula method would be non-trivial. It is also not obvious how to combine the linear Diophantine equation method or the elementary formula method with general computational logic systems.

The C-A term rewriting system methods of this paper are on the one hand generalizations of the Ballantyne and Lankford [1979] solution of the finitely presented commutative semigroup (FPCS) uniform word problem, and on the other hand special cases of more general C-A term rewriting system methods developed by Lankford and Ballantyne [1977]. These methods are easily incorporated into general computational logic systems, such as with blocked permutative narrowing and resolution, see Lankford and Ballantyne [1979]. Our solution of the FPAG uniform word problem with these methods solves part of the stumbling block problem mentioned by Bergman [1978]. We believe similar methods will solve other important parts of the stumbling block problem, such as the uniform word problem for finitely presented C-A rings.

BACKGROUND

It is known that the word problem for free Abelian groups is decidable by the complete C-A term rewriting system below, see Lankford and Ballantyne [1977] and Lankford [1979].

R1. $[x1] \longrightarrow [x]$

R2. $[xx^{-1}] \longrightarrow [1]$

R3. $[1^{-1}] \longrightarrow [1]$

R4. $[(x^{-1})^{-1}] \longrightarrow [x]$

R5. $[(xy)^{-1}] \longrightarrow [x^{-1}y^{-1}]$

In the language of term rewriting systems, an FPAG consists of the above five C-A rewrite rules plus a finite number of ground (variable free) rewrite rules $[L_i] \longrightarrow [R_i]$ where L_i and R_i are terms constructed from the group operations and constants. In the usual language of uniform word problems, the constants are the generators, and the equations $L_i = R_i$ are the relations. In general a presentation may have generators which do not occur in any of the relations, but we ignore them here (because they can be collected together using commutativity and associativity and kept in normal form by R1 - R5).

In the case of FPCS it was obvious that the C-A term rewriting systems were Noetherian when ordered by a vector lexicographic order. It is natural to conjecture a similar result for FPAG, but we must be careful. For example, $[a] \longrightarrow [a^{-1}]$ is not Noetherian, and there are non-ground rules. Evidently we require a norm which is compatible with certain lexicographic orders. Let $F_1 = 1$, $F_a = 1$ for any constant a, $F(x,y) = x + y$, $F_{-1}(x) = x^2$, $|\,[v]\,| = 1$ for any variable symbol v, and define $|\,[t]\,|$ as in Lankford [1979]. It follows that if [u] is an immediate reduction of [t] by R1 - R5, then $|\,[t]\,| \geq |\,[u]\,|$ with equality possible only for immediate reductions using R3 or R4. And because R3 and R4 decrease the number of occurrences of -1's, it follows that R1 - R5 is Noetherian. That R1 - R5 is Noetherian is already known from Lankford [1979], but the norm used there is not compatible with lexicographic orders.

MAIN RESULTS

Lemma 1 Suppose the term algebra contains a finite number of constants ordered in some manner: $a_1, \ldots, a_k$. For any ground C-A congruence class [g] which is irreducible relative to R1 - R5, let m_i denote the number of occurrences of a_i^{-1} in g, n_i the number of occurrences of a_i in g, and $V_1([g]) = (m_1, \ldots, m_k, n_1, \ldots, n_k)$. If R is a C-A term rewriting system consisting of R1 - R5 and ground rules $[1] \longrightarrow [R]$ which are irreducible relative to R1 - R5 and satisfy $V_1([L]) > V_1([R])$ in the vector lexicographic order, then R is Noetherian.

Proof Suppose there were an infinite sequence $[t_1] \longrightarrow [t_2] \longrightarrow [t_3] \longrightarrow \ldots$. For any ground rule $[L] \longrightarrow [R]$ of R, $|\,[L]\,| \geq |\,[R]\,|$, i.e., the norm and vector lexicographic order are compatible. Thus it follows that $|\,[t_i]\,| \geq |\,[t_{i+1}]\,|$. Without loss of generality assume $|\,[t_i]\,| = |\,[t_{i+1}]\,|$, hence the only rules used in the immediate reductions are R3, R4, or ground. Because the ground rules are irreducible relative to R1 - R5, there is no infinite subsequence of applications of R3. The norms are equal, so R4 applies only to subterms of the form[1] x^{-n} where $n \geq 2$ and x is a variable symbol or constant. Clearly there can be no infinite subsequence of applications of R4 to subterms of the form x^{-n} where x is a variable symbol. When x is a constant the only rules besides R4 which apply to a subterm of the form x^{-n} are those of the form $[x] \longrightarrow [c]$, $[x^{-1}] \longrightarrow [c]$, or $[x^{-1}] \longrightarrow [c^{-1}]$ where c is a constant. The number of occurrences of -1's in these subterms is nonincreasing and strictly decreases with each application of R4. And because no new subterms of the form x^{-n} are introduced (there are no rules of the form $[x] \longrightarrow [c^{-1}]$, and the ground rules are irreducible relative to R1 - R5), there can be no infinite subsequence of applications of R4.

---

[1] The form x^{-n} denotes $(\ldots((x^{-1})^{-1})^{-1}\ldots)^{-1}$ rather than $(x^{-1})^n$

So we now assume that the given infinite sequence of immediate reductions consists entirely of ground rule applications, and all t_i are ground. (This does not mean that the $[t_i]$ are necessarily irreducible relative to R1 - R5.) When x is a constant and $n \geq 2$ let us define the excess -1's of x^{-n} as $n - 1$. The excess -1's of a ground [g] is the sum of the excess -1's of subterms of g of the form x^{-n} where x is a constant and $n \geq 2$. The excess -1's parameter of the $[t_i]$ is non-increasing, so without loss of generality assume the excess -1's parameter is constant. Now extend the domain of V_1 to arbitrary ground [g] as follows. A term g may be regarded as the product L(g)N(g) where L(g) is the linear part of g, that is a product of a_i^{-1}'s and a_i's, and N(g) is the nonlinear part, that is a product of terms of the form $g_j^{-n_j}$ where $n_j \geq 2$, and the leading function symbol of g_j is not -1. (The g_j may in turn be a product $L(g_j)N(g_j)$ and so on.) Let $V_1([g]) = V_1([L(g)]) + \Sigma V_1([g_j])$. It can be shown that if [h] is an immediate C-A reduction of [g] by a ground rule of R, and the excess -1's parameters of [g] and [h] are equal, then $V_1([g]) > V_1[h])$ in the vector lexicographic order. But then $V_1([t_1]) > V_1([t_2]) > V_1([t_3]) > \ldots$ which is impossible.

<u>Conjecture</u> If V_1 is replaced by $V_2[g]) = (m_1, n_1, \ldots, m_k, n_k)$, then the revised Lemma 1 holds.

<u>Theorem 1</u> If R is a Noetherian, C-A term rewriting system, and all critical pairs of R and its embeddings exist, then R is Church-Rosser iff for each critical pair X,Y, $X^* = Y^*$.

<u>Proof</u> See Lankford and Ballantyne [1977].

The completion procedure is well-known to workers in the field, and so will not be described in detail here. We just remind the reader of certain key points: equations are expressed as rewrite rules using the order described above, all rewrite rules are kept irreducible relative to the others, and equations [t] = [u]

which satisfy $t \in [u]$ are deleted. To show that the completion procedure decides the FPAG uniform word problem we must show (1) that all critical pairs exist at each round of the completion procedure, and (2) the completion procedure terminates uniformly.

Lemma 2 All critical pairs exist at each round of the completion procedure.

Proof It has recently been announced by Fages [1982] that the C-A unification procedure halts for arbitrary pairs of input terms, in which case it is immediate that all critical pairs exist at each round of the completion procedure. One can also prove Lemma 2 directly as follows. Let S_1 be R1 - R5 plus their embeddings and S_2 be the ground rules plus their embeddings. Because R1 - R5 are already known to be Church-Rosser, critical pairs need only be shown to exist between certain pairs of rewrite rules, namely, one each from S_1 and S_2 or both from S_2. This leads to 17 cases, two of which we show below.

Consider the critical pairs of the embedding $x1y \longrightarrow xy$ of R1 and the embedding $Lz \longrightarrow Rz$ of a ground rule $[L] \longrightarrow [R]$. The trick in this case is that inverses of constants in L and R can themselves be treated as constants (remember that [L] and [R] are irreducible relative to R1 - R5), and so the C-A unification procedure halts.

Consider the critical pairs of the embedding $xx^{-1}y \longrightarrow y$ of R2 and the embedding $Lz \longrightarrow Rz$ of a ground rule. In this case, following Stickel [1975], a new variable w is troduced for x^{-1} so that the unifiers of xwy and Lz can be found (again treating inverses of constants as new constants). Now consider w' and $(x')^{-1}$ where x' and x' are the assignments of x and w by one of the C-A unifiers in the complete set for xwy and Lz. Because w' and x' are products of variables, constants, and inverses of constants, the only way that w' and $(x')^{-1}$ can be C-A unifiable is for w' to be the inverse of a constant and x' to be the same constant. Thus the complete set of C-A unifiers of $xx^{-1}y$ and Lz is a subset of the complete set of C-A unifiers of xwy and Lz.

Lemma 3 After a current set of rules is reduced to irreducible form by the completion procedure, the new irreducible set consists of R1 - R5 and ground rules.

Proof It can be shown that if a fully reduced critical pair of the form [$g_1$variables],[$g_2$variables] is formed, where g_1 and g_2 are ground, then [g_1],[g_2] is also formed.

Thus when the completion procedure is iterated, except for R1 - R5, only ground rules are present after the current set is reduced to irreducible form. Of course, before passing to a subsequent critical pairs step we must again add embeddings.

Lemma 4 The C-A completion procedure halts uniformly.

Proof By the same argument as Lemma 2 of Ballantyne and Lankford [1979], if the completion procedure ran indefinitely, then there would exist an infinite set of mutually incomparable vectors, which is impossible.

Theorem 2 The uniform word problem for finitely presented Abelian groups is decidable by the C-A completion procedure.

Proof Use Lemmas 1 - 4 and Theorem 1.

Inefficiencies in the raw procedure (Theorem 2) arise from two sources: (1) the C-A unification procedure, and (2) the fact that constants can occur in equations in equivalent ways, as a constant on one side of a rule, or as the inverse of the constant on the other side of the rule. We obviate much of this inefficiency by taking a different approach. First, the uniform identitiy problem is solved by a modified C-A completion algorithm which generates Church-Rosser bases, and then the desired complete set can be computed directly from the Church-Rosser bases. Computer generated examples of the first part of this two part procedure are given. The second part is still under development.

For any C-A term rewriting system $\mathcal{R}$ let $\mathcal{E}(\mathcal{R})$ denote the set of equations obtained from $\mathcal{R}$ by deleting brackets and replacing $\longrightarrow$ by = throughout, together with the commutative and associative equations which define the congruence classes, let $[t]^*$ denote the normal form of [t] relative to R1 - R5, and let $\longmapsto$ denote an immediate C-A reduction by R1 or a ground rule. $\mathcal{R}$ is a Church-Rosser basis means $\mathcal{R}$ consists of R1 - R5 and ground rules of the form $[L] \longrightarrow [1]$, and $\mathcal{E}(\mathcal{R}) \vdash t = 1$ iff $[t]^* \longrightarrow^* [1]$.

Lemma 5 If $\mathcal{R}$ consists of R1 - R5 and ground rules of the form $[L] \longrightarrow [1]$, and [L] is irreducible relative to R1 - R5, then $\mathcal{R}$ is a Church-Rosser basis iff for each critical pair X,Y obtained by the maximum overlap method of Ballantyne and Lankford [1979] (with inverses of constants treated as constants), $[XY^{-1}]^* \longmapsto^* [1]$, and for each $[L] \longrightarrow [1] \in \mathcal{R}$, $[L^{-1}]^* \longrightarrow^* [1]$.

Proof ($\Longrightarrow$) Obvious. ($\Longleftarrow$) Clearly $[t]^* \longmapsto^* [1]$ implies $\mathcal{E}(\mathcal{R}) \vdash t = 1$. Now let $\mathcal{E}(\mathcal{R}) \vdash t = 1$. Then t = 1 has a proof of length n, that is a sequence $t \equiv t_0, t_1, \ldots, t_n \equiv 1$ where each step is obtained by one application of some axiom of $\mathcal{E}(\mathcal{R})$. We will show by induction that if t = 1 has a proof of length n, then $[t]^* \longrightarrow^* [1]$. For n = 1 we have $[t_0] \longrightarrow [1]$. If the rule which applied is R1 - R4, then $[t_0]^* = [1]$, hence $[t_0]^* \longrightarrow^* [1]$. And if the rule which applied is ground, then $[t_0] = [t_0]^*$ (because the left sides of ground rules are irreducible), so $[t_0]^* \longrightarrow [1]$. For the induction step take a proof of length n + 1. By the induction hypothesis $[t_1]^* \longmapsto^* [1]$. There are now five cases to consider: (1) $[t_0] = [t_1]$, (2) $[t_0] \longrightarrow [t_1]$ by R1 - R5, (3) $[t_0] \longleftarrow [t_1]$ by R1 - R5, (4) $[t_0] \longrightarrow [t_1]$ by a ground rule, or (5) $[t_0] \longleftarrow [t_1]$ by a ground rule. In cases (1) and (2) clearly $[t_0] \longrightarrow^* [t_1]^*$, and in case (3) $[t_0] \longrightarrow^* [t_1]^*$ because R1 - R5 is Church-Rosser. Cases (4) and (5) require rather lengthy arguments, so we will only indicate the approach with a discussion of part of case (4). Let $[L] \longrightarrow [1]$ be the ground rule used in the $[t_0] \longrightarrow [t_1]$ step. If $[L] \longrightarrow [1]$ does not interact with any rule which produces $[t_1]^*$ except $[x1] \longrightarrow [1]$ on the occurrence of 1, then $[t_0] = [L][t_1'] \longrightarrow^* [L][t_1']^*$ where $[t_1']^* = [t_1]^*$. So $[L][t_1']^* \longmapsto [1][t_1']^* \longmapsto [t_1]^*$. But $[L][t_1']^*$ may not be irreducible relative to R1 - R5. If it is not irreducible relative to R1 - R5, then it follows that only R1 and R2 apply, and that R2 applies only to constant-inverse of constant pairs. Without loss of generality let $[L][t_1']^* = [L'x_1 \ldots x_n y_1 \ldots y_n z]$, $([L][t_1']^*)^* = [L'z]$, $[t_1]^* = [y_1 \ldots y_n z] = [L_1 \ldots L_j]$ where $[L_i] \longrightarrow [1] \in R$, and $[L_1] = [y_1 \ldots y_k w]$. It follows that $[L_1^{-1}]^* \longrightarrow [1] \in R$ and overlaps with $[L] \longrightarrow [1]$ on $x_1 \ldots x_k$. The overlap may not be maximum, but it can be shown that for any overlap critical pair X,Y, $[XY^{-1}]^* \longrightarrow^* [1]$. Hence $[t_0]^* =$ $[L'x_{k+1} \ldots x_n (y_{k+1}^{-1} \ldots y_n^{-1} z^{-1})^{-1}]^* \longmapsto^* [1]$, so we are done with the non-interaction part of case (4).

If R is not a Church-Rosser basis, then R can be transformed into a Church-Rosser basis by a completion algorithm which adds rules $[M] \longrightarrow [1]$ when $[XY^{-1}]^* \longrightarrow^* [M] \neq [1]$ or $[L^{-1}]^* \longrightarrow^* [M] \neq [1]$ with $[M]$ irreducible. The algorithm terminates because there can be no infinite set of mutually incomparable vectors.

Example 1 The Abelian group presented by $a^2b^{-3}c = 1$, $a^{-3}b^2c^3 = 1$, $a^2b^2c^{-2} = 1$ is typical of our computer experience with small examples. To conserve space we drop brackets. A 70 ground rule Church-Rosser basis was generated which consisted of the following 35 rules and their reduced inverses.

G1. $a^2b^{-3}c \longrightarrow 1$	G8. $a^{-3}b^7 \longrightarrow 1$
G2. $a^{-3}b^2c^3 \longrightarrow 1$	G9. $abc^{-2} \longrightarrow 1$
G3. $a^2b^2c^{-2} \longrightarrow 1$	G10. $a^{-1}b^4c^{-5} \longrightarrow 1$
G4. $a^{-1}b^4c \longrightarrow 1$	G11. $a^{-6}b^4 \longrightarrow 1$
G5. $b^5c^{-3} \longrightarrow 1$	G12. $a^{-3}b^2c^{-3} \longrightarrow 1$
G6. $abc^{-4} \longrightarrow 1$	G13. $A^5c^{-5} \longrightarrow 1$
G7. $a^{-4}bc \longrightarrow 1$	G14. $c^6 \longrightarrow 1$

G15. $a^{-2}b^{3}c^{5} \longrightarrow 1$

G16. $a^{-1}b^{-6}c \longrightarrow 1$

G17. $a^{3}b^{3} \longrightarrow 1$

G18. $a^{-1}b^{9}c^{-2} \longrightarrow 1$

G19. $a^{5}c \longrightarrow 1$

G20. $b^{5}c^{3} \longrightarrow 1$

G21. $a^{6}bc^{-3} \longrightarrow 1$

G22. $a^{7}b^{2}c^{-1} \longrightarrow 1$

G23. $a^{9}b^{-1} \longrightarrow 1$

G24. $a^{10}c^{-4} \longrightarrow 1$

G25. $a^{2}b^{7}c \longrightarrow 1$

G26. $b^{10} \longrightarrow 1$

G27. $a^{11}bc^{-2} \longrightarrow 1$

G28. $a^{4}b^{-1}c^{5} \longrightarrow 1$

G29. $a^{15}c^{-3} \longrightarrow 1$

G30. $a^{12}b^{2} \longrightarrow 1$

G31. $a^{16}bc^{-1} \longrightarrow 1$

G32. $a^{20}c^{-2} \longrightarrow 1$

G33. $a^{21}b \longrightarrow 1$

G34. $a^{25}c^{-1} \longrightarrow 1$

G35. $a^{30} \longrightarrow 1$

Our computer experiences suggest that Church-Rosser bases may often contain large numbers of rules, even for small examples. Whether this remains true for the corresponding Church-Rosser, Noetherian, C-A term rewriting systems is unknown because we have not yet implemented an algorithm which transforms Church-Rosser bases into complete sets of C-A reductions. Regardless of the outcome of that investigation we are naturally interested in finding better Church-Rosser basis generation algorithms. With some minor changes the method of Smith [1966] can be used to generate Church-Rosser bases. Smith's algorithm accepts a matrix of integers corresponding to the exponents of a given presentation and outputs a matrix of integers which describes the FPAG in terms of its cyclic summands. The output matrix is known as the Smith normal form (SNF). Smith [1966] points out that the SNF algorithm is equivalent to transforming a given presentation into an equivalent presentation of the form $a_i = \Pi b_j^{n_{ij}}$, $b_j^{m_j} = 1$ where the a_i and b_j are distinct, the a_i are constants which occur in the given presentation, and the b_j are either constants which occur in the given presentation or new constants (introduced by the SNF algorithm). The number of equations of the first kind is bounded above by the number of distinct constants which occur in the given presentation, and the number of equations of the second kind is bounded above by the number of given relations. The corresponding C-A term rewriting system consisting of R1 - R5, the rules $[a_i] \longrightarrow [b_j^{n_{ij}}]$ and $[b_j^{m_j}] \longrightarrow [1]$ will be denoted by R_{SNF}. R_{SNF} is generally not a Church-Rosser basis, but it can be easily transformed into one by adding the rules $[b_j^{-m_j}]^* \longrightarrow [1]$. We call the Church-Rosser bases obtained in this way SNF Church-Rosser bases.

<u>Theorem 3</u> If R is an SNF Church-Rosser basis, and R' is obtained from R by replacing the rules $[b_j^{\pm m_j}] \longrightarrow [1]$ by $[b_j^{-\frac{1}{2}|m_j|}] \longrightarrow [b_j^{\frac{1}{2}|m_j|}]$ and $[b_j^{\frac{1}{2}|m_j| + 1}] \longrightarrow [b_j^{\frac{1}{2}|m_j| - 1}]$ when m_j is even, or by $[b_j^{-\frac{1}{2}(|m_j| + 1)}] \longrightarrow [b_j^{\frac{1}{2}(|m_j| - 1)}]$ and $[b_j^{\frac{1}{2}(|m_j| + 1)}] \longrightarrow [b_j^{\frac{1}{2}(|m_j| - 1)}]$ when m_j is odd, then R' is a Noetherian, Church-Rosser, C-A term rewriting system.

<u>Proof</u> The number of applications of ground rules of the first kind is finite because the number of occurrences of a_i is non-increasing, and strictly decreases with each application of a rule of the first kind. R' is thus Noetherian because any C-A term rewriting system consisting of R1 - R5 and ground rules of the second kind satisfies the hypothesis of Lemma 1. That R' is Church-Rosser follows from Theorem 1. Because of the simple structure of R' the superpositions which must be checked are simple and will not be given here.

<u>Example 2</u> Let us call the complete sets generated via Theorem 3 SNF complete sets. The ground rules of the SNF complete set for Example 1 are given below (without brackets).

G1. $b \longrightarrow a^9$	G3. $a^{16} \longrightarrow a^{-14}$
G2. $c \longrightarrow a^{-5}$	G4. $a^{-15} \longrightarrow a^{15}$

This is obviously quite an improvement over the Church-Rosser basis of Example 1.

Example 3 The following presentation was obtained by randomly selecting the last four digits of telephone numbers. The ground rules of the SNF complete set are given below (without brackets).

$$c_1^{3827} c_2^{-2223} c_3^{1934} c_4^{-3400} c_5^{4815} c_6^{-6646} c_7^{7833} c_8^{-9443} c_9^{4584} c_{10}^{-4462} = 1$$

G1. $c_3 \longrightarrow c_1^{475} c_2^{-47} c_4^{-771} c_6^{-285} c_9^{130} c_{10}^{74} d_1^{-251} d_2^{-380} d_3^{1200}$

G2. $c_5 \longrightarrow c_1^{22} c_2^{-7} c_4^{-35} c_6^{-16} c_9^{7} c_{10}^{4} d_1^{-11} d_2^{-13} d_3^{58}$

G3. $c_7 \longrightarrow c_1^{-139} c_2^{17} c_4^{226} c_6^{86} c_9^{-39} c_{10}^{-22} d_1^{75} d_2^{109} d_3^{-335}$

G4. $c_8 \longrightarrow c_1^{2} c_2^{-2} c_4^{-2} c_6^{-2} c_9 d_1 d_2 d_3^{3}$

CONCLUSIONS

Our approach has been based upon the methods of Lankford and Ballantyne [1977] because we are familiar with them. The more general methods of Peterson and Stickel [1981] could also be used to develop these results. Our investigations are incomplete, but preliminary computer experiments suggest that the SNF completion algorithm is currently the most practical method of deciding the FPAG uniform word problem with complete sets of reductions. Smith [1966] gave an example with 8 generators and 8 relations all of whose exponents are one digit which generated numbers that exceeded his machine capacity (10^8). We wonder if there are other basis generation methods which achieve a better balance between the number of rules and the exponent size. We also wonder what can be said about the theoretical complexity of SNF completion. Cardoza [1975] points out that the linear Diophantine equation method of solution of the FPAG uniform word problem is polynomial complexity. But as we have said, that method does not combine readily with general logical systems. A basis solution of the uniform word problem for finitely presented nilpotent groups has been given by Mostowski [1966a,b], so it is reasonable to ask if similar term rewriting system methods can be developed for nilpotent groups. However, it appears to be unknown whether the word problem for free nilpotent groups is decidable by some kind of complete sets of reductions. We are especially interested in extensions of these methods to finitely presented C-A rings, and believe there is a close relationship with extensions of the method of Buchberger [1979] to Z-algebras, such as by Kapur [1983].

REFERENCES

Ballantyne, A. and Lankford, D. New decision algorithms for finitely presented commutative semigroups. Louisiana Tech U., Math. Dept., report MTP-4, May 1979; J. Comput. Math. with Appl. 7 (1981), 159-165.

Bergman, G. The diamond lemma for ring theory. Advances in Math. 29 (1978), 178-218.

Buchberger, B. A criterion for detecting unnecessary reductions in the construction of Gröbner-bases. Lecture Notes in Comp. Sci. 72 (1979), Springer-Verlag, 3-21.

Cardoza, E. Computational complexity of the word problem for commutative semigroups. M.Sc. thesis, MIT, Cambridge, MA, Aug. 1978; and MAC Tech. Memo. 67, Oct. 1975.

Fages, F. (some results of Fages presented by J. P. Jouannaud at a GE term rewriting system conference, Sept. 1983)

Jacobson, N. Lectures in Abstract Algebra, II Linear Algebra, Van Nostrand, 1953.

Kapur, D. (comparison of computer generated examples at a GE term rewriting system conference, Sept. 1983)

Lankford, D. On proving term rewriting systems are Noetherian. Louisiana Tech U., Math. Dept., report MTP-3, May 1979.

Lankford, D. and Ballantyne, A. Decision procedures for simple equational theories with commutative-associative axioms: complete sets of commutative-associative reductions. U. of Texas, Math. Dept., ATP project, report ATP-39, Aug. 1977.

Lankford, D. and Ballantyne, A. The refutation completeness of blocked permutative narrowing and resolution. Proc. Fourth Workshop on Automated Deduction, Austin, Texas, Feb. 1979, W. Joyner, ed., 168-174.

Mostowski, A. On the decidability of some problems in special class of groups. Fund. Math. LIX (1966), 123-135.

Mostowski, A. Computational algorithms for deciding some problems for nilpotent groups. Fund. Math. LIX (1966), 137-152.

Peterson, G. and Stickel, J. Complete sets of reductions for some equational theories. JACM 28, 2 (Apr. 1981), 233-264.

Smith, D. A basis algorithm for finitely generated Abelian groups. Math. Algorithms 1, 1 (Jan. 1966), 13-26.

Stickel, M. A complete unification algorithm for associative-commutative functions. Advance Papers of the Fourth International Conference on Artificial Intelligence, AI Lab, MIT, Aug. 1975, 71-76.

Reidemeister, K. Einführung in die kombinatorische Topologie, Braunschweig, 1932, 50-56.

Szmielew, W. Elementary properties of Abelian groups. Fund. Math. XLI (1954), 203-271.

Canonical Forms in Finitely Presented Algebras

Philippe Le Chenadec
INRIA, Domaine de Voluceau
Rocquencourt B.P. 105
78153 Le Chesnay Cedex, France

ABSTRACT

This paper is an overview of rewriting systems as a tool to solve word problems in usual algebras. A successful completion of an equational theory, defining a variety of algebras, induces the existence of a completion procedure for the finite presentations in this variety.

The common background of these algorithms implies a unified vision of several well-known algorithms: Thue systems, abelian group decomposition, Dehn systems for small cancellation groups, Buchberger and Bergman's algorithms, while experiments on many classical groups proove their practical efficiency despite negative decidability results.

Keywords: Finitely Presented Algebras, Word Problem, Rewriting Systems, Completion Procedures.

1. Introduction

To compute in the group G defined by the generators a, b, c, and the equations abc = aa = bbb = ccccccc = 1, the mathematician alternatively works on two levels : first, he deals with a group satisfying the associativity, right unit and inverse axioms, so that he simplifies words of the form aa^{-1}, takes the inverses, etc; second, the group G allows new substitutions such that $bcab=b^{-1}$. We explain in the present paper how this mathematician, becomed informatician as he discovered that a computer makes less mistakes than himself in tedious works, will learn to his favorite computer the elementary laws of classical algebras.

In the theory of rewriting systems, the canonical system of rules associated to a variety performs the first type of algebraic operations. Such systems are known for all classical algebras [Hul80a]. This paper presents an approach for the second step. By coding the canonical system in new data and control structures, we get completion procedures for finitely presented algebras of the variety.

After a brief presentation of the rewriting systems background , we analyse each variety from simplest ones (semigroups...) to more complex ones (rings...). In abelian case, the completion always terminates as the uniform word problem is solvable. From unsolvability results, this is false in non-commutative varieties. However, many examples of complete group presentations are given, showing the practical interest of an algorithm whose behaviour is theoretically worse. Also in group theory, we show that the basic result of the important small cancellation theory can be refined with rewriting theory. In other cases, we find as special cases Thue systems (monoids), Bergman's algorithm (associative algebras), Buchberger's one (commutative algebras); these facts prove the unifying effect of the present view.

2. REWRITING SYSTEMS and the COMPLETION PROCEDURE

This section is a succint presentation, for a detailed development, we refer to [Hue80a, Hue80b], we just define the basic notions.

2.1. Definitions

Let S be a finite set, S^* is the monoid of words on S, with the concatenation and the empty word 1; $S^+=S^*-\{1\}$ is the semigroup of non-empty words.

- An equivalence relation $\equiv$ on S^* is a congruence iff

$$\forall U,V,W,W' \in S^* \; U \equiv V \Rightarrow WUW' \equiv WVW'.$$

- The word $U \neq 1$ is subword, prefix or suffix of V iff $\exists$ W,W' s.t. $V=WUW'$, $V=UW$ or $V=WU$, U is proper iff $U \neq V$.

For computer handling, algebraic expressions are coded as terms, defined by an operator domain F graded by arity $\alpha: F \to \mathbf{N}$, and a denumerable set V of variables, disjoint from F, the algebra of terms is noted $\mathbf{T}(F,V)$ and defined on $(F \cup V)^*$:

$\forall c \in F$ s.t. $\alpha(c)=0$, $c \in \mathbf{T}(F,V)$,

$\forall v \in V$, $v \in \mathbf{T}(F,V)$,

$\forall f \in F \; \forall t_1, \ldots, t_{\alpha(f)} \in \mathbf{T}(F,V)$, $f\, t_1 \cdots t_{\alpha}(f) \in \mathbf{T}(F,V)$.

For convenience, we use infix notation and parenthesis. Subterms of a term t are t subwords that are themselves terms. The set of variables occuring in t is noted $\mathbf{V}(t)$. A ground term is a term t s.t. $\mathbf{V}(t)=\phi$, their set is a $\mathbf{T}(F,V)$-subalgebra noted $\mathbf{G}(F)$. Substitutions are term endomorphisms defined by:

$\sigma(f\, t_1 \cdots t_{\alpha(f)}) = f\, \sigma(t_1) \cdots \sigma(t_{\alpha(f)})$

$\sigma(v) = v$ almost everywhere on V.

Finally, a term congruence is an equivalence relation compatible with term structure. Let us now define an equational variety and a presentation of an algebra.

An equation is a pair of terms, the congruence generated by a set E of equations is the smallest congruence $\equiv_E$ containing all pairs $(\sigma(t),\sigma(t'))$ for (t,t') in E, and σ a substitution. The quotient set $\mathbf{T}(F,V)/\equiv_E$ is then an algebra on F and V. An algebra **A** satisfies an equation t=t' of $\mathbf{T}(F,V)$ iff $\nu(t)=\nu(t')$ in **A** for all morphisms $\nu: \mathbf{T}(F,V) \to \mathbf{A}$

Definition 2.1

The variety $V(E)$ *defined by a set of equations* E *is the class of all Algebras satisfying the equations of* E.
A presentation of an algebra **A** *in* $V(E)$ *is a pair of sets* (G,R) *s.t.*

- $G \cap F \neq 0$, $\forall g \in G$, $\alpha(g)=0$.
- R is a set of equations on $\mathbf{G}(F \cup G)$.

The algebra **A** *is the quotient* $[\mathbf{G}(F \cup G)/\equiv_E]/\equiv_R$.
A *is finitely presented iff* G *and* R *are finite.*

For example, groups are defined by the operators ., 1, $^{-1}$, and the three equations $(x.y).z=x.(y.z)$, $x.(x)^{-1}=1$ and $x.1=x$, $x,y,z\in V$; while (a,b; a.a = 1, a.b = b.a ; b.(b.b) = 1) is a presentation of $\mathbb{Z}/2\mathbb{Z} \times \mathbb{Z}/3\mathbb{Z}$.

2.2. Reductions

The basic idea to compute normal forms is to reduce terms by rules which are directed equations, noted $\lambda\to\rho$, s.t. $\mathbf{V}(\rho)\subset\mathbf{V}(\lambda)$. The precise definition of reductions needs subterm replacements : if u is a t subterm, then t[u<-u'] is the term obtained from t by replacing its subterm u by u' (two distinct occurences of a term in another one are distinct subterms).

Definition 2.2

The term t *reduces in the term* t' *by the rule* $\lambda\to\rho$ *iff there is a* t*-subterm* u *and a substitution* σ *s.t.* $\sigma(\lambda)=u$ *and* $t'=t[u\leftarrow\sigma(\rho)]$.

The transitive-reflexive closure of $\to$ is noted $\to^*$. To compute one reduction step, we must find the substitution σ, this operation is called the match of two terms. We also need the unification of two terms : $\exists\,\sigma$ s.t. $\sigma(t)=\sigma(t')$? When the equality is replaced by the associative-commutative equivalence, these operations are called AC-matching and AC-unification, in all cases complete and finite algorithms are known [Pet82a, Fag84a, Liv76a].

2.3. The Completion Algorithm

We are looking for sets of rules such that a term has a unique irreducible form. This is achieved when the rules satisfy two classical properties:

- Nœtherianity, there exist no infinite reduction chains.
- Confluence, $\forall$ m,n,p $p\to^*m$ and $p\to^*n \Rightarrow \exists q\; m\to^*q$ and $n\to^*q$.

This last condition may be tested by critical pairs when the reduction is nœtherian [Hue80a]:

Definition 2.3

If the rules $\lambda\to\rho$ *and* $\mu\to\nu$ *are such that there exists a non-variable subterm* λ' *of* λ *unificable with* μ, *under the substitution* σ, *then the equation* $(\sigma(\lambda)[\lambda'\leftarrow\sigma(\nu)],\sigma(\rho))$ *is called a critical pair (c.p.) obtained by the superposition of* $\mu\to\nu$ *on* $\lambda\to\rho$.

Thus, the unification is a crucial point to compute the critical pairs. For example, the two rules $x*(x.(a*y))\to f(x)$ and $b.z\to g(z)$ give the c.p. $(b*g(a*y),f(b))$, under the sustitution {<x,b>,<z,a*y>}. These pairs are elementary divergent points in the reductions.

We now present the *Completion Algorithm*. It tries, giving a set of equations and a reduction ordering over terms, to compute a nœtherian and confluent rewriting system [Hul80a, Knu70a]. Along this study, we modify the previous notions (term structure, matching, reductions, superposition) by examining the behaviour of the completion procedure in the various varieties. However, the control structure of this algorithm will be essentially the same:

COMPLETION ALGORITHM

Input

E :*A finite set of equations,*

< :*A reduction ordering,*

Red(W,R) :*Returns a normal form of* W *under the rewriting system* R,

Super(k,R) :*Computes all critical pairs between the rule* k *and those in* R.

Init R = ϕ ; *The set of rules.*

Step [1]
 If E $\neq \phi$ Then choose a pair (M,N) $\in$ E;
 E = E - {(M,N)}; Go To [3];

Step [2]
 If all rules are superposed
 Then stop with success;
 Else If rule k is not superposed with other rules
  Then E = Super(k,R); Go To [1];

Step [3]
 (M,N) = (Red(M,R) , Red(N,R));
 If M = N Then Go To [1];
 Else If M>N Then let λ = M, ρ = N;
  If M<N Then let λ = N, ρ = M;
  Else stop with failure;
 R = {$\lambda \to \rho$} $\cup$ R; Go To [1] ▪

In this algorithm, a crucial point is to keep the rules in R interreduced, for two reasons: on one hand, the reduction speed is greatly increased; on the other hand, it is a basic assumption in the following theorem. Also, a weight function on terms ensures that no equation will be forgotten for ever. Then, the correction of the algorithm follows from a theorem due to Huet [Hue81a] :

Theorem 2.4
Let E *be a finite set of equations, let* R_∞ *be the set of rules that appears in* R *and whose members are never reduced by the other rules, then* R_∞ *is a canonical rewriting system. More precisely,*
- if R_∞ *is finite, the algorithm stops in a finite time with* R_∞ *as final set of rules.*
- otherwise, it loops undefinitely.

The two facts that 1) R_∞ is canonical and 2) if R_∞ is finite, the algorithm stops in success are of great importance. They will be used many times along this study, providing corollaries about the termination of the completion and its correction in different varieties.

To resume, let us say that the algorithm can stop in success, loop undefinitely, or stop with failure as an equation cannot be oriented without transgressing the finite termination property. This is the case for a commutative axiom: the rule $x+y \to y+x$ gives the infinite chain $a+b \to b+a \to a+b \cdots$. To handle this stumbling-block, several tools have been developped to perform reductions modulo associative-commutative equivalence classes. For our purpose, let us say that the main difference is that to each rule $\lambda \to \rho$ whose left member leading operator f is an AC-operator, we associate an extension $f(\lambda,x) \to f(\rho,x)$ where $x \in V - \mathbf{V}(\lambda) \cup \mathbf{V}(\rho)$. Just remember that complete and finite AC-matching and AC-unification algorithms are known.

After this overview of the basic notions, we now detail the completion behaviour in algebraic structures. The order of analysis is semigroups, monoids, groups, rings, modules and algebras, by increasing structure complexity. Abelian and non-abelian cases will be studied alternatively. The hierarchical aspect of these varieties allows knowledge inheritance.

3. MONOIDS and GROUPS

Until now, we made the assumption that rewriting systems were nœtherian, in fact, given a sytem, the question of its termination is undecidable [Hue78a]. In practice, orderings over terms are used in termination proofs [Jou81a, Der82a]. In this paper, we only need some classical orderings such as the words length, or length together with a lexicographic ordering. However, we shall see some more complex examples.

3.1. Semigroups and Monoids

The associativity law of a single binary operator defines the variety of semigroups. Any of the two possible orientations gives a canonical system, say $\mathbf{A} = \{(x.y).z \to x.(y.z)\}$. According to the Def. 2.1, a semigroup S is presented by a set of generators G and a set of equations E over ground terms, S is the quotient of the free semigroup on G by the congruence generated by E. We want to solve the word problem for S: are two ground terms representing the same abstract element in S ? The completion algorithm tries to solve this problem if we give it as input the rule **A** and the equations E. The following lemma details this completion:

Lemma 3.1

i) There is an isomorphism between G^+ *and* **A***-canonical terms in* $G(\{.\} \cup G)$.

ii) If the ground term rule $\alpha \in R_\infty$*, there exists an associated rule* $\beta \in R_\infty$ *with one variable :*

$\alpha: a_1.(\ldots.(a_{n-1}.a_n)\ldots) \to b_1.(\ldots.(b_{m-1}.b_m)\ldots)$

$\beta: a_1.(\ldots.(a_{n-1}.(a_n.x)\ldots) \to b_1.(\ldots.(b_{m-1}.(b_m.x)\ldots)$

$n,m \in \mathbb{N},\ a_i,b_j \in G,\ x \in V.$

iii) Superpositions without the associative rule result from a β *left member and an* α *one s.t.* $\exists\, i \in \mathbb{N}$ s.t. $a'_{i+1-j} = a_{n+1-j},\ 1 \le j \le i \le n$.

The first proposition asserts the linear structure of irreducible terms, under the correspondence: $a_1.(a_2 \cdots (a_{n-1}.a_n)\ldots) \leftrightarrow a_1 a_2 \cdots a_{n-1} a_n$. The other ones detail the possible superpositions: in ii), the associative rule on a ground term one gives a β rule, and in ii) an α rule on a β one gives an α rule. There are no other legal superposition (remember that we superpose on a non-variable subterm). Thus, we get two consequences. First, the data-structure of words is better in this case than the term one, second, all the β rules are redundant. So that we get a new algorithm, its control structure is the completion one, the following list gives the connection between old and new keywords:

- A term is now a word.
- The word U matches the word V iff U is a prefix of V.
- The rule $U \to V$ superposes on $P \to Q$ iff $\exists$ A,B,C s.t. $B \neq 1$, AB=P and BC=U, the critical pair is then (QC,AV).
- The word W reduces in W' iff $\exists\, U \to V$, A,B s.t. W=AUB and W'=AVB.

Then, if R(V) is the irreducible form of the word V under the canonical word system R, we have from theorem 2.4 :

Corollary 3.2

If (G,E) *is a finite presentation of a semigroup* S, *and* R *is a canonical system for* S, *then the subset* Σ *of* G^+ *whose elements are all* R*-irreducible words, with the law* x *defined by*

$$\forall W,W' \in \Sigma \quad WxW' = R(WW')$$

is a S *isomorphic semigroup.*

Of course, such a finite set of rules does not always exist, the word problem for semigroups beeing unsolvable [Nov55a]. The variety of monoids has the following canonical system :

$$\mathbf{M} \begin{cases} (x.y).z \rightarrow x.(y.z) \\ x.1 \rightarrow x \\ 1.x \rightarrow x \end{cases}$$

The only difference between semigroups and monoids is that the reductions must remove the constant 1 from words, as done by the two last rules in **M** : as superpositions occur on a non variable subterm and all rules are interreduced, these two rules cannot give c.p. The sets of rules on words are a generalization of Thue Systems for which only length reducing replacements are allowed. Many results can be found about such systems in [Boo82a, Boo82b] or in [Coc76a], especially in connection with language theory.

4. GROUPS

The study of non abelian groups is done in two times. The completion of groups is observed as in the previous cases. But a new fact appears: rules are symmetrized, this operation is closely related with the canonical system for groups, the second part of the present section is an analysis of this symmetrization process. We first need some technical definitions.

- Let G be a set of generators, G^{-1} is a copy of G whose elements are noted a^{-1}, $a \in G$. If $b=a^{-1} \in G^{-1}$ then b^{-1} is the generator a. If $U=u_1 \cdots u_n \in (G \cup G^{-1})^*$, $u_i \in G \cup G^{-1}$, then $U^{-1}=u_n^{-1} \cdots u_1^{-1}$, with $1^{-1}=1$.
- The word $u_i \cdots u_n u_1 \cdots u_{i-1}$ is called a cyclic permutation of U, $1 \le i \le n$.
- The length of U, $|U|$, is the number of generators in U, with $|1|=0$.

4.1. Completion

The famous canonical system for groups is the following one :

$$\mathbf{G} \begin{cases} x.1 \rightarrow x & 1.x \rightarrow x \\ x.x^{-1} \rightarrow 1 & x^{-1}.x \rightarrow 1 \\ x.(x^{-1}.y) \rightarrow y & x^{-1}.(x.y) \rightarrow y \\ 1^{-1} \rightarrow 1 & \\ (x.y).z \rightarrow x.(y.z) & \\ (x.y)^{-1} \rightarrow y^{-1}.x^{-1} & \\ (x^{-1})^{-1} \rightarrow x & \end{cases}$$

By the rules **G**3, **G**4, **G**8 and **G**9, not all words of $(G \cup G^{-1})^*$ are **G**-irreducible. Also, we need the following function, it computes the **G**-canonical forms of words:

let **G**(W) =

Case W of
1V, $aa^{-1}V$ or $a^{-1}aV$ Then **G**(V);
aV Then
Case **G**(V) of
1 Then a;
$a^{-1}U$ Then U;
Otherwise a**G**(V);
Otherwise W ▪

This function is now used to define the group reduction :

$$W \to W' \text{ iff } \exists\, U \to V,\ A,\ B \text{ s.t. } W=AUB \text{ and } W'=\mathbf{G}(AVB)$$

The function **G** computes the well-known canonical form in free groups, i.e. the reduction generated by the rules $aa^{-1} \to 1$, $a \in G \cup G^{-1}$, or by the canonical system **G**. Of course, we could merge these rules with those generated by the completion of a given group, but the previous reduction runs faster, and this fact is essential in an implementation.

We then have the equivalent of lemma 3.1, whose details are left out. The point is that new critical pairs are computed between ground term rules and **G** ones, expressed as words they give the :

Lemma 4.1
To the rule $a_1 \cdots a_n \to b_1 \cdots b_m$, $a_i, b_j \in G \cup G^{-1}$ *the completion associates the following critical pairs :*

$(a_1 \cdots a_{n-1}\ ,\ b_1 \cdots b_m a_n^{-1})$,

$(a_2 \cdots a_n\ ,\ a_1^{-1} b_1 \cdots b_m)$,

$(a_n^{-1} \cdots a_1^{-1}\ ,\ b_m^{-1} \cdots b_1^{-1})$.

As these critical pairs are closed to the group variety, we may call them canonical. Now, the group completion is completely defined. The control structure is modified, we must add one step to compute the canonical pairs of the new rules. A good heuristic is to give them the highest priority in the set E of waiting equations, and to generate them between the two first steps. The reader can now think to the corollary 4.2 versus 3.2, just add the inverse operator :

$$\text{If } W \in \Sigma \text{ then } W^{\sim 1} = R(W^{-1}).$$

If we delete the superposition step in completion, we obtain the symmetrization algorithm, whose name will become obvious later on. As in the monoid case, the group word problem is unsolvable [Nov55a] ,so the completion may not terminate. However, it halts on finite groups, as noted by Métivier [Met3.a]:

Proposition 4.2
Given a presentation of a finite group **G**, *the completion algorithm always halts in success.*

Suppose the completion does not halt, by theorem 2.4, it computes an infinite canonical system R_∞. As **G** is finite and the rules right members are R_∞-irreducible, there exists a word W, right member of infinitely many rules. Let $\Sigma=(W_i)_{i \in N}$ be the sequence of associated left members. As the rules are interreduced, W_i is not subword of W_j, $j \neq i$, moreover, all proper subwords of W_i are R_∞-irreducible. Thus, we can extract from Σ a subsequence Σ' strictly length increasing. Let $W_i=a_iV_i$, $W_i \in \Sigma'$, $a_i \in G \cup G^{-1}$, the set of all V_i is an infinite set of R_∞-

irreducible words, contradicting the finiteness of **G**. The algorithm must halt ▪

Now, the question of complexity may be asked about the completion of finite groups. Our experience shows that complex situations exist: the symmetric groups S_n have presentations whose completion gives a set of rules whose cardinality is in $O(|S_n|) \sim n!$. Moreover, we did not succeed in completing the alternating groups A_n. However, the proof of prop. 4.2 may be used to give an upper bound to $|R|$, the number of rules. Let $(\rho_k)_{k=1,\dots,n}$ be an enumeration of the right members, $\rho_k \neq \rho_l$ if $k \neq l$, then $n \leq |\mathbf{G}|$. If $L_k = \{\lambda_i \,/\, \lambda_i \to \rho_k \in R\}$, then $\sum_k |L_k| = |R|$. Let $M_k = \{\mu_i \,/\, \exists a_i \in G \cup G^{-1} \text{ s.t. } a_i\mu_i = \lambda_i \in L_k\}$, we may suppose $\mu_i \neq \mu_j$ if $i \neq j$, otherwise $a_i\mu_i =_{\mathbf{G}} a_j\mu_i =_{\mathbf{G}} \rho_k$ implies $a_i =_{\mathbf{G}} a_j$ while $a_i \neq a_j$ as the rules are interreduced. But this means that a generator was redundant. Eliminating such cases, we get $|L_k| = |M_k|$, as the words in M_k are **G**-irreducible, we have $|M_k| \leq |\mathbf{G}|$. Thus,

$$|R| = \sum_{k=1}^{n} |L_k| \leq \sum_{k=1}^{n} |\mathbf{G}| \leq |\mathbf{G}|^2$$

And we get $|R| \leq |\mathbf{G}|^2$. But this upper bound may be improved. Fix one $\mu_i \in M_k$, then μ_i appears at most in $|G \cup G^{-1}| = 2|G|$ sets M_l, otherwise two distinct rules would have the same left member. As the number of μ_i is bounded by $|\mathbf{G}|$, we have $\sum_k |M_k| \leq |G \cup G^{-1}|.|\mathbf{G}|$. In other words, $|R| \leq 2|G|.|\mathbf{G}|$, as in practice the number of generators is very small, this inequality tells us that we get the multiplication table of the group **G** in $O(|\mathbf{G}|)$ rules. However, this bound is still large as one can see in § 7.

For infinite groups, we would like some criterions restricting the number of superpositions. In the following section, we show that the symmetrization is in some cases powerful enough to solve the word problem.

4.2. Symmetrization

The first step is to observe that, in groups, the word problem is equivalent to the question : is a given word belonging to the congruence class of identity, for $W = W' \iff WW'^{-1} = 1$. Then, we can search for hypothesis s.t. the symmetrization answers to the identity class membership question. However, this study needs some words combinatorics out of the present paper scope, for a detailed presentation see [Che83a]. First, some definitions:

- A cyclically reduced word (c.r.w.) is a word whose all cyclic permutations are **G**-reduced.
- A relation is a c.r.w. congruent to identity.
- A symmetrized set of relations $R \in E$ (s.s.r.) is a set of words including the relations R, the inverses R^{-1}, and all their cyclic permutations.
- A piece is a common prefix of two distinct relations.
- A group presentation is henceforth a set of generators and a set of defining relations to which we may associate a s.s.r. A presentation satisfies C(n), $n \in \mathbf{N}$, iff every piece P of the defining relations s.s.r. is s.t. $|P| < \frac{1}{n}|W|$, the piece P belonging to the relation W.

For example, $(a,b,c\ ;\ abc, a^3, b^5, c^7)$ is a presentation of a polyhedral group; it satisfies C(2), but not C(3): a is a piece between the two first relations and $|a| = \frac{|abc|}{3}$.

The first step consists in a detailed analysis of the symmetrization algorithm under the hypothesis of length decreasing rules :

Hypothesis 1 : If $U \to V$ is a rule, then $|U| \geq |V|$.

The reason for this hypothesis lies in the fact that the present study is a detailed analysis of successive reductions, which is impossible when the words length increases by the replacements. Then, the next proposition summarizes the symmetrization's behaviour:

Proposition 4.3
Given a group presentation G, *the symmetrization always terminates.*
Let Γ *be the set of rules computed, then if* $U \to V \in \Gamma$ *:*
- *The words* UV^{-1}, $U^{-1}V$, VU^{-1} *and* $V^{-1}U$ *are relations.*
- *If* $|U|+|V|=2p+1$ *then* $|U|=p+1, |V|=p$.
- *If* $|U|+|V|=2p$ *then* $|U|=p=|V|$ *or* $|U|=p+1, |V|=p-1$.

If G satisfies C(2) *then*
- *The set* $S=\{UV^{-1}, VU^{-1} / U \to V \in \Gamma\}$ *is the s.s.r. of defining relations.*
- *If* $PQ \in S$ *and* $|P|>|Q|$ *then* $P \overset{*}{\to}_\Gamma Q$.
- *A non confluent critical pair results from a superposition on a piece between two relations.*

Let us remember that the algorithm takes a nœtherian ordering as input, so that the reductions always terminates, it is of course always possible to find such an ordering: for example a lexicographical one based on length and a total ordering of generators. Thus, as the reduction is nœtherian and only a finite number of words can appear in rules, it follows that the symmetrization halts in a finite number of steps. The remaining assertions follows from elementary manipulations on words. Now, the name of symmetrization appears clearly, this algorithm brokes the defining relations of a finite group presentation by computing what is called in literature a s.s.r. [Lyn77a]. As example, the fundamental group of a double torus, defined by $(A,B,C,D;ABCDA^{-1}B^{-1}C^{-1}D^{-1})$, gives the following Γ set:

$$\begin{cases} DCBA \to ABCD \\ BCDA^{-1} \to A^{-1}DCB \\ B^{-1}A^{-1}DC \to CDA^{-1}B^{-1} \\ DA^{-1}B^{-1}C^{-1} \to C^{-1}B^{-1}A^{-1}D \\ D^{-1}C^{-1}B^{-1}A^{-1} \to A^{-1}B^{-1}C^{-1}D^{-1} \\ B^{-1}C^{-1}D^{-1}A \to AD^{-1}C^{-1}B^{-1} \\ BAD^{-1}C^{-1} \to C^{-1}D^{-1}AB \\ D^{-1}ABC \to CBAD^{-1} \end{cases}$$

In Prop. 4.3, the two words in the definition of S are necessary : in the previous example, the defining relation is of type VU^{-1} by rule 1, but not of type UV^{-1}, while $BCDA^{-1}B^{-1}C^{-1}D^{-1}A$ is of second type by the rule 2, but not of first type. Note that we have made the assumption that G satisfies C(2). As for the length hypothesis, this is necessary in order to localize the possible reductions of a word. The second step is an analysis of the critical pairs associated to a piece. Thus, we have two rules whose left members have the non-empty word B as prefix and suffix, B is a subword of a piece, say ABC, between the distinct relations computed from these rules according to Prop. 4.3, that is, we have the configuration **C**:

$$\mathbf{C} \begin{cases} \mathbf{k}: \ \alpha AB \to \beta C^{-1} \\ \mathbf{l}: \ BC\gamma \to A^{-1}\delta \end{cases} \quad B \neq 1 \text{ and } \beta^{-1}\alpha \neq \gamma\delta^{-1}$$

The superposition of rules **k** and **l** in B gives the c.p. $(\alpha AA^{-1}\delta, \beta C^{-1}C\gamma)$, and after G-reductions, $(\alpha\delta, \beta\gamma)$. Thus, we get an elementary point of divergence in the reductions. But this is false in reductions of words congruent to identity : the word $\alpha\delta\gamma^{-1}\beta^{-1}$ is Γ-reducible under hypothesis C(2) for in this case $|\delta\gamma^{-1}| > |ABC|$ which proves the Γ-reducibility by Prop. 4.3 . In fact, the two words of the c.p., although beeing **G**-irreducible, may be Γ-reducible, and we do not want this possibility: in this case we cannot assert anything about the c.p. Let us examine this eventuality. As the rules are interreduced, there exists a rule $\mathbf{m}:\mu\nu\rightarrow\tau$ and words β',γ' with $\beta=\beta'\mu$ and $\gamma=\nu\gamma'$. Now, suppose that the word μ is not a piece between rules **k** and **m**, we can write the identity in $(G\cup G^{-1})^*$ of the associated relations : $C^{-1}B^{-1}A^{-1}\alpha^{-1}\beta'\equiv\nu\tau^{-1}$. But this identity implies that the left member in **l** would be **G**-reducible in the subword $C\nu$, contradicting the irreducibility of rules members. Thus the words μ and ν are pieces. Clearly, this is impossible if we suppose the :

Hypothesis 2: All presentations satisfy C(4).

We have just prove the

Lemma 4.4
Under Hyp. 2, all critical pairs are, eventually after some **G***-reductions, in* **G** Γ*-normal forms.*

We gave the proof of this lemma because it provides a good and succint example of what are the demonstrations of the next lemmata : an assumption on existence of a reduction gives a configuration impossible under our precise hypotheses or under the general features of a rewriting sytem, the identification of relations beeing constantly used. Indeed, we know the canonical form of the c.p. The main part of this second step is to examine carefully the reduction of a c.p. in a *context*:

Definition 4.5
A context is a **G** Γ*-irreducible word* μM *(left context) or* $T\tau$ *(right context) that reduces a member of a c.p. when concatenated to it,* M *(resp* T*) giving an eventually* **G***-reduction, and* μ *(resp.* τ*), left member proper prefix (resp. suffix) of a* Γ *rule, giving a* Γ*-reduction.*

For example, if $\mu\lambda\rightarrow\rho\in\Gamma$, we have the following reductions:

$$\mu M\beta\gamma \xrightarrow{*}_{G} \mu\lambda X \rightarrow_{\Gamma} \rho X$$

The words $\beta\gamma$ and $\alpha\delta$ beeing symmetric, we restrict our attention to the reductions of $\beta\gamma$, searching on one hand confluence conditions in the reductions of $\mu M\beta\gamma$ and $\mu M\alpha\delta$ (resp. $\beta\gamma T\tau$ and $\alpha\delta T\tau$). Then, we get two configurations:

$$\mathbf{L}\left\{\begin{array}{lcl} \mathbf{k}: \alpha AB & \rightarrow & M^{-1}\mu_1\nu_1^{-1}\beta' C^{-1} \\ \mathbf{l}: BC\sigma\gamma' & \rightarrow & A^{-1}\delta \\ \mathbf{m}: \mu\mu_1 & \rightarrow & \nu\rho\nu_1 \\ \mathbf{n}: \rho\beta'\sigma & \rightarrow & \tau \\ \textbf{pieces}: & \beta', & ABC,\ \mu_1\nu_1^{-1},\ \sigma,\ \rho. \end{array}\right. \qquad \mathbf{R}\left\{\begin{array}{lcl} \mathbf{k}: \alpha AB & \rightarrow & \beta'\mu C^{-1} \\ \mathbf{l}: BC\gamma'\sigma_1^{-1}\tau_1 T^{-1} & \rightarrow & A^{-1}\delta \\ \mathbf{m}: \tau_1\tau & \rightarrow & \sigma_1\nu\sigma \\ \mathbf{n}: \mu\gamma'\nu & \rightarrow & \rho \\ \textbf{pieces}: \gamma', & ABC, & \mu,\ \sigma_1^{-1}\tau_1,\ \nu. \end{array}\right.$$

Their signification appears in the following proposition:

Proposition 4.6

In the reduction of $\mu M\beta\gamma$ (resp. $\beta\gamma T\tau$), we have

Case 1: β (resp. γ) is absorbed by M (resp. T) in a G-reduction, then, in the context, the c.p. conflues.

*Case 2: There is no piece between rule **k** (resp. **l**) and **m** (resp **n**), then, in the context, the c.p. conflues.*

*Case 3: The reduction of $\mu M\beta\gamma$ (resp. $\beta\gamma T\tau$) gives a word $\nu\beta'\gamma$ (resp. $\beta\gamma'\sigma$) G-irreducible, Γ-reducible only by the rule **n** in the configuration **L** (resp. **R**).*

The interesting facts of this proposition are 1) many pieces exist in the configurations, 2) in case of non confluence, the words β' or γ', beeing pieces, are not equal to 1, in other words the second component of the original c.p. has not been affected by the reductions. Under this tricky result lies in fact the heart of what is called the small cancellation theory, as we are now going to explain it. The last step in our study needs two assumptions:

Hypothesis 3: The set Γ is s.t. the rules **n** in **R** and **L** does not exist.

Assumption 4: $\exists$ W,W' s.t. $W \overset{*}{\to}_{G\Gamma} 1$, $W \overset{*}{\to}_{G\Gamma} W'$ and $W' \neq 1$ is **G**Γ–irreducible.

Nœtherianity and finiteness of Γ implies that the set

$$\Theta(W)=\{ Z \ / \ W \overset{*}{\to}_{G\Gamma} Z,\ \exists\, Z_1,Z_2 \text{ s.t. } Z \to_\Gamma Z_i,\ 1 \notin \mathrm{Irred}(Z_2),\ \mathrm{Irred}(Z_1)=\{1\}\}$$

where Irred(V) is the set of all **G**Γ-irreducible forms of V, is non-empty. A minimal word Y in $\Theta(W)$ with respect to the reduction ordering is then of the form $U\alpha ABC\beta V$ with U,V **G**Γ-irreducible and $U=\mu_n M_n \cdots \mu_1 M_1$, $V=T_1\tau_1 \cdots T_m\tau_m$, where the $\mu_i M_i$ and $T_j\tau_j$ are those from Def. 4.5. This word Y is a minimal point of strong divergence in the reductions from W, strong because our choice of $\Theta(W)$ is such that after two reductions of a word Z in $\Theta(W)$ we get two reduction dags totally disconnected, and this divergence implies that the reductions are done in a non-confluent c.p., that is on a piece. We have the following scheme:

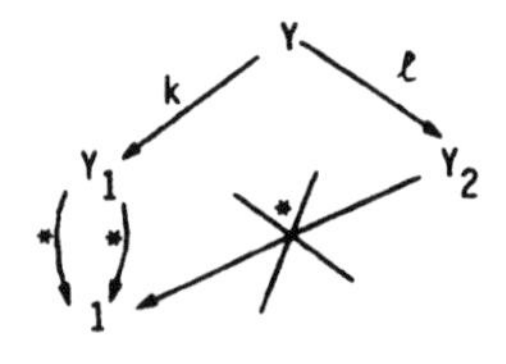

Y_1 reduces only on 1

Y_2 can't reduce on 1

<u>i.e. there is no confluence of Y_1 and Y_2</u>

Then, as $\mathrm{Irred}(Y_1) = \{1\}$, we may choose a precise chain of reductions starting at Y_1. Under the hypothesis 3, and because of non-confluence of Y_1 an Y_2, the Case 3 in prop. 4.6 details the first reductions of Y_1: $Y_1 \overset{*}{\to} (\prod_{n}^{2}[\mu_i M_i)\nu\beta'\gamma'\sigma(\prod_{2}^{m}[T_j\tau_j)$.

We would like to prove the falsity of assumption 4, we are done if we could prove that Y_1 reduces to a non-empty word **G**Γ-irreducible. The shorter way to achieve this goal is to reduce Y_1 in such a manner that appears a propagation of the irreducible words β'_i and γ'_j of Prop. 4.6, according to the structure of U and V. And in fact, this propagation is possible as explained above: $Y_1 \overset{*}{\to}_{G\Gamma} \beta'_{n-1} \cdots \beta'_2\gamma'_2 \cdots \gamma'_{m-1} = Y_3$. Is this last word reducible? In looking for a reduction of $\beta'_1\gamma'_1$, we get the following last configurations:

$$\mathbf{LR}_Y^1 \begin{cases} \mathbf{k}: \alpha AB & \to M^{-1}\mu_1\nu_1^{-1}\beta' C^{-1} \\ \mathbf{l}: BC\varepsilon\gamma''\sigma_1^{-1}\tau_1 T^{-1} & \to A^{-1}\delta \\ \mathbf{m}: \mu\mu_1 & \to \nu\nu_1 \\ \mathbf{n}: \tau_1\tau & \to \sigma_1\sigma \\ \mathbf{p}: \xi_1\beta'\varepsilon\xi_2 & \to \omega \\ \mathbf{pieces}: ABC,\ \beta', & \varepsilon,\ \mu_1\nu_1^{-1},\ \sigma_1^{-1}\tau_1. \end{cases}$$

$$\mathbf{LR}_Y^2 \begin{cases} \mathbf{k}: \alpha AB & \to M^{-1}\mu_1\nu_1^{-1}\beta''\varepsilon C^{-1} \\ \mathbf{l}: BC\gamma'\sigma_1^{-1}\tau_1 T^{-1} & \to A^{-1}\delta \\ \mathbf{m}: \mu\mu_1 & \to \nu\nu_1 \\ \mathbf{n}: \tau_1\tau & \to \sigma_1\sigma \\ \mathbf{q}: \xi_1\varepsilon\gamma'\xi_2 & \to \omega \\ \mathbf{pieces}: ABC,\ \gamma', & \varepsilon,\ \mu_1\nu_1^{-1},\ \sigma_1^{-1}\tau_1. \end{cases}$$

And the corresponding $\mathbf{LL}_Y^{1,2}$, $\mathbf{RR}_Y^{1,2}$ for $\beta'_i\beta'_{i-1}$ and $\gamma'_{j-1}\gamma'_j$. So that, we assert the

Hypothesis 5: Γ is s.t. for all Y, the configurations $\mathbf{LL}_Y^{1,2}$, $\mathbf{LR}_Y^{1,2}$, and $\mathbf{RR}_Y^{1,2}$ do not exist.

Consequently, Y_1 reduces in $Y_3 \neq 1$ and $\mathbf{G}\Gamma$-irreducible, but this contradicts the definition of Y, hypothesis 3 & 5 implies the falsity of assumption 4, and:

$$W \xrightarrow{*}_{G\Gamma} 1 \implies \mathrm{Irred}(W) = \{1\}$$

The conclusion follows from an elementary result from group theory, direct consequence of the normal closure definition:

If W belongs to the congruence class of the empty word, then there exists W' of the form $\prod_1^k T_i R_i T_i^{-1}$ where $R_i \in S$, the s.s.r. of defining relations, obtained from W by insertion and deletion of subwords aa^{-1} or $a^{-1}a, a \in G$, i.e. by G-reductions. Obiously, $W' \xrightarrow{*}_{G\Gamma} 1$, and the following scheme illustrates the last step of our study :

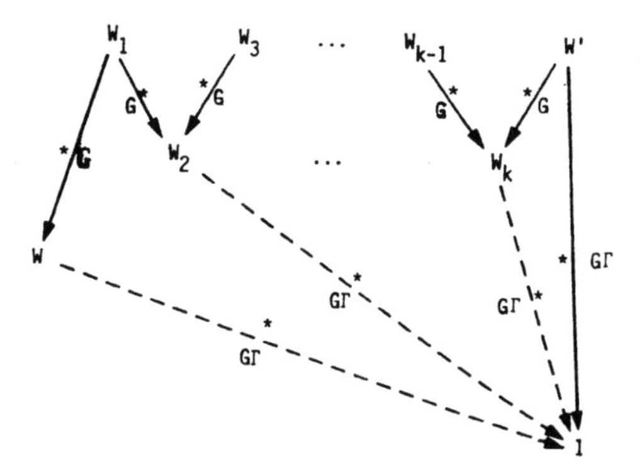

Theorem 4.7
If for a presentation of a group G *satisfying C(4), there exists a set of rules* Γ *satisfying Hyp.3 & 5, then*

$$W =_G 1 \Rightarrow W \to^* 1.$$

Thus, the nice properties of Γ (cf prop. 4.3) allows, in the study of a restricted confluence over 1, the localization of a strong divergence and then, exhibits the configurations that may prevent this confluence. However, the Hyp. 5 is quantified over all Y words. A finite condition is used in the small cancellation theory which uses, together with the condition C(n), the following one:

T: $\forall R_1, R_2, R_3 \in S$ s.t. R_i, R_{i+1} and R_3, R_1 are not inverses,

one of the products R_1R_2, R_2R_3 and R_3R_1 is **G**–irreducible.

Corollary 4.8 (Dehn)
If a group satisfies C(6), *or* C(4) *and* T, *then it has a solvable word problem.*

In the case C(6), Hyp.3 is satisfied because the rules **n** cannot exist, its left member, concatenation of three pieces, would be shorter than its right one. On the other hand, C(4) is Hyp.2, and for example the configuration **L** gives the triple $(\beta'^{-1}\nu_1\mu_1^{-1}M\alpha ABC\ ,\ C^{-1}B^{-1}A^{-1}\delta\gamma'^{-1}\sigma^{-1}\ ,\ \sigma\tau^{-1}\rho\beta')$ in contradiction with T, so that Hyp.3 is satisfied in both cases. $\mathbf{LR}^2_Y$ gives the triple $(\beta'^{-1}\nu_1\mu_1^{-1}M\alpha ABC,\ C^{-1}B^{-1}A^{-1}\delta T\tau_1^{-1}\sigma_1\gamma''^{-1}\varepsilon^{-1},\ \varepsilon\xi_2\omega^{-1}\xi_1\beta')$, impossible if T holds; while C(6) implies that $M\alpha \to^*_{G\Gamma} \mu'\nu'^{-1}\beta' C^{-1}B^{-1}A^{-1}$, i.e. the words Y_1 and Y_2 conflues: under C(6) the configuration **LR** and others may exist, but are not associated to a word Y, Hyp. 5 holds in both cases.

The group (A,B,C ; ABC=CBA) gives a counter-example and an example. Indeed, it gives two Γ sets. The first one is:

$$\Gamma \left\{ \begin{array}{llll} CBA & \to ABC & & \\ A^{-1}CB & \to BCA^{-1} & ABCA^{-1} & \to CB \\ B^{-1}A^{-1}C & \to CA^{-1}B^{-1} & BCA^{-1}B^{-1} & \to A^{-1}C \\ C^{-1}B^{-1}A^{-1} & \to A^{-1}B^{-1}C^{-1} & CA^{-1}B^{-1}C^{-1} & \to B^{-1}A^{-1} \\ B^{-1}C^{-1}A & \to AC^{-1}B^{-1} & BAC^{-1}B^{-1} & \to C^{-1}A \\ C^{-1}AB & \to BAC^{-1} & & \end{array} \right.$$

Then the rules 2,8,7,9 and 8 in this order are a $\mathbf{LR}_Y$ configuration with

$Y = ABABABCA^{-1}CBCA^{-1}B^{-1}A^{-1}B^{-1}C^{-1}B^{-1}C^{-1}B^{-1}C^{-1}$ and

$Irred(Y) = \{1\ ,\ ABABABCBCA^{-1}CA^{-1}B^{-1}A^{-1}B^{-1}C^{-1}B^{-1}C^{-1}B^{-1}C^{-1}\}$.

Thus this Γ-set does not solve the word problem. The reader may check that it has four **L** and four **R** configurations, equivalent modulo a permutation of the letters; however, they do not provide non-confluent words. The other Γ set is:

$$\Gamma \left\{ \begin{array}{ll} ABC & \to CBA \\ A^{-1}CB & \to BCA^{-1} \\ CA^{-1}B^{-1} & \to B^{-1}A^{-1}C \\ C^{-1}B^{-1}A^{-1} & \to A^{-1}B^{-1}C^{-1} \\ B^{-1}C^{-1}A & \to AC^{-1}B^{-1} \\ BAC^{-1} & \to C^{-1}AB \end{array} \right.$$

This symmetrized set is in fact canonical, consequently it satisfies the theorem 4.7. This group is not C(6), but C(5), and the triple

$(AC^{-1}B^{-1}A^{-1}CB, B^{-1}A^{-1}CBAC^{-1}, CBAC^{-1}B^{-1}A^{-1})$ shows that T does not hold. Thus, this group is not a small cancellation one, while the present proof applies to it.

Historically, this corollary appears first with Dehn [Deh12a], about the fundamental surface groups. The algorithm used to solve the word problem just replaces every subword greater than half a member of a s.s.r. by the remainding relation. Usual proofs uses graphs associated to words and Euler's Formula [Lyn77a]. Greendlinger [Gre60a] proved it by the only use of G-reductions. Bücken [Buc79a] initiated the present approach, while our work details the symmetrization and, by localization, exhibits the basic configurations.

The main advantage of the present proof is, by localization of a strong divergence in reductions, to give structural conditions more general than the usual metric ones.

5. ABELIAN CASE

5.1. Semigroups and Monoids

In non-commutative cases, we examined the normal forms under the variety canonical system. Now, this is achieved by the associative-commutative equivalence of terms. We pointed out in § 2 that the commutative law could not be oriented into a rule without loosing the nœtherianity. But a coherent theory of AC rewriting and completion has been developped [Liv76a, Sti81a, Hul80a, Fag84a]. All this background is assumed in this section, we give the main applications in finitely presented commutative algebras.

A term of an abelian semigroup or monoid on the set $G=\{g_1, \ldots, g_p\}$ is, in accordance with the previously mentionned theory and with the usual notation is the flat structure $g_{i_1}+\cdots+g_{i_j}$, whose *one* AC-equivalent term is $t = g_{i_1}+(g_{i_2}+(\cdots+g_{i_j})\cdots)$. As we did in the non abelian case, we now choose a new data structure. In the AC-unification, an ordering of the flat structure is taken from a total ordering of the operator domain. Such a total ordering on G gives us the new structure : we represent terms by abelian words $(a_1, \ldots, a_p)$ in $\mathbb{N}^p$, a_j beeing the occurence number of g_j in t. In other words, we choose for the free semigroup on G (resp. monoid) the concrete representation $\mathbb{N}^p-\{0\}$ (resp. $\mathbb{N}^p$). This deduction may bee juged too formal as we find the trivial model of free semigroups, but we think that the present paper justifies this technical and formal work. We also need some definitions :

- Abelian words addition, component by component, deduced from the AC-term structure.
- The partial order $\ll$: $M=(a_1, \ldots, a_p) \ll N=(b_1, \ldots, b_p)$ iff $a_i < b_i$, $i=1,\ldots,p$, where $<$ is the ordering on integers.
- $\max(M,N) = (\max(a_1,b_1), \ldots, \max(a_p,b_p))$ (resp. min).

The ordering $\ll$ is nœtherian and compatible with the addition. The embedding rules now replace the β-rules of § 2, their variables allow the subterm reduction. Once more, the completion in monoids has the usual control structure with the following definitions:

- A term is an abelian word.
- The word M matches N iff $M \ll N$, the matching is $N-M = (b_1-a_1, \ldots, b_p-a_p)$. M reduces in N by the rule $\lambda\to\rho$ iff $\lambda \ll M$, and then $N = (M-\lambda)+\rho$.
- The rules $\lambda\to\rho$ and $\sigma\to\tau$ superpose iff $\min(\lambda,\sigma) \neq 0$, and we may consider only one critical pair: $P = (\max(\lambda,\sigma)-\rho, \max(\lambda,\sigma)-\tau)$, the others beeing confluent if P is introduced as a rule, i.e. they are less general.

In the variety of semigroups, we have the following result [Cli67a], obtained by induction using $\ll$:

Theorem 5.1
Every finitely generated commutative semigroup is finitely presented.

In other words, in an infinite set of abelian words of finite length, the set of $\ll$-minimal words is finite. Then, observing that all abelian words introduced by superposition are not greater than the common max of the defining equations, we have [Bal79a]:

Proposition 5.2
On a finite presentation, the semigroup completion terminates.
On a finitely generated semigroup, defined by an infinite set of equations, the completion computes in an infinite time a finite canonical set of rules.

This result is of course true for the monoids. In other words, we find the well-known fact that the uniform word problem is solvable in these varieties.

5.2. Groups

The canonical system for abelian groups is

$$\mathbf{GA}\begin{cases} x+0 & \to x \\ x+(-x) & \to 0 \\ x+(-y)+y & \to x \\ -0 & \to 0 \\ -(-x) & \to x \\ -(x+y) & \to (-x)+(-y) \end{cases}$$

The previous structure of abelian words is still used. But now, with the the inverse operator, we work in the free abelian group $\mathbf{Z}^p$. The rule **GA3**, embedding of **GA2**, may be used via superpositions to control the behaviour of the completion in abelian groups. For example, the rule $a+b \to c$ superposed on **GA3** with the substitution $\sigma = \{(x,a),(y,b)\}$ gives the c.p. $(a, c-b)$, so that we may restrict the left members to only one generator, and consequently, there is no reason to keep the waiting equations as a list E of abelian word pairs rather than a list of abelian words, the difference between the pair members. Now, in step 3 of the completion, the choice of a word to create a new rule **k** is based on the minimal coefficient a_i in absolute value among those in E:

$$\mathbf{k} \ : \ |a_i|\ g_i \ \to \ \sum_{j \neq i} \varepsilon a_j g_j \qquad \varepsilon = \text{opposite sign of } a_i.$$

Then we look for the next minimum, but this minimum will decrease if all waiting words are reduced by the rule **k**. Also, all words in E are reduced. If we loop on this operation, there exist a step such that the last generator introduced as left member, say g_j, has disappeared from E. Note that when a left member is reducible, the associate rule is reintroduced as a word in E. Then we get two cases:

- $|a_j|=1$, the generator g_j was redundant.
- $|a_j| \neq 1$, again here we have two cases. The simpler one occurs when the right member is null, that is, the generator g_j generates the cyclic group $\mathbf{Z}/\,|a_j|\mathbf{Z}$. In the last one, with **GA3**, we create the rule:

$$\mathbf{k}' \ : \ |a_j|(g_j - \sum_{i \neq j} c_i g_i) \ \to \ \sum_{i \neq j} d_i g_i$$

with $b_i = c_i * \varepsilon a_j + d_i$, the integer division of right coefficients by the left one.

This operation implies the interesting fact that all d_i are now smaller than a_j in absolute value, so that the minimum in E will decrease if the rule **k'** is of the same type than the other ones, i.e. if we introduce a new generator g'_j:

$$1 : g_j \rightarrow g'_j + \sum_{i \neq j} c_i g_i$$

Then the rules **k** and **k'** are reduced, they give the same word reintroduced in E, and the minimum decreases.

The algorithm halts necessarily : when a new rule is created, the minimum in E decreases strictly or a generator disappears from E. This study shows that, to a given generator g, original or not, only three cases may happen:

1) It does not appear in left members, it generates an infinite cyclic group as $n*g$ is in canonical form, $\forall n \in \mathbf{N}$. Let G_1 be the set of all these generators.
2) It appears in a left member (such a rule is unique), with a coefficient equal to one. Then it was dependent from the remaining generators.
3) It appears in a left member with a coefficient $c>1$. The right member is then null and the generator g generates a finite cyclic groupc $\mathbf{Z}/c\mathbf{Z}$. Let G_2 be the set of those generators.

We have proved the

Theorem 5.3
The group G is the direct product of $|G_1|$ infinite cyclic groups, and $|G_2|$ finite groups whose cardinals are given by the rewriting system R, wich reduces a word given in the primitive generators into this cyclic decomposition of G.

The reader who is familiar with abelian groups has noticed that the reductions in E and the introduction of a new generator correspond respectively to the elementary row and column operations on the matrix E. The interest of the present proof of abelian group decomposition is that the set of rules R keeps trace of what usually appears as some row and column magic on matrices. In fact, the completion halts without superpositions between ground rules.

To conclude this section, let us say that the abelian monoid completion was first studied by J.M. Hullot and G. Huet (unpublished manuscript), and D.A. Smith [Smi66a] published a detailed algorithm for abelian group decomposition from which the present section is inspired.

6. RINGS, MODULES and ALGEBRAS

6.1. Rings

The canonical system for non-abelian rings is [Hul80a]:

$$\mathbf{AU}\left\{\begin{array}{llll}
x+0 & \rightarrow x & & \\
(-x)+x & \rightarrow 0 & x+(-y)+y & \rightarrow x \\
1.x & \rightarrow x & (x.y).z & \rightarrow x.(y.z) \\
x.1 & \rightarrow x & x.(y+z) & \rightarrow (x.y)+(x.z) \\
-0 & \rightarrow 0 & (x+y).z & \rightarrow (x.z)+(y.z) \\
-(-x) & \rightarrow x & -(x+y) & \rightarrow (-x)+(-y) \\
x.0 & \rightarrow 0 & x.(-y) & \rightarrow -(x.y) \\
0.x & \rightarrow 0 & (-x).y & \rightarrow -(x.y)
\end{array}\right.$$

The rules are partitionned in two sets, the right ones are needed in the symmetrization, while the left ones only define the new data-structure. The **AU**-canonical forms are the usual ones for polynoms, i.e. the orientation of the distributive law expresses the polynomial linearization. Thus, the data-structure is the composition of the two previous ones, first we have an abelian group, such that the first level of structure will be abelian words, second, this abelian structure is infinitely generated by the words from G^*. The order on G previously used in abelian words is now replaced by the lexicographic ordering on monomials from G^*. We can speak of the heading monomial. From the addition structure, we get a symmetrization, but still more complex than the abelian group one. However we can choose the heading monomial as left member of the rules. And, still with the rule **AU9**, the symmetrization gives the rules:

$$\mathbf{k} : (p+1)m \rightarrow (-p)m+\sum_j \beta_j m_j \ , \ p \in \mathbf{N}$$

$$\mathbf{l} : (p+1)m \rightarrow (-p+1)m+\sum_j \beta_j m_j \ , \ p \in \mathbf{N}^*.$$

according to the parity of m's coefficient. Then we have two cases for the embedding rules:

1) If p=0, then we have the classical embedding with associativity (cf. § 3).
2) Otherwise, two c.p. with the distributive rules create the new β-rules :

$$\mathbf{k}' : (p+1)xmy \rightarrow (-p)xmy+\sum_j \beta_j(xm_jy) \ , \ x,y \in V \ (\text{resp. } \mathbf{l}').$$

This last rule allows the reduction of a monomial such that k.m'mm", where k>p in a polynomial P. Such a set of rules is symmetrized : all critical pairs between them and **AU** are solved. Of course, in ring completion as in the other ones, we only need the rules of type **l** or **k**. The next lemma details the c.p.:

Lemma 6.1
The two rules

$$\mathbf{k}: \alpha U \rightarrow P$$

$$\mathbf{l}: \beta V \rightarrow Q \ , \quad \alpha,\beta \in \mathbf{N}^*, \ U,V \in G^*, \ P,Q \in \mathbf{Z}\langle G\rangle.$$

superpose iff $\exists$A,B,C B$\neq$1, AB=U, BC=V, *the c.p. is then*

$$((\max(\alpha,\beta)-\alpha)\,ABC+PC \ , \ (\max(\alpha,\beta)-\beta)\,ABC+AQ).$$

The description of ring completion is then achieved. The algorithm may not terminate. Observe that we *choose* a precise symmetrization, the following *canonical system* shows that other more complex situations exist:

$$\Delta \begin{cases} A+B \rightarrow AC+D \\ -A+-B \rightarrow -D+-(AC) \\ D+-A \rightarrow B+-(AC) \\ A+-D \rightarrow -B+AC \\ B+-D \rightarrow -A+AC \end{cases}$$

Here is an open question: is there a canonical system such as Δ for which no *symmetrized* canonical system exist? In other words, is our symmetrization choice, reasonable for the complexity point of view, complete or uncomplete?

We briefly talk about abelian rings. The data-structure is now two superposed abelian word structures. Thus, the completion is deduced from the previous and monoid ones. We have a symmetrization that enables the choice of the left member. Taking an degree and lexicographical ordering, this left member may be the heading monomial. Then, as for the abelian monoids, we compute the c.p. with respect to the min-max functions on monomials:

Lemma 6.2
The two rules

k: $\alpha U \to P$

l: $\beta V \to Q$, $\alpha,\beta \in \mathbb{N}^*$, $U,V \in G^*$, $P,Q \in \mathbb{Z}<G>$.

superpose iff $\min(U,V) \neq 1$, *the null degree monomial, the c.p. is then*

$$((\max(\alpha,\beta)-\alpha)\max(U,V)+[\max(U,V)\div U].P ,$$

$$(\max(\alpha,\beta)-\beta)\max(U,V)+[\max(U,V)\div V].Q).$$

In this lemma, as for abelian monoids, a new operation $\div$ appears, corresponding to a reduction step. This operation is of course associated to the division modulo an ideal, the reduction computes an equivalent term modulo a sub-structure, sub-monoid, normal subgroup or two-sided ideal in the present case. As the coefficients and monomials introduced by the superposition are not greater than the common max in the defining equations, the algorithm halts in a finite number of steps. Of course, other reduction orderings may be used.

6.2. Modules and Algebras

These algebraic structures are generalizations of abelian groups, Z-modules, and rings, Z-algebras. In consequence, we just point out the main differences.

First of all, the scalar ring is supposed to have a canonical system: the normal form of $\alpha.m$, α scalar, needs α's one. In fact, this assumption is not necessary, as we have a surjective map from the scalar ring in the module, not always an injective one, but this hypothesis worths nothing while the explanations will be clearer. Then, the data-structure are the same as previously where scalars now replace integers. The superposition between two rules occurs on the scalars, the monomials or on both ones. The scalar match and reduction is done as described in § 6.1, in place of the max-min operations on integers.

The Module completion always terminate : fix one generator g, create a new rule whose left member is $\alpha.g$ until g disappears from the waiting equations etc... until the last generator. This is the same control structure as in the abelian group completion. Let us say that more than one rule with g in its left member may exist, and that if the scalar ring has sufficient properties, the algorithm may be improved; for example, on a division ring, the abelian group algorithm may be used, the minimum beeing defined by the division absolute value.

Finally, in algebras over a ring, the completion runs as well. In abelian case, it terminates. In the other case, recall that we choose a precise symmetrization, but other less efficient ways exist. These two cases are studied in the literature. G.M. Bergman [Ber78a] presents the reductions in non-commutative rings. On an example, he runs the Knuth-Bendix procedure. The reader will find in this paper many applications of the diamond lemma. In 1965, B. Buchberger also discovers the completion of commutative multivariate polynomial rings [Buc81a], providing many interesting examples of polynomial completion. These two authors restrict the left members to unitary monomials. We have seen that this restriction is unnecessary. However, when the scalar ring is the field $\mathbb{Q}$, this restriction can be used to extend the completion: the scalar division is used to keep only unitary monomials in the left members. The reader will observe that it is an extension of the classical Knuth-Bendix algorithm, as the field theory is not equational ($\forall x \neq 0, xx^{-1}=1$). The computer is however supposed to have a decision procedure on rationals!

7. COMPLETE GROUP PRESENTATIONS

We investigated this variety because no systematic completions was done for groups, as it was the case for other varieties. Some complete systems for classical groups are given, when such a system exists, many others can be found. Also, the one presented has the smaller number of rules. Thus, we prefer to call such sytems complete presentations rather than canonical ones as they are not unique. The defining presentations were found in [Cox72a], the complete ones with a system written in Maclisp. For an exhaustive enumeration of the rules, see our thesis [Che83a].

7.1. Surface Groups

The defining presentation of a p holes torus is $(A_1, \ldots, A_{2p} ; A_1 \cdots A_{2p} = A_{2p} \cdots A_1)$. The completion gives the system T_p of 2p rules whose length is 2p:

$$
T_p \left\{ \begin{array}{lcl}
A_n A_{n-1} \ldots A_1 & \rightarrow & A_1 A_2 \ldots A_n \\
A_2 \ldots A_n A_1^{-1} & \rightarrow & A_1^{-1} A_n \ldots A_2 \\
A_2^{-1} A_1^{-1} A_n \ldots A_3 & \rightarrow & A_3 \ldots A_n A_1^{-1} A_2^{-1} \\
\ldots & & \\
A_{2k} \ldots A_n A_1^{-1} \ldots A_{2k-1}^{-1} & \rightarrow & A_{2k-1}^{-1} \ldots A_1^{-1} A_n \ldots A_{2k} \\
A_{2k}^{-1} \ldots A_1^{-1} A_n \ldots A_{2k+1} & \rightarrow & A_{2k+1} \ldots A_n A_1^{-1} \ldots A_{2k}^{-1} \\
\ldots & & \\
A_n^{-1} \ldots A_1^{-1} & \rightarrow & A_1^{-1} \ldots A_n^{-1} \\
A_2^{-1} \ldots A_n^{-1} A_1 & \rightarrow & A_1 A_n^{-1} \ldots A_2^{-1} \\
\ldots & & \\
A_{2k}^{-1} \ldots A_n^{-1} A_1 \ldots A_{2k-1} & \rightarrow & A_{2k-1} \ldots A_1 A_n^{-1} \ldots A_{2k}^{-1} \\
A_{2k} \ldots A_1 A_n^{-1} \ldots A_{2k+1}^{-1} & \rightarrow & A_{2k+1}^{-1} \ldots A_n^{-1} A_1 \ldots A_{2k} \\
\ldots & & \\
A_{n-1} \ldots A_1 A_n^{-1} & \rightarrow & A_n^{-1} A_1 \ldots A_{n-1}
\end{array} \right.
$$

The termination of T_p is proved with a lexicographical ordering s.t. $A_{2p} > A_{2p}^{-1} > A_{2p-2} > A_{2p-2}^{-1} > \cdots > A_2 > A_2^{-1} > A_1 > A_1^{-1} > A_3 > A_3^{-1} > \cdots > A_{2p-1} > A_{2p-1}^{-1}$.

The group (A,B,C ; ABC=CBA) used in § 4.2 has a symmetrized set which is also canonical. But its nœtherianity does not follow from a classical ordering. It belongs to the family $\perp_p$ defined by $(A_1, \ldots, A_{2p+1} ; A_1 \cdots A_{2p+1} = A_{2p+1} \cdots A_1)$. Here is a complete presentation having 2p+1 rules, the words length is 2p+1:

$$
\perp_p \left\{
\begin{array}{lcl}
A_n \cdots A_1 & \rightarrow & A_1 \cdots A_n \\
A_2 \cdots A_n A_1^{-1} & \rightarrow & A_1^{-1} A_n \cdots A_2 \\
A_2^{-1} A_1^{-1} A_n \cdots A_3 & \rightarrow & A_3 \cdots A_n A_1^{-1} A_2^{-1} \\
\cdots & & \\
A_{2k} \cdots A_n A_1^{-1} \cdots A_{2k-1}^{-1} & \rightarrow & A_{2k-1}^{-1} \cdots A_1^{-1} A_n \cdots A_{2k} \\
A_{2k}^{-1} \cdots A_1^{-1} A_n \cdots A_{2k+1} & \rightarrow & A_{2k+1} \cdots A_n A_1^{-1} \cdots A_{2k}^{-1} \\
\cdots & & \\
A_1^{-1} \cdots A_n^{-1} & \rightarrow & A_n^{-1} \cdots A_1^{-1} \\
A_1 A_{2p+1}^{-1} \cdots A_2^{-1} & \rightarrow & A_2^{-1} \cdots A_{2p+1}^{-1} A_1 \\
A_3^{-1} \cdots A_n^{-1} A_1 A_2 & \rightarrow & A_2 A_1 A_n^{-1} \cdots A_3^{-1} \\
\cdots & & \\
A_{2k-1} \cdots A_1 A_n^{-1} \cdots A_{2k}^{-1} & \rightarrow & A_{2k}^{-1} \cdots A_n^{-1} A_1 \cdots A_{2k-1} \\
A_{2k+1}^{-1} \cdots A_n^{-1} A_1 \cdots A_{2k} & \rightarrow & A_{2k} \cdots A_1 A_n^{-1} \cdots A_{2k+1}^{-1} \\
\cdots & & \\
A_{2p-1} \cdots A_1 A_{2p+1}^{-1} A_{2p}^{-1} & \rightarrow & A_{2p}^{-1} A_{2p+1}^{-1} A_1 \cdots A_{2p-1} \\
A_n^{-1} A_1 \cdots A_{2p} & \rightarrow & A_{2p} \cdots A_1 A_n^{-1}
\end{array}
\right.
$$

The nœtherianity of $\perp_p$ follows from the fact that each member of $G \cup G^{-1}$ appears in one an only one rule as prefix. The rules having the same length, we may restrict our attention to the reduction chains of words having the same length. Then, one shows by induction on $|U|$ that the reductions $PU \rightarrow^{*}_{\perp_p} P^{-1}V$, with $|U|=|V|$ and P left member prefix of length n-1 , are impossible; it follows that if $|U|=|V|$, one never has $\rho_i U \rightarrow^{*}_{\perp_p} \lambda_j V$, ρ_i (resp. λ_j) right member of a rule (resp. left). Thus the reductions must halt as a prefix may be reduced only once time. Note that T_p and $\perp_p$ are symmetrized sets of rules (all c.p. are solved by the symmetrization).

The non-orientable surface groups are defined by $(A_1, \ldots, A_p ; A_1^2 \cdots A_p^2 = 1)$. There are two cases: p odd or even.

p=2n

$$
R_p \left\{
\begin{array}{l}
A_k^{-1} A_{k-1}^{-2} \ldots A_{k-n}^{-2} \rightarrow A_k A_{k+1}^{2} \ldots A_{k+p}^{2} \\
A_k^{-2} A_{k-1}^{-2} \ldots A_{k-n+1}^{-2} A_{k-n}^{-1} \rightarrow A_{k+1}^{2} \ldots A_{k+n}^{2} A_{k+n+1} \\
\quad \cdots \\
A_k^{2} \ldots A_{k+n-1}^{2} A_{k+n} \rightarrow A_{k-1}^{-2} \ldots A_{k-n+1}^{-2} A_{k-n}^{-1} \\
A_k A_{k+1}^{2} \ldots A_{k+n}^{2} \rightarrow A_k^{-1} A_{k-1}^{-2} \ldots A_{k-n+1}^{-2} \\
A_k^{-2} \ldots A_{k-n+2}^{-2} A_{k-n+1}^{-1} A_{k-n+2}^{2} \ldots A_{k+1}^{2} \rightarrow A_{k+1}^{2} \ldots A_{k+n}^{2} A_{k+n+1}^{-1} A_{k+n}^{-2} \ldots A_{k+2}^{-2} \\
A_k^{-1} A_{k-1}^{-2} \cdots A_{k-n+1}^{-2} A_{k-n+1}^{2} \cdots A_{k-1}^{2} A_k \rightarrow A_k A_{k+1}^{2} \cdots A_{k+n-1}^{2} A_{k+n} A_{k+n-1}^{-2} \cdots A_{k+1}^{-2} A_k^{-1}
\end{array}
\right.
$$

p=2n+1

$$R_p \begin{cases} A_k^{-1}A_{k-1}^{-2}\dots A_{k-n}^{-2} \rightarrow A_kA_{k+1}^{2}\dots A_{k+p}^{2} \\ A_k^{-2}A_{k-1}^{-2}\dots A_{k-n+1}^{-2}A_{k-n}^{-1} \rightarrow A_{k+1}^{2}\dots A_{k+n}^{2}A_{k+n+1} \\ \dots \\ A_k^{2}\dots A_{k+n}^{2} \rightarrow A_{k-1}^{-2}\dots A_{k-n}^{-2} \\ A_kA_{k+1}^{2}\dots A_{k+n}^{2}A_{k+n+1} \rightarrow A_k^{-1}A_{k-1}^{-2}\dots A_{k-n+1}^{-2}A_{k-n}^{-1} \\ A_k^{-2}\dots A_{k-n+1}^{-2}A_{k-n}A_{k-n+1}^{2}\dots A_k^{2} \rightarrow A_{k+1}^{2}\dots A_{k+n}^{2}A_{k+n+1}A_{k+n}^{-2}\dots A_{k+1}^{-2} \\ \dots \\ A_k^{-1}A_{k-1}^{-2}\cdots A_{k-n+1}^{-2}A_{k-n}^{-1}A_{k-n+1}^{2}\cdots A_k^{2}A_{k+1} \rightarrow A_kA_{k+1}^{2}\cdots A_{k+n}^{2}A_{k+n-1}^{-1}A_{k+n}^{-2}\cdots A_{k+2}^{-2}A_{k+1}^{-1} \end{cases}$$

where $n=[p/2]$, $A_i = A_{i+p}$ if $i=1-p,\dots,0$ and $A_i=A_{i-p}$ if $i=p+1,\dots,2p$. The nœtherianity follows from a lexicographical ordering with $A_1^{-1} > \cdots > A_p^{-1} > A_1 > \cdots > A_p$. There are 6p rules in R_p which is not a symmetrized set: the two last rules types result from c.p.

7.2. Symmetric Groups

Many finite presentations of the symmetric group S_n are known. We succeeded in completing some of them, especially the following one:

$$S_n \begin{cases} R_i = (i\ i{+}1) & i=1,\dots,n-1 \\ R_i^2 = 1 & i=1,\dots,n-1 \\ R_iR_j = R_jR_i & i \leq j-2 \\ (R_iR_{i+1})^3 = 1 & 1 \leq i \leq n-2 \end{cases}$$

The completion gives the n^2-2n+2 rules:

$$\mathcal{S}_n \begin{cases} R_i^{-1} \rightarrow R_i & \\ R_i^2 \rightarrow 1 & i=1\dots n-1 \\ R_iR_j \rightarrow R_jR_i & j \leq i-2 \\ R_iR_{i-1}\cdots R_jR_i \rightarrow R_{i-1}R_iR_{i-1}\cdots R_j & j<i \end{cases}$$

Let $1 = R_0$ in $\mathcal{S}_n$, then for each rule the integer made by the concatenation of the left members generators indices is greater than the right member one, thus the system is nœtherian.

7.3. Polyhedral Groups

The polyhedral group $G_{l,m,n}$ is defined by $(A,B,C\ ;\ ABC, A^l, B^m, C^n)$. The finite polyhedral groups are the rotation groups of the five regular polyhedrons of the 3-space:

Tetrahedron

$$(2,3,3)\left\{\begin{array}{l} CC \to C^{-1} \\ BB \to B^{-1} \\ B^{-1}B^{-1} \to B \\ C^{-1}C^{-1} \to C \\ C^{-1}B^{-1} \to BC \\ B^{-1}C^{-1} \to CB \\ BCB \to C^{-1} \\ CBC \to B^{-1} \\ C^{-1}BC^{-1} \to CB^{-1}C \\ CBC^{-1} \to B^{-1}C \\ C^{-1}BC \to CB^{-1} \\ BCB^{-1} \to C^{-1}B \\ B^{-1}CB \to BC^{-1} \\ B^{-1}CB^{-1} \to CB^{-1}C \\ BC^{-1}B \to CB^{-1}C \end{array}\right.$$

Octahedron (Cube)

$$(2,3,4)\left\{\begin{array}{l} BB \to 1 \\ B^{-1} \to B \\ A^{-1} \to BABAB \\ AAA \to BABAB \\ ABABA \to B \\ ABAABA \to BAAB \\ AABAAB \to BAABAA \end{array}\right.$$

Icosahedron

(Dodecahedron)

$$(2,3,5)\left\{\begin{array}{l} BBB \to 1 \\ AA \to 1 \\ A^{-1} \to A \\ B^{-1} \to BB \\ BABABAB \to ABBA \\ BBABB \to ABABABA \\ BABABBABAB \to ABBABABBA \\ BABBABABBABA \to ABABBABABBAB \end{array}\right.$$

The termination of $G_{2,3,3}$ follows from a lexicographical ordering defined by $B^{-1}>C^{-1}>C>B$, the generator A beeing eliminated. For the others, we use the Knuth-Bendix weight function on words defined by

Cube: $\pi(A^{-1})=5, \pi(B^{-1})=\pi(A)=\pi(B)=1$, with a lexicographic ordering $A^{-1}>B^{-1}>A>B$ when the weights are equal.

Icosaedron: $\pi(B^{-1})=6, \pi(B)=3, \pi(A^{-1})=\pi(A)=1$, and $A^{-1}>B^{-1}>A>B$.

The remaining groups seem to have a complete presentation of less than forty rules: we succeed in completing them for $l,m,n<20$. The rules are regular but a general form is hard to give due to the presence of three parameters implying many subcases.

7.4. The seventeen plane groups

These are the famous 17 geometric groups that cover the plane from a basic pavement. All of them are completed. We present here two of them whose completion gives interesting termination problems. The first one is p31m in the notation of [Cox72a], and is defined by $A^2=B^2=C^2=(AB)^3=(BC)^3=(CA)^3=1$:

$$\text{p31m}\left\{\begin{array}{l} C^{-1} \to C \\ A^{-1} \to A \\ B^{-1} \to B \\ BB \to 1 \\ CC \to 1 \\ AA \to 1 \\ CBC \to BCB \\ BAB \to ABA \\ ACA \to CAC \end{array}\right.$$

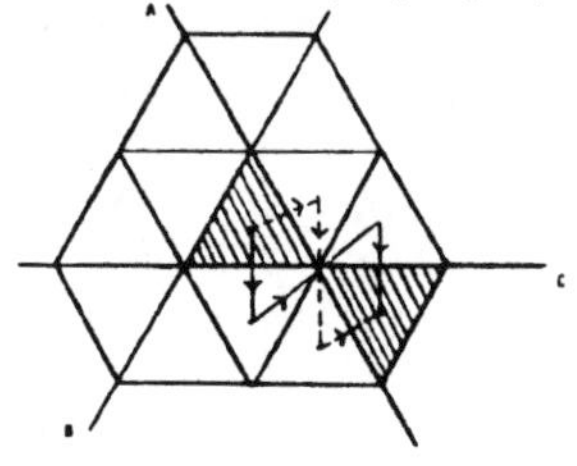

Pavement of plane by p31m, A,B,C are three symetries. Arrows represent the rule ACA → CAC.

Its termination proof is essentially$\perp_p$'s one. In an infinite reduction chain, as the rules do not increase the words length, this length becomes constant. Indeed, only the three last rules may be used, analysing the prefixes, one may see that p31m is nœtherian. The last one is p3m1, defined by $A^2=B^3=(AB^{-1}AB)^3=1$. The last equation gives only one rule in the following locally confluent set:

$$\text{p3m1}\left\{\begin{array}{l} A^{-1} \to A \\ AA \to 1 \\ BB \to B^{-1} \\ B^{-1}B^{-1} \to B \\ AB^{-1}ABA \to B^{-1}ABAB^{-1}AB \end{array}\right.$$

If R is the last rule and Q the set of other ones, it is easy two show that the reduction $\to_Q^* . \to_R . \to_Q^*$ is nœtherian. But we do not know if $\to_{p3m1}$ halts. The point is that, in the rule R, each proper subword of the left member is subword of the right one.

8. CONCLUSION

Along the way from a symbolic algorithm to various applications, we found several procedures that may be called numerical: abelian group decomposition, completion over rational scalars are the main ones. This is perhaps the first conclusion that symbolic and numerical algorithms are not so far one from the other. Moreover, the formal approach leads to a unified view of many algorithms, this fact is of course theoretically pleasant, but let us say that in practice this is of great importance: we wrote a system running on various Lisp (Maclisp, Franzlisp, LeLisp). In this program, only two distinct completion procedures exist: one for abelian groups, the other for the remainding cases, all functions are overloaded according to the current variety. Together with some recent developments on inheritance languages, this tells us what could be the future formal computation systems: a dag structure of parametrized theories, to which are associated distinct semantics for common procedures, together with more specific functions.

The third main point is the power of the method. It has been clearly proved for polynomials by Buchberger et all[Buc82a]; in groups, we hope that the reader is convinced by § 7 despite the negative decidability results. Also, let us say that coding a set of equations in data and control structure through canonical systems is clearly an important way of program synthesis.

References

Bal79a. A.M. Ballantyne and D.S. Lankford, *New Decision Algorithms for Finitely Presented Commutative Semigroups*, Report MTP-4, Department of Mathematics, Louisiana Tech. U (May 1979).

Ber78a. George M. Bergman, *The Diamond Lemma for Ring Theory*, Advances in Mathematics, 29,2 (pp. 178-218) (Aug. 1978).

Boo82b. Ronald V. Book, *Confluent and Other Types of Thue Systems*, Journal of the ACM, Vol 29, No. 1 (January 1982).

Boo82a. Ronald V. Book, Matthias Jantzen, and Celia Wrathall, *Monadic Thue Systems*, Theoretical Computer Science 19, p. 231-251, North-Holland Publishing (1982).

Buc81a. B. Buchberger, "H-Bases and Grobner-Bases for Polynomial ideals ," CAMP Publ. No 81 -2.0, Johannes Kepler Universitat, Austria (Fevrier 1981).

Buc82a. B. Buchberger, "Miscellaneous results on Grobner-Bases for polynomial ideals II," CAMP Publ. No 82 - 23.0, Johannes Kepler Universitat, Austria (Juin 1982).

Buc79a. Hans Bucken, *Reduction Systems and Small Cancellation Theory*, Proc. Fourth Workshop on Automated Deduction, pp. 53-59 (1979).

Che83a. Philippe Le Chenadec, *Formes canoniques dans les algèbres finiment présentées*, Thèse de 3ème cycle, Univ. d'Orsay (Juin 1983).

Cli67a. A.H. Clifford and G.B. Preston, *The algebraic theory of semigroups, Vol II*, Amer. Math. Soc. Providence, Rhode Island (1967).

Coc76a. Y. Cochet, *Church-Rosser congruences on free semigroups*, Colloq. Math. Soc. Janos Bolyai: Algeb. theory of Semigroups. 20,51-60 (1976).

Cox72a. H.S.M. Coxeter and W.O.J Moser, *Generators and Relations for Discrete Groups*, Springer-Verlag (1972).

Deh12a. M. Dehn, *Transformation der Kurve auf zweiseitigen Flache*, Math Ann. 72, 413-420 (1912).

Der82a. Nachum Dershowitz, *Orderings for Term-rewriting Systems*, Theoretical Computer Science 17, pp. 279-301 (1982).

Fag84a. F. Fages, *Associative-Commutative Unification*, 7th Conference on Automated Deduction, Napa Valley California (May 1984).

Gre60a. M. Greendlinger, *Dehn's Algorithm for the Word Problem*, Communications on Pure and Applied Mathematics, 13, pp. 67-83 (1960).

Hue80a. G. Huet, *Confluent Reductions: Abstract Properties and Applications to Term Rewriting Systems.*, JACM 27,4, pp. 797-821 (Oct. 1980).

Hue81a. G. Huet, *A Complete Proof of Correctness of the Knuth-Bendix Completion Algorithm*, JCSS 23,1, pp. 11-21 (Aug. 1981).

Hue78a. G. Huet and D.S. Lankford, *On the Uniform Halting Problem for Term Rewriting Systems*, Rapport Laboria 283, IRIA (Mars 1978).

Hue80b. G. Huet and D. Oppen, *Equations and Rewrite Rules: a Survey*, In Formal Languages: Perspectives and Open Problems, Ed. Book R., Academic Press (1980).

Hul80a. J.M. Hullot, *Compilation de Formes Canoniques dans les Théories Equationnelles*, Thèse de 3ème cycle, U. de Paris Sud (Nov. 80).

Jou83a. J.P Jouannaud, H. Kirchner, and J.L. Remy, *Church-Rosser properties of weakly terminating equational term rewriting systems*, Centre de Recherche en Informatique de Nancy, R-004 (1983).

Jou81a. Jean-Pierre Jouannaud, Pierre Lescanne, and Fernand Reinig, *Recursive decomposition ordering*, Rapport GRECO No 2 (1981).

Knu70a. D. Knuth and P. Bendix, *Simple Word Problems in Universal Algebras*, In Computational Problems in Abstract Algebra Ed. Leech J., Pergamon Press, pp. 263-297. (1970).

Liv76a. M. Livesey and J. Siekmann, *Unification of Bags and Sets*, Internal Report 3/76, Institut fur Informatik I, U. Karlsru~he (1976).

Lyn77a. R.C. Lyndon and P.E. Schupp, *Combinatorial Group Theory*, Springer-Verlag (1977).

Met3.a. Yves Metivier, *Systèmes de réécriture de termes et de mots*, Thèse de 3ème cycle d'enseignement supérieur, No d'ordre 1841, Université de Bordeaux I (26 Mai 1983.).

Nov55a. P.S. Novikov, *On the algorithmic unsolvability of the word problem in group theory*, Trudy Mat. Inst. Steklov 44,143 (1955).

Pet82a. Gerald E. Peterson and Mark E. Stickel, "Complete systems of reductions using associative and/or commutative unification," Technical note 269, SRI International (October 1982).

Smi66a. David A. Smith, *A basis algorithm for finitely generated abelian groups*, Mathematical Algorithms Vol I,1 (Jan. 1966).

Sti81a. M.E. Stickel, *A Complete Unification Algorithm for Associative-Commutative Functions*, JACM 28,3 pp 423-434 (1981).

TERM REWRITING SYSTEMS AND ALGEBRA

Pierre LESCANNE
CRIN
Campus Scientifique
BP 239
54506 Vandoeuvre-les-Nancy, FRANCE

ABSTRACT: This paper presents two ideas for proving theorems in algebra. First the equivalence of two presentations of algebras is verified by attempting to make them equivalent to the same noetherian and confluent set of rewrite rules. Second the Knuth-Bendix procedures that performs that conevergence often fails by generating too big or non directable rules. We propose to tame it by an iterative approach.

1. INTRODUCTION

In this paper we report experiments that we have done using the REVE implementation of the Knuth-Bendix procedure refered as KBP here. The method is based on the convergent set uniqueness theorem [1] [2] [12]. This theorem says that given an ordering and a set E of equations there exists a unique irreducible and convergent (i.e., noetherian and confluent) set of rules compatible with the ordering that is a decision procedure for the theory presented by E. Irreducible means that left-hand and right-hand sides cannot be reduced by the other rules. Compatible means that s --> t implies s > t. The method for proving the equivalence of two sets of axioms, let say E and F is as follows. KBP is run on E with an ordering then KBP is run on F with the same ordering. If they provide the same convergent set (up to a renaming of the variables) then E and F are equivalent which means that E and F present the same theory.

People who have experimented with the Knuth-Bendix completion procedure know it is not always obvious to get a convergent set of rules from any set of equations. In this paper we explained features we added to the Knuth-Bendix completion procedure presented by Huet in [5]. We refer to it as KBP. The main characteristic of KBP is that it can be iterated which means that after a failure, KBP can be restarted with the input produced by the previous iteration. It is a part of the term rewriting system laboratory named REVE, so let say a word on how REVE directs equation to build a noetherian i.e. uniformely terminating set of term rewrite rules. We claim that the proof of term rewriting system uniform termination is the main problem that any implementation of this procedure has to solve. We call this termination the Noether property to make the difference with the termination of the Knuth-Bendix procedure. For that REVE uses a simplification ordering, the Recursive Decomposition Ordering with status (RDOS). The RDOS is based on a precedence i.e., an ordering on symbols. RDOS is well-founded and can easily be made stable for subtitutions of terms to variables. So if a set of rules is oriented by using a RDOS, then it is noetherian.

In the experiments we conducted with REVE we wanted to prove some theorems in algebra, namely that a given set of axioms actually presents a variety of algebras, especially in group theory. Using an unrestricted KBP two kinds of problems arise.

- The program generates longer and longer equations and starts getting much slower. However no regularity occurs and no loop seems to have been entered. For instance, starting with medium size equations, production of ten line equations is common even among the first ones.

- The program generates an equation it is enable to direct with the tool it has e.g. RDOS fails to propose an orientation. It can have made previously a choice on the precedence which seemed reasonable at the time it was done, but which is inadequate for the current equation.

The solution we propose was already sketched in Knuth and Bendix paper [9], it works as follows. Instead of using KBP to produce directly the desired noetherian set of rules, we use KBP ability to generate "useful" consequences of the input set of equations. By "useful" consequence, we mean either a compact rule, short or strongly reducing, or a new operator that expresses an interesting fact about the theory. For example, in group with right division, the property "x / x == y / y" has to be exploited, to create a new constant which allows the procedure to make further simplifications. When KBP eventually fails, these "useful" consequences have not to be lost, but have to be inserted in the input set of equations of a new activation of KBP, with the hope that, this time, it will finish successfully. Our approach is rather experimental, but precisely the experiments that will be given in this paper show its effectiveness.

Let us say a little more about the method we propose. Let us start with a set A of equations and let us call KBP with the possible restriction that all the equations that exceed a certain size are discarded. Suppose the procedure stops after generating a set B of rules. This may happen either after a failure on the orientation of an equation or after exploring all the possibilities allowed by the restriction we have imposed. Consider the result B not as a set of rules but as a set of equations, any orientation is irrelevent because the current ordering will be forgotten and replaced subsequently by another. We have no reason to believe that B is equivalent to A, we know only that all the equations in B are consequences of A. Therefore adding B to A makes obvioulsy an equivalent system. However we have serious reasons to hope that if we feed KBP with the set A + B it will have more chances to produce a noetherian and confluent set of rewrite rules. Notice that during the process KBP can generate new operations as explained in [9] for example, this feature is nicely implemented in REVE. If the resulting set is the same as a set known to present or decide a theory, the method provides a computer proof of an algebraic theorem. The convergent set uniqueness theorem says that shall be the case if in both cases the same precedence is chosen. As an example, let us shortly present, in this introduction, an interesting result we got. We proposed to REVE the Higman and Neuman presentation of groups [4]:

x / ((((x /x) / y) / z) / (((x / x) / x) / z)) == y

After two iterations, REVE generates the following noetherian and confluent set of rewrite rules:

x / x -> b
x / b -> x
b / y -> c(y)
c(x / y) -> y / x
(x / c(y)) / y -> x
c(b) -> b
(x / y) / c(y) -> x
c(c(x)) -> x
x / (y / z) -> (x / c(z)) / y

The perspicacious reader will recognize "b" as the identity and "c" as the inverse. Actually this set of rules is produced by KBP, starting with the equations

```
x * (y * z) == (x * y) * z
a * x == x
b(x) * x == a
x / y == x * b(y)
```

and using an appropriate instance of the RDOS. The fourth equation is just a definition of the right division. Thus we may claim that REVE proposed the first computer proof of Higman and Neuman result. If we compar Higman and Neuman's proof with that proposed by REVE we may make the following comments. The human proof analyzes subtly the operating group of translations of the considered algebra and uses extensively second order reasonings. REVE performs only long and obvious computations based on replacement of terms by equal terms. Using uniformly terminating term rewriting systems avoids loops. That could not be done by a human but all the steps can be easily followed by a patient reader.

2. REVE AND THE RECURSIVE DECOMPOSITION ORDERING

Before we propose an account on the experiments we conduct, we would like to give more information on the Knuth-Bendix completion procedure (KBP) and on the Recursive Decomposition Ordering with Status (RDOS) implemented in REVE.

2.1 Special features of the Knuth-Bendix Completion Procedure

Compared with a common KBP the procedure implemented in REVE has essentially three features in addition to its algorithm to prove Noether property.

First, it is possible to set a limit on the size of the equations that are kept during the activation of the procedure. If an equation exceeds the size, it is discarded. The user has also the possibility of throwing away equations that cannot be directed. A flag is set when at least one rule was discarded. Then the final set of rules produced by KBP will no longer be guaranteed to be equivalent to the input set of equations. This facility will not be used in the following experiments but was useful during previous tests done before those presented here.

Second, when KBP fails to direct an equation but recognizes that the set of variables of the left-hand side contains variables that do not occur in the right-hand side and vice-versa and there is no hope for an orientation, it proposes to introduce a new operator which arity is the size of the intersection of these two sets. Then it creates two new rules using the new operator.

Third, when KBP fails on directing an equation, the user is always asked whether he or she wants to keep the current set of rules, in order to continue the experiment with a new KBP and another precedence or whether he or she rather wants to restart with the initial data. In the first case or when KBP terminates after inspection of all the rules and after discard of some of them, the initial set of equations is always appended to the result. By this mean the set of equations is maintained equivalent to the input. Notice that in REVE-2 [3] these features are a full part of the system and not only a facility added to it.

2.2 Survey of the Recursive Decomposition Ordering with Status

REVE uses the Recursive Decomposition Ordering with Status. More details on this ordering and a formal presentation can be found in [7] and [11]. We give here only a flavor of the RDOS. A full definition would require a long development. Nevertheless we hope that the following examples will allow the reader to become familiar with its use. RDOS is based on a precedence, i.e., a partial quasi-ordering on the symbols. Thus terms can be declared as equivalent or as comparable. The precedence is used to induce a quasi-ordering on terms. For instance, if * > i, then

after the informal principle "a term decreases, when its lighter symbols go up, and its heavier symbols go down". In the following when no relation among two symbols is expressed, that means the precedence says nothing about these symbols. This method is refined to prove for example

with ! > &, using all the power of the RDOS. Informally, ~ goes down and & is replaced by a lighter symbol namely !. To make the RDOS stable by substitution to variables, the precedence has to forbid any variable to be comparable with any other variable and any symbol. Obviously no precedence can help RDOS to direct

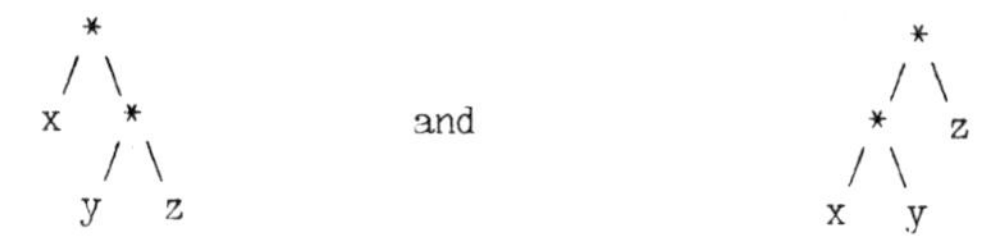

The solution is to associate a status with each operator. If the status is right-to-left, e.g. for *, that means that one has to start to weigh the rightmost direct subterm first, y * z and z in the previous example at the top. In addition, RDOS includes tests that check that the other terms are not too big. For example RDOS can easily decide

$$x \; / \; (y \; / \; z) > (x \; / \; b(z)) \; / \; y$$

with a right-to-left status for / and a precedence that says that b and / are equivalent. Indeed y is less than y / z, x / b(z) is not too big. Indeed it is less than x / (y / z) because b(z) is less than y / z, and obviously x is less than the whole term. Notice that in most of the cases the precedence is not set before running KBP, but when KBP cannot direct an equation. In such a case, REVE proposes pairs of symbols that could be added to the precedence. When the answer can be decided by REVE, KBP continues, otherwise the user makes choices.

In the following we will indicate UC running times for a VAX implementation. They have to be taken as order of magnitude. In its current state, indeed REVE as a prototype does not present optimized algorithm everywhere and values were taken on versions of the software different from this presented in [10].

3. TWO PRELIMINARY EXAMPLES.

We present first two completions solved by REVE and never done before. They do not use specific features of KBP, but only the power of RDOS.

3.1 Groups with right division

Let us start with

```
x * (y * z) == (x * y) * z
e * x == x
i(x) * x == e
x / y == x * i(y)
```

Let us set the precedence to e < / < * and / equivalent to i. The status of * and / is right-to-left. After seventeen seconds, REVE produced the following noetherian and confluent set of rewrite rules.

```
x / x -> e
x / e -> x
e / y -> i(y)
i(x / y) -> y / x
(x / i(y)) / y -> x
i(e) -> e
(x / y) / i(y) -> x
i(i(x)) -> x
x / (y / z) -> (x / i(z)) / y
x * y -> x / i(y)
```

This set of ten rules is a completely new way to decide group theory using the rewrite rule method. Until now, only the set of ten rules (eleven if the definition of the right division is added) presented in [9] was known and often called "the" canonical systems for group theory. The previous one is another canonical system. Let us call the former "classical". It is possible to think that the non classical system was found because of the power of the RDOS. The next example will justify this assertion.

3.2 Taussky presentation of groups

In their paper Knuth and Bendix study an axiomatisation of groups as proposed by Taussky (Example 11 in [9]). They show that groups can be presented as the algebras that satisfy the following equations. The reader who is not familiar with this example is invited to read Knuth and Bendix paper first or to accept these five equations

```
x * (y * z) == (x * y) * z
e * e == e
x * i(x) == e
f(x * y, x) == g(x * y, y)
f(e, x) == x
```

For technical reasons related to their ordering, Knuth and Bendix make f ternary. RDOS does not require such assumption. A way to direct these equations is to set the status of * to be "right-to-left" and the precedence to be

```
e < * < i < f < g
```

After 32 seconds, REVE generates the "classical" set of rules for groups plus the two rules.

$$g(x, y) \to f(x, x * i(y))$$
$$f(e, x) \to x$$

The key point is probably when REVE generates the rule $e * x \to x$ as the 27th consequence. In their experiment Knuth and Bendix fail to complete the given set of equation, even after adding a new monadic operator. They say "Actually it is not hard to see that [...] axioms 1 through 5 cannot be completed to a finite set of reductions" (probaly because of their ternary f). Thus their discussion at the end of their example 11 become unfounded. Notice that the precedence and the status proposed here are not the only ones that make KBP to terminate successfully.

4 GROUPS AS BINARY ALGEBRAS

Interesting experiments on the Knuth-Bendix completion procedure can be done on axiomatisation of groups as binary algebras i.e. algebras that contain a unique binary operation we write /. They are called "groupoids" by Higman and Neuman. This operation is interpreted as the right division and is usually enough to characterize groups. Notice that identity is not introduced, but can be easily defined as the common value of x / x for all the instanciations of x after an identity $x / x == y / y$ is proved. The inverse is introduced similarly.

4.1 Groups as binary algebras with one axioms

The unique axiom is

$$x / ((((x /x) / y) / z) / (((x / x) / x) / z)) == y$$

In the experiment we proposed, we started with this equation and no restriction. Soon REVE cannot direct an equation it generates. This equation of the form $L(x, y, z) == R(x, y, u)$ allows REVE to introduce a new operator $a(x, y)$. After 40 rules are generated, REVE proposes the equation

$$a(x, (x / x)) == y / y$$

and introduces a constant it calls b and the rule

$$x / x \to b.$$

After 67 rules, from an equation of the form $LL(y, z, u) == a(x, y)$, REVE introduces the monadic function c and the rule

$$a(x, y) \to c(y)$$

The 144th rules is

$$c(y / x) \to x / y.$$

After 382 rules and 49 minutes, REVE deduces the equation

$$x / (y / z) == (x / (c(w) / z)) / (y / c(w))$$

that cannot be directed. For instance setting a status for / makes the ordering unable to direct the rule 144. Then we restart with the following precedence

$$b < / \text{ and } c \sim / \text{ and } c < a.$$

The status of / is set to "right-to-left". After 3 1/2 minutes, REVE generates the set of equations mentioned in the introduction, plus the useless rule,

a(x, y) -> c(y).

4.2 Groups as binary algebras with two axioms

Another interesting exercice in algebra using REVE can be to prove that the two following axioms present actually the variety of groups.

(x / x) / ((y / y) / y) == y
(x / z) / (y / z) == (x / y).

Running KBP on this input, REVE generates an equation of the form

L(x, y) == (z / z) / y

which suggests the monadic operator a and the rule

(x / x) / y -> a(y).

Later REVE generates the equation

(x / a(y)) / y == x / (z / z)

and the rule

x / (y / y) -> b(x)

and later the rule

b(x) -> x.

To direct the equation

(x / y) / (z / (y / u)) == (x / u) / z

the status of / is set to "right-to-left". But previous choices do not make possible to direct an equation like

a(y / x) == x / y.

A second iteration is necessary, with a precedence as above. The equation

(z / (u / y)) / (x / u) == z / (x / y)

that cannot be directed is discarded. The equation

x / x == y / y

is generated and the identity is introduced. Eventually a set of equations is generated. A third iteration is made necessary to check the confluence of the set generated previously. The durations of the three iterations of KBP are respectively 3.9 s., 12 s., and 5 s.

4.3 Groups as associative algebras with left and right division

A group can also be presented as an associative binary algebra with a left division and a right division. The following axioms are satisfied.

$$x * (y * z) == (x * y) * z$$
$$x \backslash (x * y) == y$$
$$(x * y) / y == x.$$

The problem turns out to be not as easy as we believed first. Because we have no reason to prefer an operator to another, we make all the operators equivalent and we set their status to "right-to-left". After about 150 equations, REVE generates the equation $x / x == y / y$ and proposes to introduce a constant that it calls a. Ten rules later it generates the equation

$$(x \backslash y) / y \;==\; a / x.$$

That suggests to introduce a new monadic operator b and the rule

$$a / x \rightarrow b(x).$$

At that time REVE sets the precedence to $b < a$ and $b < /$. Therefore REVE is unable to direct the equation $b(y / x) == x / y$. Thus after stopping KBP a new precedence is declared. For instance, $a < / < \backslash < *$ and $b \sim /$ and the status of $\backslash$ is now set to "left-to-right". KBP terminates and generates the 9 axioms already seen in 3.1 for the groups with right division plus the definition of * and /,

$$x * y \rightarrow x / b(y)$$
$$x \backslash y \rightarrow b(x) / b(y).$$

The first iteration took about 17 minutes and the second 80 seconds. Notice that a precedence like $a < * < b < / < \backslash$ at the beginning of the second iteration generates the "classical" axioms plus the "definition" of / and $\backslash$.

5. CONCLUSION

In this paper, we presented REVE experiments in algebra and showed how an iterative KBP can be used in REVE to generate noetherian and confluent term rewriting systems. A case we did not talk about is when the second iteration needs a new unification algorithm because during the first iteration an equation that cannot be directed was generated. For instance, the commutativity or an equation like

$$(x / y) / z == (x / z) / y$$

that occurs often in presentation of commutative groups [6]. When such unification algorithms will be available, we will try this kind of experiments. On another hand, during a discussion with Gerard Huet we thought that instead of throwing the rules away it could be possible to keep them for a further use in the same iteration of KBP. For example, if an equation was eliminated because it was too big, may be a new rule will reduce it or a lack of equation will make necessary to relax the condition on the size. Same for equations that cannot be directed, a new rule can reduce one side and allows REVE to direct it. Informally in REVE the "trash can" should be replaced by a "refrigerator". A new completion procedure based on these ideas was designed and proved by Helene Kirchner (private communication) and Randy Forgaard has implemented these features in REVE 2.

Acknowledgments: I would like to acknowledge my colleagues of MIT Randy Forgaard, John Guttag and Jeannette Wing for their help when I started these experiments, and my colleagues of the group Eureca at Nancy later on. During this work I was partially by the Agence de l'Informatique (grant 82/767) and the Greco Programmation.

6. REFERENCES

[1] BUTLER G., LANKFORD D., "Experiments with computer implementations of procedures which often derive decision algorithms for the word problem in abstract algebras" Louisiana Tech U. Math Dept. Ruston LA 71272 Dept MTP-7 Aug. 1980.

[2] DERSHOWITZ N., MARCUS L., "Existence and Construction of Rewrite Systems," Memo, Information Sciences Researsh Office, The Aerospace Corporation, El Segundo California USA, (August 1982).

[3] FORGAARD R., "A Program for Generating and Analizing Term Rewriting Systems" Master's Thesis, MIT Lab. for Computer Science, Cambridge Massachusetts USA (1984) (To appear)

[4] HIGMAN G., NEUMAN B. H., "Groups as groupoids with one law," Publ. Math. Debrecen. 2 (1952), 215-221.

[5] HUET G., "A Complete Proof of Correctness of Knuth-Bendix Completion Algorithm," J. Comp. Sys. Sc., 23 (1981), 11-21.

[6] JEANROND H. J., "Deciding unique termination of permutative rewriting systems: Choose your term algebra carefully" 5th Conf. on Automated Deduction, Lecture Notes in Computer Science 87 (1980), 335-355.

[7] JOUANNAUD J-P., LESCANNE P., REINIG F., "Recursive Decomposition Ordering," Conf. on Formal Description of Programming Concepts II, North-Holland (D Bjorner Ed.) (1982), 331-346.

[8] KIRCHNER H., "Current Implementation of the general completion algorithm" Private Communication.

[9] KNUTH D.E., BENDIX P.B., "Simple Word Problems in Universal Algebras," in Computational Problems in Abstarct Algebra, Ed. Leech J., Pergamon Press (1969), 263-297.

[10] LESCANNE P., "Computer Experiments with the REVE Term Rerwiting System Generator" 10th POPL Conf. Austin Texas (1983).

[11] LESCANNE P., "Uniform Termination of Term Rewriting Systems: Recursive Decomposition Ordering with Status" 9th Coll. on Trees in Algebra and Programming (Bordeaux March 1984).

[12] METIVIER B., "About the rewriting systems produced by the Knuth-Bendix Completion Algorithm," Inf. Proc. Ltrs 16 (1983), 31-34.

Termination of a Set of Rules Modulo a Set of Equations[1]

Jean-Pierre Jouannaud[2]
Centre de Recherche en Informatique de NANCY and GRECO Programmation
Campus Scientifique, BP 239, 54506 Vandoeuvre les Nancy CEDEX, FRANCE.

Miguel Munoz[2]
Universidad de Valencia, c/o Dr Moliner, Burjasot, Valencia, SPANES.

Abstract

The problem of termination of a set R of rules modulo a set E of equations, called E-termination problem, arises when trying to complete the set of rules in order to get a Church-Rosser property for the rules modulo the equations. We first show here that termination of the rewriting relation and E-termination are the same whenever the used rewriting relation is E-commuting, a property inspired from Peterson and Stickel's E-compatibility property. More precisely, their results can be obtained by requiring termination of the rewriting relation instead of E-termination if E-commutation is used instead of E-compatibility. When the rewriting relation is not E-commuting, we show how to reduce E-termination for the starting set of rules to classical termination of the rewriting relation of an extended set of rules that has the E-commutation property. This set can be classicaly constructed by computing *critical pairs* or *extended pairs* between rules and equations, according to the used rewriting relation. In addition we show that different orderings can be used for the starting set of rules and the added critical or extended pairs. Interesting issues for further research are also discussed.

---

[1]Research supported in part by Agence pour le Developpement de l'Informatique under contract 82/767 and for another part by Office of Naval Research under contract N00014-82-0333.

[2]Part of this work was done while the second author was visiting the Centre de Recherche en Informatique de Nancy and another part while the first author was visiting the Stanford Research Institute, Computer Science Laboratory, 333 Ravenswood Avenue, Menlo Park, CA 94025, USA.

1 Introduction

Term Rewriting Systems (TRS in short) are known to be a very general and major tool for expressing computations. In addition, they have a very simple **Initial Algebra semantics** whenever the result of a computation does not depend on the choice of the rules to be applied (the so called **Confluence** property). When a TRS is not confluent, it can be transformed into a confluent one, using the Knuth and Bendix **completion procedure** [Knuth and Bendix 70].

The Knuth and Bendix completion procedure is based on using equations as **rewrite rules** and computing **critical pairs** when left members of rules overlap. If a critical pair has distinct irreducible forms, a new rule must be added and the procedure recursively applies until it maybe stops. This procedure requires the **termination** property of the set of rules, which can be proved by various tools. A full implementation of these techniques is described in [Lescanne 83]. When the termination property is not satisfied, the set of axioms can be splitt into two parts: those causing non termination are used as equations while the others are used as rewrite rules. In that case, critical pairs must be computed between rules and between rules and equations. In addition to a *complete unification algorithm* for the theory expressed by these equations, the process requires the *termination* property of the set of rules *modulo* the set of equations, called **E-termination**. This means that infinite sequences are not allowed for the relation obtained by composition of the equality generated by the equations with one step rewriting using the rules. The theory for such completion algorithms was developped successively by [Lankford and Ballantyne 77a], [Lankford and Ballantyne 77b], [Lankford and Ballantyne 77c], [Huet 80], [Peterson and Stickel 81], [Jouannaud 83] and [Jouannaud and Kirchner 84]. The last paper gives a very general completion algorithm for such mixed sets of rules and equations based on two abstract properties called **E-Confluence** and **E-Coherence.**

Implementers of these algorithms are faced to this problem of proving E-termination, that was never adressed before, except for a recent work [Dershowitz, Hsiang, Josephson and Plaisted 83, Plaisted 83]. Those authors propose two techniques for solving the problem in the particular case of *associative commutative* theories, both based on **transformations of terms**. Unlike these authors, we don't want to transform terms, because we don't want to carry on many representations of a same term that are to be updated accordingly. However, our general purpose is the same: to *reduce* the E-termination problem to the ordinary termination problem, in order to apply well known and powerfull techniques based on **simplification orderings** such as the **Recursive Path Ordering** [Dershowitz 82] or the **Recursive Decomposition Ordering** [Jouannaud, Lescanne and Reinig 83]. This reduction can be obtained through a process of transforming left and right hand sides of rules, or by **adding new rules** to the starting set. The first method proposed in [Dershowitz, Hsiang, Josephson and Plaisted 83] actually uses these two techniques at the same time. Our purpose in this paper is to investigate the power of the second technique in the general case of arbitrary mixed sets of rules and equations and show how these added rules are related to the so called **extended rules** added during the run of an E-completion procedure [Peterson and Stickel 81, Jouannaud and Kirchner 84].

Sections 2 and 3 introduce the classical background about Term Rewriting. E-termination is defined in section 4. We show in section 5 that E-termination and termination of the rewriting relation are the same whenever the used rewriting relation is **E-commuting**. More precisely, we show that Peterson and Stickel's Church-Rosser results can be obtained by requiring termination for this relation instead E-termination if E-commutation is used instead of E-compatibility as defined in [Peterson and Stickel 81]. When the rewriting relation is not E-commuting, we show in section 6 how to reduce E-termination for the starting set of rules to termination of the rewriting relation for an extended set of rules that has the E-commutation property. This set can be classicaly constructed by computing *critical pairs* or *extended pairs* between rules and equations, according to the used rewriting relation. In addition, we show that different orderings can be used for ordering on one hand the starting set of rules, on the other hand the critical or extended pairs. Finally, we show that Peterson and Stickel's associative-commutative extensions ensure E-commutation as well as E-compatibility which proves that our approach is well suited to associative-commutative theories.

2 Term Rewriting Systems

Definition 1: Given a set X of variables and a graded set F of function symbols, T(F,X) denotes the **free algebra** over X. Elements of T(F,X), called **terms**, are viewed as *labelled trees* in the following way: A term t is a partial application of N_+^* into F$\cup$X such that its domain $\mathcal{D}$(t) satisfies:
the empty word ϵ is in $\mathcal{D}$(t) and αi is in $\mathcal{D}$(t) iff α is in $\mathcal{D}$(t) and i $\in$[1,arity(t(α))].
$\mathcal{D}$(t) is the set of **occurrences** of t, $\mathcal{G}$(t) the subset of **ground occurrences** (i.e. non variable ones), $\mathcal{V}$(t) the set of variables of t, t/α the subterm of t at occurrence α, t[α←t'] the term obtained by replacing t/α by t' in t and #(x,t) the number of occurrences of x in t. A term t is **linear** iff #(x,t) =1 for any x in $\mathcal{V}$(t). □

Definition 2: We call **axiom** or **equation** any pair (t,t') of terms and write it t=t'. The **equality** $=^E$ or $|\text{-}*\text{-}|^E$ is the smallest congruence which is generated by the set E of axioms. Let $|\text{-}n\text{-}|^E$ denote n steps of the elementary one step E-equality $|\text{-}|^E$ and $|\text{-}+\text{-}|^E$ the transitive closure of $|\text{-}|^E$. An equation t=t' is said to be **linear** if t and t' are both linear. □

Definition 3: **Substitutions** σ are defined to be endomorphisms of T(F,X) with a finite domain $\mathcal{D}(\sigma)$ = {x | σ(x) $\neq$ x}. An **E-match** from t to t' is a substitution σ such that t' $=^E$ σ(t). σ is simply called a **match** if E is empty. An **E-unifier** of two terms t and t' is a substitution σ such that σ(t) $=^E$ σ(t'). If E is empty there exists a **most general unifier** called **mgu** for any pair (t,t') of terms that has unifiers. All unifiers are actually instances of the mgu. If E is not empty, a basis for generating the set of unifiers by instanciation, whenever one exists, is called **complete set of unifiers** (**csu** in short). □

Many theorical problems arise in equational theories that can be approached by the use of *rewrite rules* i.e. directed equations, or more generally by the use of *mixed sets of rules R and equations E*. We define the rewriting relation in the general case and specialize it for the standard case where equations are not used in the rewriting relation.

Definition 4: A **rewrite rule** is a pair of terms denoted **l→r** such that $\mathcal{V}$(r) is a subset of $\mathcal{V}$(l). The **rewriting relation** $\to^{R,E}$ (or more simply $\to^R$ if E is empty) is defined as follows:
t $\to^{R,E}$ t' iff there exist an occurrence α of $\mathcal{G}$(t), a rule l→r of the set R of rewrite rules and a substitution σ such that t/α $=^E$ σl and t' = t[α←σ(r)].
$\text{-}*\!\to^{R,E}$ (resp. $\text{-}+\!\to^{R,E}$) denotes the reflexive transitive (resp. transitive) closure of $\to^{R,E}$ and is

called the **derivation relation**. $\text{-n}\rightarrow^{R,E}$ denotes n step rewritings with $\rightarrow^{R,E}$. A term t is said **reducible** if t $\rightarrow^{R,E}$ t' for some t', else it is said in **normal form**. □

Notations: In the following, i, j, k denote natural numbers, x, y, z denote variable symbols, a, b, c and e denote constant symbols, f and h denote function symbols, s and t denote terms, λ, μ and ν denote occurrences, ϵ denotes the empty occurrence, and $\lambda\mu$ denotes the concatenation of occurrences λ and μ. Letters σ, τ, θ and η denote substitutions, σt or σ(t) denote the instanciation of t by σ, composition of substitutions σ and τ is witten $\sigma\tau$, l=r and g=d denote equations, l$\rightarrow$r and g$\rightarrow$d denote rules or specify the way an equation is used. Each one of these notations can be indexed by a natural number or by a letter denoting a natural number.
Finally, composition of relation other than substitutions is denoted by a bold dot. □

Notice that R,E-reducibility is *decidable* whenever the matching problem is decidable in the theory E. This is always the case if the theory E is empty. When the theory E is not empty, we will use either the relation $\rightarrow^{R}$ or the relation $\rightarrow^{R,E}$. We will speak about **rewritings** for $\rightarrow^{R}$ and **E-rewritings** for $\rightarrow^{R,E}$.

It must be well understood that the two relations $\rightarrow^{R,E}$ and $=^{E}\circ\rightarrow^{R}$ are different: If t $\rightarrow^{R,E}$ t' at occurrence u, the E-equality steps may apply in t only at occurrences that are suffixes of u. If t $=^{E}\circ\rightarrow^{R}$ t', then the E-equality steps may apply at any place in t. The two relations are therefore the same only if $\rightarrow^{R,E}$ rewrites at the top. This is illustrated by the following example where + is a binary infix operator with the properties:
associativity used as an equation: (x+y)+z = x+(y+z), left identity used as a rule: e+x $\rightarrow$ x
Now consider the term t = (y+e)+z. By associativity, t =E y+(e+z) and this last term rewrites to y+z. However, t is in normal form for $\rightarrow^{R,E}$ because its subterm (y+e) is obviously in normal form and the whole term itself cannot be matched with the left member of rule using associativity since neither (y+e)+z nor y+(e+z) are instances of e+x.

A usually required property for a rewriting relation is *Confluence*, which expresses that the result of a computation does not depend on the choice of the rules to be applied. This property has to do with the so called *Church-Rosser property*, which says roughly that the word problem can be solved by mean of checking normal forms for equality or E-equality.

Definition 5: Let R be a set of rules and E a set of equations. Let =A be the congruence generated by both equations and rules used both ways. Then we say that the rewriting relation $\rightarrow$ (either $\rightarrow^{R}$ or $\rightarrow^{R,E}$) is:

1. E-Church-Rosser iff $\forall t_1,t_2$ such that $t_1 =A\ t_2$, $\exists t_1',t_2'$ such that t_1 -*$\rightarrow$ t_1', t_2 -*$\rightarrow$ t_2' and $t_1' =^{E} t_2'$.

2. E-confluent iff the property holds for t_1 and t_2 such that $\exists$s, s -*$\rightarrow$ t_1 and s -*$\rightarrow$$t_2$.

3. Locally E-confluent iff the property holds for t_1 and t_2 such that $\exists$s, s $\rightarrow$ t_1 and s$\rightarrow$$t_2$. □

If the rewriting relation is terminating, then we can choose for t_1' and t_2' the normal forms of t_1 and t_2.

When E is empty, the rewriting relation is necessarily $\rightarrow^R$. In that case, Church-Rosser and Confluence are the same property [Huet 80]. This is not the case anymore when E is not empty, as shown by the following example, where + is a binary infix operator with the following properties:
commutativity used as an equation: x+y =y+x, left identity used as a rule: e+x → x
Confluence is obviously satisfied, however x+e and x are equal under the whole theory, are both in normal form and are not equal under commutativity. Note that the problem disappears if we allow commutative rewritings, because x+e rewrites now to x. But this is not the case in general [Peterson and Stickel 81, Jouannaud 83, Jouannaud and Kirchner 84]: If we look back at our example with associativity, (y+e)+z and y+z were also equal under the whole theory, both were in normal form for the E-rewriting relation, however they were not equal under associativity.

In order to have the E-Church-Rosser property, another relation must actually be introduced to express that two E-equal terms must have E-equal normal forms. This property can be either the *E-commutation* property defined in section 4 the *E-compatibility property* of [Peterson and Stickel 81] or the more general *E-coherence property* defined first in [Jouannaud 83] and further improved in [Jouannaud and Kirchner 84].

3 Termination

Definition 6: A TRS R is **terminating** if it does not exist an infinite sequence of terms such as $t_0 \rightarrow^R t_1 \rightarrow^R t_2 \rightarrow^R \dots \rightarrow^R t_n \dots$. □

The termination problem for TRS was studied by [Manna and Ness 70] who introduced *reduction orderings*:

Definition 7: An ordering > on terms is a **reduction ordering** iff it is **compatible** with the operations of the term Algebra i.e.: t > t' implies f(... t ...) > f(... t' ...) □

The main property of reduction orderings is that they contain a rewriting relation $\rightarrow^R$ whenever they contain the pairs (σg,σd) for any rule g→d in R and any substitution σ [Manna and Ness 70]. Therefore, they allow to prove the termination of the $\rightarrow^R$ rewriting relation, whenever they are well founded.
An important class of such orderings is given by the so-called **Polynomials orderings**. In order to compare terms, these orderings compare polynomials in their variables. The key point is to guess a polynomial that will make each left hand side of rule greater than the corresponding right hand side. This clearly requires some expertise. See [Huet and Oppen 80] for a discussion and accurate references.

A very important class of reduction orderings is introduced in [Dershowitz 82]:

Definition 8: A **simplification ordering** is a reduction ordering > that has the **subterm property**, i.e. f(... t ...) > t. □

The main property of simplification orderings is that they contain the embedding relation, which is actually the smallest (in a set theoretic sense) simplification ordering:

Definition 9: Let s and t be two terms. s is said to be **embedded** in t and we write s emb t iff one of the following conditions hold:
(1) s is a variable of $\mathcal{V}(t)$
(2) $s = f(\ldots\ s_i \ldots)$, $t = f(\ldots\ t_i \ldots)$ and s_i emb t_i for any i
(3) s emb t_i for t_i a subterm of t. □

Embedding can just be seen as a way to map injectively the nodes of s on the nodes of t, with respect to the topology of s. But the map is not surjective in general: when it is the case, then s and t are identical. Let us draw an example:

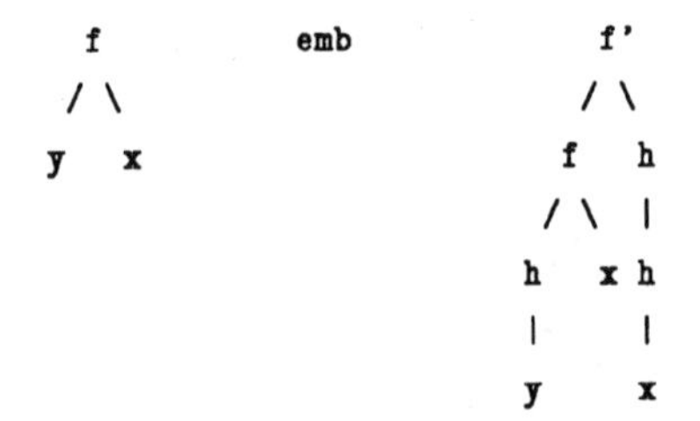

Several interesting simplification orderings have been designed these past years, mainly the *Knuth and Bendix ordering* [Knuth and Bendix 70], the *path of subterm ordering* [Plaisted 78], the *recursive path ordering* [Dershowitz 82] [Kamin and Levy 80] and the *recursive decomposition ordering* [Jouannaud, Lescanne and Reinig 83] [Lescanne 84]. All these orderings extend to terms a precedence defined on function symbols. The precedence can be buit automatically for the last one, which was first done in [Lescanne 83] and then improved in [Choque 83].

Before closing this section, we note that the problem of termination of the relation $\rightarrow^{R,E}$ has never been adressed. We will see in the following that this is one of the main issues for the E-termination problem that we define now.

4 E-Termination

Our problem is slightly different from the problem of termination because we allow equality steps between rewritings:

Definition 10: A TRS R is **E-terminating** iff it does not exist an infinite sequence of the form $t_0 =^E t_0' \rightarrow^R t_1 =^E t_1' \rightarrow^R \ldots =^E t_n' \rightarrow^R t_{n+1} \cdots$. □

Note that we can use either $\rightarrow^R$ or $\rightarrow^{R,E}$ in the previous definition, since $=^E\circ\rightarrow^{R,E}$ and $=^E\circ\rightarrow^R$ are the same relation. Remind that the relation $=^E\circ\rightarrow^R$ is strictly more powerfull than the relation $\rightarrow^{R,E}$ itself, because it allows E-equalities to take place *anywhere* in the term to be rewritten, whereas $\rightarrow^{R,E}$ allows E-equalities to take place only *under* the occurrence where the term is rewritten. The same happens with the relation $=^E\circ\rightarrow^{R,E}$ since it is equal to $=^E\circ\rightarrow^R$. Let us give an additional example with our favorite binary infix operator +, and the following properties:

associativity used as equation: (x+y)=z = x + (y+z)
simplifiability used as a rule: $x + (- x) \to e$
Then the term $x + ((- x) + y)$ is in normal form for $\to^{R,E}$, since neither one of his subterms nor itself is an instance (modulo associativity) of $x+(-x)$. However, the whole term is equal under associativity to $(x + (- x)) + y$ and rewrites to $e + y$.

Let now $\to^{R/E}$ be the relation $=^{E}.\to^{R}.=^{E}$. This relation on terms simulates the rewriting relation induced by $\to^{R}$ in E-congruence classes. It is clear that the E-termination problem is nothing else but the termination of $\to^{R/E}$. These remarks will be freel used in what follows.

Let us now start with some general comments about E-termination that show simple restrictions on the set of equations.
First $\mathcal{V}(g) = \mathcal{V}(d)$ for any equation g=d. Else there are obviously infinite loops as we may instanciate the extra variable by anything, for instance a left member l of some rule $l \to r$, then rewrite the term using the rule and then coming back to the starting term by applying the starting equation twice, in order to first erase r, then obtain l again.
A second important remark is that E-termination cannot be satisfied whenever there are some equations of the form x=t where x has several occurrence in t, because in that case l is E-equal to a term with several occurrences of l. One can rewrite one of these and start the process again. This is actually the case with any instance of such an axiom whith x replaced by a non ground term. In the following, we assume that E does not contain such axioms.

We are primarily interested in proving termination of TRS generated by the generalized Knuth and Bendix completion procedure [Jouannaud and Kirchner 84]. This procedure completes the set of rules in order to get two main properties, namely E-confluence and E-coherence. In the following, we introduce and use a property stronger than E-coherence, called E-commutation, that allows to reduce E-termination to termination of the rewriting relation.

5 E-Termination of E-Commuting Rewriting Relations

In this section, we prove abstract results for an arbitrary rewriting relation $\to$. These results are applied in the next section to the case where the rewriting relation is $\to^{R}$ or $\to^{R,E}$. In the following, the rewriting relation $\to$ is supposed to satisfy the property: $\to^{\mathbf{R}} \subseteq \to \subseteq \to^{\mathbf{R/E}}$.
As a consequence,
$=^{E}\circ\to^{R}\circ=^{E}$, $=^{E}\circ\to\circ=^{E}$, $=^{E}\circ\to^{R/E}$, $=^{E}\circ\to^{R/E}\circ=^{E}$ and $\to^{R/E}$ are the same relation.

Definition 11: A rewriting relation $\to$ is
E-commuting with a set E of equations iff for any s', s and t such that $s' =^{E} s \to^{+} t$, then $s' \to^{+} t' =^{E} t$ for some t'.
locally E-commuting, whenever the property holds for $s' \vdash\!\dashv^{E} s \to t$. □

Notice that we require at least one $\to$ step from s' before to get t'. This is coherent with our goal to prove E-termination: If R is E-terminating, then s' must be different from t' , else there would be the following cycle: $t =^{E} t' = s' =^{E} s \to t$, therefore $t \to^{R/E} t$ since $=^{E}\circ\to$ is included into $\to^{R/E}$.

E-commutation differs from *E-compatibility* by rewriting from s' until a term t' is found to be E-equal to t, instead rewriting once from s' to a term $s^\bullet$ and then rewriting from t until a term $t^\bullet$ is found to be E-equal to $s^\bullet$. On the other hand, *E-coherence* allows both kinds of rewritings and is therefore more general than either one.

The importance of the E-compatibility relation was shown by Peterson and Stickel: Assuming E-compatibility, E-Church-Rosser and E-confluence are equivalent. This can easily be shown by induction on the length of the proof that s $=_{R \cup E}$ t in the E-church-Rosser definition. Let us now point out the main importance of E-commutation for the E-termination problem:

Theorem 12: [Munoz 83] Let R be a set of rules and E a set of equations. Assume the rewriting relation $\rightarrow$ is terminating and E-commuting. Then R is E-terminating.

proof: by noetherian induction on $\rightarrow$. Let $t_0 \rightarrow^{R/E} t_1 \text{-*}\!\rightarrow^{R/E} t_n$ a derivation issued from t_0. By definition of R/E, there exists a t_0' such that $t_0 =^E t_0' \rightarrow^R t_1$, therefore $t_0 =^E t_0' \rightarrow t_1$ by definition of $\rightarrow$. Now, by E-commutation there exists a $t_0^\bullet$ such that $t_0 \text{-+}\!\rightarrow t_0^\bullet =^E t_1$. As $t_0^\bullet =^E t_1$, the rest of the derivation starting from t_1 is also a R/E derivation starting from $t_0^\bullet$. By induction hypothesis, it must be finite, because $t_0^\bullet$ is a proper son of t_0 for $\rightarrow$. □

Note that there is no need for R to be E-confluent.
Proving the same property with local E-commutation instead of E-commutation requires little more work.

Lemma 13: Assume $\rightarrow$ is locally E-commuting.
Then $t_0 =^E t_0' \rightarrow t_1$ implies $t_0 \text{-+}\!\rightarrow\!\circ\text{-*}\!\rightarrow^{R/E} t_1$.

proof: by induction on n if t_0 |-n-| t_0'. The basic n=0 case is straightforward.
Let now t_0 $|\text{-n-}|^E$ t |-| $t_0' \rightarrow t_1$. By local E-commutation, we have t $\text{-+}\!\rightarrow\!\circ\!=^E t_1$, thus $t \rightarrow t' \text{-*}\!\rightarrow\!\circ\!=^E t_1$ for some t'. By induction hypothesis, $t_0 \text{-+}\!\rightarrow\!\circ\text{-*}\!\rightarrow^{R/E}$ t', thus $t_0 \text{-+}\!\rightarrow\!\circ\text{-*}\!\rightarrow^{R/E} t_1$, using the fact that $\rightarrow$ is included into $\rightarrow^{R/E}$. □

Theorem 14: Let R a set of rules and E a set of equations. Assume that the rewriting relation is terminating and locally E-commuting. Then R is E-terminating.

proof: Almost the same proof as for theorem 12. What differs is that we use the lemma instead of E-commutation. This allows us to construct from t_0 a derivation of the form $t_0 \text{-+}\!\rightarrow t_0^\bullet \text{-*}\!\rightarrow^{R/E} t_1 \text{-*}\!\rightarrow^{R/E} t_n$, and we are done. □

As previously, there is no need for R to be E-confluent or locally E-confluent.
From these results, we can obtain a new Church-Rosser result for Equational Term Rewriting Systems using Termination instead of E-Termination:

Theorem 15: Let R be a set of rules and E a set of equations. Assume that $\rightarrow$ is terminating. Then $\rightarrow$ is E-confluent and E-commuting (thus E-Church-Rosser) iff it is locally E-confluent and locally E-commuting.

proof: For the only if part, we remark that global properties imply local ones and that E-Church-Rosser is true since on one hand E-commutation implies E-Coherence and on the other hand E-coherence and E-confluence imply E-Church-Rosser [Jouannaud and Kirchner 84].
For the if part, we first use theorem 14 to prove E-termination. Then we can prove E-commutation from local E-commutation by noetherian induction on the relation $(\rightarrow^{R/E})_{mult}$, which is the extension of $\rightarrow^{R/E}$ to multisets of terms (see [Jouannaud and Lescanne 82] for a discussion about multiset extensions of orderings):
Let $t_0 \rightarrow t_0' \dashrightarrow^{*} t_0''$ and $t_0 \mid\!-\!\mid^{E} t_1 \ldots \mid\!-\!\mid^{E} t_n$. By local E-commutation, $t_1 \dashrightarrow^{+} t_1' =^{E} t_0'$. By induction hypothesis applied to the multiset $\{t_1,\ldots,t_n\}$ which is strictly smaller than the starting multiset, we get $t_n \dashrightarrow^{+} t_n' =^{E} t_1'$. As terms along the proof that $t_0' =^{E} t_1' =^{E} t_n'$ are all proper sons of t_0 for the relation $\rightarrow^{R/E}$, the multiset of these terms is smaller than the multiset $\{t_0\}$ itself, thus than the starting multiset. By induction hypothesis, $t_n' \dashrightarrow^{*} t_n'' =^{E} t_0''$ and we are done. E-confluence can now be obtained easily from local E-confluence and E-coherence by noetherian induction on $\rightarrow^{R/E}$. □

Let us point out that multiset induction allows simple and elegant proofs. This technique was introduced in [Jouannaud and Kirchner 84] for proving very general Church-Rosser results using local E-coherence instead of local E-commutation.

We already know how to prove termination of the standard rewriting relation $\rightarrow^{R}$, using a reduction ordering $>$ such that $\sigma(g) > \sigma(d)$ for any rule $g \rightarrow d$ and any substitution σ. This enables us to prove E-termination of R, provided it is E-commuting.

Example 1: [Huet 80] Let 0 and 1 be constants, Exp be a unary function symbol and + and $*$ be binary infix function symbols. Assume + and $*$ are both associative and commutative. Rules are:
$x+0 \rightarrow x$, $0+x \rightarrow x$, $x*1 \rightarrow x$, $1*x \rightarrow x$, $Exp(0) \rightarrow 1$ and $Exp(x+y) \rightarrow Exp(x)*Exp(y)$.
It is easy to check that the standard rewriting relation is locally commuting. Moreover, it is terminating as shown by a recursive path ordering [Dershowitz 82] with $Exp > *$ and $Exp > 1$ the precedence on function symbols. By theorem 14, R is terminating modulo associativity and commutativity of + and $*$. Actually, R is also locally E-confluent and we could use directly theorem 15 for proving its Church-Rosser property.

In order to prove termination of the rewriting relation $\rightarrow^{R,E}$, we simply need a reduction ordering $>$ that satisfies: $t > \sigma(d)$ for any rule $g \rightarrow d$, any substitution σ and any t such that $t =^{E} \sigma(g)$. As a matter of fact, such an ordering will contain the $\rightarrow^{R,E}$ rewriting relation and thus prove its termination provided it is well founded.

Example 2: Let f be a binary symbol and - be a unary symbol with the following properties:
$f(x,f(-x,y)) \rightarrow y$ and $f(f(y,-x),x) \rightarrow y$ used as rules.
$--x=x$ and $-f(x,y)=f(-y,-x)$ used as equations.
This equational theory is introduced in [Jouannaud, Kirchner C. and H. 81] and studied in [Kirchner C. and H. 82] where R,E is proved to be E-commuting. Since R,E must reduce the

number of f symbols, it is also terminating. Therefore R is E-terminating. Note that the same argument used for proving termination of R,E actually proves termination of R/E.

An interesting open problem now arises: the design of well-suited reduction orderings for proving termination of R,E. We expect it to be easier than the design of reduction orderings working in E-congruence classes, as it is done in [Dershowitz, Hsiang, Josephson and Plaisted 83, Plaisted 83], that is reductions orderings for proving termination of R/E. By the way, these orderings can also be used for proving termination of R,E since R,E is included into R/E.

6 E-Termination of non E-Commuting Rewriting Relations

The main idea of this section is quite simple: Since the set R of rules is not E-commuting, we first compute the smallest E-commuting set of rules that contains R and then prove the termination of the rewriting relation associated with this new set of rules. In addition, we will show that we can use two different proofs for R and the added rules. This will give us a much more powerfull technique.

Definition 16: A relation $>$ is said to be:
E-commuting with a rewriting relation $\rightarrow$ iff for any s',s and t such that $s' =^E s \dashrightarrow^+ t$, then $s' > t' =^E t$ for some t'.
semi-locally E-commuting if the property holds for $s' =^E s \rightarrow t$.
locally commuting if it holds for $s' \vdash\dashv^E s \rightarrow t$. □

E-commuting relations play the same role as E-commuting rewriting relations. Notice that we obtain our previous definition if $>$ is taken to be $\dashrightarrow^+$. In the following, we assume $>$ to be an ordering. Notice that E-commutation and semi-local E-commutation can actually be proved equivalent by a simple induction on the length of the derivation. This will be freely used in what follows by calling E-commutation what in fact is semi-local E-commutation. We can now adapt theorem 12 to the case of E-commuting orderings, provided they are assumed well founded. Our goal is actually to relax the hypothesis that $>$ is well founded and assume that it contains the embedding relation.

Theorem 17: [Munoz 83] Assume that $>$ is E-commuting with $\rightarrow$ and contains the embedding relation. Then $\rightarrow^R$ is E-terminating.

proof: By contradiction. Assume there exist an infinite chain for $\rightarrow^{R/E}$ issued from t_0: $t_0 =^E t_0' \rightarrow^R t_1 =^E t_1' \rightarrow^R t_2 \ldots t_n =^E t_n' \rightarrow^R t_{n+1} \ldots$. Then, applying E-commutation as many times as needed, we construct an infinite chain for $>$: $t_0 > t_1'' > \ldots > t_n'' > t_{n+1}'' \ldots$. By Kruskal's theorem, there exist two terms t_i and t_j with $i <_N j$ in the sequence such that t_i is embedded in t_j. It follows from hypothesis that $t_i < t_j$ or $t_i = t_j$. But $t_i > t_j$ by transitivity of the ordering $>$, which gives a contradiction. □

Note that there is no need for $>$ to be a well founded ordering: what is important is that it contains the embedding relation. A same trick was actually used in [Dershowitz 82] when proving that simplification orderings can be used to prove termination. This theorem can now be applied to simplification orderings, because they contain the embedding relation. We can thus suppose that $>$ is an extension of such an ordering.

Let us now remark that our definition of E-commutation collects two different notions:

- Let us assume that the E-equality step in the E-commutation definition is empty i.e. s' = s $\dashrightarrow^{+}$ t. In that case, we obtain from the definition that there exists a t' such that s' $>$ t' $=^E$ t. A reasonable restriction is to assume t' = t for that special case, as for the empty theory. This means that the rewriting relation $\rightarrow$ is included into the commuting ordering, which is easily obtained by requiring that the simplification ordering or the E-simplification ordering orients the rules: as seen previously it will contain the whole rewriting relation.

- Let us assume now that s'$\neq$s and s' $=^E$ s $\rightarrow$ t, which implies that ww[s' $\rightarrow$ t' $=^E$ t] for some t'. This is what is called E-commutation in the following. This E-commutation notion has as its purpose to ensure E-termination provided termination is true.

Theorem 17 can now be restated in the two following ways:

Theorem 18: [Munoz 83] Let $>$ be a simplification ordering such that σ(g) $>$ σ(d) for any rule g$\rightarrow$d in R and any substitution σ. Then $\rightarrow^R$ is E-terminating if $>$ is contained into an ordering $>$' E-commuting with $\rightarrow^R$.

Theorem 19: Let $>$ be an E-simplification ordering such that t $>$ σ(d) for any rule g$\rightarrow$d in R, any substitution σ and any t such that t $=^E$ σ(g). Then R is E-terminating if $>$ is contained into an ordering $>$' E-commuting with $\rightarrow^{R,E}$.

proofs: We verify hypotheses of theorem 17. $>$' contains $>$ thus the embedding since $>$ is a simplification ordering. Using the hypotheses, $>$ contains the rewriting relation $\rightarrow^R$ or $\rightarrow^{R,E}$ depending upon theorem 18 or 19. Therefore $>$' has the same property. As $>$' is E-commuting, we capture the full power of our previous definition. □

This result can now be applied in the following way: Let us start with a *simplification ordering* that orients the instances of rules of R (up to E-equality for the instance of the left member in the case where $\rightarrow^{R,E}$ is our rewriting relation). This ordering can be for example the Recursive Path Ordering or the Recursive Decomposition Ordering if the rewriting relation is $\rightarrow^R$. Now try to extend this ordering into an *E-commuting reduction ordering*. This will be the difficult step in practice and we now study various ways for achieving it.

The technique we propose relies on the two following main remarks: First of all, the commutation relation is related to some kind of *critical pairs* i.e. can be checked on a set of pairs of terms (t_0,t_1) that we can try to orient with the extended ordering. Second, usual simplification orderings are *monotonic* with respect to the precedence on function symbols they use. A first simple idea is thus to *increase*, if necessary, the *precedence* on function symbols in order to orient the *critical pairs*.

However, the ordering used to compare the critical pairs has no need to be the same than the ordering used to compare the rules themselves. What is required is that it must be an extension, and we can actually imagine various ways to build such an extension in practice, based for example on a lexicographic way of using the first ordering then another one. This will provide a much more powerfull technique than increasing the precedence only.

The previous techniques reduce the problem of E-termination of a set of rules to the problem of termination of an extended set of rules. We now split our discussion into two sections, according to the two different rewriting relations we can use.

6.1 $\rightarrow^R$ Rewritings

Definition 20: Given two terms t' and t, we say that t' **overlaps** t at occurrence λ in $\mathcal{G}(t)$ if t' and t/λ are unifiable . If σ is their most general unifier, $\sigma(t)$ is called the **overlaping term**. Given now two rules $g \rightarrow d$ and $l \rightarrow r$ such that g overlaps l at occurrence λ with the substitution σ, we call **critical pair** the pair of terms $<\sigma r, \sigma l[\lambda \leftarrow \sigma d]>$. Let SCP($g \rightarrow d, l \rightarrow r$) be the set of all critical pairs obtained by overlaping of g on l. □

Notice that any term overlaps with itself at the top. Such overlapings produce trivial critical pairs that we don't consider here. Notice also that a critical pair is not symmetric: this is due to the fact that the overlaping operation itself is not symmetric. More precisely, the overlaping term σl rewrites both ways as follows: $\sigma l \rightarrow \sigma r$ and $\sigma l \rightarrow \sigma l[\lambda \leftarrow \sigma d]$. The first rewriting uses the rule $l \rightarrow r$ whereas the second one uses the rule $g \rightarrow d$.

With critical pairs is associated the so called:

Critical pair lemma [Huet 80]: Let $t \rightarrow t_1$ at the top with the rule $g \rightarrow d$ and $t \rightarrow t_2$ at occurrence λ with the rule $l \rightarrow r$. If λ is in $\mathcal{G}(g)$, then there exist a critical pair (p,q) in SCP($l \rightarrow r, g \rightarrow d$) and a substitution σ such that $t_1 = \sigma p$ and $t_2 = \sigma q$. □

We now introduce an operation over sets of rules that we call *Critical Pair*:

Definition 21: We say that a set of rules is closed with respect to a set E of equations under **Critical Pair operation** iff for any rule $l \rightarrow r$ of R and any equation g=d of E:
- for any (p,q) in SCP($l \rightarrow r, g \rightarrow d$) or in SCP($l \rightarrow r, d \rightarrow g$), $p \xrightarrow{+}{}^{R} \circ =^{E} q$.
- for any (p,q) in SCP($g \rightarrow d, l \rightarrow r$) or in SCP($d \rightarrow g, l \rightarrow r$), $q \xrightarrow{+}{}^{R} \circ =^{E} p$. □

Note that the previous condition is satisfied if $p \rightarrow q$ (resp. $q \rightarrow p$) is in R. Actually, this definition means that we want the E-commutation property to be true at least for the critical pairs of the rules with the equations. As usually, it will be then satisfied for all possible cases.

Let now **CP(R,E)** be a (smallest) set of rules that contains R and is closed under Critical Pair operation. Such a (non unique) set can be obtained by adding the rules $p \rightarrow q$ or $q \rightarrow p$ each time the previous property is not satisfied. Generally, we can't expect it to be finite, but let us postpone this problem.

Theorem 22: Let R be a set of left linear rules and E a set of linear equations. Then R is E-terminating if there exist:

•a simplification ordering $>$ such that $\sigma(l) > \sigma(r)$ for any rule $l \rightarrow r$ in R and any substitution σ.

•a reduction ordering $>'$ that contains $>$ and such that $\sigma(l) >' \sigma(r)$ for any rule $l \rightarrow r$ in CP(R,E) - R and any substitution σ. □

proof: We have to prove the commutation of $>'$ with $\rightarrow^R$. Let us first remark that $>'$ contains $>$ thus $\rightarrow^R$, because $>$ is a reduction ordering that contains the instances of rules in R. For the same reason, $>'$ contains $\rightarrow^{CP(R,E)-R}$, thus $\rightarrow^{CP(R,E)}$, therefore $\text{-+}\rightarrow^{CP(R,E)}$ by transitivity. It is therefore sufficient to prove E-commutation of $\text{-+}\rightarrow^{CP(R,E)}$ with $\rightarrow^R$. Since $\rightarrow^{CP(R,E)}$ contains $\rightarrow^R$, this property follows from the E-commutation of $\text{-+}\rightarrow^{CP(R,E)}$ with $\rightarrow^{CP(R,E)}$. This last property is proved now:

Let $t_1 |\text{-n-}|^E\ t \rightarrow^{CP(R,E)} t_2$. We prove the property by induction on n.

If n is greater than 1, the property is easily proved by induction.

If n=0, the property is obvious.

Let us now deal with the n=1 case. Assume that $t/\lambda = \sigma(g)$ and $t_1 = t[\lambda \leftarrow \sigma(d)]$ for an equation g=d in E and that $t/\mu = \sigma(l)$ and $t_2 = t[\mu \leftarrow \sigma(r)]$ for a rule $l \rightarrow r$ in CP(R,E). In order to have the same substitution for both rewritings, we classicaly assume without loss of generality that g and l have disjoint sets of variables. We discuss different cases according to the respective positions of λ and μ:

(1) If λ and μ are disjoint, then the equality step and the rewriting step commute.

(2) If λ is a prefix of μ, let us say $\mu = \lambda\mu'$, we distinguish two cases:

(a) If μ' belongs to $\mathcal{G}(g)$, then by the critical pair lemma, there exist a critical pair (p,q) in SCP($l \rightarrow r, g \rightarrow d$) such that $t_1 = \sigma'p$ and $t_2 = \sigma'q$ for a substitution σ'. Then $t_1 = \sigma'p > t_2 = \sigma'q$ or $t_1 = \sigma'p >\circ=^E \sigma'q = t_2$ by hypotheses and definition of CP(R,E).

(b) If μ' does not belong to $\mathcal{G}(g)$, then $t_1 \text{-m}\rightarrow t_1' |\text{-m'-}| t_2$, where m is the number of occurrences of the variable x at occurrence ν in g such that $\mu' = \nu\mu''$. As equations are supposed to be linear, m=1 and the result is true.

(3) If μ is a prefix of λ, the proof works as previously and uses the fact that rules in R are supposed to be left linear, which implies that rules in CP(R,E) are left linear too, since equations are linear. This is actually a property of unification. This ends the proof of our theorem. □

Example 3: Let us use example 1 again, with the new rules

$(x+y)*z \rightarrow (x*z)+(y*z)$ and $z*(x+y) \rightarrow (z*x)+(z*y)$.

By adding $*>+$ to the precedence, the recursive path ordering will orient all our rules. Let us now check what are rules to be added in order to have E-commutativity. Actually, they are infinitely many, but all have a same form that enable us to prove their termination with the same recursive path ordering. Let us give some of them:

$((x+y)+z)*z' \rightarrow (x*z')+((y+z)*z')$, $(x+y)*(z*z') \rightarrow ((x*z)+(y*z))*z'$ and

$(x'*(x+y))*z \rightarrow x'*((x*z)+(y*z))$.

6.2 $\rightarrow^{R,E}$ REWRITINGS

The previous technique is rather restrictive, especially for the kind of rules that can be handled. In order to relax this restriction, we now use the $\rightarrow^{R,E}$ rewriting relation of Peterson and Stickel. This relation was introduced by these authors to solve the problem of non left linearity of rules, and a new notion of critical pairs:

Definition 23: Given two terms t' and t, t' **E-overlaps** t at occurrence λ in $\mathcal{G}(t)$ with a *complete set of E-unifiers* Σ iff $\sigma(t') =^E \sigma(t/\lambda)$ for any substitution σ in Σ. Given two rules $l \rightarrow r$ and $g \rightarrow d$ such that l E-overlaps g at occurrence λ with a complete set Σ of E-unifiers, we call **complete set of E-critical pairs** of $l \rightarrow r$ on $g \rightarrow d$ at occurrence λ to the set of pairs $\{<\sigma(d),\sigma(g[\lambda \leftarrow r])> \mid \text{for any } \sigma \text{ in } \Sigma\}$, and **extended pair** of $g \rightarrow d$ on $l \rightarrow r$ at occurrence λ to the pair $<g[\lambda \leftarrow l],g[\lambda \leftarrow r]>$. In the following, **SEP**$(g \rightarrow d, l \rightarrow r)$ denotes the set of all extended pairs of $g \rightarrow d$ on $l \rightarrow r$. □

Extended pairs were first introduced in [Peterson and Stickel 81] for the associative commutative case, and as above in [Jouannaud and Kirchner 84]. Notice that the substitution σ is not involved in the computation of the extended pair. On the other hand, for any unifier σ in Σ, we have: $\sigma(g[\lambda \leftarrow l]) = \sigma g[\lambda \leftarrow \sigma l] = \sigma g[\lambda \leftarrow \sigma(g/\lambda)] = \sigma g$ If g=d is an equation of E, then σg $|\text{-}|^E$ σd at the top. It follows that if σg rewrites at the top with $\rightarrow^{R,E}$, then σd rewrites actually to the same term. Therefore, if the left member of the extended pair rewrites at the top to a term t, then the left member of any corresponding critical pair will rewrite at the top to the corresponding instance of t. This is very important for practice because it allows to avoid the computation of the complete set of unifiers of l and g/λ: we only need to know wether the two terms unify or not.

With extended pairs is associated the

E-extended pair lemma: Let t $|\text{-}|^E$ t_1 at the top with the equation $g \rightarrow d$ of E and t $\rightarrow^{R,E}$ t_2 at occurrence λ in $\mathcal{G}(g)$ with the rule $l \rightarrow r$ of R. Then there exist an extended pair (p,q) in $SEP(l \rightarrow r, g \rightarrow d)$ and a substitution σ such that $t_1 =^E \sigma p$ and $t_2 =^E \sigma q$.

proof: Same as the proof of *E-critical pairs lemma 1* in [Jouannaud 83]. □

As previously, extended pairs define a closure operation:

Definition 24: We say that a set of rules is closed with respect to a set E of equations under **Extended Pair operation** iff for any rule $l \rightarrow r$ of R and any equation g=d of E:
•for any (p,q) in $SEP(l \rightarrow r, g \rightarrow d) \cup SEP(l \rightarrow r, d \rightarrow g)$, p $\rightarrow^{R,E}$ p' at the top and p' $\text{-}^*\!\!\rightarrow^R \circ =^E$ q
Let **EP(R,E)** be the smallest set of rules that contains R and is closed under Extended Pair operation. □

As previously, p must be rewritten at least once. This is the case if $p \rightarrow q$ is in R.

Theorem 25: Let R be a set of rules and E a set of linear equations. Then R is E-terminating if there exist

•a simplification ordering $>$ such that $t > \sigma(d)$ for any rule $g \rightarrow d$ in R and any substitution σ such that $t =^E \sigma(g)$
•a reduction ordering $>'$ that contains $>$ and such that $t > \sigma(q)$ for any extended pair (p,q) in EP(R,E) and any substitutions σ such that $t =^E \sigma(p)$.

proof: The proof is nearly the same as the proof of theorem 3, except that $\rightarrow^R$ or $\rightarrow^{CP(R,E)}$ reductions are replaced by $\rightarrow^{R,E}$ or $\rightarrow^{EP(R,E),E}$ reductions. Notice that the starting relations contained in these orderings are no the instances of rules anymore, but that E-equalities can affect the instance of the left member of rule. Thus, by commutation with the operations of the algebra, these orderings contain the relations $\rightarrow^{R,E}$ or $\rightarrow^{EP(R,E),E}$. Accordingly, the lemma that we have to prove states that $\text{-+}\!\rightarrow^{EP(R,E),E}$ is E-commuting with $\rightarrow^{EP(R,E),E}$. As previously, the proof is by induction and then by case analysis for the n=1 case. We use the same notations, but now $t \rightarrow^{EP(R,E),E} t_2$ and $t/\mu =^E \sigma(l)$.
The case where λ and μ are disjoint is once more straightforward.
If μ is a prefix of λ, then $t_1/\mu =^E t/\mu =^E \sigma(l)$. Therefore $t_1 \rightarrow^{EP(R,E),E} t_2$ and there is no need of extended pairs for this case.
If λ is a prefix of μ, let us say $\mu=\lambda\mu'$, we distinguish two cases:
•If μ' belongs to $\mathcal{G}(g)$, then by extended pair lemma, there exist a substitution σ and an extended pair (p,q) such that $t_1 =^E \sigma p$ and $t_2 =^E \sigma q$. If $p \rightarrow q$ is in EP(R,E), then $t_1 \rightarrow^{EP(R,E),E} \sigma q =^E t_1$. If $p \rightarrow^{EP(R,E),E} p'$ at the top and $p' \text{-*}\!\rightarrow^{EP(R,E),E} \circ\!=^E q$ then $t_1 \rightarrow^{EP(R,E),E} \sigma(p') \text{-*}\!\rightarrow^{EP(R,E),E} \circ =^E \sigma q =^E t_2$.
•If μ' does not belong to $\mathcal{G}(g)$, then the diagram commutes, provided equations are linear. □

This last result provides an easy way of proving E-termination orderings, provided the set of extended pairs is finite. Let us show that it is actually the case for the important case of associative commutative theories.

Theorem 26: The set EP(R,AC) is finite. More precisely:
$EP(R,AC) = R \cup \{f(x,l) \rightarrow f(x,r) \mid \#(x,l) = 0$ and f is top function symbol of l.$\}$

This result is already proved in [Peterson and Stickel 81], because it is one of the basements of their AC-completion algorithm. By the way, we can see that extended pairs ensure the commutation property as well as the compatibility property: in fact they ensure an even stronger property, their common instance.

Example 4: The same previous example can be reused with rewriting modulo associativity and commutativity. Rules $x+0 \rightarrow x$ and $x*1 \rightarrow x$ can now be discarded, since they are associative commutative instances of others. We have now exactly three extended pairs:
$(x+y)*(z+z') \rightarrow ((x*z)+(y+z))*z'$, $(x'*(x+y))*z \rightarrow x'*((x+z)+(y+z))$ and $x'*((x+y)*z) \rightarrow ((x'*x)+(x'*y))*z$.
To prove E-termination of the starting set R of rules, we first prove termination of the relation R,E. This can be done by taking the following lexicographic ordering:
$t<t'$ iff lexicographically (1) $t =AC\ t^{\blacksquare}$ emb t' (2) + and 0 have less occurrences in t than in t'.
(1) is well-founded since embedding is a size-decreasing relation and =AC is a size-preserving

one. (2) is clearly well-founded, therefore $<$ is well founded. Now $<$ contains embedding by (1) and R satisfies the requirements of theorem 25. We are left to construct an extension $>'$ of $>$ that orients the extended rules. This can be done lexicographically with an ordering that counts the number of + symbols, each one with a multiplicity 2^h where h is its depth in the tree associated with the term.

7 Conclusion

This work is an attempt to clarify the E-termination problem and show how it is related to the classical termination problem and the E-commutation property that already arised in [Dershowitz, Hsiang, Josephson and Plaisted 83] in a rather magic way. Rather than providing new orderings to solve the problem, we give here some interesting ways of addressing it. In addition, we have shown precisely what are the problems to be addressed before providing practical and efficient tools for E-termination proofs.
Let us note that we can imagine to use mixted rewriting relations by splitting the whole set R of rules into a first subset Rl of left linear ones and a second subset Rnl, that must contain all non left linear rules of R and maybe some left linear rules too. This technique was used in [Jouannaud 83, Jouannaud and Kirchner 84] to improve the efficiency of rewritings as well as the Knuth and Bendix completion. It is clear from the previous proofs that our results can be easily adapted to this case.
Finally, a last question arises: How to design a completion algorithm for mixed sets of rules and equations that is based on termination of the rewriting relation instead E-termination? First, we must ensure the E-commutation property instead of the E-coherence property as in [Jouannaud and Kirchner 84]. This can be achieved by adding systematically extended pairs each time an equation and a rule overlapp. This is however not sufficient to conclude that E-termination follows from termination all along the completion process, since the E-commutation property will be ensured at the end of the completion and not during the course of the algorithm. On the other hand, E-commutation will be true at least for those rules whose critical pairs are already processed and we can expect that it is sufficient for the completion process to be sound.

Acknowledgments: The authors acknowledge Nachum Dershowitz, Jieh Hsiang and Pierre Lescanne for fruitfull discussions about this problem and Helene Kirchner and Jose Meseguer for carefully reading this draft and suggesting many improvements.

References

[Choque 83] Choque, G.
Calcul d'un ensemble complet d'incrementations minimales pour l'ordre recursif de decomposition.
Technical Report, CRIN, Nancy, France, 1983.

[Dershowitz 82] Dershowitz, N.
Ordering for Term-rewriting Systems.
Journal of Theoretical Computer Science 17(3):279-301, 1982.
Preliminary version in 20th FOCS, 1979.

[Dershowitz, Hsiang, Josephson and Plaisted 83]
Dershowitz, N., Hsiang,J., Josephson, N. and Plaisted,D.
Associate-Commutative Rewriting.
Proceedings 10th IJCAI, 1983.

[Huet 80] Huet, G.
Confluent Reductions: Abstract Properties and Applications to Term Rewriting Systems.
Journal of the Association for Computing Machinery 27:797-821, 1980.
Preliminary version in 18th FOCS, IEEE, 1977.

[Huet and Oppen 80]
Huet, G. and Oppen, D.
Equations and Rewrite Rules: A Survey.
In Book, R. (editor), *Formal Language Theory: Perspectives and Open Problems*, . Academic Press, 1980.

[Jouannaud 83] Jouannaud, J.P.
Church-Rosser Computations with Equational Term Rewriting Systems.
Technical Report, CRIN, Nancy, France, January, 1983.
Preliminary version in Proc. 5th CAAP, 1983, to appear in Springer Lecture Notes in Computer Science.

[Jouannaud and Kirchner 84]
Jouannaud, J.P. and Kirchner, H.
Completion of a set of rules modulo a set of Equations.
Technical Report, SRI-International, 1984.
Preliminary version in Proceedings 11th ACM POPL Conference, Salt Lake City, 1984, submitted to the SIAM Journal of Computing.

[Jouannaud and Lescanne 82]
Jouannaud, J.P., Lescanne, P.
On Multiset Ordering.
Information Processing letters 15(2), 1982.

[Jouannaud, Kirchner C. and H. 81]
Jouannaud, J.P., Kirchner, C. and Kirchner, H.
Algebraic manipulations as a unification and matching strategy for equations in signed binary trees.
Proceedings 7th International Joint Conference on Artificial Intelligence, Vancouver , 1981.

[Jouannaud, Lescanne and Reinig 83]
Jouannaud, J.-P., Lescanne, P., Reinig, F.
Recursive Decomposition Ordering.
In *IFIP Working Conference on Formal Description of Programming Concepts II*, . North-Holland, 1983, edited by D. Bjorner., 1983.

[Kamin and Levy 80]
Kamin, S. and Levy, J.J.
Attempts for generalizing the recursive path ordering.
1980.
Unpublished draft.

[Kirchner C. and H. 82]
Kirchner, C. and Kirchner, H.
Resolution d'equations dans les Algebres libres et les varietes equationelles d'Algebres.
PhD thesis, Universite Nancy I, 1982.

[Knuth and Bendix 70]
Knuth, D. and Bendix, P.
Simple Word Problems in Universal Algebra.
In J. Leech (editor), *Computational Problems in Abstract Algebra*, . Pergamon Press, 1970.

[Lankford and Ballantyne 77a]
Lankford, D. and Ballantyne, A.
Decision Procedures for Simple Equational Theories with Commutative Axioms: Complete Sets of Commutative Reductions.
Technical Report, Univ. of Texas at Austin, Dept. of Mathematics and Computer Science, 1977.

[Lankford and Ballantyne 77b]
Lankford, D. and Ballantyne, A.
Decision Procedures for Simple Equational Theories with Permutative Axioms: Complete Sets of Permutative Reductions.
Technical Report, Univ. of Texas at Austin, Dept. of Mathematics and Computer Science, 1977.

[Lankford and Ballantyne 77c]
Lankford, D. and Ballantyne, A.
Decision Procedures for Simple Equational Theories with Associative Commutative Axioms: Complete Sets of Associative Commutative Reductions.
Technical Report, Univ. of Texas at Austin, Dept. of Mathematics and Computer Science, 1977.

[Lescanne 83] Lescanne P.
Computer Experiments with the REVE Term Rewriting Systems Generator.
In *Proceedings, 10th POPL,* . ACM, 1983.

[Lescanne 84] Lescanne, P.
How to prove termination? An approach to the implementation of a new Recursive Decomposition Ordering.
Proceedings 6th CAAP, Bordeaux , 1984.

[Manna and Ness 70]
Manna, Z. and Ness, N.
On the Termination of Markov Algorithms.
Proceedings of the third Hawaii Conference on System Sciences , 1970.

[Munoz 83] Munoz, M.
Probleme de terminaison finie des systemes de reecriture equationnels.
PhD thesis, Universite Nancy 1, 1983.

[Peterson and Stickel 81]
Peterson, G. and Stickel, M.
Complete Sets of Reductions for Some Equational Theories.
JACM 28:233-264, 1981.

[Plaisted 78] Plaisted, D.
A Recursively Defined Ordering for Proving Termination of Term Rewriting Systems.
Technical Report R-78-943, University of Illinois, Computer Science Department, 1978.

[Plaisted 83] Plaisted, D.
An associative path ordering.
Technical Report, University of Illinois, Computer Science Department, 1983.

ASSOCIATIVE-COMMUTATIVE UNIFICATION

François Fages

CNRS, LITP 4 place Jussieu 75221 Paris Cedex 05,
INRIA Domaine de Voluceau Rocquencourt, 78153 Le Chesnay.

ABSTRACT

Unification in equational theories, that is solving equations in varieties, is of special relevance to automated deduction. Recent results in term rewriting systems, as in [Peterson and Stickel 81] and [Hsiang 82], depend on unification in presence of associative-commutative functions. Stickel [75,81] gave an associative-commutative unification algorithm, but its termination in the general case was still questioned. Here we give an abstract framework to present unification problems, and we prove the total correctness of Stickel's algorithm.

The first part of this paper is an introduction to unification theory. The second part is devoted to the associative-commutative case. The algorithm of Stickel is defined in ML [Gordon, Milner and Wadsworth 79] since in addition to being an effective programming language, ML is a precise and concise formalism close to the standard mathematical notations. The proof of termination and completeness is based on a relatively simple measure of complexity for associative-commutative unification problems.

1. Unification in equational theories

1.1. Equational theories

We assume well known the concept of an algebra $\mathbf{A} = \langle A, F \rangle$ with A a set of elements (the carrier of $\mathbf{A}$) and F a family of operators, given with their arities. More generally, we may consider heterogeneous algebras over some set of sorts, but all the notions considered here carry over to sorted algebras without difficulty, and so we will forget sorts and even arities for simplicity of notation. With this provision, all our definitions are consistent with [Huet and Oppen 80].

We denote by $T(F)$ the set of (ground) terms over F. We assume that there is at least one constant (operator of arity 0) in F so that this set is not empty. We also assume the existence of a denumerable set of variables V, disjoint from F, and denote by $T(F,V)$ the set of terms with variables over F and V. When F and V are clear from the context, we abbreviate $T(F,V)$ as T and $T(F)$ as G (for ground). We denote terms by M,N,..., and write V(M) for the set of variables appearing in M.

We denote by **T** (resp. **G**) the algebra with carrier T (resp. G) and with operators the term constructors corresponding to each operator of F.

The *substitutions* are all mappings from V to T, extended to T, as endomorphisms of **T**. We denote by S the set of all substitutions. If $\sigma \in S$ and $\mathrm{M} \in T$, we denote by σM the application of σ to M. Since we are only interested in substitutions for their effect on terms, we shall generally assume that $\sigma x=x$ except on a finite set of variables $D(\sigma)$ which we call the ***domain*** of σ by abuse of notation. Such substitutions can then be represented by the finite set of pairs $\{<x,\sigma x> \mid x\in D(\sigma)\}$. We define the ***range*** $R(\sigma)$ of σ as :

$$R(\sigma) = \bigcup_{x\in D(\sigma)} V(\sigma x).$$

We say that σ is ***ground*** iff $R(\sigma)=\phi$. The composition of substitutions is the usual composition of mappings: $(\sigma o \rho)x=\sigma(\rho x)$. And we say that ***$\sigma$ is more general than ρ*** : $\sigma\leq\rho$ iff $\exists\eta\ \eta o\sigma=\rho$.

An equation is a pair of terms M=N. Let E be a set of equations (axioms), we define the ***equational theory presented by E*** as the finest congruence over T containing all pairs σM=σN for M=N in E and σ in S. It is denoted by $\underset{E}{=}$. An equational theory presented by E is ***axiomatic*** iff E is finite or recursive.

An algebra **A** is a ***model*** of an equation M=N if and only if νM=νN as elements of A. for every assignment ν (i.e. mapping from V to A extended as a morphism from **T** to **A**). We write **A** |= M=N. **A** is a model of an equational theory E iff **A** |= E for every E in E. We denote by $M(E)$ the class of models of E, which we call the ***variety*** defined by E.

E-equality in T is extended to substitutions by extensionality:

$$\sigma \underset{E}{=} \rho \text{ iff } \forall x\in V\ \sigma x \underset{E}{=} \rho x.$$

We write for any set of variables V:

$$\sigma \overset{V}{\underset{E}{=}} \rho \text{ iff } \forall x\in V\ \sigma x \underset{E}{=} \rho x.$$

In the same way, ***σ is more general than ρ in E over*** V,

$$\sigma \overset{V}{\underset{E}{\leq}} \rho \text{ iff } \exists\eta\ \eta o\sigma \overset{V}{\underset{E}{=}} \rho.$$

The corresponding equivalence relation on substitutions is denoted by $\overset{V}{\underset{E}{\equiv}}$; i.e. $\sigma \overset{V}{\underset{E}{\equiv}} \rho$ iff $\sigma \overset{V}{\underset{E}{\leq}} \rho$ and $\rho \overset{V}{\underset{E}{\leq}} \sigma$. We will omit V when V=$V$, and E when E=ϕ.

1.2. E-unification

1.2.1. Historical

Let E be an equational theory. A substitution σ is a ***E-Unifier*** of terms M and N if and only if σM $\underset{E}{=}$ σN.

Hilbert's tenth problem (solving of polynomial equations over integers, called Diophantine equations) is the unification problem in arithmetic. Livesey, Siekmann, Szabo and Unvericht [79] have proved that Associative-Distributive unification is undecidable, and thus that the undecidability of Hilbert's tenth problem [Matiyasevich 70, Davis 73] does not rely on a specific property of integers.

We denote by U_E the set of all E-unifiers of M and N:

$$U_E\,(M,N) = \{\sigma\in S \mid \sigma M \underset{E}{=} \sigma N\}.$$

Axiomatic equational theories are semi-decidable, and U_E is always recursively enumerable, but of course we are mostly interested in a generating set of the E-unifiers (called Complete Set of E-Unifiers by Plotkin [72], and denoted by CSU_E), from which we can generate U_E by instantiations. Or better by a basis of U_E (called Complete Set of Minimal Unifiers and denoted by μCSU_E) satisfying the minimality condition $\sigma \neq \sigma' \Rightarrow \sigma \not\leq_E^V \sigma'$.

So we shall make the difference between ***unification procedures*** which enumerate a CSU_E (the exhaustive enumeration procedure in semi-decidable theories enumerates U_E entirely), ***unification algorithms***, which always terminate with a finite CSU_E, empty if terms are not unifiable, and ***minimal unification procedures or algorithms*** which compute a μCSU_E.

Unification has been for the first time studied in first order languages (the case $E=\phi$) by Herbrand [30]. In his thesis he gave an explicit algorithm to compute a most general unifier. However the notion of unification really grew out of the work of the researchers in automatic theorem-proving, since the unification algorithm is the basic mechanism needed to explain the mutual interaction of inference rules. Robinson [65] gave the algorithm in connection with the resolution rule, and proved that it indeed computes a most general unifier. Independently, Guard [64] presented unification in various systems of logic. Unification is also central in the treatment of equality [Robinson and Wos 69, Knuth and Bendix 70]. Implementation and complexity analysis of unification is discussed in [Robinson 71], [Venturini-Zilli 75], [Huet 76], [Baxter 77], [Paterson and Wegman 78] and [Martelli and Montanari 82]. Paterson and Wegman give a linear algorithm to compute the most general unifier.

First order unification was extended to infinite (regular) trees by Huet [76], who showed that a single most general unifier exists for this class, computable by an almost linear algorithm. This problem is relevant to the implementation of PROLOG like programming languages [Colmerauer 72,82, Fages 83].

In the context of higher order logic, the problem of unification was studied by Gould [66], who defined "general matching sets" of terms, a weaker notion than that of CSU. The existence of a unifier is shown to be undecidable in third order languages in [Huet 73], and at second order by Goldfarb [81]. The general theory of CSU's and μCSU's in the context of higher order logic is studied in [Huet 76, Jensen and Pietrzykowski 77].

Unification in equational theories has been first studied by Plotkin [72] in the context of resolution theorem provers to build-up the underlying equational theory into the rules of inference. In this paper Plotkin conjectured that there existed an equational theory E where a μCSU_E did not always exist. Theorem 1 in the next chapter proves this conjecture.

Further interest in unification in equational theories arose from the problem of implementing programming languages with "call by patterns", such as QA4 [Rulifson 72]. Associative unification (finding solutions to word equations) is a particularly hard problem. Plotkin [72] gives a procedure to enumerate a μCSU_A (eventually infinite), and Makanin [77] shows that the word equation problem is decidable. Stickel [75,81] and independently Livesey and Siekmann [76,79], give an algorithm for unification in presence of associative-commutative operators. However its termination in the general case was still questionned since some recursive calls are made on terms having a bigger size than the initial terms. Theorems 3 and 4 in this paper prove the termination and the completeness in the general case. Siekmann [78] studied the general problem in his thesis, especially the extension of the AC-unification algorithm to idempotence and identity. Lankford [79,83] gave the extension to a unification procedure in Abelian group theory.

In the class of equational theories for which there exists a canonical term rewriting system (see [Huet and Oppen 80]), Fay [79] gives a universal procedure to enumerate a CSU_E. It is based on the notion of "narrowing", as defined in [Slagle 74]. Hullot [80] gives a similar procedure and a sufficient termination criterion, further generalized in [Jouannaud and Kirchner 83]. Siekmann and Szabo [82] investigate the domain of regular canonical term rewriting systems in order to find general minimal unification procedures, but we show in [Fages and Huet 83] that even in this framework μCSU_E may not exist.

Termination or minimality of unification procedures is much harder to obtain than completeness. However the main applications of unification in equational theories to the generalizations of the Knuth and Bendix algorithm, such as in [Peterson, Stickel 81 and Hsiang 82], are covered by the associative-commutative unification algorithm.

1.2.2. Definitions

Let $M,N \in \mathcal{T}$, $V=\mathcal{V}(M) \cup \mathcal{V}(N)$ and W be a finite set of "protected variables". S is a *Complete Set of E-Unifiers of* M *and* N *away from* W if and only if :

a) $\forall \sigma \in S \;\; \mathcal{D}(\sigma) \subseteq V$ and $\mathcal{R}(\sigma) \cap W = \phi$ (purity)

b) $S \subseteq \mathcal{U}_E(M,N)$ (correctness)

c) $\forall \rho \subseteq \mathcal{U}_E(M,N) \; \exists \sigma \subseteq S \;\; \sigma \overset{V}{\underset{E}{\leq}} \rho$ (completeness)

Furthermore S is a *complete Set of Minimal E-Unifiers of* M *and* N *away from* W if additionally :

d) $\forall \sigma, \sigma' \in S \;\; \sigma \neq \sigma' \Rightarrow \sigma \overset{V}{\underset{E}{\not\leq}} \sigma'$ (minimality)

Remark that the most general unifiers in first order languages are μCSU_ϕ reduced to one element. The reason to consider W is that in many algorithms, unification must be performed on subterms, and *it is necessary to separate the variables introduced by unification from the variables of the context*. It is the case for instance for resolution in equational theories [Plotkin 72], and for the generalization of the Knuth and Bendix completion procedure in congruence classes of terms [Peterson and Stickel 81]. It is easy to show that there always exists a CSU_E away from W, by taking all E-unifiers satisfying a).

We may add to the definition of CSU_E :

d') $\forall \sigma, \sigma' \in S \;\; \sigma \neq \sigma' \;\Rightarrow\; \sigma \overset{V}{\underset{E}{\neq}} \sigma'$ (non-congruency)

Such CSU_E still always exist but we loose the property that if $\mathcal{U}_E$ is recursively enumerable then there exists a recursively enumerable one. For example, in undecidable axiomatic equational theories $\mathcal{U}_E$ is recursively enumerable but in general the CSU_E satisfying d') are not.

1.2.3. Existence of basis of the E-unifiers

It is well known that *there may not exist a finite* CSU_E. For instance a*x=x*a in the theory where * is associative [Plotkin 72]. When there exists a finite CSU_E, there always exists a minimal one, by filtering out redundant elements. But it is not true in general :

Theorem 1 (non-existence of basis) : In some first order equational theory E there exist E-unifiable terms for which there is no μCSU_E.

Proof : The proof is in [Fages and Huet 83], where it is shown that there may be infinite strings of E-unifiers more and more general. And thus that minimality d) may be incompatible with completeness c). The example is $E=\{f(0,x)=x\ ,\ g(f(x,y))=g(y)\}$, for terms g(x) and g(a). ∎

However when a μCSU_E exists, it is unique up to $\overset{V}{\underset{E}{\equiv}}$.

Theorem 2 (unicity of basis) : Let M and N be two terms, U_1 and U_2 be two μCSU_E of M and N. There exists a bijection $\varphi : U_1 \to U_2$ such that $\forall\sigma\in U_1\ \ \sigma \overset{V}{\underset{E}{\equiv}} \varphi(\sigma)$.

Proof : $\forall\sigma\in U_1\ \exists\rho\in U_2\ \ \rho \overset{V}{\underset{E}{\leq}} \sigma$ since U_2 is complete. We pick-up one such ρ as $\varphi(\sigma)$.
$\forall\sigma'\in U_2\ \exists\rho'\in U_1\ \ \rho' \overset{V}{\underset{E}{\leq}} \sigma'$. We pick-up one such ρ' as $\psi(\sigma')$.
Thus $\forall\sigma\in U_1\ \ \psi(\varphi\sigma) \overset{V}{\underset{E}{\leq}} \sigma$ so $\psi(\varphi\sigma)) = \sigma$ by minimality,
$\varphi(\sigma)) \overset{V}{\underset{E}{\leq}} \sigma \overset{V}{\underset{E}{\leq}} \varphi(\sigma)$ i.e. $\sigma \overset{V}{\underset{E}{\equiv}} \varphi(\sigma)$. ∎

2. ML as a programming language for the term structure

ML is an applicative language with *exceptions* in the line of ISWIM [Landin 66], POP2 [Burstall, Collins and Popplestone 71], GEDANKEN [Reynolds 70]. Functions are "first class citizens", the type inference mechanism allows *functionals* at any order, and *polymorphic* operators. *Primitive types* are : void, bool, int and string, and *type operators* are : cartesian product ×, sum + and function →. Type variables are denoted by *, **, ... The declaration of an *abstract type* consists in the definition of constructors and external functions. For instance, the polymorphic list type, ** list = void + (* × * list)*, is predefined with the constructors **nil** and **cons** (noted . in infix), and the destructors **hd** : ** list→ ** and **tl** : ** list→ * list*. Variables can be structured in lists and pairs, bindings are then made by *matching*. We refer to the LCF manual [Gordon, Milner and Wadsworth 79] for the syntax and semantics of ML.

The composition of functions is an infix operator **o** : *(**→ ***)×(*→ **)→ *→ ****.
It could be defined by :
mlinfix 'o';;
let (f o g) x = f (g x) ;;

The curried function **map** : *(*→ **)→ * list→ ** list* returns the list of the function applications to a list of arguments. It is equivalent to :
letrec map f l = if l = nil then nil else f (hd l) . (map f (tl l));;

Particularly important is **itlist** : (*→**→**)→* *list*→**→** which iterates the application of a function to a list of arguments, by composing the results :
itlist f $[l_1;...;l_n]$ x = (f l_1 (f l_2 $\cdots$ (f l_n x) $\cdots$)) = ((f l_1)o(f l_2)o...o(f l_n)) x. It is defined by :
letrec itlist f l x = if l=nil then x else f (hd l) (itlist f (tl l) x);;
For example **flatten** : ** list list→ * list* which eliminates the first level parentheses in a list can be defined by :
let flatten l = itlist append l nil;;
We generalize itlist to the iterative application of a curried function on two lists of arguments, with :
letrec itlist2 f k l x = if k=nil then x else f (hd k) (hd l) (itlist2 f (tl k) (tl l) x);;

We will not detail the abstract types related to the term structure, we keep previous notations related to them, and define the function **op** : $T \to F$ which returns the head symbol of a term, and **largs** : *T→T list* which returns the list of the arguments.

```
abstype F = ...;;
abstype V = ...;;
absrectype T = F × T list + V
  with ... and op M = ... and largs M = ... and isvar M = ... and equal(M,N) = ...;;
lettype S = T→T;; mlinfix '←';; letrec x←M = ...;;
```

Since substitutions are functions from terms to terms, the composition of substitutions is exactly the composition of functions **o**, and the identity function **I** serves as the identity substitution.

If we suppose defined the function **occurs** : *T×T→bool* which recognizes if a variable occurs in a term, the unification algorithm of Robinson [65] may be defined by the function **uni** : *T×T→S* with :

```
letrec uni(M,N) =
   if isvar M then if equal(M,N) then I
      else if occurs(M,N) then fail else M←N
   else if isvar N then if occurs(N,M) then fail else N←M
   else if op M≠op N then fail
      else (itlist2 unicompound (largs M) (largs N) I)
and unicompound A B σ = uni (σ A,σ B) o σ;;
```

3. AC-unification

3.1. Connection with the solving of linear homogeneous diophantine equations

A binary function + is associative and commutative iff it satisfies (in infix notation) :

$$\begin{cases} x+y = y+x \\ (x+y)+z = x+(y+z) \end{cases}$$

The set of those function symbols is denoted by F_{AC}. We will not consider them as symbols of variable arity in order to stay in the term algebras formalism, but we define the function **largAC** : *T→T list* which returns the list of the AC-arguments, that is the list of the arguments after the elimination of parentheses on the head symbol if it is AC. For example with M = (x+(x+y))+(f(a+(a+a))+(b+c)) where $+\in F_{AC}$, we have op M = +, largs M = [x+(x+y) ; f(a+(a+a))+(b+c)] and largAC M= [x ; x ; y ; f(a+(a+a)) ;b ; c].

Let two terms M and N beginning with the same AC head symbol to be unified. Stickel [81] proved that the elimination of common arguments, pair by pair, does not change U_{AC}(M,N). For example the unification of M = (x+(x+y))+(f(a+(a+a))+(b+c)) and N = ((b+b)+(b+z))+c, where $+\in F_{AC}$, is equivalent to the unification of the arguments lists [x;x;y;f(a+(a+a))] and [b;b;z], obtained by eliminating the common arguments b and c. If a variable is eliminated in this operation, it must be added to the set of context variables W.

Unification of (non-ordered) lists of arguments is the problem of solving equations in the free abelian semigroup, which is isomorphic to the solving of homogeneous linear diophantine equations $\sum a_i x_i = \sum b_j y_j$ over $\mathbb{N}$-{0}. Thus for any AC-unification problem, we associate such an equation by associating integer variables to distinct arguments, with their multiplicity as coefficient. For instance $2x_1+x_2+x_3 = 2y_1+y_2$ in the example. In return the solutions to the diophantine equation induce the unifiers of the lists elements, that are still to unify with the effective arguments.

Stickel [76] and Huet [78] give an algorithm to solve homogeneous linear diophantine equations, which enumerates a basis of solutions, by backtracking with a certain bound on the value of the variables, and elimination of the redundant solutions. The best bound, is double :

1) $\forall i\ x_i \leq \max_k b_k\ ,\ \forall j\ y_j \leq \max_l a_l$

2) $\forall i\ \forall j\ \ x_i \leq \dfrac{\text{lcm}(a_i,b_j)}{a_i}$ or $y_j \leq \dfrac{\text{lcm}(a_i,b_j)}{b_j}$

The basis of solutions can be represented as a matrix with as many columns as variables in the equation, and as many lines as solutions in the basis. For example the equation $2x_1+x_2+x_3 = 2y_1+y_2$ admits as solutions basis the matrix :

	x_1	x_2	x_3	y_1	y_2
s_1	0	1	1	1	0
s_2	0	0	2	1	0
s_3	0	2	0	1	0
s_4	1	0	0	1	0
s_5	0	0	1	0	1
s_6	0	1	0	0	1
s_7	1	0	0	0	2

Any linear combination of the seven elements of the basis is a solution to the equation. However, because the absence of a zero in the unification problem, we must consider all subsets of the solutions basis, with the constraints that the sum of the coefficients in a column must be non null, and equal to 1 if the corresponding term is not a variable.

Hullot [80] gives a method for a *constrained enumeration of partitions*, in order to reduce the high complexity of this computation. We refer to his thesis for details of this technique which is crucial in an implementation. In the example, among the $2^7=128$ solutions, only 6 are to be considered. Let us examine $\{s_4,s_5,s_6,s_7\}$: $x_1 = s_4+s_7$, $x_2 = s_6$, $x_3 = s_5$, $y_1 = s_4$, $y_2 = s_5+s_6+s_7$. We deduce a unifier from it, by associating to each solution s_i a new variable u_i :

$$\sigma = \{x \leftarrow b+u_7\ ,\ z \leftarrow f(a+(a+a))+(y+(u_7+u_7))\ ,\ u_6 \leftarrow y\ ,\ u_5 \leftarrow f(a+(a+a))\ ,\ u_4 \leftarrow b\}.$$

3.2. The algorithm in ML

The unification of non-ordered lists of arguments, described in the previous chapter, may be defined with the function **unilist** : *T list×T list×F→T list×(T list list)* as :

let unilist (lM,lN,f) = let lA,matrix,n = dio (elimcom (lM,lN))
in lA , trad (f , lA , partit (matrix,n));;

where **elimcom** : *T list×T list→T list×T list* eliminates common terms to the two lists of arguments; **dio** : *T list×T list→T list×(int list list)×int* solves the diophantine equation associated to the two lists of arguments, and returns the simplified list of arguments with variable arguments first, the matrix of integer solutions that forms a basis (with variable arguments columns first) together with the number of variable arguments; **partit** : *(int list list)×int→(int list list list)* enumerates all the partitions of the basis satisfying the constraint that the sum of the coefficients in a column must be non null, and greater than 1 for the non-variable arguments (last columns); and where **trad** : *F×T list×(int list list list)→(T list list)* expresses each solution attached to a partition as a list of terms built on f and new variables, that are still to unify with the effective arguments in lA.

We give the AC-unification algorithm as the ML definition of the function **uniAC** : *T×T→S list* which returns as a list a finite CSU_{AC} of its arguments, empty if they are not unifiable :

```
letrec uniAC(M,N) =
    if isvar M then if equal(M,N) then [I]                              {case 1}
                    else if occurs(M,N) then nil                        {case 2}
                         else [M←N]                                     {case 3}
    else if isvar N then if occurs(N,M) then nil                        {case 4}
                         else [N←M]                                     {case 5}
    else if op M≠op N then nil                                          {case 6}
    else if isAC (op M) then                                            {case 7}
        let lA,lLS = unilist (largAC M,largAC N,op M)
        in (flatten (map (λ lS . itlist2 unicompound lA lS [I]) lLS))
    else (itlist2 unicompound (largs M) (largs N) [I])                  {case 8}
and unicompound A B lσ = flatten (map continue lσ)
    where (continue σ = map compose (uniAC (σ A,σ B))
          where compose ρ = ρ o σ);;
```

3.3. A measure of complexity for AC-unification problems

We give here a measure of complexity for AC-unification problems, which is at the basis of our proof of termination and completeness of uniAC. In order to define precisely this measure of complexity, we see terms as labeled trees, and we define *the set of occurrences of a term* M, $\mathcal{O}(M)$, as the set of nodes of the tree, designed by integer lists [Huet and Oppen 80]. ε is the empty list, occurrences are denoted by u, v, w, ...

$$\begin{cases} O(M) = \{\varepsilon\} \text{ if M is a variable or a constant} \\ O(f(M_1,\dots,M_n)) = \{\varepsilon\}\cup\{i.u \mid 1\le i\le n \ \&\ u\in O(M_i)\} \end{cases}$$

We denote by M/u *the subterm of* M *at occurrence* u :

$$\begin{cases} M/\varepsilon = M \\ f(M_1,\dots,M_n)/i.u = M_i/u \end{cases}$$

We are not interested in occurrences of a AC symbol inside a small homogeneous tree such as :

+
⊕ ⊕
x y ⊕ v
z u

Accordingly, we say that an occurrence $u\in\mathcal{O}(M)$ is ***admissible*** if and only if u=ε or $op(M/u)\notin F_{AC}$ or u=v.i i∈ℕ & op(M/u)≠op(M/v). In the above example, the circled occurrences of + are not admissible.

An occurrence is ***strict*** if it is not the occurrence of a variable. The set of strict and admissible occurrences of a term M is denoted by $\hat{\mathcal{O}}(M)$. For example, let M = (x+a)+(f(x)*b), F_{AC} = {+,*} and x∈*V*. The occurrence 1 is not admissible, 1.1 and 2.1.1 are occurrences of variables, therefore $\hat{\mathcal{O}}(M)$ = {ε, 1.2, 2, 2.1, 2.2}. We denote by SUBT(M) *the set of the non-variable admissible subterms of* M : SUBT(M) = {M/u | u∈$\hat{\mathcal{O}}$(M)}.

We want also to distinguish the variables having occurrences immediately under at least two different function symbols. *The set of immediate operators of a term* N *in a term* M, is the set Op(N,M) = {op M/u | u∈O(M) & ∃i∈ℕ M/u.i = N}.

The complexity of AC-unification of two terms M *and* N, is a pair (ν,τ), denoted by $C_{AC}(M,N)$, where ν is the number of distinct variables in M and N, having occurrences immediately under at least two different function symbols : $\nu = \text{card}\{x \in V \mid \text{card}(Op(x,M) \cup Op(x,N)) > 1\}$, and τ is the number of distinct non-variable admissible subterms in both M and N : $\tau = \text{card}(SUBT(M) \cup SUBT(N))$. That is τ is the number of non-variable nodes in the minimal graph of representation of M and N where AC symbols are represented as operators of variable arity.

For example, consider $+ \in F_{AC}$, $f \notin F_{AC}$, $M = (x+x)+f(y+(x+x))$ and $N = f(y)$. Only y has occurrences under + and f, and we count four admissible non-variable subterms. $C_{AC}(M,N) = (1,4)$. If + was not AC then $C_{AC}(M,N)$ would be equal to (1,5).

We shall use the lexicographic ordering on complexities :

$$(\nu,\tau) < (\nu',\tau') \text{ iff } \nu < \nu' \text{ or } \nu = \nu' \;\&\; \tau < \tau'$$

Lemma 1 < is a nœtherian ordering.

Proof : < is the lexicographic extension of two nœtherian orderings on terms. ■

Now we express that the complexity of two admissible subterms in two terms is strictly smaller.

Lemma 2 Let $M,N \in T\text{-}V$, $u \in \hat{O}(M)-\{\varepsilon\}$, $v \in \hat{O}(N)-\{\varepsilon\}$. We have $C_{AC}(M/u,N/v) < C_{AC}(M,N)$.

Proof : Let $(\nu,\tau) = C_{AC}(M/u,N/v)$ and $(\nu',\tau') = C_{AC}(M,N)$. Obviously $\nu \leq \nu'$. Since u is in $\hat{O}(M)$, $SUBT(M/u) \subseteq SUBT(M)$. Similarly $SUBT(N/v) \subseteq SUBT(N)$. Since $u \neq \varepsilon$, M is not a subterm of M/u. Similarly N is not a subterm of N/v. Either M is not a subterm of N/v, in which case $M \notin SUBT(N/v)$, or N is not a subterm of M/u, in which case $N \notin SUBT(M/u)$. Therefore since $M \in SUBT(M)$ and $N \in SUBT(N)$ (they are not variables), we get $\tau < \tau'$. ■

Remark that in this lemma, the hypotheses on u and v are necessary. For example for $M=N=(a+b)+c$ where $+ \in F_{AC}$, $C_{AC}(M,N)=(0,1)$, $M/1=a+b$, $N/2=c$ and $C_{AC}(M/1,N/2)=(0,2)$. But 1 is not an admissible occurrence in M.

The next lemma exhibits the elementary substitutions that are composed by uniAC as it will be shown in theorem 3, and proves that they do not increase the complexity of the initial terms.

Lemma 3 : Assume M and N are two terms, W is a finite set of variables, and σ is an elementary substitution of one of the four forms :

1) $M \leftarrow N$ (resp. $N \leftarrow M$) if M (resp. N) is a variable not appearing in N (resp. M),
2) $x \leftarrow T$ where $(Op(x,M) \cup Op(x,N)) \cap (Op(T,M) \cup Op(T,N)) \neq \phi$ and $x \notin V(T)$.
3) $x \leftarrow T$ where $x \notin V(M) \cup V(N) \cup W$,
4) $x \leftarrow x_1 + \ldots + x_m + T_1 + \ldots + T_n$ where $m+n \geq 1$, $+ \in F_{AC}$, $+ \in Op(x,M) \cup Op(x,N)$, $+ \in Op(T_i,M) \cup Op(T_i,N)$, $x_i \notin V(M) \cup V(N) \cup W$ and $x \notin V(T_i)$.

We have $C_{AC}(\sigma M,\sigma N) \leq C_{AC}(M,N)$.

Proof :
Let $(\nu,\tau) = C_{AC}(M,N)$ and $(\nu',\tau') = C_{AC}(\sigma M,\sigma N)$. In case 1 $\nu'=\nu$ and $\tau'=\tau$. In case 2 $\nu' \leq \nu$. Suppose $\nu'=\nu$ and $T \notin V$. Since T is a subterm of M or N, the set of non-variable subterms of M and N is the same as the set of non-variable subterms of σM and σN. Furthermore, if some occurrence of T was admissible in M or N, then it will stay so after substitution; otherwise let $+=op(T)$, all occurrences of T in M and N are immediately under +, the same for x because $\nu'=\nu$, thus all new occurrences of T in σM and σN are still not admissible. In all cases $SUBT(\sigma M) \cup SUBT(\sigma N) = SUBT(M) \cup SUBT(N)$ and $\tau'=\tau$. Case 3 is trivial, $\nu'=\nu$, $\tau'=\tau$. In case 4 the first subcase is $m=1$, $n=0$, where trivially $C_{AC}(\sigma M,\sigma N) = C_{AC}(M,N)$. Otherwise since $x_i \notin V(M) \cup V(N)$, we have $Op(x_i,\sigma M) \cup Op(x_i,\sigma N) = \{+\}$ and since

$+\in Op(T_j,M)\cup Op(T_j,N)$, we have $\nu'\leq\nu$. Furthermore if $\nu'=\nu$, since $x\notin \mathcal{V}(T_j)$, we must have card$(Op(x,M)\cup Op(x,N))=1$, thus $Op(x,M)\cup Op(x,N)=\{+\}$, and similarly to case 2 we conclude that $\tau'=\tau$.

■

Corollary : Assume W is a finite set of variables. For any terms M and N if $\sigma=\sigma_n \circ \cdots \circ \sigma_1$ is the composition of elementary substitutions satisfying the hypotheses of lemma 3 on M, N and W for σ_1, on $\sigma_1 M$, $\sigma_1 N$ and W for σ_2, ... , on $\sigma_{n-1} \cdots \sigma_1 M$, $\sigma_{n-1} \cdots \sigma_1 N$ and W for σ_n, then $C_{AC}(\sigma M,\sigma N)\leq C_{AC}(M,N)$.

Proof : By n applications of lemma 3.

■

The last lemma expresses some kind of stability by context of these properties. It is needed in the induction steps for the proof of termination.

Lemma 4 : Assume P and Q are two terms, M and N are two subterms such that $Op(M,P)\cap Op(N,Q)\neq\phi$, and W is a finite set of variables such that $\mathcal{V}(P)\cup\mathcal{V}(Q)\subseteq W\cup\mathcal{V}(M)\cup\mathcal{V}(N)$. If σ is a substitution satisfying the hypotheses of corollary of lemma 3 on M, N and W, then $C_{AC}(\sigma P,\sigma Q)\leq C_{AC}(P,Q)$.

Proof : We just have to prove that the hypotheses of lemma 3 on σ, M, N and W are equivalent to the hypotheses of lemma 3 on σ, P, Q and W. Case 1 with M and N is identical to case 2 with P and Q. Case 2 with M and N remains the same with P and Q since M and N are subterms of P and Q. Case 3 and case 4 are also unchanged because $\mathcal{V}(P)\cup\mathcal{V}(Q)\subseteq W\cup\mathcal{V}(M)\cup\mathcal{V}(N)$.

■

3.4. An example

Let $M=x+(y+(z+f(x,y,z)))$ to be unified with $N=u+(v+(w+f(u,v,w)))$ where $+\in\mathcal{F}_{AC}$ and $f\notin\mathcal{F}_{AC}$. $C_{AC}(M,N)=(6,4)$.

This example corresponds to the case 7 of the algorithm. M and N do not share arguments, lM = [x;y;z;f(x,y,z)], lN = [u;v;w;f(u,v,w)]. The function dio solves the associated diophantine equation, $u_1+u_2+u_3+u_4=v_1+v_2+v_3+v_4$, and returns the list lA of arguments with variables first :

x y z u v w f(x,y,z) f(u,v,w)

together with the matrix of solutions (with variable arguments columns first) :

1	0	0	1	0	0	0	0
1	0	0	0	1	0	0	0
1	0	0	0	0	1	0	0
1	0	0	0	0	0	0	1
0	1	0	1	0	0	0	0
0	1	0	0	1	0	0	0
0	1	0	0	0	1	0	0
0	1	0	0	0	0	0	1
0	0	1	1	0	0	0	0
0	0	1	0	1	0	0	0
0	0	1	0	0	1	0	0
0	0	1	0	0	0	0	1
0	0	0	1	0	0	1	0
0	0	0	0	1	0	1	0
0	0	0	0	0	1	1	0
0	0	0	0	0	0	1	1

and the number of variable arguments, here 6. The function partit returns the list of all the partitions of the 16 lines of the matrix that satisfy the constraint that only variable arguments are affected by a sum of coefficients greater than 1, and no column has a null sum. Let us consider such a particular partition by

keeping only lines 1,2,3,5,6,7,9,10,11 and 16. The traduction of this solution is the list
$[x_1+(x_2+x_3);\ x_4+(x_5+x_6);\ x_7+(x_8+x_9);\ x_1+(x_4+x_7);\ x_2+(x_5+x_8);\ x_3+(x_6+x_9);\ x_{10};\ x_{10}]$, where $\{x_1,\dots,x_{10}\}$ are new variables attached to the 10 solutions of the partition. The terms of this list are then recursively unified with the terms in lA. First recursive calls are simple instanciations of x,y,z,u,v,w and x_{10}, but *the last recursive call relates to two terms of greater size than M and N, and containing simultaneously more variables and more distinct subterms*, that is : $f(x_1+(x_2+x_3),\ x_4+(x_5+x_6),\ x_7+(x_8+x_9))$ with $f(x_1+(x_4+x_7),\ x_2+(x_5+x_8),\ x_3+(x_6+x_9))$.

However the complexity is $(0,8) < C_{AC}(M,N)$. Unification goes on with $x_1+(x_2+x_3)$ and $x_1+(x_4+x_7)$ of complexity (0,2), and so on ...

3.5. Proof of total correctness

There is a notational difficulty in proofs of correctness, connected to the set W of protected variables. In the implementation W is implicit for a gensym like mechanism; this is what we expressed by "new variables" above. In the proofs below we must be more precise and give explicitly W as a third argument of uniAC.

Theorem 3 : Assume W is a finite set of variables. For any terms M and N uniAC(M,N,W) terminates.

Proof : The proof consists in showing that all recursive calls uniAC(M',N',W') in cases 7 and 8 of uniAC(M,N,W) are such that either $M'\in\mathcal{V}$, or $N'\in\mathcal{V}$, or $C_{AC}(M',N')<C_{AC}(M,N)$. In the first two cases, they terminate immediatly. For the third case, we have to show also that any σ in uniAC(M',N',W') satisfies the hypotheses of the corollary of lemma 3 on M, N and W, this by nœtherian induction on $C_{AC}(M,N)$.
In cases 3 and 5, σ is of the form 1 in lemma 3.
In case 8, let k be the arity of op(M), $\sigma_0=I$, $W_0=W$ and for $1\le i\le k$, let $M_i=\sigma_{i-1}\cdots\sigma_1\,M/i$, $N_i=\sigma_{i-1}\cdots\sigma_1\,N/i$, $W_i=(W_{i-1}\cup\mathcal{V}(\sigma_{i-1}\cdots\sigma_1 M)\cup\mathcal{V}(\sigma_{i-1}\cdots\sigma_1 N))-(\mathcal{V}(M_i)\cup\mathcal{V}(N_i))$ and σ_i be any unifier in uniAC(M_i,N_i,W_i). We prove by induction on i that $C_{AC}(\sigma_i\cdots\sigma_1\,M,\sigma_i\cdots\sigma_1\,N)\le C_{AC}(M,N)$.
By induction on i, $C_{AC}(\sigma_{i-1}\cdots\sigma_1\,M,\sigma_{i-1}\cdots\sigma_1\,N)\le C_{AC}(M,N)$. Now either $M_i\in\mathcal{V}$, or $N_i\in\mathcal{V}$, or by lemma 2 we get $C_{AC}(M_i,N_i)<C_{AC}(M,N)$.
Furthermore, by nœtherian induction σ_i satisfies the hypotheses of corollary of lemma 3 on M_i, N_i and W_i, thus also on $\sigma_{i-1}\cdots\sigma_1 M$, $\sigma_{i-1}\cdots\sigma_1 N$ and W_i by lemma 4, $C_{AC}(\sigma_i\cdots\sigma_1\,M,\sigma_i\cdots\sigma_1\,N)\le C_{AC}(M,N)$.
In case 7, let +=op(M), and lA,llS=unilist (largAC M, largAC N, W). Assume $lA=[M_1;\dots;M_m;M_{m+1};\dots;M_n]$ and $lS=[X_1;\dots;X_n]$ is an element of llS. $X_1,\dots,X_n$ are terms built on + and some new variables (not appearing in $\mathcal{V}(M)\cup\mathcal{V}(N)\cup W$). $M_1,\dots,M_n$ are variables and $M_{m+1},\dots,M_n$ are arguments not headed by +. The first n recursive calls are simple instanciations of the form 4 in lemma 3, so the complexity of M and N does not increase. The last recursive calls are performed on either one new variable and a term in lA, leading to a substitution of form 3 or 2, or either on two non-variable arguments of lA, namely $\sigma_{i-1}\cdots\sigma_1 M/j$ and $\sigma_{i-1}\cdots\sigma_1 N/k$, and the proof is the same as in case 8, or on one non-variable argument of lA and a term headed by +, thus leading to a failure.

■

Remark that if variable arguments were not instanciated first, next recursive calls could be done on two terms headed by +, and the proof of termination would be a little more complicated. For this, uniAC is nearer Livesey and Siekmann's algorithm [76,79] than Stickel's one [75,81]. They differ themselves only on this point.

Let U(M,N,W) be the finite set of substitutions returned by uniAC(M,N,W) and restricted to $V=\mathcal{V}(M)\cup\mathcal{V}(N)$.

Theorem 4 : Assume M and N are two terms, and W is a finite set of variables disjoint from $\mathcal{V}(M)\cup\mathcal{V}(N)$. U(M,N,W) is a CSU_{AC} of M and N away from W.

Proof : We show by nœtherian induction on $C_{AC}(M,N)$ that :
1) $\forall\sigma\in U(M,N,W)$ $\mathcal{D}(\sigma)\subseteq V$ and $\mathcal{R}(\sigma)\cap W=\phi$
2) $U(M,N,W)\subseteq \mathcal{U}_{AC}(M,N)$
3) $\forall\rho\in \mathcal{U}_{AC}(M,N)$ $\exists\sigma\in U(M,N,W)$ $\sigma \underset{AC}{\overset{V}{\leq}} \rho$.

Cases 1,2,4,6 are trivial.
Cases 3,5 : If M (resp. N) is a variable having no occurrence in N (resp. M), then the substitution $M\leftarrow N$ (resp. $N\leftarrow M$) satisfies the three properties.
Case 8 : Let k be the arity of op(M). We show that the set

$$S=\{\sigma_k o\ldots o\sigma_{1|V} \mid \sigma_1\in U(M/1\ ,\ N/1\ ,\ W),\ \sigma_2\in U(\sigma_1\ M/2\ ,\ \sigma_1\ N/2\ ,\ W_2),\ \cdots,$$
$$\sigma_k\in U(\sigma_{k-1}\cdots\sigma_1\ M/k\ ,\ \sigma_{k-1}\cdots\sigma_1\ N/k\ ,\ W_k)\}$$

is a CSU_{AC} of M and N away from W, with $W_0=W$ and for $1\leq i\leq k$ $W_i=(W_{i-1}\cup\mathcal{V}(\sigma_{i-1}\cdots\sigma_1 M)\cup\mathcal{V}(\sigma_{i-1}\cdots\sigma_1 N))-(\mathcal{V}(\sigma_{i-1}\cdots\sigma_1 M/i)\cup\mathcal{V}(\sigma_{i-1}\cdots\sigma_1 N/i))$.
Notice that that for any i $W\subseteq W_i$.

1) $\mathcal{D}(\sigma_k o\ldots o\sigma_{1|V})\subseteq V$, $\mathcal{R}(\sigma_k o\ldots o\sigma_{1|V})\subseteq\bigcup_{i=1}^{k}\mathcal{R}(\sigma_i)$, and by nœtherian induction $\mathcal{R}(\sigma_i)\cap W_i=\phi$. Thus $\mathcal{R}(\sigma_k o\ldots o\sigma_1)\cap W=\phi$, since $W\subseteq W_i$.
2) By nœtherian induction $\forall i$ $1\leq i\leq n$ $\sigma_i\cdots\sigma_1\ M/i \underset{AC}{=} \sigma_i\cdots\sigma_1\ N/i$,
thus $\sigma_k\cdots\sigma_1\ M/i \underset{AC}{=} \sigma_k\cdots\sigma_1\ N/i$, i.e. $\sigma_k\ o\ \cdots\ o\ \sigma_{1|V}$ is a AC-unifier of M and N.
3) Let $\rho\in \mathcal{U}_{AC}(M,N)$, M and N being AC-unifiable. We show there exists $\sigma\in U(M,N,W)$ such that $\sigma\underset{AC}{\overset{V}{\leq}}\rho$, by nœtherian induction on k.
k=0 trivial since U(M,N,W)={I}.
k>0 by nœtherian induction $\exists\lambda_{i-1}\in S$ $\lambda_{i-1}o\sigma_{i-1}o\ldots o\sigma_1 \underset{AC}{\overset{V}{=}} \rho$.
But $\lambda_{i-1}\ \sigma_{i-1}\ \cdots\ \sigma_1\ M/i \underset{AC}{=} \lambda_{i-1}\ \sigma_{i-1}\ \cdots\ \sigma_1\ N/i$,
so by induction $\exists\sigma_i\in U(\sigma_{i-1}\cdots\sigma_1\ M/i,\sigma_{i-1}\cdots\sigma_1\ N/i,W_i)$ $\sigma_i \underset{AC}{\overset{V_i}{\leq}} \lambda_{i-1}$,
with $V_i=\mathcal{V}(\sigma_{i-1}\cdots\sigma_1\ M/i)\cup\mathcal{V}(\sigma_{i-1}\cdots\sigma_i\ N/i)$.
$\mathcal{D}(\sigma_i)\subseteq V_i$ so $\sigma_i \underset{AC}{\leq} \lambda_{i-1}$ and $\sigma_i o\ldots o\sigma_1 \underset{AC}{\overset{V}{\leq}} \rho$.
Therefore $\sigma_k o\ldots o\sigma_1 \underset{AC}{\overset{V}{\leq}} \rho$.

Case 7 : Let $lA=[M_1,\ldots,M_n]$ and $llS=[lS_1,\ldots,lS_p]$ be the solutions returned by unilist (largAC M, largAC N, W). By isomorphism with the solving of the equation $\sum_{i=1}^{m'} a_i x_i = \sum_{j=1}^{n'} b_j y_j$ over $\mathbb{N}$-{0}, we prove with the same inductions than in case 8, that the AC-unifiers of the effective arguments in lA with the terms in the solutions lS in llS form a $CSU_{AC}(M,N,W)$.

■

In general U(M,N,W) is not a μCSU_{AC} of M and N, but because it is finite, it suffices to eliminate the redundant unifiers (by AC-matching) in a final pass to get one.

3.6. Extension to identity and idempotence

In [Fages 83] we describe the extension to a unification algorithm in presence of operators that may be associative (A), commutative (C), idempotent (I) with unit (U), with the only restriction that an associative operator must be commutative.

Unification of terms built over only one function symbol ACU or ACUI, is studied in [Livesey and Siekmann 76,79]. These authors show that the case 7 in the algorithm is simplified since the presence of a unit permits to consider not all subsets of the solutions base, but only the *sum*, and idempotence allows to solve the associated equation in *0, 1* rather than over integers.

However in the general case of unification in presence of operators U or I, case 6 is no longer a fail case. For example if 1 is the unit of f, f(x,y) and g(a,b) are unifiable with the CSU_U $\{\{x \leftarrow 1, y \leftarrow g(a,b)\}, \{y \leftarrow 1, x \leftarrow g(a,b)\}\}$. If f is also idempotent, $\{x \leftarrow g(a,b), y \leftarrow g(a,b)\}$ is another UI-unifier, and so on ... The cases 6 and 8 must be changed in this way.

Termination can be proved as for AC-unification, since the only introduced symbols are unit constants and new variables under the same function symbol. The combinatory explosion in the case 7 on the computation of partitions is eliminated when the AC function has a unit. The possibility of having more than one single most general U-unifier comes from the cases 8 and 6, which conversely introduces a new combinatory on the arguments, especially if both head functions are idempotent.

Conclusion

We have shown that varieties defined by a set of associative-commutative function symbols are finitary for the unification problem, by proving the termination and completeness of Stickel's algorithm. In the "formel" system, under development at INRIA, we use an implementation in Maclisp originally written by J.M. Hullot [79,80]. In this program the constrained generation of partitions, coded by the binary representation of bignums, allows to treat efficiently very large AC-unification problems.

In [Fages 83] we show that the sharing of common subterms is possible by representing terms by ordered graphs, and we extend the algorithm to unify finite and infinite rational terms modulo associativity-commutativity, identity and idempotence. In presence of Abelian group operators, the existence of a unification procedure extending the one of Lankford [83] which would terminate in the general case is an open problem.

Acknowledgements

I wish to thank Gérard Huet for his decisive help in the presentation, Guy Cousineau and Thierry Coquand for their comments.

References

Baxter L.D. The Complexity of Unification. *Ph.D. Thesis, University of Waterloo.* (1977)

Baxter L.D. The Undecidability of the Third Order Dyadic Unification Problem. *Information and Control 38, p170-178.* (1978)

Burstall R.M., Collins J.S. and Popplestone R.J. Programming in POP-2 *Edinburgh University Press.* (1971)

Colmerauer A. Prolog II, manuel de référence et modèle théorique. *Rapport interne, Groupe d'Intelligence Artificielle, Université d'Aix-Marseille II.* (Mars 1982)

Colmerauer A., Kanoui H. and Van Caneghem M. Etude et Réalisation d'un

système PROLOG. *Rapport Interne, GIA, Univ. d'Aix-Marseille Luminy.* (1972)
Davis M. Hilbert's Tenth Problem is Unsolvable. *Amer. Math. Monthly 80,3, pp. 233-269.* (1973)
Fages F. Formes canoniques dans les algèbres booléennes, et application à la démonstration automatique en logique de premier ordre. *Thèse, Université de Paris VI.* (June 83)
Fages F. Note sur l'unification des termes de premier ordre finis et infinis. *Rapport LITP 83-29.* (May 1983)
Fages F. and Huet G. Unification and Matching in Equational Theories. *CAAP 83, l'Aquila, Italy. Lecture Notes in Computer Science 159.* (March 1983)
Fay M. First-order Unification in an Equational Theory. *4th Workshop on Automated Deduction, Austin, Texas, pp. 161-167.* (Feb. 1979)
Goldfarb W.D. The Undecidability of the Second-order Unification Problem. *Theoritical Computer Science, vol. 13, pp. 225-230. North Holland Publishing Company.* (1981)
Gordon M.J., Milner A.J. and Wadsworth C.P. Edinburgh LCF. *Springer-Verlag LNCS 78.* (1979)
Gould W.E. A Matching Procedure for Omega Order Logic. *Scientific Report 1, AFCRL 66-781, contract AF19 (628)-3250.* (1966)
Guard J.R. Automated Logic for Semi-Automated Mathematics. *Scientific Report 1, AFCRL 64,411, Contract AF19 (628)-3250.*(1964)
Herbrand J. Recherches sur la théorie de la démonstration. *Thèse, U. de Paris, In: Ecrits logiques de Jacques Herbrand, PUF Paris 1968.* (1930)
Hsiang J. Topics in Automated Theorem Proving and Program Generation. *Ph.D. Thesis, Univ. of Illinois at Urbana-Champaign.* (Nov. 1982)
Huet G. The Undecidability of Unification in Third Order Logic. *Information and Control 22, pp. 257-267.* (1973)
Huet G. A Unification Algorithm for Typed λ-Calculus. *Theoretical Computer Science 1.1, pp. 27-57.* (1975)
Huet G. Résolution d'équations dans des langages d'ordre 1,2, ... omega. *Thèse d'Etat, Université de Paris VII.* (1976)
Huet G. An Algorithm to Generate the Basis of Solutions to Homogenous Linear Diophantine Equations. *Information Processing Letters 7,3, pp. 144-147.)* (1978)
Huet G. and Oppen D. Equations and Rewrite Rules: a Survey. *In Formal Languages: Perspectives and Open Problems, Ed. Book R., Academic Press.* (1980)
Hullot J.M. Associative-Commutative Pattern Matching. *Fifth International Joint Conference on Artificial Intelligence, Tokyo.* (1979)
Hullot J.M. Compilation de Formes Canoniques dans les Théories Equationnelles. *Thèse de 3ème cycle, U. de Paris Sud.* (Nov. 80)
Jensen D. and Pietrzykowski T. Mecanizing ω-Order Type Theory through Unification. *Theoretical Computer Science 3, pp. 123-171.* (1977)
Jouannaud J.P. and Kirchner C. Incremental Construction of Unification Algorithms in Equational Theories. *Proc. 10th ICALP.* (1983)
Kirchner H. and Kirchner C. Contribution à la résolution d'équations dans les algèbres libres et les variétés équationnelles d'algèbres. *Thèse de 3ème cycle, Université de Nançy.* (Mars 1982)
Knuth D. and Bendix P. Simple Word Problems in Universal Algebras. *In Computational Problems in Abstract Algebra, Pergamon Press, pp. 263-297.* (1970)
Landin P.J. The Next 700 Programming Languages. *CACM 9,3.* (March 1966)
Lankford D.S. A Unification Algorithm for Abelian Group Theory. *Report MTP-1, Math. Dept., Louisiana Tech. U.* (Jan. 1979)
Lankford D.S., Butler G. and Brady B. Abelian Group Unification Algorithms for

Elementary Terms. *Math. Dept., Louisiana Tech. U., Ruston Louisiana 71272* (1983)
Livesey M. and Siekmann J. Unification of Bags and Sets. *Internal Report 3/76, Institut fur Informatik I, U. Karlsruhe.* (1976)
Livesey M., Siekmann J., Szabo P. and Unvericht E. Unification Problems for Combinations of Associativity, Commutativity, Distributivity and Idempotence Axioms. *4th Workshop on Automated Deduction, Austin, Texas, pp. 161-167.* (Feb. 1979)
Makanin G.S. The Problem of Solvability of Equations in a Free Semigroup. *Akad. Nauk. SSSR, TOM pp. 233,2.* (1977)
Martelli A. and Montanari U. An Efficient Unification Algorithm. *ACM T.O.P.L.A.S., Vol. 4, No. 2, pp 258-282.* (April 1982)
Matiyasevich Y. Diophantine Representation of Recursively Enumerable Predicates. *Proceedings of the Second Scandinavian Logic Symposium, North-Holland.* (1970)
Paterson M.S. and Wegman M.N. Linear Unification. *J. of Computer and Systems Sciences 16, pp. 158-167.* (1978)
Peterson G.E. and Stickel M.E. Complete Sets of Reduction for Equational Theories with Complete Unification Algorithms. *JACM 28,2 pp 233-264.* (1981)
Plotkin G. Building-in Equational Theories. *Machine Intelligence 7, pp. 73-90.* (1972)
Raulefs P., Siekmann J., Szabo P., Unvericht E. A Short Survey on the State of the Art in Matching and Unification Problems. *Sigsam Bulletin 1979, Vol. 13.* (1979)
Reynolds John C. GEDANKEN - A Simple Typeless Language Based on the Principle of Completeness and the Reference Concept. *CACM 13,5, pp. 308-319.* (May 1970)
Robinson J.A. A Machine Oriented Logic Based on the Resolution Principle. *JACM 12, pp. 32-41.* (1965)
Robinson G.A. and Wos L.T. Paramodulation and Theorem Proving in First-order Theories with Equality. *Machine Intelligence 4, American Elsevier, pp. 135-150.* (1969)
Rulifson J.F., Derksen J.A. and Waldinger R.J. QA4 : a Procedural Calculus for Intuitive Reasoning. *Technical Note 73, A.I. Center, SRI, Menlo Park.* (Nov. 1972)
Siekmann J. Unification and Matching Problems. *Ph. D. thesis, Memo CSM-4-78, University of Essex,* (1978)
Siekmann J. and Szabo P. Universal Unification in Regular Equational ACFM Theories. *CADE 6th, New-York.* (June 1982)
Slagle J.R. Automated Theorem-Proving for Theories with Simplifiers, Commutativity and Associativity. *JACM 21, pp. 622-642.* (1974)
Stickel M.E. A Complete Unification Algorithm for Associative-Commutative functions. *4th International Joint Conference on Artificial Intelligence, Tbilisi.* (1975)
Stickel M.E. Unification Algorithms for Artificial Intelligence Languages. *Ph. D. thesis, Carnegie-Mellon University.* (1976)
Stickel M.E. A Complete Unification Algorithm for Associative-Commutative Functions. *JACM 28,3 pp 423-434.* (1981)
Venturini-Zilli M. Complexity of the Unification Algorithm for First-Order Expressions. *Calcolo XII, Fasc. IV, p361-372.* (October December 1975)

A Linear Time Algorithm for a Subcase of Second Order Instantiation

Donald Simon[1]
University of Texas at Austin, Texas

1. Introduction

A problem that comes up in any proof-checking system is whether or not a proof step is a valid instantiation of a lemma or theorem. Often, the lemma or theorem may include set variables and so in general can be second order. This problem is somewhat simpler than the more general problem of second order unification. Jensen and Pietrzykowski [1] and Huet [2] give semi-decision procedures for finding ω-order unifiers. The second order instantiantion problem is shown to be NP-complete in Baxter [3]. Our approach will be to find useful subcases of the second order instantiantion problem which yield to fast algorithms. This paper is a first approximation towards that goal.

We are interested in finding the solution to the following problem: Given a set of pairs of expressions, $\{(a_i , b_i \mid i \epsilon [1,n]\})$, where, for $1 \leq i \leq n$, a_i must be a first-order constant or variable, or a term of the form g(x), where g is a second order constant or variable, and x is a first-order constant or variable find a unifier α such that $\alpha a_i = b_i$, for $1 \leq i \leq n$. We will assume that for $1 \leq i \leq n$, b_i is in normal form.

In the following, c will be a first-order constant, u, v, w, x, y, and z will be first-order variables, e, e1, and e2 are first-order expressions, f, g, and h are second order constants, and γ and δ will be second order variables. $\mu|V$ means the restriction of the unifier μ to the set V.

The algorithm is broken down into three phases. In the first phase, we find a unifier for each pair of terms, independent of the other pairs. In the second phase, we merge together unifiers which contain a common second order variable. In the final phase, unifiers which share common first-order variables are merged. The result will be a set of tables of unifiers from which we can construct a unifier for the original problem by selecting one row from each table. If no such unifier exists, some table will be empty.

[1]This work was supported in part by National Science Foundation Grants MCS-8011417 and MCS-831499

2. Phase 1.

In the first phase, we define a unifier α_i which will unify a_i and b_i, for each i. In addition, for some pairs, we will have to remember a variable and this we will call v_i .

For each i, $1 \leq i \leq n$, we have 6 possible cases:

1. $a_i = c$. If $b_i = c$, then define $\alpha_i := \{\}$, otherwise halt and return "no unifier".
2. $a_i = x$. Define $\alpha_i := \{<b_i, x>\}$.
3. $a_i = f(c)$. If $b_i = f(c)$, then define $\alpha_i := \{\}$, otherwise halt and return "no unifier".
4. $a_i = f(x)$. If $b_i = f(e)$, then define $\alpha_i := \{<e, x>\}$, otherwise halt and return "no unifier".
5. $a_i = \gamma(c)$. Create a new variable, x_i, define $\alpha_i := \{<\lambda z \bullet b_i, \gamma>, <c, x_i>\}$ and define $v_i := x_i$.
6. $a_i = \gamma(x)$. Define $\alpha_i := \{<\lambda z \bullet b_i, \gamma>\}$ and $v_i := x$.

Clearly, these unifiers can differ on the values of some variables. In the next two two phases we resolve these conflicts. First, for each second order variable γ that occurs in some a_i, we will find a set of quasi-unifiers, each of which will unify all the pairs of terms containing γ. The reason that these are not true unifiers is that they may contain two pairs of the form <e1, x>, <e2, x> where x is a first-order variable. In the second merge step, we will resolve any conflicts on first-order variables and end up with a set of unifiers that will unify all the terms of the problem.

3. Phase 2.

Suppose that γ is a second order variable that occurs in some term and $\alpha_{i_1}, \alpha_{i_2}, \ldots, G[a]_{i_k}$ are the unifiers that contain γ. If $\alpha_{i_1}\gamma = \alpha_{i_2}\gamma = \ldots = \alpha_{i_k}\gamma$, then the set of unifiers for γ certainly includes the singleton $\{\{<\alpha_{i_1}\gamma, \gamma>\}\}$.

On the other hand, suppose that $\alpha_{i_j}\gamma \neq \alpha_{i_l}\gamma$ for some pair of unifiers. In this case, we will find one list of unifiers that unifies these two pairs of terms. We will then extend each of these unifiers, if possible, to unify the other pairs of terms which contain the variable γ. The main loop of this procedure is given below:

$j := 1$; $V := \{\}$;

WHILE $j < k$ AND $\alpha_{i_j}\gamma = \alpha_{i_{j+1}}\gamma$ DO [$V := V \cup \{v_{i_j}\}$; $j := j + 1$];

IF $j \geq k$ THEN RETURN $\{\{<\alpha_{i_1}\gamma, \gamma>\}\}$

ELSE [S := UNIFY-PAIR($\alpha_{i_j}\gamma(v_{i_j})$, $\alpha_{i_{j+1}}\gamma(v_{i_{j+1}})$, γ, V, $v_{i_{j+1}}$));

$j := j + 1$;

WHILE $j \leq k$ DO [S := EXTEND-UNIFIER(S, γ, $\alpha_{i_j}\gamma(v_{i_j})$, v_{i_j}); $j := j + 1$]

RETURN S];

The procedure UNIFY-PAIR(e1, e2, γ, V, v') works by finding the unifiers for (x_i, y_i, γ_i, V, v') where x_i and y_i are corresponding arguments of e1 and e2, respectively, when $x_i \neq y_i$. Since all the unifiers for different pairs of arguments must agree on the variables of V and v', we can take the intersection of the sets of unifiers restricted to these variables. This yields a set of unifiers which can be extended to unifiers for the original problem.

UNIFY-PAIR(e1, e2, γ, V, v')

If either e1 or e2 is a first-order variable or constant,

or e1 and e2 have different heads or different numbers of arguments

then return $\{\{<\lambda z \bullet z, \gamma>, <e2, v'>\} \cup \{<e1, v> : v \in V\}\}$,

else $e1 = f(x_1, x_2, \ldots, x_r)$ and $e2 = f(y_1, y_2, \ldots, y_r)$. In this case,

for each i, $1 \leq i \leq r$, such that $x_i \neq y_i$, we create a new variable γ_i and define

$D_i :=$ UNIFY-PAIR(x_i, y_i, γ_i, V, v').

We also define

$$E := \bigcap_{1 \leq i \leq r,\ x_i \neq y_i} \{\nu | (V \cup \{v'\}) : \nu \in D_i\}.$$

We will extend each $\mu \in E$ to a unifier for the original problem as follows:

Suppose $\mu \in E$. For each i, $1 \leq i \leq r$, define δ_i by

$$\delta_i := \begin{cases} \lambda z \bullet x_i & \text{if } x_i = y_i \\ \nu\ \gamma_i \text{ where } \nu \in D_i \text{ such that } \nu | (V \cup \{v'\}) = \mu & \text{if } x_i \neq y_i. \end{cases}$$

The extension of μ is $\{<\lambda z \bullet f(\delta_1(z), \delta_2(z), \ldots, \delta_r(z)), \gamma>\} \cup \mu$.

To this set of unifiers, we add the unifier

$\{<\lambda z \bullet z, \gamma>, <e2, v'>\} \cup \{<e1, v> : v \in V\}$,

and return this set as the result.

To extend a unifier μ to include a new pair $\gamma(v) \rightarrow e2$, all we need do is to compare $(\mu\ \gamma)v$ to e2 and find the subterm of e2 that corresponds to v in $(\mu\ \gamma)v$.

EXTEND-UNIFIER(S, γ, v, e)

Set S' := {}.

For each unifier $\mu \in$ S, find ν := EXTEND-UNIFIER+(($\mu\ \gamma$)v, e, v).

If ν is "No extension", do nothing, otherwise add $\mu \cup \nu$ to S'.

RETURN S'.

Calling EXTEND-UNIFIER+(δ(v), e, v) will find a unifier ν, such that $\delta(\nu\ v)) = e$, provided such a unifier exists.

EXTEND-UNIFIER+(e1, e2, v)

If either e1 or e2 is a first-order variable or constant,

then if e1 = e2 then return {}

else if e1 = v then return {<e2, v>}

else return "No extension"

else if e1 and e2 have different heads or different numbers of arguments,

then return "No extension"

else e1 = $f(x_1, x_2, \ldots, x_r)$ and e2 = $f(y_1, y_2, \ldots, y_r)$. In this

case, define H_i := EXTEND-UNIFIER+(x_i, y_i, v) for $1 \leq i \leq r$.

If any H_i is "No extension", then return "No extension".

If all the H_i's are empty, then return {}. Otherwise, take

$$H := \bigcap_{1 \leq i \leq r,\ H_i \neq \{\}} H_i .$$

If H is empty, return "No extension", otherwise return H.

Let's look at a simple example.

Suppose that γ occurs in the following four pairs:

$\gamma(v)$ --> f(g(h(a)), h(a))

$\gamma(w)$ --> f(g(h(a)), h(a))

$\gamma(u)$ --> f(g(h(b)), h(b))

$\gamma(v)$ --> f(g(c), c).

In the first phase, we find four unifiers and record four variables.

$\alpha_1 := \{<f(g(h(a)), h(a)), \gamma>\}$ $\quad v_1 := v$

$\alpha_2 := \{<f(g(h(a)), h(a)), \gamma>\}$ $\quad v_2 := w$

$\alpha_3 := \{<f(g(h(b)), h(b)), \gamma>\}$ $\quad v_3 := u$

$\alpha_4 := \{<f(g(c), c), \gamma>\}$ $\quad v_4 := v$.

Since $\alpha_1\gamma(v) = \alpha_2\gamma(w)$, the call to UNIFY-PAIR will be UNIFY-PAIR(f(g(h(a)), h(a)), f(g(h(b)), h(b)), γ, {v, w}, u). The algorithm will recurse on the arguments and find that UNIFY-PAIR(g(h(a)), g(h(b)), γ_1, {v, w}, u) has unifiers

{{<λz•g(h(z)), γ_1>, <a, v>, <a, w>, <b, u>},

{λz•g(z), γ_1>, <h(a), v>, <h(a), w>, <h(b), u>},

{<λz•z, γ_1>, <g(h(a)), v>, <g(h(a)), w>, <g(h(b)), u>}}

and that UNIFY-PAIR(h(a), h(b), γ_2, {v, w}, u} has the unifiers

{{<λz•h(z), γ_2>, <a, v>, <a, w>, <b, u>},

{<λz•z, γ_2>, <h(a), v>, <h(a), w>, <h(b), u>}}.

By restricting these unifiers to the set {v, w, u} and taking the intersection we find two partial unifiers,

{{<a, v>, <a, w>, <b, u>}, {<h(a), v>, <h(a), w>, <h(b), u>}}.

The extension of the two unifiers gives us the set

{{<λz•f(g(h(z)), h(z)), γ>, <a, v>, <a, w>, <b, u>},

{<λz•f(g(z), z), γ>, <h(a), v>, <h(a), w>, <h(b), u>}},

to which we add the unifier

{<λz•z, γ>, <f(g(h(a)), h(a)), v>, <f(g(h(a)), h(a)), w>, <f(g(h(b)), h(b)), u>}

and return this set as the result of UNIFY-PAIR.

Next, we extend this set of unifiers to include the fourth pair. This entails three calls to EXTEND-UNIFIER+, one for each unifier. EXTEND-UNIFIER+(f(g(h(v), h(v)), f(g(c), c), v)) will return "No extension" since no extension exists for h(v) and c. EXTEND-UNIFIER+(f(g(v, v), f(g(c), c), v)) returns the extension {<c, v>} and EXTEND-UNIFIER+(v, f(g(c, c), v)) returns {<f(g(c, c), v>}).

Thus, the final result from the second phase is the list of quasi-unifiers

{{<λz•f(g(z), z), γ>, <h(a), v>, <h(a), w>, <h(b), u>, <c, v>},

{<λz•z, γ>, <f(g(h(a)), h(a)), v>, <f(g(h(a)), h(a)), w),

<f(g(h(b)), h(b)), u>, <f(g(c), c), v>}}.

We should note that these unifiers conflict on the value of variable v. This fact will be caught in the third phase, when we merge the unifiers by their common first-order variables.

The intersection in the procedure UNIFY-PAIR can be done in time linear to the size of the input expressions e1 and e2. To show this, we make use of the following theorem:

Theorem 1: If μ and ν are unifiers such that $\mu\ \gamma(\mathrm{v1}) = \nu\ \gamma(\mathrm{v1}) = \mathrm{e1} \neq \mathrm{e2} = \mu\ \gamma(\mathrm{v2}) = \nu\ \gamma(\mathrm{v2})$, then there exists a function ϕ, such that either $\phi(\mu\ \mathrm{v1}) = \nu\ \mathrm{v1}$, $\phi(\mu\ \mathrm{v2}) = \nu\ \mathrm{v2}$, and $\mu\ \gamma = (\nu\ \gamma) \circ \phi$, or $\phi(\nu\ \mathrm{v1}) = \mu\ \mathrm{v1}$, $\phi(\nu\ \mathrm{v2}) = \mu\ \mathrm{v2}$, and $\nu\ \gamma = (\mu\ \gamma) \circ \phi$).

The proof of this theorem appears in the appendix. The first result of this theorem is that there are a linear (proportional to the depths of e1 and e2) number of unifiers, and the list of unifiers $\mu_1, \mu_2, \ldots, \mu_p$ can be ordered so that $\mu_i v'$ is a subterm of $\mu_{i+1} v'$ for $1 \leq i < p$. To see this, suppose that the first case, i.e., $\phi(\mu\ v1) = \nu\ v1$ holds. Since $(\nu\ \gamma)(\nu\ v1) \neq (\nu\ \gamma)(\nu\ v2)$, $\nu\ \gamma$ cannot be a constant function, and so $(\nu\ v1)$ must be a subterm of e1. Also, $\nu\ v1 \neq \nu\ v2$, so ϕ cannot be a constant function and hence, $(\mu\ v1)$ must be a subterm of $(\nu\ v1)$. Therefore, the images of v1 under the unifiers can be ordered by the subterm relation and all are subterms of e1.

A second result of the theorem is that if there is an ordered list of all unifiers, $\mu_1, \mu_2, \ldots, \mu_p$, which unify $\gamma_1(v1)$ and x_1, and unify $\gamma_1(v2)$ and y_1, and an ordered list of all unifiers $\nu_1, \nu_2, \ldots, \nu_q$, which unify $\gamma_2(v1)$ and x_2, and $\gamma_2(v2)$ and y_2, such that $\mu_k v1 = \nu_l v1$ and $\mu_k v2 = \nu_l v2$, then for all $i < k$, there is a $j < l$, such that $\mu_i v1 = \nu_j v1$, and $\mu_i v2 = \nu_j v2$. By the theorem, there is a ϕ such that $\phi(\mu_i v1) = \mu_k v1$ and $\phi(\mu_i v2) = \mu_k v2$. So, $\{<(\nu_l \gamma_2 \circ \phi, \gamma_2>, <\mu_i v1, v1>, <\mu_i v2, v2>\})$ will unify $\gamma_2(v1)$ and x_2, and $\gamma_2(v2)$ and y_2. This unifer appears before ν_l in the ordering since $\mu_i\ v1$ is a subterm of $\mu_k\ v1 = \nu_l\ v1$. Thus to find the intersection of two lists of unifiers, all we need do is to check that the first unifiers in the lists agree on the variables and then compare the corresponding ϕ functions.

By modifying UNIFY-PAIR in the following manner, the algorithm takes linear time: UNIFY-PAIR will return two results: a list of unifiers $\mu_1, \mu_2, \ldots, \mu_k$ and a list of functions $\phi_1, \phi_2, \ldots, \phi_{k-1}$ such that $\phi_i(\mu_i v') = \mu_{i+1} v'$ for $1 \leq i < k$. To do the intersection, we take the first unifier in each D_i, restrict each one to the set $V \cup \{v'\}$ and check if they are all equal to one another. If so, we extend the unifier as above and add it to the list E. Instead of comparing the second unifier in each D_i to each other, we can compare the first function in each function list to each other and so forth. The new algorithm is given below.

UNIFY-PAIR-1(e1, e2, γ, V, v')

If either e1 or e2 is a first-order variable or constant,

or e1 and e2 have different heads or different numbers of arguments

then return {{<λz•z, γ>, <e2, v'>} ∪ {<e1, v> : v ∈ V}} and {},

else e1 = $f(x_1, x_2, \ldots, x_r)$ and e2 = $f(y_1, y_2, \ldots, y_r)$. In this case,

for each i, $1 \leq i \leq r$, such that $x_i \neq y_i$, we create a new variable γ_i and define

D_i := the list of unifiers returned by UNIFY-PAIR-1(x_i, y_i, γ_i, V, v')

and

G_i := the list of functions returned by UNIFY-PAIR-1(x_i, y_i, γ_i, V, v').

We also set E and F to {}. Let i_0 be such that $1 \leq i_0 \leq r$ and $x_{i_0} \neq y_{i_0}$.

(We know that such an i exists, since UNIFY-PAIR-1 is only called when e1 and e2 are different.)

If $D_{i,1}|V \cup \{v'\} = D_{i_0,1}|V \cup \{v'\}$, for $x_i \neq y_i$, then we add the extension

of $D_{i_0,1}|V \cup \{v'\}$ to E and do the following loop:

j := 1;

WHILE $G_{i,j} = G_{i_0,j}$, for $x_i \neq y_i$, {add $G_{i_0,j}$ to F and the extension

of $D_{i_0,j}|V \cup \{v'\}$ to E;

j := j + 1}.

Finally, we add $\mu\gamma$ to the set of functions, F, where μ is the last unifier added to E, and we add the unifier

{<λz•z, γ>, <e2, v'>} ∪ {<e1, v> : v ∈ V} to E

and return these two lists as the result.

To see that the entire procedure UNIFY-PAIR is linear, we first note that since y_i = $(G_{i,k} \circ G_{i,k-1} \circ \ldots \circ G_{i,1}(D_{i,1} v'))$, each occurence of a symbol in y_i corresponds exactly to a symbol in one of the $G_{i,j}$'s or $D_{i,1}$ v'.

Second, after performing the intersection, we will throw away (i.e., never examine again) all the $G_{i,j}$'s except when $i = i_0$. We can then assign the cost of the comparision of $G_{i,j}$ and $G_{i_0,j}$ to $G_{i,j}$ and we see that the total number of comparisions is equal to total number of subterms in all the y_i 's, i.e. the size of e2. Thus, UNIFY-PAIR takes only take linear time.

Similarly, we can modify EXTEND-UNIFIER to execute in linear time. The new procedure will be given the set of unifiers and the set of functions and will return a new set of unifiers and a new set of functions. We will begin with the last unifier in the set and extend it. Since the last unifier always contains <$\lambda z \bullet z$, γ>, it can always be extended by adding the pair <e, v>. The other extensions will be found, not by comparing $(\mu_j\gamma)v$ to e, but rather $(\phi_j\ v)$ to a subterm of e, where the ϕ_j's are in the function list.

EXTEND-UNIFIER-1(S, G, γ, v, e)

Call the last unifier in S, μ_k. We set S' := $\{\mu_k \cup$ <e, v>$\}$, G' := {}, and e' := e.

For each function ϕ_j in G, starting from the end, and working backwards,

find EXTEND-UNIFIER+(ϕ_j(v), e', v), and if an extension ν exists,

add ϕ_j to G' and $(\mu_j \cup \nu)$ to S' and set e' := $(\nu\ v)$. If no extension exists,

halt and add $(\phi_1 \circ \phi_2 \circ \ldots \circ \phi_j)$ to G'.

Return S' and G'.

To see how this routine works, consider the invariant $(\mu_j\gamma)e' = e$. This is true before the loop when $\mu_j = \mu_k = \lambda z \bullet z$ and e' = e. After one execution of the loop, assuming an extension is found, $\phi_j\ (\nu\ v) = e'$, by the definition of EXTEND-UNIFIER+. Since $(\mu_j\gamma) = (\mu_{j+1}\gamma) \circ \phi_j$ and $(\mu_{j+1}\gamma)e' = e$ (the induction hypothesis), $(\mu_j\gamma)\ (\nu\ v) = (\mu_{j+1}\gamma) \circ \phi_j(\nu\ v) = (\mu_{j+1}\gamma)e' = e$. At the end of loop, we define e' := $(\nu\ v)$. Thus, after the execution of the loop, $(\mu_j\ \gamma)e' = e$. Therefore, EXTEND-UNIFIER-1 finds the proper extension. Since it also examines each subterm of e at most once, it runs in linear time.

4. PHASE 3.

After we have merged the unifiers found in the first phase by their common second order variables, we are left with a list of quasi-unifiers for each second order variable. In addition, there will also usually be some first-order unifiers left from Cases 2, 4, and 5 in phase 1. In this phase, we will merge all the unifiers on their first-order variables. We can think of these lists as tables, where each unifier is a row, and the columns corresponds to the variables in the unifiers. The first-order unifiers have only one row. If a variable, say v, occurs in two or more columns, we want to merge the tables containing those columns on the common values of v, and delete any rows for which the value of v does not occur in all the v columns. In database terminology, we need to do a natural join of the tables.

The previous example written as a table looks like:

γ	v	w	u	v
$\lambda z \bullet f(g(z), z)$	h(a)	h(a)	h(b)	c
$\lambda z \bullet z)$	f(g(h(a)), h(a))	f(g(h(a)), h(a))	f(g(h(b), h(b))	f(g(c), c)

The only columns that need to be merged are the v columns. However, there is no row in which the values of v match, so the merged table is empty, and there is no unifier.

Again we note that the rows are ordered by the subterm relation on the columns corresponding to first-order variables. If we could check for equality in constant time, the obvious algorithm for finding the intersection of sorted lists would give a linear time algorithm. Here again, we can use the ϕ functions to get a linear time algorithm. To illustrate the algorithm, we suppose that we want to merge two columns together. The columns are $x_1, x_2, \ldots, x_r$ and $y_1, y_2, \ldots, y_s$ and we have sets of functions ϕ_i, and ψ_i such that $\phi_i(x_i) = x_{i+1}$, for $1 \leq i < r$, and $\psi_j(y_j) = y_{j+1}$, for $1 \leq j < s$.

> For each column, we find the depth of every element.
>
> We also create two new variables v and w.
>
> Starting at the bottom row of each table (the elements having the smallest depth), we move up the columns independently until we come to rows whose elements have the same depth, say row i_1 and row j_1. We compare the elements and if they are equal then we merge the rows together. We will call the result of this test "last-test". We continue to move up the columns until again, we come to rows, i_2 and j_2, whose elements have the same depth. We call
>
> COMPARE-TERMS-OF-SAME-DEPTH($(\phi_{i_2-1} \circ \phi_{i_2-2} \circ \ldots \circ \phi_{i1})$v,
>
> $(\psi_{j_2-1} \circ \psi_{j2-2} \circ \ldots \circ \psi_{j_2})$w,
>
> v, w, last-test, x_{i_1}, y_{j_1})
>
> and if this returns "TRUE" we merge the rows. In any case, we set last-test to this result, and continue up the rows in this manner, until all the rows have been examined.

COMPARE-TERMS-OF-SAME-DEPTH(e1, e2, v, w, last-test, x1, x2) assumes that x1 and x2 have the same depth and the result of comparing x1 and x2 is stored in last-test. It returns the result of comparing $(\lambda v \bullet e1)x1$ and $(\lambda w \bullet e2)x2$.

COMPARE-TERMS-OF-SAME-DEPTH(e1, e2, v, w, last-test, x1, x2)
There are six cases to consider:

1. e1 = v and e2 = w. Return the results of the last comparision (last-test).
2. e1 = v and e2 ≠ w. Do COMPARE-TERMS-OF-SAME-DEPTH+(e2, w, x1).
3. e1 ≠ v and e2 = w. Do COMPARE-TERMS-OF-SAME-DEPTH+(e1, v, x2).
4. Either e1 or e2 is another first-order variable, or a first-order constant. If e1 = e2, return "TRUE", otherwise, return "FALSE".
5. e1 and e2 are not first-order variables or constants and have different heads or numbers of variables. Return "FALSE".
6. e1 = $f(x_1, x_2, \ldots, x_r)$ and e2 = $f(y_1, y_2, \ldots, y_r)$. For $1 \leq i \leq r$, call COMPARE-TERMS-OF-SAME-DEPTH(x_i, y_i, v, w, last-test, x1, x2). In any of these is "FALSE", return "FALSE", otherwise, return "TRUE".

COMPARE-TERMS-OF-SAME-DEPTH+(e1, u, e2)

If e1 = u, return "FALSE", (In this instance, x1 and x2 occur in the larger terms at

different heights, hence the larger terms cannot be equal)

else if e1 or e2 is a first-order variable or constant,

return "TRUE" if e1 = e2, "FALSE" otherwise,

else if e1 and e2 have different heads or numbers of arguments, return "FALSE"

else e1 = $f(x_1, x_2, \ldots, x_r)$ and e2 = $f(y_1, y_2, \ldots, y_r)$.

For $1 \leq i \leq r$, call

COMPARE-TERMS-OF-SAME-DEPTH+(x_i, u, y_i). In any of these is "FALSE",

return "FALSE", otherwise, return "TRUE".

By doing the intersection this way, whenever we compare two subterms, we are sure that one of them has never been involved in a comparision before. Thus the algorithm is linear. The full algorithm for any number of columns is the obvious extension of this.

5. Implementation.

The algorithm has been encoded in UCILSP on a DEC-20.

6. Acknowledgements.

I would like to thank Dr. Woody Bledsoe, Larry Hines, Ernie Cohen, and Natarajan Shankar for their useful discussions of the problem. Also, I would like to thank the referees for their suggestions.

Proof of theorem:

In the following, the term "constant function" means a lambda expression of the form $\lambda z \bullet c$ where z does not appear in the term c.

First, we need two lemmas:

Lemma 2: If $x \neq y$, and $f(x) = f(y)$, then f is a constant function.

The proof is by induction on the body of f.

Basis: Suppose that the body of f is atomic. Hence $f = \lambda z \bullet z$ or $f = \lambda z \bullet c$. In the first case, $x = f(x) = f(y) = y$, which is false. Therefore f is constant.

Induction step: Suppose that $f = \lambda z \bullet h(r_1(z),\ldots,r_m(z))$ and the lemma is true for $r_1,\ldots,r_m$. Then $h(r_1(x),\ldots,r_m(x)) = f(x) = f(y) = h(r_1(y),\ldots,r_m(y))$. Thus, for j in $[1,m]$, $r_j(x) = r_j(y)$ and by induction r_j is a constant function, hence f is also constant.

Lemma 3: If $f(x) = g(x)$, $f(y) = g(y)$, and $f(x) \neq f(y)$ then $f = g$.

Proof of lemma 3:

Basis: Suppose that the body of f is atomic. Then either $f = \lambda z \bullet z$ or $f = \lambda z \bullet c$ where c is a constant other than z. The second case is impossible, since then $f(x) = c = f(y)$.

Basis: Suppose that the body of g is atomic. Either $g = \lambda z \bullet z$ or $g = \lambda z \bullet d$. In the second case $f(x) = g(x) = d = g(y) = f(y)$, which is false. Thus $f = g$.

Induction step: Suppose that $g = \lambda z \bullet h(g_1(z),\ldots,g_n(z))$. $x = f(x) = g(x) = h(g_1(x),\ldots,g_n(x))$. Hence for i in $[1,n]$, $g_i(x)$ is a subterm of x. Therefore g_i must be a constant function for i in $[1,n]$. Thus g is a constant function and a contradiction arises.

Induction step: Suppose that $f = \lambda z \bullet h(f_1(z),\ldots,f_n(z))$ and that the lemma is true for $f_1,\ldots,f_n$. Now we induct on the body of g.

Basis: If g is atomic, then $f = g$ follows analogously to the case when f is atomic and g is non-atomic.

Induction step: $g = \lambda z \bullet h'(g_1(z),\ldots,g_{n'}(z))$. Since $h(f_1(x),\ldots,f_n(x)) = f(x) = g(x) = h'(g_1(x),\ldots,g_{n'}(x))$ $h = h'$, $n = n'$, and for i in $[1,n]$, $f_i(x) = g_i(x)$. Similarly, since $f(y) = g(y)$, $f_i(y) = g_i(y)$ for i in $[1,n]$. If $f_i(x) \neq f_i(y)$, then by induction $f_i = g_i$. Otherwise, since $x \neq y$, and $f_i(x) = f_i(y)$, by Lemma, 1 f_i must be a constant function. Similarly, g_i

is a constant function, and f_i and g_i must be the same function. In either case $f_i = g_i$. Hence $f = g$.

Theorem:

If $a \neq b$

$f(x1) = a$, $f(y1) = b$,

$g(x2) = a$, and $g(y2) = b$.

then there exists a ϕ such that either

$x1 = \phi(x2)$, $y1 = \phi(y2)$, and $g = f \circ \phi$,

or $x2 = \phi(x1)$, $y2 = \phi(y1)$, and $f = g \circ \phi$.

The proof follows from induction on the bodies of f and g.

Basis: Suppose that the body of f is atomic. Therefore, $f = \lambda z \bullet z$ or $f = \lambda z \bullet c$ where c is a constant other than z. The second case cannot occur, since it implies that $a = f(x1) = f(x2) = b$. In the first case, define ϕ to be g.

Induction step: Suppose that $f = \lambda z \bullet h(f_1(z),...,f_n(z))$ and that the theorem holds for f_1, f_2, ..., f_n. We now induct on the body of the function g.

Basis: The body of g is atomic. $g \neq \lambda z \bullet c$ as above. Hence, $g = \lambda z \bullet z$, so define ϕ to be f.

Induction step. Suppose that $g = \lambda z \bullet h'(g_1(z),...,g_{n'}(z))$ and that the theorem holds for $g_1, g_2,...,g_{n'}$. Since $f(x1) = g(x2)$, $h(f_1(x1),...,f_n(x1)) = h'(g_1(x2),...,g_{n'}(x2))$, so $h = h'$, $n = n'$, and for i in $[1,n]$, $f_i(x1) = g_i(x2)$. Similarly, since $f(y1) = g(y2)$, for i in $[1,n]$ $f_i(y1) = g_i(y2)$. Since $f(x1) \neq f(x2)$, there exists j in $[1,n]$ such that $f_j(x1) \neq f_j(y1)$. Let jo be such a j. We define $a_{jo} := f_{jo}(x1)$ and $b_{jo} := f_{jo}(y1)$. Then

$a_{jo} \neq b_{jo}$,

$f_{jo}(x1) = a_{jo}$, $f_{jo}(y1) = b_{jo}$,

$g_{jo}(x2) = a_{jo}$, and $g_{jo}(y2) = b_{jo}$.

Hence, by the induction hypothesis, there exists a function ϕ_{jo} such that either

$x1 = \phi_{jo}(x2)$, $y1 = \phi_{jo}(y2)$, and $g_{jo} = f_{jo} \circ \phi_{jo}$,

or $x2 = \phi_{jo}(x1)$, $y2 = \phi_{jo}(y1)$, and $f_{jo} = g_{jo} \circ \phi_{jo}$.

We define $\phi := \phi_{jo}$ and claim that this function satisfies the theorem's conclusion. We examine two cases, based on the value of ϕ. **Case 1:** $x1 = \phi(x2)$, $y1 = \phi(y2)$, and $g_{jo} = f_{jo} \circ \phi$. First, we note that since $f(x1) \neq f(y1)$, $x1 \neq y1$, and since $g(x2) \neq g(y2)$, $x2 \neq y2$.

We first show that for all i in $[1,n]$, $g_i = f_i \circ \phi$.

Case 1A: $f_1(x1) = f_1(y1)$. By lemma 1, f_1 is a constant function.
Similarly, since $g_1(x2) = f_1(x1) = f_1(y1) = g_1(y2)$,
g_1 must be the same constant function.
Therefore $g_1 = f_1 \circ \phi$.

Case 1B: $f_1(x1) \neq f_1(y1)$. In this case, define $a_1 := f_1(x1)$
and $b_1 := f_1(y1)$. Then

$a_1 \neq b_1$
$f_1(x1) = a_1$, $f_1(y1) = b_1$,
$g_1(x2) = a_1$, and $g_1(y2) = b_1$.

By the induction hypothesis, there exists a ϕ_1 such that either

$x1 = \phi_1(x2)$, $y1 = \phi_1(y2)$, $g_1 = f_1 \circ \phi_1$
or $x2 = \phi_1(x1)$, $y2 = \phi_1(y1)$, $f_1 = g_1 \circ \phi_1$.

In the first case, $\phi_1(x2) = x1 = \phi(x1)$ and
$\phi_1(y2) = y1 = \phi(y2)$. Thus by Lemma 2, $\phi_1 = \phi$ and
$g_1 = f_1 \circ \phi$. In the second case, $x2 = \phi_1(x1)$. Since ϕ_1
cannot be constant ($\phi_1(x1) = x2 \neq y2 = \phi_1(y1)$),
x1 must be a subterm of x2. Similarly, since
$x1 = \phi(x2)$, x2 must be a subterm of x1.
Therefore $x1 = x2$ and $\phi_1 = \phi = \lambda z \bullet z$.
Hence $g_1 = f_1 = f_1 \circ \phi$.

In either case, $g_1 = f_1 \circ \phi$. Hence

$g = \lambda z \bullet h(g_1(z),\ldots,g_n(z))$
$= \lambda z \bullet h(f_1(\phi(z)),\ldots,f_n(\phi(z))$
$= f \circ \phi$.

Case 2: $x2 = \phi(x1)$, $y2 = \phi(y1)$, and $f_{jo} = g_{jo} \circ \phi$. This case is analogous to Case 1.

Finally, if we take $f := (\mu\ \gamma)$, $x1 := (\mu\ v1)$, $y1 := (\mu\ v2)$, $g := (\nu\ \gamma)$,
$x2 := (\nu\ v1)$, $y2 := (\nu\ v2)$, $a := e1$, and $b := e2$, we have the desired
result.

References

[0] T. Pietrzykowski, "A Complete Mechanization of Second Order Type Theory", (1973), JACM 20, 333-364.

[1] D. C. Jensen and T. Pietrzykowski, "Mechanizing ω-Order Type Theory Through Unification", (1976) Theorectical Computer Science 3, 123-171.

[2] G. Huet, "A Mechanization of Type Theory", (1975), Theorectical Computer Science 1, 25-58.

[3] L. D. Baxter, "The Complexity of Unification," (1976), Doctoral Thesis, Dept. of Computer Science, University of Waterloo, Waterloo, Ontario.

A NEW EQUATIONAL UNIFICATION METHOD : A GENERALISATION OF MARTELLI-MONTANARI'S ALGORITHM

Claude Kirchner

Centre de Recherche en Informatique de NANCY
Campus scientifique. BP 239 Vandoeuvre les Nancy
Cedex FRANCE

and

GRECO-Programmation

ABSTRACT

We address here the problem of unification in equational theories. A new unification method for some equational theories is given. It is a generalization of the Martelli and Montanari's algorithm and is based on a decomposition-merging-normalization process. We prove that for the class of decomposable theories this method gives a complete set of unifiers for any equation.

We apply the general results to the MINUS theories that contain axioms like -(-x) =x and -(f(x,y))=f(-y,-x) and we give an original unification algorithm for this type of theory.

1 INTRODUCTION

Solving equations in a theory E is a very fundamental problem especially in computer science. This notion appears in artificial intelligence with Robinson as the central step of the inference rule called resolution [ROB,65]. E-unification (i.e. unification in the theory E) is now a basic point in the generalized Knuth and Bendix algorithm [P&S,81], [J&K,84], [KIR,84], where the critical pairs are computed modulo some set of axioms. It also has important applications in automatic theorem proving.

This research was supported in part by Agence pour le Developpement de l'Informatique, under contract 82/767.

Solving an equation t1=t2 or equivalently finding an unifier of the terms t1 and t2 in the theory E is finding, if it exists, a substitution (i.e. an assignment of a term to every variable) which makes the two terms equal in the theory E.

In the case where there is no axiom there exists a unique simplest substitution (or solution), but when we consider theories with at least one axiom, the unicity of the simplest solution disappears. Plotkin [PLO,72] has introduced the notion of complete set of E-unifiers, which is in fact a basis of the set of E-solutions of the equation in the theory E.

Many algorithms have been found in order to compute the simplest substitution in the case where there is no axiom, see [ROB,72], [HUE,76], [P&W,78], [C&B,83], [FAG,83] and Martelli and Montanari in [M&M,82]. But finding complete set of E-unifiers is a very hard problem. Many important equational theories have been studied in the past, and specific unification algorithms are known for some of them, including commutativity, associativity [PLO,72], idempotence [R&S,78], associativity and commutativity [STI,81], [FAG,84], associativity, commutativity and idempotence [L&S,77] and abelian group theory [LAN,79]. All these algorithms are complete.

On the other hand, Fay [FAY,79] described a complete unification algorithm for the class of equational theories that possess a confluent and noetherian term rewriting system as defined by Knuth and Bendix [K&B,70]. This method relies on using the narrowing process defined by Lankford [LAN,75]. Hullot [HUL,80] described an improved version of Fay's algorithm, which both allows avoiding many useless computations and gives sufficient conditions for termination. These results have been generalized in [JKK,83] to equational term rewriting systems ([J&K,84]). But one drawback of the narrowing process is that it often does not terminate.

We propose here a E-unification method, based on a generalization of the Martelli and Montanari's algorithm to some kind of equational theories, the decomposable theories.

Before stating the method with all its details, we first sketch the main ideas.

A very simple and natural idea is to decompose the problem of E-unification into many simpler ones. We decompose an equation into a system of simpler ones, in such a way that the set of E-solutions is preserved. But this decomposition does not work with any kind of equations. For example if E is {f(x,y)=f(y,x)} then the two systems of equations {f(a,b)=f(b,x)} and {(a=b), (x=b)} do not have the same set of E-solutions.

We introduce a subset of the function symbols set, for which the decomposition is correct in the theory E. For example, if g is a symbol which does not appear on top of any axiom of E, then any equation g(t1,t2)=g(t1',t2') has the same set of E-solutions as the system {(t1=t1'),(t2=t2')}.

The decomposition of an equation gives two kinds of equations

* either x=t where x is a variable,
* or t1=t2 where further decomposition is not possible.

On one hand we assume there is a known way of solving equations x = t in the theory; that is we are able to give a complete set of E-unifiers for such an equation. On the other hand, we suppose that there exists in the theory E a mechanism which allows us to transform the equation t1=t2 into another one x=t, for a variable x of t1 or t2, which has the same set of E-solutions.

With such theories we are able to decompose equations in simpler ones. We then have to regroup all the conditions imposed by the equations on the same variable. This operation, called merging, yields to work with multiequations, that is a multiset of terms for which a substitution which makes all the terms equal in the theory E has to be found.

For example, if, in a theory E, we have to solve the system {(x=t1), (x=t2)}, it is equivalent to solving the multiequation x=t1=t2 and so, we then have to apply the decomposition process on t1 and t2.

The basic idea is to repeat the operations of decomposition and merging on the system of multiequations coming from the initial equation in order to obtain equations of the form x=u where x is a variable and u a term. But what happens with the multiequations x=t1=t2 where t1 and t2 don't have the same top symbol? To allow the continuation of the decomposition we have to provide a transformation of the equation t1=t2 into an equation y=t where y is a variable and t a term such that t1=t is decomposable. Such a transformation is called normalization.

The main interest of this method is to give tools to decompose unification problems in simpler ones, and to provide a finite description of a complete set of E-solutions even when this set in infinite. This comes from the fact that we don't use the form of the E-solutions of the equations x=t during the decomposition-merging-normalization process, but only in the final step, where we deduce a complete set of E-unifiers of the initial equation, from the final system. The example developed in the last section shows this property.

In the next section we give basic definitions. In section 3 we prove the main result of this paper: decomposition-merging-normalisation method, applied to the decomposable theories, yields a complete set of E-unifiers for any equations. Section 4 deals with an important example: the MINUS theories.

2. PRELIMINARIES

Our definitions and notations are consistent with those of G.Huet and D.Open [H&O,80], J.P.Jouannaud C.Kirchner and H.Kirchner [JKK,83].

Definitions 1: Given a set V of variables and a graded set F of function symbols, M(F,V) denotes the free algebra (or free magma) over V. Elements of M(F,V) are called terms. Terms can be viewed as labelled trees in the following way: a term t is a partial application of N* into FuX such that its domain Dom(t) satisfies:

(1) e ∈ Dom(t) where e is the empty word.

(2) m ∈ Dom(ti) => im C Dom(f(... ti ...)) for all i in [1,n=arity(f)].

Dom(t) is called set of occurrences of t, O(t) denotes the subset of non variable occurrences of Dom(t), V(t) denotes the set of variables of t, t/m the subterm of t at occurrence m, and t[m<-t'] the term obtained by replacing t/m by t' in t.

Definitions 2: Substitutions s are defined to be endomorphisms of M(F,V) with a finite domain D(s). We use I(s) to denote the set of variables occurring in the terms s(x) for all variable x in the domain of s. A substitution s is denoted by {(x1\t1),...,(xn\tn)}.

We denote by s¦W the restriction of the substitution s to the subset W of X.

We denote by < the subsumption preorder on M(F,V) defined by: t<t' iff t'= s(t) for a substitution s called a match from t to t'.

Composition of substitutions s and r is denoted by s.r .

Given a subset X of V, we define s<s'[X] iff s'=s".s [X] for some substitution s". X=V is omitted.

Definitions 3: We call an axiom or equation any pair {t,t'} of terms and write it t=t'. The A-equality =A is the smallest congruence closed under instantiation and generated by a finite set A of axioms.

We are interested in solving equations in equational theories, that is in finding substitutions s such that s(t) =A s(t'). To define a basis of the set of solutions, we extend the preorder on substitutions to equational theories.

Definition 4: We define t <A t' iff t'=A s(t) for some substitution s called A-match from t to t'. Given a subset X of V, we define s <A s' [X] iff s' =A s".s [X] for some substitution s".

Definitions 5: Given an equational theory =A, two terms t and t' are said to be A-unifiable iff there exists a substitution s such that s(t) =A s(t'). s is also called an A-solution of the equation t=t'. S is a complete set of A-unifiers of t and t' away from W such that X=V(t)UV(t')⊆W iff:

(1) for all s in S, D(s)⊆X and I(s)∩W = {} (no conflicts between variables)

(2) for all s in S, s(t) =A s(t')

(3) for all unifiers s', there exists an s in S such that s <A s'[X].

In addition S is said to be minimal if it satisfies the further condition :

(4) for all s and s' in S, s <A s' implies s=s'.

We denote by SU(t,t',A) the set of all A-unifiers of t and t', by CSU(t,t',A) a complete set of A-unifiers of t and t' and by CMSU(t,t',A) a complete and minimal set of A-unifiers of t and t'. A may be omitted when known or empty.

An A-unification algorithm is complete if it generates a complete set of A-unifiers. This set is not required to be finite here. In what follows, we write "finite" algorithm for a terminating and finite sets generating algorithm.

Many theoretical problems arise in equational theories (word problem,...) that can be approached by the use of rewrite rules, that is "one-way-equations". Working with rules requires good properties, as shown by Knuth and Bendix.

Definitions 6: A term rewriting system R is a set of pairs {gk->dk | 1=<k=<n} such that V(dk)$\subseteq$V(gk). We say that a term t R-reduces at occurrence m to a term t' using the rule gk -> dk and write t --->R[m,k] t' iff there exists a match **s** from **gk** to t/m and t' = t[m <- s(dk)]. We may omit R or [m,k].

We denote by -*->R the derivation relation, that is the reflexive transitive closure of --->R and by =R the generated equational theory.

An irreducible term for --->R is said in R-normal form. NF(t) is a normal form of t, that is a term t' in R-normal form s.t. t -*->R t'.

Definition 7: A term rewriting system R is said to be:

(1) noetherian iff --->R is terminating

(2) confluent iff for all t, t1, t2 such that t -*-> t1 and t -*-> t2, there exists a term t' such that t1 -*-> t' and t2 -*-> t'.

Confluent and noetherian term rewriting systems (called convergent) provide a decision procedure for equational theories because t =R t' iff t and t' have the same R-normal form.

3 THE GENERAL CASE

We want to generalize the method of Martelli and Montanari in order to provide unification algorithms for classes of equational theories. We always assume that the considered equational theory E is associated with a convergent rewriting system R.

This section is illustrated by the example of a theory on M(F,X) with a set of function symbols F={c,-,a,b} and arity(c)=2, arity(-)=1, arity(a)=arity(b)=0. It will be denoted Minus and the axioms are the very common ones:

$$-(-x) = x \quad ; \quad -(c(x,y)) = c(-y,-x).$$

Due to lack of space many proofs are omitted; they can be found in [KIR,83c].

The fundamental concept is here the notion of equivalent system of equations. We first study this notion.

3.1 E-equivalent systems of equations

Definition 8: Two equations t=t' and t1=t1' are E-equivalent, which is denoted by t=t' <=> t1=t1' , iff they have the same set of solutions:

$$SU(t,t',E)=SU(t1,t1',E).$$

Lemma 1: For any term rewriting system R: if t1 -*->R t1' and t2 -*->R t2' then t1=t2 <=> t1'=t2'.

So that we consider only convergent term rewriting systems, we suppose now that all the equations which are considered are in R-normal form, i.e. have their two terms in R-normal form.

Definition 9: A system of equations is a multiset of equations. An E-solution of a non empty multiset M of terms is a substitution s such that for any element u and v of M we have s(u) =E s(v). An E-solution of a system of equations S={(ti=ti')i=1,...,n} is a substitution s such that for any i in 1,...,n we have s(ti) =E s(ti'). The systems of equations S and S' are E-equivalent iff they have the same set of E-solutions.

In order to allow the decomposition of a system of equations in an E-equivalent one as in the case of the empty theory, we introduce a new class of theories, the Fd-Dec-theories.

Definition 10: A theory E is a Fd-Dec-theory iff

* we can associate with E a convergent term rewriting system R,
* there exists a subset Fd of F such that, for any symbol f in Fd, for any term t=f(t1,...,tn), NF(f(t1,...,tn)) = f(NF(t1),...,NF(tn)).

  We denote by top-Fd-term each term which top symbol belongs to Fd.

In such cases, we can decompose an equation into a system of simpler ones.

Lemma 2: If E is a Fd-Dec-theory, for any f and f' in Fd, t=f(t1,...,tn) t'=f'(t1',...,tp'), the equation t=t' has no solution if $f \neq f'$ and otherwise is E-equivalent to the system {(ti=ti')i=1,...,n}.

Iterating the decomposition process leads to three kinds of equations:

a) x=t with $x \in V$, $t \notin V$ and $x \notin V(t)$

b) x=t with $x \in V$, $t \notin V$ and $x \in V(t)$

c) f(t1,...,tn) = f'(t1',...,tp') with f or f' not in Fd.

In case a), the substitution (x\t) is the minimal E-solution of the equation.

In case b), according to the theory E there can exist a E-solution. For example, in Minus, the equation x=-x admits as an E-solution (x\c(a,-a)). We suppose in the following that a method for solving this type of equation is known in the theories considered.

In case c), it is necessary to transform such an equation into an E-equivalent one, for which the decomposition process can be iterated. The idea is to find operations on the equations which allow us to transform the equation t=t' in an E-equivalent one of the form x=t".

For example, in Minus, the equation c(x,a) = -y is E-equivalent to y=c(-a,-x).

The following definition introduces the class of theories for which we can extend the Martelli and Montanari's approach.

Definition 11: A theory E is a Fd-D-theory or a decomposable theory iff

* it is a Fd-Dec-theory

* any equation x=t is solvable (we can decide if the equation has solutions or not and find a complete set of E-unifiers of them, which domain is exactly {x}).
* any equation t=t' with t a top-Fd-term and t' a non variable term is such that:
  * either it has no E-solution if V(t)uV(t') = {}
  * either is E-equivalent to an equation of the form x=t" for a variable x in V(t)uV(t') and a top-Fd-term t".

Example: Minus is a (F\{-})-D-theory as it will be proved in the next paragraph.

3.2 Multiequations

In this section we introduce and study an extension of the multiequation notion in Martelli and Montanari's work.

Definition 12: A multiequation e is a multiset of terms structured as follow:

* V(e) the set of all variable terms of e,
* P(e) the multiset of terms in e which are not top-Fd-terms,
* T(e) the multiset of terms in e which are top-Fd-terms.

e will also be denoted e=(V(e)=P(e)=T(e)).

Example: for Minus {x, y}={-x, -z}={c(x,y)} is a multiequation, with V(e)={x,y}, P(e)={-x, -z}, T(e)={c(x,y)}. It will be also denoted when there is no ambiguity by x=y=-x=-z=c(x,y)

Definition 13: * A system of multiequations is a multiset of multiequations.

* An E-solution of a multiequation e is a substitution s such that for any elements u and v of e we have s(u) =E s(v).
* s is a E-solution of the system of multiequation M iff s is an E-solution of every multiequation in M. The set of all the E-solutions of the system M will be denoted by SU(M,E).
* The systems of multiequations M and M' are E-equivalent iff SU(M,E)=SU(M',E).

We shall now study the transformations which allow us to simplify a system of multiequations.

3.3 Decomposition of multiequations systems

The basic idea is that, if e has a solution then the terms of T(e) have a common part and a frontier which is defined now.

Before the definition, which is quite technical, we illustrate these notions with an example in the Minus theory. With the use of the tree notation for more clarity, let t1 and t2 such that:

```
t1 =     c      and  t2 =    c   .   Their  common part is  CP[{t1,t2}] =  c
        / \                 /  \                                           / \
       c   y               c    c                                         c  {y}
      / \                 / \  / \                                       / \
     -   a               c   a x  y                                     -   a
     |                  / \                                           { | }
     x                 u   v                                            x
```

```
and FR[{t1,t2}] = {(- =  c ), (y = c  )}.
                   |    / \        / \
                   x   u   v      x   y
```

Definition 14: Let M be a non empty multiset of terms, M={t1, ..., tn}.

- The common part CP[M] of M is the term on FuP(M(F,V)) and
- The frontier FR[M] of M is the system of multiequations,

  recursively given by:

  Case * for all i in 1,...,n ti(e)=f with f in Fd of arity p,

  then CP[M] = f(CP[{t1/1,...,tn/1}],... CP[{t1/p,...,tn/p}])

  and

  FR[M] = {(FR[t1/j,...,tn/j]}j=1,...,p

  * there exists i in 1,...,n such that ti is a variable or is not a top-Fd-term,

    then CP[M] = {ti} and FR[M] = M

  * in all the other cases the common part and the frontier of M do not exist.

It is more convenient to see the common part as a term. So with the next definition we introduce a representation of the common part.

Definition 15: To the mapping CP[M] we associate a term Cp[M] which is also named common part of M defined by:

+ CP[M] exists iff Cp[M] does, and when it exists,

+ for any m in D(Cp[M]),

* CP[M](m)∈F => Cp[M](m)=CP[M](m)

* CP[M](m)={u} => Cp[M]/m=u

Example: with the previous example we get Cp[{t1,t2}] = c(c(-x,a),v)

Now what is the effect of the decomposition on the set of E-solutions of an multiequation? To answer we first study the decomposition effect on the E-solutions of a multiset of terms.

Lemma 3: If s is an E-solution of the multiset of terms M then for any t of M, s(t) =E s(Cp[M]), and s is an E-solution of FR[M].

Proof: by structural induction on Cp[M].[]

We can now introduce the notion of decomposition of an equation.

Definition 16: Given a multiequation e, the decomposition of e denoted Dec(e) is the system of multiequations: Dec(e) = {V(e)=P(e)=Cp[T(e)]} U FR[T(e)] defined when Cp[T(e)] exists.

Example: In the Minus theory if e=(x=y=-x=t1=t2) where t1 and t2 are the terms of the previous example, then:

Dec(e) = {(x=y=-x=c(c(-x,a),v))} U {({}={-x}={c(x,v)}), ({y}={}={c(x,v)})}

The main result of this paragraph can now be stated.

Proposition 1: For any multiequation e such that T(e) is not empty, when the common part of T(e) exists, then the multiequation e=(V(e)=P(e)=T(e)) is E-equivalent to Dec(e) else, e has no E-solution.

Proof: Let s be an E-solution of e.

Let x be in V(e). We have s(x) =E s(p) =E s(t) for any p and t in P(e) and T(e) respectively. By the previous lemma we have:

* for any t in T(e), s(t) =E s(Cp[T(e)]) and thus s is an E-solution of V(e)=P(e)=Cp[T(e)].
* s is an E-solution of FR[T(e)].

Therefore, s is an E-solution of Dec(e).

Conversely, if s is an E-solution of Dec(e), by construction of Dec(e) it is also an E-solution of e. []

Definition 17: The development or the decomposition of a system of multiequations S is a system of multiequations Dec(S) obtained from S by replacing an equation e of S by Dec(e) when it exists. Formally: Dec(S) = (S\{e}) U Dec(e).

The previous results allows to state:

Corollary 1: For any system of multiequations S, if there exists e in S such that Dec(e) does not exist then S has no E-solutions, otherwise S and Dec(S) are E-equivalent.

3.4: Merging of a multiequations system.

In order to group together the hypotheses on the same variable, we now have to merge the multiequations of a system which have variable part not disjoint.

Definition 18: The merging Merg(e,e') of two multiequations e and e' such that V(e) and V(e') are not disjoint, is the multiequation defined by:

Merg(e,e') = ((V(e)UV(e') = P(e)UP(e') = T(e)UT(e')).

Example: In the theory Minus, if e=({x, y} = {-x, -z} = {c(a,x), c(a,a))} and e'=({x} = {} = {c(a,a)}) then Merg(e,e') = ({x, y} = {-x, -z} = {c(a,x), c(a,a), c(a,a)})

It is easy to prove that if e and e' are two multiequations such that V(e) and V(e') are not disjoint, then the system {e,e'} is E-equivalent to Merg(e,e').

We now extend this notion to a system of multiequations.

Definition 19: For a system of multiequations S, the *merging* of S is the system denoted Merg(S) obtained from S by replacing any pair of multiequations that can be merged by their merging.

With these definitions we have the following result: for any system of multiequation S, S and Merg(S) are E-equivalent.

3.5: Normalization of a multiequations system

We introduce in this paragraph a new notion, called normalization, which is needed by the fact that we have to transform some multiequations (for example x=-x=c(a,b) in Minus) in order to allow the continuation of the process of decomposition-merging.

In order to explain the problem, we first give an example in the Minus theory.

Let e=(x=-x=c(a,b)). We cannot decompose this equation because T(e) is reduced to one element. But on the other hand, we cannot consider this equation as "fully" decomposed, because there are elements in both P(e) and T(e). We can thus remark that -x=c(a,b) is E-equivalent to x=c(-b,-a), as it is proved in section 4. Thus e, which can be seen as the system {(x={}=c(a,b)), (x={}=c(-b,-a))} is E-equivalent to x=c(a,b)=c(-b,-a) and this multiequation allows us to continue the decomposition process.

We now formalize this idea. First we have to prove the E-equivalence between a multiequation e and the system issued from this transformation.

Lemma 4: Any multiequation e = (V(e)=P(e)=T(e)) with |P(e)|+|T(e)|>1

* either has no E-solution,
* or is E-equivalent to a multiequation system denoted Trans(e) composed of multiequations e' such that P(e')={} or |P(e')|=1 and T(e')={}.

Proof: We always consider that P(e)≠{}.

** If V(e) = {}: Let t and t' be in e. Either V(t)uV(t')={} and e has no E-solution since E is a Fd-D-theory or, let x be a variable in V(t)uV(t'). By hypothesis t=t' <=> x=t" where t" is a top-Fd-term and the problem is reduced to the case where V(e) is not empty.

** If $V(e) \neq \{\}$ then

we first consider the case where $T(e)$ is not empty. Thus if t is a term of $T(e)$, e is E-equivalent to the system $\{(V(e)=\{\}=T(e))\} \cup \{(p=t) \mid$ for each p in $P(e)\}$. But since E is supposed to be a Fd-D-theory by hypothesis, any equation $p=t$ either is E-equivalent to an equation $x=t'$ where x belongs to $V(p) \cup V(t)$ and t' is a top-Fd-term, or has no E-solution; we denote by $R(e)$ the set: $\{t' \mid (x=t') <=> (p=t)$ for each p in $P(e)\}$. So e is E-equivalent to the system $S=\{(V(e)=\{\}=T(e))\} \cup \{(x=\{\}=t') \mid$ t' element of $R(e)\}$. We take in this case $Trans(e)=Merg(S)$.

If $T(e)$ is empty there are by our hypothesis at least two terms in $P(e)$. Let p be any term of $P(e)$. The multiequation e is thus E-equivalent to the system $\{(V(e)=\{p\}=\{\})\} \cup \{p=p'' \mid p''$ is an element of $P(e)$ different of $p\}$. Since E is an Fd-D-theory, any equation of the form $p=p''$ is E-equivalent to an equation of the form $x=t$ where t is a top-Fd-term. So e is E-equivalent to the system $S = \{(V(e)=\{p\}=\{\})\} \cup \{(x=t) \mid$ $x=t <=> p=p'$ for any element p'' of $P(e)$ different of $p\}$, and by merging S, we obtain a multiequation system S' whose elements e are such that $P(e)$ is empty $\{\}$ or $T(e)$ is not empty. On S' which we can apply the first part of the proof. [ ]

Definition 20: For any multiequation e, we define the *normalized multiequation* $Nor(e)$ by: $Nor(e)$ = Case * $P(e) = \{\}$ then e

* $T(e) = \{\}$ and $|P(e)| = 1$ then e

* else if $Trans(e)$ exists then $Trans(e)$

else $Nor(e)$ does not exists.

An immediate consequence of the previous lemma is the following:

Corollary 2: Any multiequation e either has no E-solution, either is E-equivalent to its normalized form $Nor(e)$.

Definition 21: A system of multiequations S is said *normalized* iff

* all the multiequations are normalized (that is $Nor(e) = e$) and,

* S is merged.

Proposition 2: Any system of multiequations S either has no E-solution, either can be transformed into an E-equivalent and normalized one, denoted Nor(S), which is obtained by normalization of each multiequation in the merged system Merg(S).

3.6 The cycle detection.

As in any E-unification algorithm, we have to detect the cycles that can occur in the potential E-solution. Here we have two possibilities: we can detect such eventual cycles "a posteriori", when the potential E-unification substitution is determined or we can try to detect them as in the Martelli and Montanari algorithm, as soon as possible.

* In the first case a topological sort can be used to solve the problem.

* In the second case, one method is to give an ordering on the multiequations.

The algorithm, given in the next paragraph, uses the second and more elaborated method, but it can be easily modified to deal with the first one. Thus the following condition about the E theory is not essential for the application of our method.

Definition 22: Let $<$ be the relation on the multiequations set defined by: $e<e'$ $<=>$ there exists x in V(e), there exists t in P(e')UT(e'), such that $x \in V(t)$. Let $<+$ be the transitive closure of $<$.

Definition 23: A Fd-D-theory is strict iff the relation $<+$ on the set of multiequations is such that for any system of multiequations S, if S has an E-solution then the restriction of $<+$ to S is a strict ordering. (One never has $e <+ e' <+ e$ for any equation e and e' distinct of e, in S).

Such an ordering allows us to structure the system of multiequations:

Definition 24: Let S be a normalized system of multiequations. We associate with S the pair (Q,R) called ordered pair of multiequation systems, defined by:

* Q is a finite sequence of multiequations (ei)i=1,...k such that for any i

and j in 1,...,k:

a) $e_i < e_j$ is false.

b) $|P(e_i) \cup T(e_i)| < 2$

* $R = S \setminus Q$.

Q is, in fact, the beginning of the E-solution of S. R is what remains to be solved. By definition there is no cycle in Q.

Proposition 3: Let E be a strict Fd-D-theory, if the ordered pair of multiequation systems (Q,R) has an E-solution and if R in not empty, then there exists a multiequation e of R maximal for <+.

3.7 The E-unification algorithm.

We now give the E-unification algorithm. Its principle is to repeat the steps of decomposition merging and normalization and to put in Q the multiequations which are solved.

```
Initialization: Q := {} and e := (t=t') where t and t' are supposed to be in
                                        normal form for the convergent rewri-
                                        ting system associated with E.

                R := IF Nor(e) exists THEN Nor(e) ELSE failure ENDIF

WHILE R is not empty DO

     IF there exists e in R maximal for <+

     THEN CASE (1) T(e)={} THEN Q <- Q+{e}            (add e in tail of Q)

                                R <- R\{e}            (remove e of R)

               (2) Cp[T(e)] do not exists THEN Stop   (clash of symbols)

               (3) FR[T(e)] contains a multiequation x=t which has no E-solution

                   THEN failure

               (4) ELSE R := R\{e}UFR[T(e)]

                   IF Nor(R) exists THEN R := Nor(R) ELSE failure ENDIF

                   Q := Q + (V(e)={}=Cp[T(e)])

          END CASE

     ELSE Failure    (cycle)       END IF     END WHILE
```

At the end of the algorithm, either there is a failure and the set of E-solutions is empty (there is a cycle, a clash of function symbols, or an equation of an E-equivalent system has no E-solution) or the set of E-solutions is determined by Q. It is sufficient to take the multiequations of Q in order and to compose their E-solutions. That comes from the fact that since each equation x=t has, by our hypothesis, a complete set of E-unifiers whose domain is exactely $\{x\}$, if e1, e2,...,en are the multiequations of Q and if s1, s2,..., sn are E-solutions of these multiequations, then the composition sn.sn-1 ...s2.s1 will be an E-solution of Q and thus of R.

The next algorithm formalizes this last step.

SUBST(Q) (we have here R={} and Q=(ei)i=1,...n; if T(ei) is not empty we denote by ti its unique element. We denote by o the substitution composition)

SU <- {Id}

FOR i=1 TO n DO

CASE (1) T(ei)=P(ei)={} THEN (let y in V(ei))

$$SU \leftarrow \bigcup_{s \in SU} \Big[\mathop{o}_{x \in V(ei),\, x \neq y} (x \backslash y) \Big] \; o \; s$$

(2) T(ei)≠{} THEN (we have also P(ei)={} ; let ti be the unique element of T(ei))

$$SU \leftarrow \bigcup_{s \in SU} \Big[\mathop{o}_{x \in V(ei)} (x \backslash ti) \Big] \; o \; s$$

(3) P(ei)≠{} THEN (we have also T(ei)={}. Let pi be the unique element of P(ei). We denote by Z a complete set of E-solutions of the equation x=pi)

$$SU \leftarrow \bigcup_{r \in Z} \Big[\bigcup_{s \in SU} \Big[\mathop{o}_{y \in V(ei)} (y \backslash r(x)) \Big] \; o \; s \Big]$$

ENDCASE
ENDFOR

Theorem 1: Let E be a strict Fd-D-theory, for two terms t and t', if the above algorithms applied to e=(t=t') terminate with success, they return a complete set of E-solutions of t and t'. Furthermore if we are able to give for any equation x=t a minimal complete set of E-unifiers, then the algorithm returns a minimal complete set of E-solutions for the equation t=t'.

4 A COMPLETE UNIFICATION ALGORITHM FOR THE MINUS THEORY.

In this section we see how the above results can be applied to the MINUS theories.

A G-MINUS theory on M(F,V) is defined by V a denumerable set of variables, F a set of function symbols which contain the symbol "-" of arity 1, G a non empty subset of elements of F\{-} with a non-zero arity and the following set of axioms, denoted MINUS:

A1 : -(-x) = x

A2,f : -(f(x1,...,xn)) = f(-xn,...,-x1) for any f in G.

The theory Minus defined in the previous section is a {c}-MINUS theory.

These axioms are common and appear in group theory and signed trees theory ([KKJ,81]) for example.

The complete proofs of the results given in this section can be found in [KIR,83c] or [K&K,82].

We have first to find a confluent term rewriting system associated to the set of axioms MINUS. Then we prove that MINUS is a strict (F\{-})-D-theory. The results of the previous section allow then to give a complete (and in this case minimal) set of MINUS-unifiers. We will see that this set can be infinite.

4.1 The confluent term rewriting system associated to MINUS.

We denote by M-i-n-u-s-> the set of rules obtained by orienting the MINUS axiom's from left to right.

Lemma 5: M-i-n-u-s-> is noetherien.

Proof: we use a recursive path ordering < as defined by N.Dershowitz [DER,82], such that, for all f in F\{-}: f < - . []

Proposition 4: M-i-n-u-s-> is convergent.

Proof: It is very easy to check with the use of a Knuth and Bendix algorithm implementation [LES,83]. []

We can take the normal form of any term. For example NF(-f(x,g(a,-b),-k(x)))=

f(k(x),g(b,-a),-x) if f, g, k are elements of G.

4.2 MINUS is a (F\{-})-D-theory.

MINUS is a (F\{-})-Dec-theory since the axiom's structure allow to prove:

Lemma 6: For any term t1,...,tn and for any symbol f of F\{-} we have NF(f(t1,...,tn))=f(NF(t1),...,NF(tn)).

We now have to study the equations of the form x=t. If t if different of -x, the result is obvious, otherwise it is more difficult.

Lemma 7: Let x be a variable and t be a term which belongs to M(F,V)\{-x}. The equation x=t has a minimal complete set of MINUS-unifiers S: S = {} if x belongs to V(t), and S = {(x\t)} otherwise.

In order to study the set of MINUS-solutions of the equation x=-x, we have to introduce some technical definitions and results.

Definition 25: We denote by Mir(F) the following subset of M(F,V):

Mir(F) = {t | NF(t) = t and NF(-t) = t}

Example: If G={cons, car, cdr} with arity(cons)=2 and arity(car) = arity(cdr)=1 cons(u,-u), car(cons(cdr(x) cdr(-x))) are elements of Mir(F).

This set of MINUS-solutions is too big, since for example, the solution (x\cons(cons(y,-y),cons(y,-y))) can be generated with the simpler one (x\ cons(x,-x)). So, in order to be able to describe a minimal complete set of MINUS-unifiers of the equation x=-x, we introduce a minimal subsets of Mir(F) denoted MIR(F,X) where X is an infinite subset of the set of variables V. MIR(F,X) allows building all the terms of Mir(F) by instantiation.

Definition 26: Let X be a subset of V whose elements are denoted by x(i,j) with i and j integer. for all i=1,2,... let be:

Mir(0,F,X) = {t| t=f(x(1,0),...,x(p,0),-x(p,0),...,-x(1,0) with f in G and of even arity 2p}

Mir(i,F,X) = {t| t=f(x(1,i),...,x(p,i),t',-x(p,i),...,-x(1,i)) with f in G and of odd arity 2p+1 and t' element of Mir(i-1,F,X)}

$MIR(F,X) = \bigcup_{n \in N} Mir(i,F,X)$

Example: With the same notations as in the previous example;

cons(x,-x) and car(cdr(car(cons(y,-y)))) are elements of MIR(F,X) with $\{x,y\} \subset X$. But cons(cons(x,-x),cons(x,-x)), or cons(car(x),car(-x)) are not elements of any MIR(F,X).

Remarks * There is no variable or signed variable (i.e. term of the form -x for some variable x) in Mir(F).

* If G contains only odd arity symbols, then Mir(F) is empty.
* If G contains only even arity symbols, then MIR(F,X) is finite.
* If G contains at least one symbol of even arity, then Mir(F) is not empty
* If G contains at least one symbol of even arity and one symbol of odd arity, then MIR(F,X) is infinite.
* {(x\t) with $t \in$ Mir(F)} is the set of MINUS-solutions of the equation x=-x.

Proposition 5: the equation (x=-x) has for its minimal complete set of MINUS-unifiers the set {(x\t) | $t \in$ MIR(F,X)} for some set of variables X.

Proof: technical verification of the definition of a minimal complete set of unifiers. []

This result shows that in general an equation t=t' has an infinite minimal complete set of MINUS-unifiers. Thus we can see here one particular interest of our method since the MINUS-solution of the equations of form x=t are not necessary during the process of decomposition-merging-normalization, but is only required at the end of this process in order to express the form of the solutions. This allows finite description of infinite complete set of solutions. This property is fully used for example in [K&K,82] for the resolution of equations in the signed trees theory.

It remains to be proven that any equation t=t', with t=-x and t a non variable term, is MINUS-equivalent to an equation x=t". It is a consequence of the following property:

Lemma 3: Any equation t=t' is MINUS-equivalent to NF(-t)=NF(-t').

So any equation of the form -x=t where x is a variable is MINUS-equivalent to NF(-(-x))=NF(-t) which is of the required form x=t". That yields the following result: the MINUS theories are (F\{-}) decomposable theories.

Furthermore the MINUS theories are strict.

We are now able to state the main result of this section:

Theorem 2: The MINUS theories are strict (F\{-}) decomposable theories.

4.3 The minimal complete sets of MINUS-unifiers.

We apply the general results to the MINUS theories:

Theorem 3: Let E be a G-MINUS-theory,

* if there are only even or only odd arity symbols in G, then any equation has a finite minimal complete set of MINUS-unifiers, which is given by the algorithm SUBST. These theories are finitary in the sense of J.Siekmann and P.Szabo [S&S,82].

* if there is at least one odd arity and one even arity symbol in G, then some equations have infinite minimal complete set of MINUS-unifiers which are finitely described by the algorithms DEC-NOR and SUBST. These theories are infinitary in the sense of [S&S,82].

We conclude by giving an example of the resolution of an equation in the MINUS theory.

Example: Let

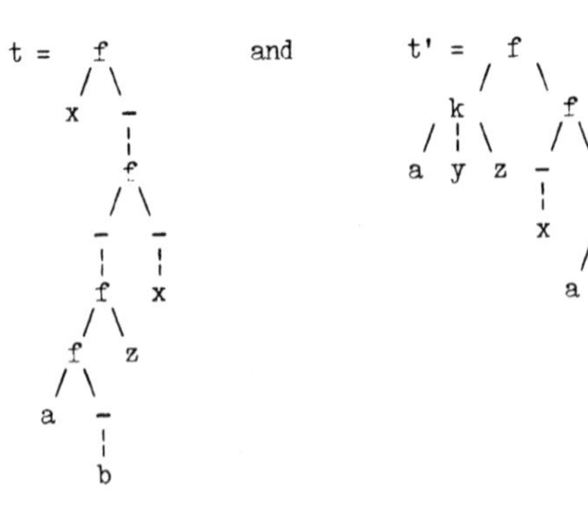

with F={-,f,a,b,k} G={f,k} and x,z,y,z' variables. What are the MINUS-solutions of the equation t=t'? First we normalize t and t':

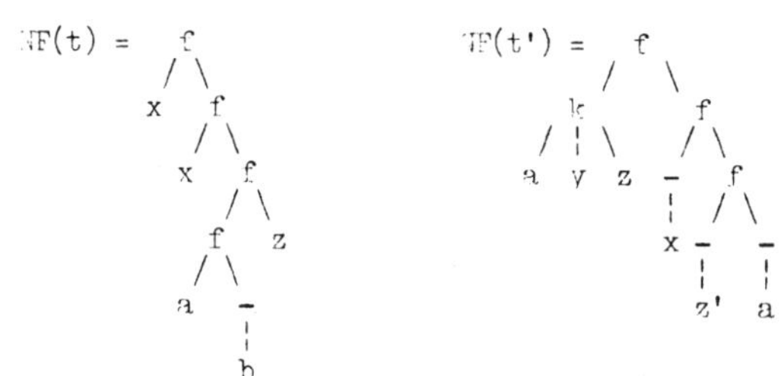

R is initialized with { (NF(t)=NF(t')) } and Q with {}.

After a first decomposition we obtain the following system:

```
x=   k                                     x=   k     =     k
   / ¦ \                                      / ¦ \       / ¦ \
  a  y  z                                    a  y  z     -  -  -
                                                         ¦  ¦  ¦
x=-x                                                     z  y  a
                 which can be normalized in  z=-a
z=-a

-z'=  f                                    z'=  f
     / \                                       / \
    a   -                                     b   -
        ¦                                         ¦
        b                                         a
```

The first and third multiequations are maximal for <*, so we get the new system

```
Q={(z'=  f  ),(x=   k   )}  and  R={(z = - = - = -),(y=-y)}
        / \        / ¦ \                 ¦   ¦   ¦
       b   -      -  y  z                a   a   a
           ¦      ¦
           a      z
```

which gives

```
Q={(z'=  f  ),(x=   k   ),(z= -),(y=-y)}   and R={}
        / \        / ¦ \      ¦
       b   -      -  y  z     a
           ¦      ¦
           a      z
```

Applying SUBST to Q gives the following minimal complete set of MINUS-unifiers of the terms t and t':

```
S = { s  ¦ s = (y\r)(z\-)(x\   k   )(z'\   f  ) with r in MIR(F,X)}  for  some
                       ¦      / ¦ \        / \
                       a    -z  y  z      b  -a
```

set of variables X.

We think that the formalism developped here is in fact the "good" one. We have treated the case of the D-theories but we are working to extend the method to a wider class of theories including permutative theories as for example associative-commutative theories.

I would like to thank P.Lescanne, J.L. Remy, K. Yelick and especially Helene Kirchner and J.P. Jouannaud for helpfull discussions and pertinent remarks.

REFERENCES

[C&B,83] CORBIN J. and BIDOIT M. "Pour une rehabilitation de l'algorithme d'unification de Robinson".
Note a l'academie de sciences de Paris. T.296, Serie I, p 279. (1983).

[DER,82] DERSHOWITZ N.: "Orderings for term-rewriting systems".
Theorical Computer Science 17-3. (1982).

[FAG,83] FAGES F.: "Formes canoniques dans les algebres booleennes, et application a la demonstration automatique en logique de premier ordre".
These de 3ieme cycle. Universite Paris 6. (1983).

[FAG,84] FAGES F.: "Associative-Commutative Unification"
Proceedings 7th CADE, Napa Valley, 1984.

[FAY,79] FAY M.: "First order unification in equational theory"
Proc. 4th Workshop on Automata Deduction Texas. (1979).

[H&O,80] HUET G. and OPPEN D.C.: "Equations and rewrite rules: a survey" in "Formal Languages: Perspectives and open problems"
Ed. Book R., Academic Press. (1980).

[HUE,76] HUET G.: "Resolution d'equation dans les langages d'ordre 1,2..., "
These d'etat, Universite Paris VII. (1976).

[HUL,80] HULLOT J. M.: "Canonical forms and unification"
Proc. 5th Workshop on Automated Deduction Les Arcs. (1980).

[JKK,83] JOUANNAUD J.P. KIRCHNER C. KIRCHNER H.: "Incremental construction of unification algorithms in equationnal theories".
Proc. of the 10th ICALP. LNCS 154, p 361. (1983).

[J&K,84] JOUANNAUD J.P. KIRCHNER H.: "Completion of a set of rules modulo a set of equations".
Proc. of the 11th ACM conference on principles of programming languages, Salt Lake City. (1984).

[KIR,83] KIRCHNER C.: "A new unification method in equational theories and its application to the MINUS theories".
Internal report CRIN 83-R-xxx. (1983).

[KIR,84] KIRCHNER H.: "A general inductive completion algorithm and application to abstract data types".
Proceedings 7th CADE, Napa Valley, 1984.

[K&B,70] KNUTH D. BENDIX P.: "Simple word problems in universal algebras" in "Computational problems in abstract algebra"
Leech J. ed. Pergamon Press, pp 263-29.7 (1970).

[KKJ,81] KIRCHNER C., KIRCHNER H., JOUANNAUD J.P.: "Algebraic manipulations as a unification and matching strategy for linear equations in signed binary trees". Proc. IJCAI 81 Vancouver. (1981).

[K&K,82] KIRCHNER C. KIRCHNER H.: "Resolution d'equations dans les algebres libres et les varietes equationnelles".
These de 3eme cycle, Universite NANCY 1. (1982)

[LAN,75] LANKFORD D.S.: "Canonical inference".
Report ATP-32. University of Texas. (1975).

[LAN,79] LANKFORD D.S.: "A unification algorithm for abelian group theory"
Report MTP-1. Math. dep., Louisiana Tech. U. (1979).

[LES,83] LESCANNE P.: "Computer experiments with the REVE term rewriting system generator".
Proc. of the 10th POPL conference. (1983).

[L&S,77] LIVESEY M. and SIEKMANN J.: "Unification of sets"
Report 3/76, Institut fur Informatik 1, Univ. Karlsruhe. (1977).

[M&M,82] MARTELLI A. and MONTANARI U.: "An efficient unification algorithm"
Transactions on Programming Languages and Systems Vol. 4, Number 2. p 258. (1982).

[PLO,72] PLOTKIN G.: "Building in equational theories"
Machine Intelligence 7, pp 73-90. (1972).

[P&W,78] PATERSON M.S. and WEGMAN M.N.: "Linear unification"
J. of Computer and Systems Sciences 16 pp 158-167. (1978).

[ROB,65] ROBINSON J.A.: "A machine oriented logic based on the resolution principle"
J.ACM 12, pp 32-41. (1965).

[ROB,71] ROBINSON J.A.: "computational logic: the unification computation".
Machine Intelligence 6. (1971).

[R&S,78] RAULEFS P. and SIEKMANN J.:"Unification of idempotent functions"
Report, Institut fur Informatik 1, Univ. Karlsruhe. (1978).

[S&S,82] SIEKMANN J. and SZABO P.: "Universal unification and a classification of equational theories".
Proc. of the 6th CADE. LNCS 138, p 369. (1982).

[STI,81] STICKEL M.E.: "A unification algorithm for associative-commutative functions"
J.ACM 28, no.3, pp 423-434. (1981).

A CASE STUDY OF THEOREM PROVING BY THE KNUTH-BENDIX METHOD: DISCOVERING THAT $x^3 = x$ IMPLIES RING COMMUTATIVITY

Mark E. Stickel
Artificial Intelligence Center
SRI International
Menlo Park, California 94025

Abstract

An automatic procedure was used to discover the fact that $x^3 = x$ implies ring commutativity. The proof of this theorem was posed as a challenge problem by W.W. Bledsoe in a 1977 article. The only previous automated proof was by Robert Veroff using the Argonne National Laboratory - Northern Illinois University theorem-proving system. The technique used to prove this theorem was the Knuth-Bendix completion method with associative and/or commutative unification. This was applied to a set of reductions consisting of a complete set of reductions for free rings plus the reduction $x \times x \times x \rightarrow x$ and resulted in the purely forward-reasoning derivation of $x \times y = y \times x$. An important extension to this methodology, which made solution of the $x^3 = x$ ring problem feasible, is the use of cancellation laws to simplify derived reductions. Their use in the Knuth-Bendix method can substantially accelerate convergence of complete sets of reductions. A second application of the Knuth-Bendix method to the set of reductions for free rings plus the reduction $x \times x \times x \rightarrow x$, but this time with the commutativity of multiplication assumed, resulted in the discovery of a complete set of reductions for free rings satisfying $x^3 = x$. This complete set of reductions can be used to decide the word problem for the $x^3 = x$ ring.

1. Introduction

We present here the solution of a challenge problem using the Knuth-Bendix method. The problem—to prove that, if $x^3 = x$ in a ring, then the ring is commutative—was offered as a challenge for theorem-proving programs by W.W. Bledsoe in 1977 [1]. Jacobson [6] proved a generalization of this problem: if every element x in a ring satisfies an equation of the form $x^{n(x)} = x$ for $n(x) > 1$, then every element has finite additive order and the ring is commutative. The only previous solution of the $x^3 = x$ challenge problem (the special case of Jacobson's theorem where $n(x) = 3$) by an automated theorem prover was done by Robert Veroff [21] using the powerful Argonne National Laboratory - Northern Illinois University (ANL-NIU) theorem-proving system.

This research was supported by the Defense Advanced Research Projects Agency under Contract N00039-80-C-0575 with the Naval Electronic Systems Command. The views and conclusions contained in this document are those of the author and should not be interpreted as representative of the official policies, either expressed or implied, of the Defense Advanced Research Projects Agency or the United States government. APPROVED FOR PUBLIC RELEASE. DISTRIBUTION UNLIMITED.

Our proof has the novel feature that all reasoning was forward reasoning with the program never having been told that the objective was proving $x \times y = y \times x$.

The solution of this problem is a substantial success for the Knuth-Bendix completion method that had already shown promise in solving less difficult problems such as completing sets of reductions for various algebras like free groups and rings. It suggests that the Knuth-Bendix completion method is a very effective method for deriving consequences from equational theories whose equations can be treated as reductions. That a 21 step proof (not counting reduction steps) for this difficult problem could be found after generating only 135 reductions is quite impressive.

The technique used was the Knuth-Bendix completion method using associative-commutative unification for addition and incomplete associative unification for multiplication. The program attempted to find a complete set of reductions beginning with a complete set of reductions for free rings plus the reduction $x \times x \times x \rightarrow x$. The program predictably failed to complete the set of reductions. Commutativity of multiplication, a consequence of $x^3 = x$, prevents there being a complete set of reductions unless commutativity of multiplication is assumed. However, the program did discover the commutative equation $x \times y = y \times x$ in the attempt, thus proving the theorem. The program also later succeeded in proving the same result with comparable effort by using ordinary, rather than associative, unification for multiplication and building associativity in with the reduction $(x \times y) \times z \rightarrow x \times (y \times z)$.

After the derivation of the commutativity of multiplication, we used the program again to attempt to complete the same set of reductions with associative-commutative unification used for both addition and multiplication. This resulted in the discovery of a complete set of reductions for free rings satisfying $x^3 = x$. This complete set of reductions can be used to provide a simple effective procedure to decide the word problem for such rings. Two words w_1 and w_2 are equivalent in the theory of free rings satisfying $x^3 = x$ if and only if $w_1\!\downarrow\, = w_2\!\downarrow$ (taking account of the associativity and commutativity of addition and multiplication) where $t\!\downarrow$ denotes the result of fully reducing t by the complete set of reductions. Equivalently, w_1 and w_2 are equivalent if and only if $(w_1 + (-w_2))\!\downarrow\, = 0$. Comer [2] proves the more general result that the theory of rings satisfying $x^n = x$ for $n > 1$ is decidable. However, our result (for the $n = 3$ case) is easier to apply.

We will not review here the Knuth-Bendix method or its extension using associative and/or commutative unification. The Knuth-Bendix method is described by Knuth and Bendix [7], Lankford [8, 9], and Huet [3], among others. Associative and/or commutative unification are treated by Siekmann [17, 18], Livesey and Siekmann [13, 14], and Stickel [19, 20] and extension of the Knuth-Bendix method to incorporate associativity and/or commutativity is treated by Lankford and Ballantyne [10, 11, 12] and Peterson and Stickel [15, 16]. In addition to the publications cited above, Hullot [5] presents numerous examples of the use of the Knuth-Bendix method with and without special treatment of associativity and/or commutativity.

The two most important differences in the use of the Knuth-Bendix method for this problem, as compared to previous work by Peterson and Stickel [15, 16], are the use of cancellation laws to simplify reductions and a better pair-evaluation function to order matching reductions.

Sections 2 and 3 describe the use of cancellation laws and the pair-evaluation function in the Knuth-Bendix method. Section 4 describes the proof of the $x^3 = x$ ring problem and Section 5 compares our approach with Veroff's. Section 6 discusses the complete set of reductions for free rings satisfying $x^3 = x$. Section 7 gives suggestions for improving the performance of the program.

2. Cancellation Laws

The most significant addition to the Knuth-Bendix method using associative and/or commutative unification that we made to solve the $x^3 = x$ ring problem is the use of cancellation laws to simplify derived equations. This addition made the solution feasible; we have so far failed to solve the $x^3 = x$ ring problem without the cancellation laws. It is certain that the effort required to do so would greatly exceed that used with the cancellation laws. We expect that cancellation laws can be widely used in the Knuth-Bendix method in the future to substantially accelerate convergence of complete sets of reductions.

To use the cancellation laws, we add the reductions $(x + y = x) \rightarrow (y = 0)$ and $(x + y = x + z) \rightarrow (y = z)$ that may be applicable to the entire derived equation, not just its subterms as is the case for all the other reductions.

With the single exception of the additive identity reduction $x+0 \rightarrow x$ that the cancellation laws are not permitted to reduce, the cancellation laws never reduce an equation of nonidentical terms to an equation of identical terms. Thus, any critical pair from which a reduction can be derived can lead instead to a simpler reduction if a cancellation law is applicable. This simpler reduction is more powerful than the original because it, plus $x + 0 \rightarrow x$, can reduce the original reduction to an identity.

Besides being more powerful, the simpler reduction has the further advantage that matching its left-hand side with the left-hand side of other reductions to generate new critical pairs will result in fewer, less complex equations and thus create less work for the program.

We ran two problems with and without using the cancellation laws. The first problem is completing the set of reductions for free commutative rings with unit element starting from the equations $0 + x = x$, $(-x) + x = 0$, $1 \times x = x$, and $x \times (y + z) = (x \times y) + (x \times z)$. The second problem is to show that rings satisfying $x^2 = x$ are commutative. This is similar to, but much simpler than, the $x^3 = x$ ring problem. The relative simplicity of the $x^2 = x$ ring problem is suggested by the fact that in $x^2 = x$ rings, $-x = x$; in $x^3 = x$ rings, $-x = 5x$. The benefit of using the cancellation laws in the $x^3 = x$ ring problem is much greater than for these simpler problems.

Problem	Pairs Matched	Equations Simplified	Reductions Created
ring completion w.o. cancel	222	428	31 (9 retained)
ring completion with cancel	115	289	21 (9 retained)
$x^2 = x$ ring problem w.o. cancel	124	268	23 (13 retained)
$x^2 = x$ ring problem with cancel	96	165	19 (13 retained)

Modifying derived equations instead of immediately transforming them into reductions provides important added variability in the Knuth-Bendix method to give it greater efficiency or additional capabilities. The use of cancellation laws is an example of the former. An example of the latter is the modification of the Knuth-Bendix method to do induction proofs in equational theories with constructors (see, for example, Huet and Hullot [4]). There, if an equation of two terms headed by the same constructor with n arguments is derived, the equation is replaced by n equations equating the arguments.

3. Pair-Evaluation Function

In the Knuth-Bendix method, it is necessary to select which pair of reductions to match next to derive new critical pairs. If the selection algorithm is poor, the completion process may diverge, even though a complete set of reductions exists. There is no way of insuring that the completion process will not diverge unnecessarily, but selecting pairs with small combined left-hand sides works well in practice.

Our implementation of the pair-selection process involves maintaining a list of all pending pairs sorted in ascending order according to an evaluation function. The evaluation function we have used in the past is simply the sum of the number of symbols in the two left-hand sides. However, the $x^3 = x$ ring problem is much more difficult than previous problems to which the Knuth-Bendix method has been applied, and this simple evaluation function was not adequate for easily solving it. An attempt to solve the $x^3 = x$ ring problem managed to get within one step of discovering $x \times y = y \times x$, but failed to select the right pair of reductions to match next after several days of computing.

The problem was the discovery of large numbers of reductions like $x^2yxyx^2yx \rightarrow xyx$. Such reductions have relatively few symbols on the left-hand side and hence were given preference for matching. However, matching these reductions with other reductions (such as the distributivity reductions) often resulted in a large number of equations that were slow to simplify. Simply counting symbols in the left-hand side of the reduction did not reflect the greater complexity (after simplification to sum-of-products form) of products when a variable is instantiated to a sum compared to the complexity of sums when a variable is instantiated. The solution is to use an evaluation function that gives less preference to products.

The evaluation function adopted to remedy this problem is $V(\lambda_1) + V(\lambda_2)$ where λ_1 and λ_2 are the left-hand sides of the two reductions to be matched and V is defined over the constant

0, all variables x, and terms $t, t_1, \ldots, t_n$ as

$$\begin{aligned} V(0) &= 2, \\ V(x) &= 2, \\ V(-t) &= 5 \times V(t), \\ V(t_1 + \cdots + t_n) &= V(t_1) + \cdots + V(t_n), \\ V(t_1 \times \cdots \times t_n) &= V(t_1) \times \cdots \times V(t_n). \end{aligned}$$

This is a natural evaluation function for ring theory problems because the value of a ring sum is simply the integer sum of the values and the value of a ring product is simply the integer product of the values. The value 2 used for constants and variables is the smallest positive integer that is not the additive or multiplicative identity.

The only part of the definition of V that seems contrived for the purpose of the $x^3 = x$ ring problem is the definition of $V(-t)$ as $5 \times V(t)$. This value reflects the fact that $-t = 5t$ is a consequence of of $x^3 = x$—a fact discovered in the earlier attempt to solve the problem. The choice of $V(-t) = 5 \times V(t)$ as opposed to other reasonable definitions for V is inconsequential because the reduction $-x \to 5x$ is discovered quite early in the completion process. All other occurrences of $-$ are then eliminated, and the evaluation function for negated terms plays no further role.

4. The Proof

The appendix lists the proof of ring commutativity from $x^3 = x$. The proof has been cleaned up slightly and unused inferences are not shown. The program did not use exponentiation or multiplication by a constant, so $3xv^2$ is our shorthand for the program's $(x \times v \times v) + (x \times v \times v) + (x \times v \times v)$.

Ring axioms were provided as reductions (1)-(11). These reductions, plus the assumptions of associativity and commutativity for $+$ and associativity for $\times$, constitute a complete set of reductions for free rings. We did not allow matching pairs of ring axiom reductions because this is a complete set of reductions and no new reductions could be created. Reduction (12), not shown, invoked the cancellation laws allowing the derivation of $x = y$ from $x + z = y + z$ and $x = 0$ from $x + z = z$ or $z = x + z$. Reduction (13) is the hypothesis that $x^3 = x$.

Each of the intermediate steps in the derivation is a derived reduction. In effect, reductions are created by forming an expression to which both parent reductions are applicable. The results of applying the two reductions are set equal and fully simplified. If the result is not an identity, it is saved as a new reduction. After each reduction, there may be a line *simplifying* x *by* y (when there is not, it means the reduction is exactly an equation formed from matching the parent reduction left-hand sides) where x is the equation formed from the two parent reductions and y is the list of simplifiers used to simplify it (*cancel* and *distrib* refer to use of the cancellation and distributivity laws). The reduction numbers $(n)^e$, $(n)^{e1}$, etc., refer to embedded forms of reduction (n). For the reduction (n) $\lambda \to \rho$ where λ is headed by the function symbol f, if f is associative-commutative then $(n)^e$ is the reduction $f(u,\lambda) \to f(u,\rho)$; if f is associative then $(n)^{e1}$ is the reduction $f(u,\lambda) \to f(u,\rho)$, $(n)^{e2}$ is the reduction $f(\lambda,v) \to f(\rho,v)$, and $(n)^{e3}$ is the reduction $f(u,f(\lambda,v)) \to f(u,f(\rho,v))$.

A useful perspective is to consider, for example, the derivation of (14) as the simplification of $x^3 = x$ with x instantiated by $y+z$ (i.e., $(y+z)^3 = y+z$) by applying distributivity to $(y+z)^3$.

At the completion of the proof, only 135 reductions had been created of which 52 were retained. The economy of the Knuth-Bendix procedure in number of retained results is demonstrable from the fact that these 135 reductions were the result of simplifying 9,013 equations derived from matching 988 pairs of reductions. Most of the remaining equations were simplified to identities and discarded; a few were simplified to equations like $3xy = 3yx$ that could not be converted into reductions and were also discarded. Total time was about 14.3 hours (including garbage-collection time) on a Symbolics LM-2 LISP Machine. This reflects slowness of the simplification procedure on numerous lengthy terms and could be greatly reduced.

5. The Veroff Proof

The only previous computer proof of the $x^3 = x$ problem was done by Robert Veroff in 1981 using the ANL-NIU theorem-proving system [21]. His solution required an impressively fast 2^+ minutes on an IBM 3033. It is interesting to compare the approaches taken in these two proofs. Both rely heavily on equality reasoning. The process of fully simplifying equations with respect to a set of reductions is just demodulation. The Knuth-Bendix method's means for deriving equations from pairs of reductions is similar to the paramodulation operation used in the ANL-NIU prover. Cancellation laws were also used by the ANL-NIU prover. Despite such similarities in approach, solution by the Knuth-Bendix method required less preparation of the problem. The Knuth-Bendix program was given only the 11 reductions for free rings, the reduction $x \times x \times x \to x$, declarations of associativity and commutativity, and a reduction for the cancellation laws. The ANL-NIU program was provided with a total of 60^+ clauses, including the negation of the theorem (their proof was goal-directed whereas our program derived ring commutativity as a result of pure forward reasoning, attempting to complete a set of reductions). Some clauses expressed information about associativity and commutativity, which are handled by declarations in the Knuth-Bendix program. A large number were present to support a polynomial subtraction inference operation—e.g., to derive $a + (-c) = 0$ from $a + b = 0$ and $b + c = 0$. Comparable operations are implicit in the Knuth-Bendix method, which can infer $a = c$ by matching the reductions $x + a + b \to x$ (the embedded form of $a + b \to 0$) and $y + b + c \to y$ (the embedded form of $b + c \to 0$).

6. A Complete Set of Reductions

After the discovery that multiplication is commutative, the problem can be run again to derive a complete set of reductions for free rings satisfying $x^3 = x$. This can then be used to provide a decision procedure for the word problem for such rings. Assuming the associativity and commutativity of addition and multiplication, reductions 1, 2, 3, 5, 6, 7, 9, and 10 comprise a complete set of reductions for free commutative rings. Attempting to complete the set of reductions consisting of these reductions plus $x \times x \times x \rightarrow x$ resulted in the discovery that the reductions marked by • in the proof comprise a complete set of reductions (assuming finite termination) for free rings satisfying $x^3 = x$. During this computation, 120 pairs of reductions were matched and 2,121 equations simplified; this resulted in 18 reductions, 8 of which were retained to form the complete set of reductions. The cancellation laws were necessary for deriving this result as well.

To verify that this set of reductions derived by the program is actually a complete set of reductions, it must be proved that the set of reductions has the finite termination property. This can be done using a polynomial complexity measure over terms as used by Lankford [8, 9]. The polynomial complexity measure $\|\cdot\|$ is defined over the constant 0, all variables x, and terms $t, t_1, \ldots, t_n$ as

$$\begin{aligned}
\|0\| &= 2, \\
\|x\| &= 2, \\
\|-t\| &= 7 \times \|t\| + 1, \\
\|t_1 + \cdots + t_n\| &= \|t_1\| + \cdots + \|t_n\| + n - 1 \quad \textit{i.e.}, \ \|t_1 + t_2\| = \|t_1\| + \|t_2\| + 1, \\
\|t_1 \times \cdots \times t_n\| &= \|t_1\| \times \cdots \times \|t_n\|.
\end{aligned}$$

It can be shown that all the reductions in the final set (and also the deleted intermediate reductions from which they were derived) are complexity reducing according to this measure. Thus the set of reductions has the finite termination property and is complete.

7. Performance Improvement

Simplification accounts for most of the time spent by the program. Either increasing the speed of simplification or reducing the number of equations that need to be simplified could substantially improve the program's performance.

Probably the most serious defect of the simplification procedure is that the subsumption algorithm, used to determine if a term is an instance of the left-hand side of a reduction and form the matching substitution, is not incremental. When matching a pair of terms, the subsumption algorithm returns a set of all most general substitutions instantiating the first term to match the second, instead of returning a single substitution and generating additional substitutions on demand. The large number of number of substitutions so produced, and the fact that the simplification procedure requires only a single substitution, result in much wasted effort.

The use of a nonincremental subsumption algorithm was especially bad in two instances. Note that subsuming $t_1 + \cdots + t_n$ by $x + y$ where $+$ is associative-commutative results in the production of $2^n - 2$ substitutions. All these substitutions will be produced as an intermediate result when attempting to reduce $t_1 + \cdots + t_n = s$ by the cancellation-law reduction $(x + y = x + z) \rightarrow$

$(y = z)$ or to reduce $(t_1 + \cdots + t_n) \times s$ by the distributivity reduction $(x+y) \times z \rightarrow (x \times z)+(y \times z)$. Given the lengthy terms sometimes simplified, and the extensive use of distributivity, we found it necessary to use LISP code to perform the cancellation-law and distributivity simplifications.

Subsuming $t_1 \times \cdots \times t_n$ by $x \times y$ where $\times$ is associative but not commutative is less costly, but still results in the production of $n-1$ substitutions, and is another source of inefficiency. Adoption of an incremental subsumption algorithm, especially with "look ahead" for recognizing that a substitution for a pair of subterms will not be extendable to match the whole terms, could greatly speed the simplification process.

An obvious addition to the Knuth-Bendix method to reduce the number of equations generated for this and similar examples is to recognize symmetries. Although multiplication is not assumed to be commutative, each reduction using multiplication has a symmetric variant, e.g.,

(3) $x \times (y + z) \rightarrow x \times y + x \times z$ and
(4) $(x + y) \times z \rightarrow x \times z + y \times z$,

(7) $x \times 0 \rightarrow 0$ and
(8) $0 \times x \rightarrow 0$,

(10) $x \times (-y) \rightarrow -(x \times y)$ and
(11) $(-x) \times y \rightarrow -(x \times y)$.

If a_1, a_2 and b_1, b_2 are pairs of reductions that are symmetric variants, it is not necessary to match each a_i with each b_j. It would be sufficient to match a_1 with b_1 and b_2, provided that the symmetric variants of derived reductions are also added.

Winkler [22] also offers a criterion for rejecting matches that could be used to reduce the number of equations generated.

Acknowledgments

Mabry Tyson provided much useful advice and assistance in this work. Dallas Lankford was extremely helpful in providing suggestions and reference material for this paper. Richard Waldinger also gave many helpful comments on an earlier draft of the paper. The strong encouragement by these people was very important to me in writing this paper.

References

[1] Bledsoe, W.W. Non-resolution theorem proving. *Artificial Intelligence 9*, 1 (August 1977), 1-35.

[2] Comer, S.D. Elementary properties of structures of sections. *Boletin de la Sociedad Matematica Mexicana 19* (1974), 78-85.

[3] Huet, G. A complete proof of the correctness of the Knuth-Bendix completion algorithm. *Journal of Computer and System Sciences 23*, 1 (August 1981), 11-21.

[4] Huet, G. and J.M. Hullot. Proofs by induction in equational theories with constructors. *Proceedings of the 21st IEEE Symposium on the Foundations of Computer Science,* 1980.

[5] Hullot, J.M. A catalogue of canonical term rewriting systems. Technical Report CSL-113, Computer Science Laboratory, SRI International, Menlo Park, California, April 1980.

[6] Jacobson, N. Structure theory for algebraic algebras of bounded degree. *Annals of Mathematics 46,* 4 (October 1945), 695–707.

[7] Knuth, D.E. and P.B. Bendix. Simple word problems in universal algebras. In Leech, J. (ed.), *Computational Problems in Abstract Algebras,* Pergamon Press, 1970, pp. 263–297.

[8] Lankford, D.S. Canonical algebraic simplification. Report ATP-25, Department of Mathematics, University of Texas, Austin, Texas, May 1975.

[9] Lankford, D.S. Canonical inference. Report ATP-32, Department of Mathematics, University of Texas, Austin, Texas, December 1975.

[10] Lankford, D.S. and A.M. Ballantyne. Decision procedures for simple equational theories with commutative axioms: complete sets of commutative reductions. Report ATP-35, Department of Mathematics, University of Texas, Austin, Texas, March 1977.

[11] Lankford, D.S. and A.M. Ballantyne. Decision procedures for simple equational theories with permutative axioms: complete sets of permutative reductions. Report ATP-37, Department of Mathematics, University of Texas, Austin, Texas, April 1977.

[12] Lankford, D.S. and A.M. Ballantyne. Decision procedures for simple equational theories with commutative-associative axioms: complete sets of commutative-associative reductions. Report ATP-39, Department of Mathematics, University of Texas, Austin, Texas, August 1977.

[13] Livesey, M. and J. Siekmann. Termination and decidability results for string-unification. Memo CSM-12, Essex University Computing Center, Colchester, Essex, England, August 1975.

[14] Livesey, M. and J. Siekmann. Unification of A+C-terms (bags) and A+C+I-terms (sets). Interner Bericht Nr. 5/76, Institut für Informatik I, Universität Karlsruhe, Karlsruhe, West Germany, 1976.

[15] Peterson, G.E. and M.E. Stickel. Complete sets of reductions for some equational theories. *Journal of the Association for Computing Machinery 28,* 2 (April 1981), 233–264.

[16] Peterson, G.E. and M.E. Stickel. Complete systems of reductions using associative and/or commutative unification. Technical Note 269, Artificial Intelligence Center, SRI International, Menlo Park, California, October 1982. To appear in M. Richter (ed.) *Lecture Notes on Systems of Reductions.*

[17] Siekmann, J. String-unification, part I. Essex University, Cochester, Essex, England, March 1975.

[18] Siekmann, J. T-unification, part I. Unification of commutative terms. Interner Bericht Nr. 4/76, Institut für Informatik I, Universität Karlsruhe, Karlsruhe, West Germany, 1976.

[19] Stickel, M.E. Mechanical theorem proving and artificial intelligence languages. Ph.D. Dissertation, Department of Computer Science, Carnegie-Mellon University, Pittsburgh, Pennsylvania, December 1977.

[20] Stickel, M.E. A unification algorithm for associative-commutative functions. *Journal of the Association for Computing Machinery 28,* 3 (July 1981), 423–434.

[21] Veroff, R.L. Canonicalization and demodulation. Report ANL-81-6, Argonne National Laboratory, Argonne, Illinois, February 1981.

[22] Winkler, F. A criterion for eliminating unnecessary reductions in the Knuth-Bendix algorithm. CAMP-LINZ Technical Report 83-14.0, Ordinariat Mathematik III, Johannes Kepler Universität, Linz, Austria, May 1983.

Appendix: Proof that $x^3 = x$ Implies Ring Commutativity

Following is the proof that $x^3 = x$ implies ring commutativity.

The program was first run with + declared to be associative-commutative and $\times$ declared to be associative. After 988 pairs were matched, 9,013 equations simplified, and 135 reductions created (52 retained), the commutativity of $\times$ was derived.

The program was then run again with both + and $\times$ declared to be associative-commutative. After 120 pairs were matched, 2,121 equations simplified, and 18 reductions created (8 retained), a complete set of reductions for the $x^3 = x$ ring was discovered. The 8 reductions in the complete set are marked by •.

•(1) $0 + x \to x$,

(2) $(-x) + x \to 0$,

•(3) $x(y + z) \to xy + xz$,

(4) $(x + y)z \to xz + yz$,

(5) $-0 \to 0$,

(6) $-(-x) \to x$,

•(7) $x0 \to 0$,

(8) $0x \to 0$,

(9) $-(x + y) \to (-x) + (-y)$,

(10) $x(-y) \to -(xy)$,

(11) $(-x)y \to -(xy)$,

•(13) $x^3 \to x$,

(14) $zyz + yz^2 + y^2z + z^2y + zy^2 + yzy \to 0$ — from (3) and (13),
simplifying $(y + z)^2y + (y + z)^2z = y + z$ by cancel, distrib, (13),

(18) $yzw + zyw + wyz + wzy + ywz + zwy \to 0$ — from (3) and (14),
simplifying $w(y + z)w + (y + z)w^2 + (y + z)^2w + w(y + z)^2 + (y + z)w(y + z) + w^2y + w^2z = 0$ by distrib, (14),

•(21) $6z \to 0$ — from (13) and (14),
simplifying $5z^3 + z = 0$ by (13),

(22) $3z^2 + 3z \to 0$ — from (13) and (14),
simplifying $3z^9 + 2z^6 + z^2 = 0$ by (13),

(24) $3yz + 3zy \to 0$ — from (3) and (22),
simplifying $2(y + z)^2 + (y + z)y + (y + z)z + 3y + 3z = 0$ by distrib, (22),

•(28) $-r \to 5r$ — from $(2)^e$ and $(21)^e$,

•(29) $3v^2 \to 3v$ — from $(21)^e$ and $(22)^e$,

•(30) $3xv^2 \to 3xv$ — from (3) and (29),
simplifying $x(3v) = xv^2 + x(2v^2)$ by distrib,

(31) $3v^2z \to 3vz$ — from (4) and (29),
simplifying $(3v)z = (2v^2)z + v^2z$ by distrib,

(33) $3xux + 3ux \to 0$ — from $(13)^{e1}$ and (24),
simplifying $3x^2ux + 2ux^3 + ux = 0$ by (13), (31),

(40) $3xux \to 3ux$ — from $(21)^e$ and $(33)^e$,

(48) $yzy + zy^2 + y^2z + 3yz \to 0$ — from $(18)^e$ and (21),
simplifying $0 = 4yzy + 4zy^2 + 4y^2z$ by (21), (30), (31), (40),

(60) $y^2t + ty^2 + yty \to 3yt$ — from $(21)^e$ and $(48)^e$,

(66) $2yty + 2y^2t \to 3yt + ty^2$ — from $(21)^e$ and $(48)^e$.
simplifying $3yt + ty^2 = 5yty + 5y^2t$ by (24), (31), (40),

(80) $xux^2 + x^2ux \to 2ux$ — from $(13)^{e1}$ and (60),
simplifying $x^2ux + ux + xux^2 = 3xux$ by cancel, (40),

(82) $y^2uy^2 + uy^2 + yuy \to 3uy$ — from $(13)^{e1}$ and (60),
simplifying $y^2uy^4 + uy^2 + yuy^5 = 3yuy^4$ by (13), (30), (40),

(115) $2y^2sy \to ysy^2 + ys$ — from $(13)^{e2}$ and (66),
simplifying $2y^2sy + y^3s + ys = 3y^2s + ysy^2$ by cancel, (13), (31),

(118) $y^2sy^2 \to sy^2$ — from $(13)^{e2}$ and (66),
simplifying $2y^5sy + y^6s + y^2s = 3y^5s + y^4sy^2$ by cancel, (13), (66),

(119) $2uy^2 + yuy \to 3uy$ — from (82),
simplifying it by (118),

(133) $xux^2 \to ux$ — from $(13)^{e1}$ and (119),
simplifying $ux^3 + ux + xux^2 = 3ux^2$ by cancel, (13), (30),

(135) $x^2ux \to ux$ — from (80),
simplifying it by cancel, (133),

(***) $sy = ys$ — from (115),
simplifying it by cancel, (133), (135).

A NARROWING PROCEDURE FOR THEORIES WITH CONSTRUCTORS

L. Fribourg

Laboratoires de Marcoussis - C.G.E.
91460 Marcoussis - France [1]

ABSTRACT

This paper describes methods to prove equational clauses (disjunctions of equations and inequations) in the initial algebra of an equational theory presentation. First we show that the general problem of validity can be converted into the one of satisfiability. Then we present specific procedures based on the narrowing operation, which apply when the theory is defined by a canonical set of rewrite rules. Complete refutation procedures are described and used as invalidity procedures. Finally, a narrowing procedure incorporating structural induction aspects, is proposed and the simplicity of the automated proofs is illustrated through examples.

1. Introduction

In order to prove a formula in a theory with constructors, a method is classically used, based on a principle of induction called structural [Bu,Au,BM].
In the late few years, a new method has appeared which proves equations without explicit induction [Mu,Go,HH]. Underlying our work is the attempt to extend such inductionless induction to prove (or disprove) a first-order logic formula F, and not only an equation.
To that goal, we show at § 2 that the problem of validity in the initial algebra for a set E of equations can be converted into a problem of equality-satisfiability (satisfiability modulo the axioms of equality). The conversion holds when F is an equational clause (disjunction of equations and inequations) and when E satisfies a certain principle of definition (a generalization of the principle introduced in [HH]).
Afterwards, the validity problem rises under its satisfiability form, and we study the case of E defining a canonical set of rewrite rules. In such a case, narrowing is a powerful rule of inference [Sl,La].
In §3-4, we introduce complete refutation procedures based on narrowing and resolution, which constitute efficient invalidity procedures. Furthermore, since they are complete, these procedures give validity results when they stop without generating the empty clause. In order to make the narrowing procedures stop more often, we introduce in §5, the notion of narrowing hypotheses and we show how to use them to simulate application of induction schemes. This finally provides us with a sound validity procedure based on narrowing, and we illustrate the simplicity of the automated proofs through several examples.

---

[1] This work was done when the author was at L.I.T.P.

2. E-validity in theories with constructors

2.1. E-validity of an equational clause

Let Σ be a set of operator symbols graded by an arity α and let V be a set of variables disjoint from Σ.
Let T be the set of terms constructed from operators in Σ and variables in V.
Let G be the set of ground terms (i.e. containing no variables) in T.

An *equational clause* in T is a disjunction of positive and negative literals (the disjunction connective is denoted '$\mathbf{v}$'). A positive literal is a word of the form '$t_1=t_2$' where t_1,t_2 are two terms in T and '=' is the equality symbol. A negative literal is the negation $\sim(t_1=t_2)$ of a positive literal (which will be sometimes denoted $t_1 \neq t_2$).

Let us consider a set E of equations and let $I(\Sigma,E)$ be the standard model defined by E - the *initial algebra*. The symbol $=_E$ will be used in a classic way to designate the finest Σ-congruence over T which, for any equation (t=u) in E and any substitution σ, contains $(\sigma(t)=\sigma(u))$.

definition

Let C be a clause. C is *E-valid* iff C is valid in $I(\Sigma,E)$.
C is *E-invalid* iff it is not E-valid.

Remarks

Beware that E-validity is distinct from the validity in all the models of E (see [HO]).
Notion of E-validity generalizes to the case where E is a set of Horn equational clauses (equational clauses with at most one equation), since the initial model still exists in this case .

proposition 2.1.1

Let C be the clause $\mathbf{v}_{k=1,\dots,p}\ t_k=t'_k\ \ \mathbf{v}_{l=1,\dots,q}\ u_l \neq u'_l$.
C is E-valid iff for any ground substitution ϑ : $V \to G$,
either $\vartheta t_i =_E \vartheta t'_i$, for some i $(1 \leq i \leq p)$ or $\sim(\vartheta u_j =_E \vartheta u'_j)$, for some j $(1 \leq j \leq q)$.

2.2. Theories with constructors

For simplicity of notation, we asume theories to be one-sorted, but all the results carry over to many-sorted theories without difficulty. In theories with constructors [HH], the signature is partitioned as $\Sigma = C \uplus D$. The members of C are called *constructors* and the members of D are called *definite operators*. We assume that there is at least one symbol of constant constructor.
GC is the set of ground terms formed solely from constructors. GCV is the set formed solely from constructors and variables.

Our aim is to express the E-validity property in terms of satisfiability, for which procedures of semi-decision such as resolution have been largely studied. In other words, can the problem of E-validity be reduced to a problem of satisfiability (i.e. existence of a model) ? We have a positive answer to that question, when there exists only one satisfying model - the initial algebra. This leads us to consider a set **D** of E-valid clauses which "lock" the initial algebra, so far as no other model satisfies $E \cup$ **D**.

First we consider clause sets which discriminate two distinct members of *GC*, then we show that such sets have the "locking" property.

The framework of this method is more general than the Huet-Hullot 's framework. We extend their principle of definition [HH], in the following way :

definition

A set E of equations satisfies the ***principle of definition with discriminating*** (or E defines D over C with discriminating) if :

(1) there exists a mapping $\Psi : G \to GC$ such that, for any t in G, $\Psi(t)$ is a member of GC equal to t modulo E.

(2) there exists a set **D** of E-valid clauses, such that for any term t,u in G, $\Psi(t)$ and $\Psi(u)$ are distinct iff $\mathbf{D} \cup \{\Psi(t)=\Psi(u)\}$ is unsatisfiable.

Ψ is called the ***constructor mapping*** of E and **D** is called the ***discriminating set*** of E for Ψ.

example

Let us consider the natural number theory; we suppose that the constructor symbols are 0,1 and +, and that $\psi(t)$ is of the form 0, 1+1, or 1+(1+(...1)...)) for any ground term t. Then the following set is a discriminating set :

$\{\ z+x \neq z+y \ \ \mathbf{v} \ \ x=y,\ \ 1+x \neq 0,\ \ 0 \neq 1+x\ \}$

In the definition above, the property (1) is the same as its counterpart in [HH].
In return, the property (2) is a generalization of its counterpart in [HH]. According to [HH] indeed, terms in GC are equal modulo E iff they are identical. In that case, ψ is determined in an unique way and associates to any t in G the unique member of GC equal modulo E. It can be shown that in such a case, the following set **C** is a discriminating set for E :

definition

The ***compatibility set of clauses*** - denoted **C** - is the union of the two following sets of clauses :

$\mathbf{C_1} = \{\ x_i=y_i \ \ \mathbf{v} \ \ c(x_1,\dots,x_k) \neq c(y_1,\dots,y_k)$ / for i=1,...,k and every k-ary (k>0) constructor operator c in C }

$\mathbf{C_2} = \{\ c(x_1,\dots,x_k) \neq d(y_1,\dots,y_l)$ / for every k-ary operator c and every l-ary operator d in C, with d distinct from c }

Remark

Note that $\mathbf{C_1}$ is the dual form of the functional substitutivity axioms, and that 0-ary (constant) constructor symbols are involved in $\mathbf{C_2}$.

THEOREM 2.2.1

Let E be a set of equations defining D over C with a discriminating set **D**, and C an equational clause on T.
C is E-valid iff $E \cup \mathbf{D} \cup \{C\}$ is equality-satisfiable.

proof

(<=) Let us suppose that $E \cup \mathbf{D} \cup \{C\}$ is equality-satisfiable.
Then there exists an interpretation J which equality-satisfies $E \cup \mathbf{D} \cup \{C\}$. Let C be of the form $\mathbf{v}_{k=1,\dots,p}\ t_k=t'_k\ \ \mathbf{v}_{l=1,\dots,q}\ u_l \neq u'_l$. Let ϑ be a ground substitution : $V \to G$. J

satisfies ϑC, hence either $J \models \vartheta t_i = \vartheta t'_i$, for some i $(1 \le i \le p)$ or $J \models \sim(\vartheta u_j = \vartheta u'_j)$, for some j $(1 \le j \le q)$.
Now $\vartheta t_i, \vartheta t'_i, \vartheta u_j, \vartheta u'_j$ are in G. Let $\tau_i, \tau'_i, \upsilon_j, \upsilon'_j$ be their respective images in GC by the constructor mapping of E. $\vartheta t_i, \vartheta t'_i, \vartheta u_j, \vartheta u'_j$ are respectively equal modulo E to $\tau_i, \tau'_i, \upsilon_j, \upsilon'_j$ (by property (1) of the principle of definition). As J satisfies E, we have :
either $J \models \tau_i = \tau'_i$ (a) or $J \models \sim(\upsilon_j = \upsilon'_j)$ (b).
Since J satisfies D, (a) implies that τ_i and τ'_i are identical (by property (2) of the principle of definition). On the other hand, (b) implies that υ_j and υ'_j are distinct, J being an equality-model. Hence either : $\tau_i = \tau'_i$ (a') or $\sim(\upsilon_j = \upsilon'_j)$ (b'). So, by substitutivity of an equal by an equal modulo E,
either $\vartheta t_i =_E \vartheta t'_i$ (a") or $\sim(\vartheta u_j =_E \vartheta u'_j)$ (b").
Thus for any ground substitution ϑ, there exists an integer i $(1 \le i \le p)$ or j $(1 \le j \le q)$ such that : $\vartheta t_i =_E \vartheta t'_i$ or $\sim(\vartheta u_j =_E \vartheta u'_j)$; hence C is E-valid (by proposition 2.1.1).

(=>) Let us suppose that C is E-valid. Then C is valid in $I(\Sigma, E)$. now, by property (2) of the principle of definition, $I(\Sigma, E)$ satisfies **D**. Furthermore $I(\Sigma, E)$ is an equality-model satisfying E (by definition). So $\{C\} \cup \mathbf{D} \cup E$ is satisfied by $I(\Sigma, E)$, and thus is equality-satisfiable. ▪

The E-validity problem is thus converted into the equality-satisfiability problem for sets of clauses.

Remark

Note that theorem 2.2.1 still holds when E is a set of Horn equational clauses (the proof is the same as the one above).

The equality-satisfiability problem is semi-decidable and can be treated with a standard method of resolution with paramodulation [Lo]. Thus, any complete method of refutation by resolution with paramodulation will stand as an E-invalidity procedure, as soon as E satisfies our principle of definition. Furthermore, such procedures show satisfiability when they stop without having generated the empty clause (see [Jo] for satisfiability-oriented strategies).
In the following, we consider theories defined by canonical sets of rewrite rules. In that case, narrowing [Sl,La] is a powerful optimization of paramodulation. Our concern is to build narrowing-based procedures which on the one hand generate shorter E-invalidity proofs and on the other hand produce more often E-validity proofs (i.e. do not loop for ever).

3. The basic E-invalidity procedure

3.1. Hypotheses on E

Henceforth, we make the following hypotheses on E :

(H0) E defines a canonical term rewriting system, where no left-hand side is a variable.

(H1) For any term t in G, the E-normal form denoted t^* is a term of GC.

The hypotheses (H0) and (H1) ensure that the property (1) of the principle of definition is satisfied, the constructor mapping Ψ being the mapping which associates the normal form t^* with any closed term t.

Given a canonical set E, a sufficient criterion for (H1)to hold, is :

(HB) For every f in D, there is in E a set of equations whose left-hand sides are of the form $f(s_i^1,\ldots,s_i^k)$ $(1\leq i\leq p)$ and the set $\{S_1,\ldots,S_p\}$
- where S_i denotes the k-tuple $(s_i^1,\ldots,s_i^k)$ - is a ***base*** for C in the sense that : for every k-tuple of ground terms in $GC(t_1,\ldots,t_k)$, there exist q, with $1\leq q\leq p$, and a substitution σ, such that, for every j, $1\leq j\leq k$, we have $t_j = \sigma(s_q^j)$.

Further sufficient criteria are developed in [HH,Th]. See also [Ka], in case of E being a set Horn equational clauses.

In addition, we make the following hypothesis :

(H2) For every t,u in GC, $t =_E u$ iff $t = u$.

The hypothesis (H2) is exactly the property (2) of the Huet-Hullot's principle of definition. With (H0-H1), it ensures that t^* is the unique member of GC equal to t, modulo E. As previously remarked, (H2) ensures that our principle of definition is verified, by taking **C** as a discriminating set.
Moreover, it can be seen that the hypothesis (H2), coupled with (H0-H1), implies (HB).

In the following, the results will often hold under more general conditions than (H2); we shall precise what actually are the useful hypotheses requested.
The E-validity problem will be treated through techniques of refutation by resolution and narrowing. We assume that the reader is familar with the notions of resolution [Lo] and narrowing [Sl,La]. A selection function φ is given which chooses from each clause a single literal to be resolved upon in that clause (see [KK]) or to be narrowed. The constructor inequations $c(x_1 \ldots x_k)\neq c(y_1 \ldots y_k)$ are assumed to be the φ-selected literal for the clauses of $\mathbf{C_1}$. The only resolvents D' considered are *selected* binary resolvents of a clause D with a clause of $\mathbf{C_1} \cup \mathbf{C_2} \cup \{x=x\}$. In resolution against $\mathbf{C_2}$ or against $\{x=x\}$, D' is obtained from D through instanciation and deletion of the φ-selected literal. In resolution against $\mathbf{C_1}$, D' is obtained from D through instanciation and replacement of the φ-selected equation by an equation between subterms. This new equation is assumed to be the φ-selected literal in D'.
For a clause D, D^* denotes its E-normal form. We say that a clause D^+ is a *selected-normal form* (resp. *selected-reduced form*) of D if D^+ is a reduced form of D and if $\varphi(D^+)$ is a normal form (resp. reduced form) of $\varphi(D)$. The symbol $*_{sel}$ will stand for the operation of normalizing literals of a clause except the φ-selected one.

Throughout, s.b.r will stand for 'selected binary resolution against a clause of $\mathbf{C} \cup \{x=x\}$' .

3.2. ground refutation

THEOREM 3.1
Let G be a ground clause on GC. G is E-invalid iff there exists an input refutation Δ_{gr} :

$\{G_0, G_1, ..., G_n\}$ of $\mathbf{C} \cup \{G\} \cup \{x=x\}$, obtained by s.b.r. (without merging) and such that :

- $G_0 = G$ and $G_n = \square$
- $G_{i+1} = G_i - \varphi(G_i)$, for $i=1,...,n$,
  and G_{i+1} is deduced from G_i by application of a linear sequence of s.b.r.

The deduction of G_{i+1} from G_i can be depicted by the following diagram :

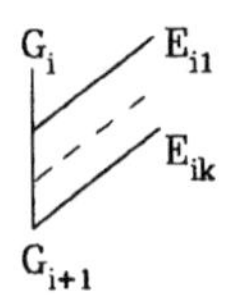

where $E_{i1},...,E_{ik} \in \mathbf{C} \cup \{x=x\}$

proof
(<=) If there exists such a refutation Δ_{gr} ,then $\mathbf{C} \cup G \cup \{x=x\}$ is unsatisfiable (by soundness of resolution) ; hence $\mathbf{C} \cup G$ is equality-unsatisfiable and so is $\mathbf{C} \cup G \cup E$. This implies that G is E-invalid (by theorem 2.2.1).
(=>) Suppose that G is E-invalid. $\varphi(G)$ contains only terms in GC, and so is either an equation t=u between terms in GC (case a), or an inequation $t \neq u$ (case b). Since G is E-invalid and is under E-normal form, then necessarily, t and u are distinct in case a, as they are identical in case b. In the latter case, G can be resolved against x=x, the resolvent being the sought clause G_1. In case a, either : t and u begin with distinct constructors (subcase a1) and then G can be resolved against a clause of $\mathbf{C_2}$,the resolvent being the sought clause C_1 ; or t and u begin with the same constructor (subcase a2), in which case there is a resolvent of G with $\mathbf{C_1}$ whose φ-selected literal is an equation between distinct sides, and the process iterates until the leading symbols of the φ-selected equation are distinct (subcase a1).
In all the cases, the φ-selected literal is eliminated through a finite sequence of s.b.r. (without merging).
The process iterates until the empty clause is produced. ▪

Remarks

The above mentioned kind of deduction applies over GCV and can be used as a decision procedure for E - validity or invalidity over GCV.
The general property underlying theorem 3.1, is the following :
let us consider the hypothesis
(HD) there exists a discriminating set $\mathbf{D}$ of E-valid clauses such that,
for any terms t,u in G, t^* and u^* are distinct iff $\mathbf{D} \cup \{t^*=u^*\}$ is unsatisfiable.
Then, under (H0) (H1) (HD), any normalized clause G^* on GC is E-invalid
iff $\{G^*\} \cup \mathbf{D} \cup \{x=x\}$ is unsatisfiable.

3.3. 1st level basic procedure

The theorem 3.1 applies to ground clauses on GC. Now, we are going to lift that theorem to the first level and extend it to all the clauses over T. This is achieved by means of the ***selected-narrowing*** operation. Let us recall that an E-narrowing on a clause C is a paramodulation of an equation $l=r$ of E into a non variable subterm t unifiable with l, followed by E-normalization. We only use selected E-narrowings, which are E-narrowings where t belongs to the φ-selected literal. The φ-selected literal within a narrowing (resp. instanciation) of a clause C is assumed to be the descendant of the φ-selected literal in C. Throughout, s.n. will stand for 'selected E-narrowing'.

lemma 3.1

Let D be a clause, D' an instance of D under a substitution ϑ.

There exists a finite (may be empty) sequence of s.n. taking D to a clause D^n and there exists a substitution ϑ' such that $\vartheta' D^n$ is a selected-normal form of D'.

The proof is the same as the proof given by Lankford [La,p.23], except that in our case reductions involve only the φ-selected literals .

THEOREM 3.2

Let C be a clause. The clause C is E-invalid iff there exists a refutation Δ of C : $\{C_0,B_1,C_1,\ldots,B_i,C_i,\ldots,B_n,C_n\}$ obtained by s.n. and s.b.r., and there is a ground refutation Δ_{gr} : $\{G_0,G_1,\ldots,G_n\}$ satisfying :

1) for $i=1,\ldots,n$, B_i is deduced from C_{i-1} by applying a finite (may be empty) sequence of s.n.

2) for $i=1,\ldots,n$, C_i is deduced from B_i by applying a linear sequence of s.b.r.

3) for $i=1,\ldots,n$, there exist two ground substitutions ϑ_i and ϑ_i' such that :

3.1) $(\vartheta_i C_{i-1})^* = G_{i-1}$

3.2) $(\vartheta'_i B_i)$ is a selected-normal form of $(\vartheta_i C_{i-1})$

4) C_0 = C and C_n = $\square$

proof

(<=) Suppose there exists such a refutation Δ. This implies that $E \cup \{C\} \cup \mathbf{C}$ is equality-unsatisfiable (by soundness of s.n. and s.b.r.). Now $I(\Sigma,E)$ is an equality-model which satisfies E. It can be seen that $I(\Sigma,E)$ also satisfies **C**. Consequently, C cannot be satisfied in $I(\Sigma,E)$, and thus is E-invalid.

(=>) Suppose that C is E-invalid , then there is a ground instance G of C such that G is E-invalid, hence such that G* is E-invalid. Now, G* is a ground clause on GC (by (H1)), so there is a ground refutation Δ_{gr} of G* (by theorem 3.1). Each portion of the deduction $\{G_{i-1},G_i\}$ can be lifted to the 1st level into a linear s.b.r. deduction $\{B_i,C_i\}$, according to the classical resolution lifting lemma (factorization is not involved because there is no merging at the ground level). The deduction from C_{i-1} of a clause B_i having G_i as an instance is obtained through selected narrowing (see lemma 3.1).

The relations between the portions of deduction are depicted by the following diagram.

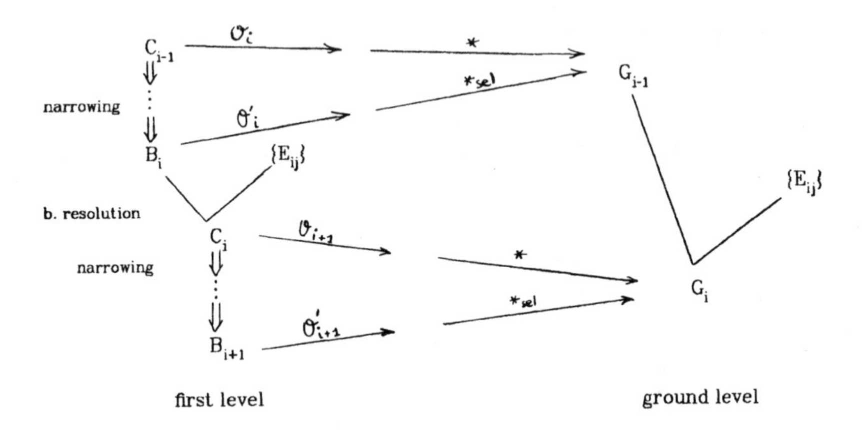

Remark

As theorem 3.2 applies under the (H2) hypothesis, an analogous theorem holds under the (HD) hypothesis (the sequences of s.b.r being no longer linear in general).

Let us now introduce our basic refutation procedure which roughly consists in computing all the resolvents and narrowings derived from C. For a clause D, R(D) denotes the set of clauses from D by s.b.r. and N(D) denotes the set of clauses derived from D by s.n.

PROC1

S_k is the current set of clauses, OLD_k is the set of ancient clauses.
Initially $k = 0$, $S_k = \{C\}$, $OLD_k = \phi$

1) if $S_k = \phi$, then stop "proof"
else select a clause D in S_k .

2) a) if $D \in OLD_k$, set $S_k = S_k - \{D\}$, go to 1.
b) if D subsumed by a clause of **C** $\cup \{x=x\}$, set $S_k = S_k - \{D\}$, go to 1.
c) if D is $\square$, stop "disproof"
d) otherwise set $S_k = S_k \cup R(D) \cup N(D) - \{D\}$
set $OLD_k = OLD_k \cup \{D\}$
set $k = k+1$, go to 1.

The selection of a clause D at step 1 requests a complete strategy (breadth-first search plan [see Lo], or selection fairness [HH]), in order to ensure the termination of PROC1 when applied to an E-invalid clause C.
The deletion of older or subsumed clauses (steps 2.a,2.b) is clearly a safe optimization.

It follows from theorem 3.2 :

THEOREM 3.3

If PROC1 stops with "disproof", then the input clause C is E-invalid.
If PROC1 stops with "proof" or does not terminate, then C is E-valid.

4. Optimizations of the basic procedure

4.1. Equations between sides beginning with constructors

Let us now optimize the procedure PROC1 by preventing the computation of narrowings on special clauses. The clauses considered have an equation, as a φ-selected literal, where each side begins with a constructor. There are two possible cases (assuming that the φ-selected literal is the rightmost one) :

(j) $D = (D',c(t_1...t_k)=c(u_1...u_k))$, with c as a k-ary (k>0) constructor symbol.

(jj) $D = (D',c(t_1...t_k)=d(u_1...u_l))$, with c and d as distinct constructor symbols.

Let Υ_1 be the mapping which associates the set $\{(D',t_i=u_i)\}_{i=1,...,k}$ with a clause D of the j-form. And let Υ_2 be the mapping which associates the set $\{D'\}$ with a clause D of the jj-form.

Remark

For a j-clause D (resp. jj-clause D), $\Upsilon_1(D)$ (resp. $\Upsilon_2(D)$) is the set R(D) of the resolvents of D.

proposition

Let C be an E-invalid clause. Let Δ:{ C, $\{(B_i, C_i)\}_{i=1,...,n}$} be a refutation of C and Δ_{gr} : $\{G_0,G_1,...,G_n\}$ be its associated ground refutation.

There exists a refutation Δ': {C, $\{(B'_i,C'_i)\}_{i=1,...,n}$} associated with Δ_{gr} without any application of r.n. to a j- or jj-clause.

The proof is just sketched out.

If, during the sequence of narrowing taking C_{i-1} to B_i, there is a narrowing N_i which is a j-clause (resp. jj-clause), then there is a clause N_{i+1} of $\Upsilon_1(N_i)$ (resp. $\Upsilon_2(N_i)$) which is in relation with the clause G_i of the ground refutation. The narrowing sequence taking N_i to B_i can thus be eliminated :

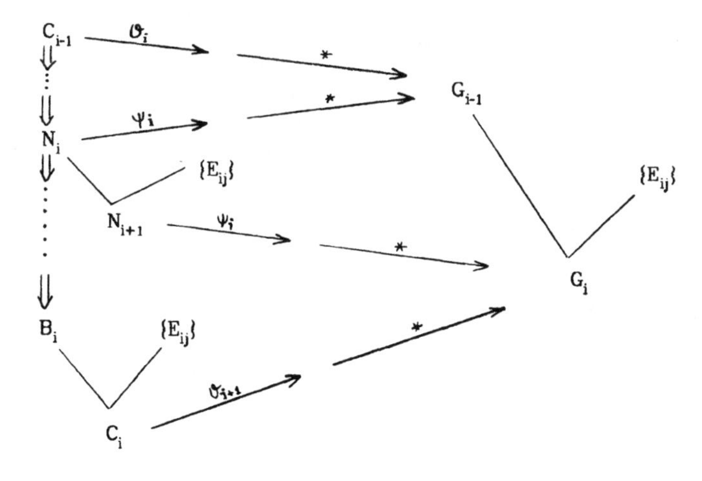

As there exists a refutation without any narrowing either on a j-clause or on a jj-clause, we can rightfully replace a clause C by the set of its resolvents that is $\Upsilon_1(C)$ or

Υ_2(C). Throughout Υ(D) denotes the set of clauses obtained by mapping a clause D with Υ_1 or Υ_2, and Υ(S) denotes the union of sets Υ(D) for every clause D in a set S.

definition

the procedure PROC1' is obtained from PROC1 by replacing at step 1
"select a clause D in S" by
"select a clause D in S; if D is a j- or jj-clause, then set $S = S \cup \Upsilon(D) - \{D\}$, go to 1".

proposition

If PROC1' stops with "disproof", then the input clause C is E-invalid.
If PROC1' stops with "proof" or does not terminate, then C is E-valid.

Remarks

The modification introduced with Υ in PROC1 corresponds to the modification made by Huet-Hullot in the Knuth-Bendix algorithm [HH].
Resolution with a clause D is only performed in PROC1' when φ(D) is of the form : $x=c(t_1 \ldots t_k)$ or x=y, with $x,y \in V$.

4.2. Narrowing heuristics

Let us consider, for instance, the application of PROC1' to the clause
C1 : 0+x=x, with : E = {E1: x+0=x, E2: x+succ(y)=succ(x+y)},
0 and succ being the constructor operators of the theory. After the first run of PROC1', we have :
S = N(C1) = { 0=0, succ(0+x)=succ(x) } and OLD = {C1}. The clause (0=0) is then selected and eliminated by subsumption. The clause (succ(0+x)=succ(x)) is then selected, transformed into (0+x=x) and finally rejected as a clause of OLD. The set S is now empty and thus C1 is declared to be E-valid.

Actually, the procedure has simulated a proof by induction on the variable x. This kind of inductionless induction method raises a problem because all the induction schemes are attempted with no regards for a former scheme being already achieved. From there comes the failure of PROC1' in proving the associativity of +, for instance. To solve this problem, we prove now the completeness of a restriction of narrowing performed on certain occurrences only.

definition

Let C be a clause and t a non variable subterm at the occurrence o in the selected literal of C.
The ***narrowing class*** of C at the occurrence o is the set of all the E-narrowings of C obtained by replacement of the subterm t at occurrence o.

definition

An ***innermost term*** t in a clause C is a term of the form $f(t_1,\ldots,t_k)$ such that f is a defined symbol of D, and such that all the t_i ($1 \le i \le k$) are only formed of constructors and variables.
An ***innermost narrowing class*** of C is a narrowing class of C at an occurrence o of an innermost term.

Remarks

An innermost narrowing class is always non empty.

For any clause C having a non empty narrowing class, there is a (non empty) innermost narrowing class.

lemma 4.2

Let C be a clause on T. Let ϑ be a ground substitution and C' the instance ϑC.
For any innermost narrowing class Cs of C, there exist a narrowing N of Cs and a clause C'' such that :

- C'' is an instance of N
- C'' is a reduced form of C'
- φ(C'') is a reduced form of φ(C').

proof
Since Cs is innermost, there is a subterm t at occurrence o in C, of the form $f(t_1...t_k)$, unifiable with a left-hand side in E and such that $t_1,...,t_k$ are composed only of constructors and variables.
Let ϑ^* be the normal substitution of ϑ (for any variable x, $\vartheta^*(x) = (\vartheta x)^*$). ϑ^* is a ground substitution which replaces every variable x by a ground irreducible subterm in GC (by (H1)) . The terms ϑ^*t_i are hence in GC, for every i=1,...,k. Now, the subset $\{f(s_i^1...s_i^k)/\ 1\leq i\leq p\}$ of the left-hand sides of E-equations defining f, is such that $\{S_1...S_p\}$ forms a base of C (since (H0-H1-H2) imply (HB)) ; so there is an integer q ($1\leq q\leq p$) and a substitution σ such that : $\vartheta^*t_j = \sigma s_q^j$ for j=1,...,k . Let $l=r$ be the equation having $f(s_q^1...s_q^k)$ as a left-hand side. We have $\vartheta^*t = \sigma l$. Hence t and l are unifiable. Let $\tau \cup \lambda$ be the most general unifier. There exists a substitution ξ such that : $\vartheta^* = \xi\tau$ and $\sigma = \xi\lambda$. Now a right narrowing of $l=r$ on C at occurrence o applies and gives the clause N : $(\tau C)^*$. Let C'' be the clause ϑ^*C. C'' is an instance of τC under the substitution ξ. Consequently, the sequence of reductions which normalizes τC, also applies to C'' and gives a clause C''_0. It can be seen that $(\tau C)^*$ has C''_0 as an instance under an irreducible substitution η_0^* and that C''_0 is a reduced form of C''. Therefore $(\tau C)^*$ and C''_0 are respectively the expected clauses N and C''. ▪

The process can be iterated if the clause C_0: $(\tau C)^*$ can itself be narrowed. Then for any innermost class Cs_1 of C_0, there exists a narrowing taking C_0 to a clause C_1. We thus successively generate clauses C_i and C''_i (i=0,...,p) by sequences of narrowing such that C_i has C''_i as an instance under an irreducible substitution η_i^* and such that C''_i is a selected-reduced form of C''_{i-1} . Whatever the innermost class Cs_i chosen at step i for C_{i-1} is, the process stops at the step m where $\varphi(C''_m)$ is irreducible (and such that no defined symbol remains in φC_m). The construction is depicted by the following diagram.

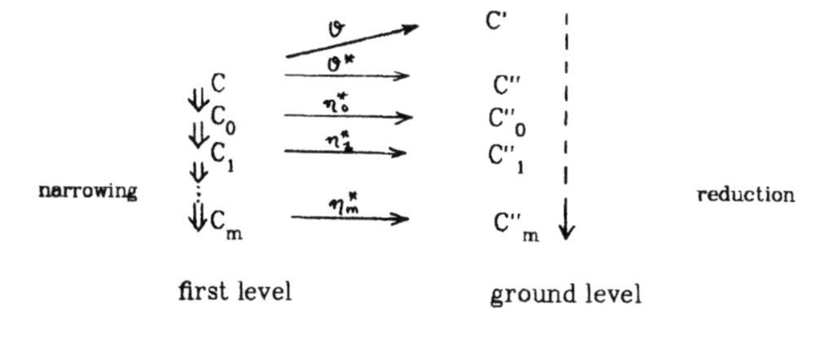

This property can be seen as a transposition of lemma 3.1, when we restrict ourselves to the narrowing computation of a single (innermost) class. Coupled with the completeness theorem 3.1 and the resolution lifting lemma, it allows us to replace in PROC1, at step 2.d
"$\mathcal{S}$:= S ∪ R(D) ∪ N(D) - {D}" by
"$\mathcal{S}$:= S ∪ R(D) ∪ $\mathcal{C}s$ - {D} , for any innermost class $\mathcal{C}s$ of D, if any".
Then we can prove the validity of the addition at step 1 of the test for j- or jj-clause, exactly in the same way as for PROC1.
Finally we can state :

THEOREM 4.1
Let PROC1" be the procedure obtained from PROC1' by replacing , at step 2.d,
"$\mathcal{S}$:= S ∪ R(D) ∪ N(D) - {D}" by
"$\mathcal{S}$:= S ∪ R(D) ∪ $\mathcal{C}s$ - {D} ,for any innermost class $\mathcal{C}s$ of D, if any".
If PROC1" stops with "disproof", then the input clause C is E-invalid.
If PROC1" stops with "proof" or does not terminate, then C is E-valid.

In terms of structural induction, the innermost narrowing classes used in PROC1" correspond to *induction schemes*. Therefore, at step 2.d, the innermost class can be selected according to the criteria used in structural induction for choosing among various induction schemes (see [Au],[BM]).

Similar strategies to that of PROC1" apply, if (HB) holds besides (H0-H1), and if there is a procedure $\mathbf{p}_{inv}$ of E-invalidity for the literals on *GCV*. The general strategy is expressed as follows :
"Perform innermost narrowing on the φ-selected literal until no symbol of defined function remains ; then eliminate the φ-selected literal through $\mathbf{p}_{inv}$, perform narrowing on the new φ-selected literal and so on until the empty clause is derived. "

4.3. Embodiment of ordinary clauses

So far, we have only considered ***equational*** clauses. However, our procedures can apply to a clause C containing a k-ary ***relation*** symbol R distinct from '='. The positive literal $R(t_1...t_k)$ can indeed be converted into an equation $R(t_1...t_k)$=true , where 'true' is a new constructor. Likewise, $\sim R(t_1...t_k)$ can be converted into $R(t_1...t_k)$=false . If there is a set E of equations which defines {R} on {true,false}, then the procedures apply as well to C.

We give a list of simple theorems borrowed from [BM], and proved by PROC1" :

- ***Constructors*** : {0,succ} {nil,cons} {true,false}
- ***Axioms*** :
  x + 0 = x
  x + succ(y) = succ(x+y)
  x × 0 = 0
  x × succ(y) = x × y + x
  lessp x 0 = false
  lessp 0 succ x = true
  lessp succ x succ y = lessp x y
  diff 0 x = 0
  diff x 0 = x
  diff succ x succ y = diff x y
  append nil y = y

append (cons x1 x2) y = cons x1 (append x2 y)
half 0 = 0
half succ 0 = 0
half succ(succ x) = half x
even 0 = true
even succ 0 = false
even succ(succ x) = even x

- *Theorems* :

LESSP.EQUAL	x = y **v** lessp x y = true **v** lessp y x = true
LESSP.DIFF	lessp y x = true **v** diff x y = 0
ASSOCIATIVITY.OF.PLUS	x + (y + z) = (x + y) + z
COMMUTATIVITY.OF.PLUS	x + y = y + x
TIMES.ZERO	0 × x = 0
ASSOCIATIVITY.OF.APPEND	append (append x y) z = append x (append y z)
DOUBLE.HALF	even x = false **v** succ(succ 0) × half x = x

These theorems have been processed by the program SEC [Fr] - a part of the system FORMEL developed at INRIA, which embodies PROC1''.

5. E-validity procedure

PROC1'' has the nice completeness property, and it simulates a weak form of structural induction principle. However, PROC1'' is not powerful enough to prove difficult theorems, for it makes use neither of lemmas nor of induction hypotheses. We study now extensions of PROC1'' incorporating these missing aspects through two strategies :

- hypothesis decomposition
- subsumptive rewriting .

5.1. Induction hypotheses

5.1.1. Narrowing hypotheses

We would like to rewrite a narrowing N of a clause C, using some kind of induction hypotheses. Rather than to consider directly C as an induction hypothesis (see [Bi]), we introduce the notion of narrowing hypotheses for N.
In keeping with [Au], we have :

definition

For s and t in T, s is an ***immediate substructure*** of t iff t is the result of applying a constructor to s and possibly some other term.
$<_P$ denotes the transitive closure of the immediate substructure relation.
For any k-tuples $(s_1,\ldots,s_k)$, $(t_1,\ldots,t_k)$ in T^k,
$(s_1,\ldots,s_k) <_P (t_1,\ldots,t_k)$ iff $s_1 = t_1$ and...and $s_{i-1} = t_{i-1}$, and $s_i <_P t_i$ for some i $(1 \leq i \leq k)$.
We say that $(s_1,\ldots,s_k)$ is a ***predecessor*** of $(t_1,\ldots,t_k)$ if $(s_1,\ldots,s_k) <_P (t_1,\ldots,t_k)$.

Remark

Given a k-tuple $(t_1,\ldots,t_k)$, all the variable of its predecessors also appear in $(t_1,\ldots,t_k)$ (shared variables).

definition

Let C be a clause, $l=r$ an equation in E, and o an occurrence of a non variable subterm t in C, such that :

t and l are unifiable, with m.g.u. $\sigma = \{x_i \leftarrow t_i\}_{i=1,...,k}$.

Let N be the narrowing $(\sigma C)^*$ of $l=r$ into C at occurrence o.

The ***predecessor set*** of σ is the subset of substitutions defined by :

$Pd(\sigma) = \{ \{x_i \leftarrow s_i\}_{i=1,...,k}$ /for any k-tuple $(s_1...s_k)$ predecessor of $(t_1...t_k)\}$.

The ***set of narrowing hypotheses*** of N is the set of clauses defined by :

$\Theta(N) = \{\sigma' C \ / \ \sigma' \in Pd(\sigma)\}$

Remark

The variables of terms s_i in clauses of $\Theta(N)$ appear in the terms t_i of σC .

proposition 5.1

Let C be a clause, and $Cs = \{N_i\}_{i=1,...,p}$ an innermost narrowing class of C.

Let U_i $(1 \le i \le p)$ be implications of the form :

$(P_{ij}$ E-valid , for all $P_{ij} \in \Theta(N_i)$) ==> (N_i E-valid)

If every U_i $(1 \le i \le p)$ is true, then C is E-valid.

proof

Let us suppose that, for every i $(1 \le i \le p)$, U_i is true and let us show that C is E-valid.

Let σ be a given ground substitution. Cs is obtained from E-narrowings into C at an occurrence o of an innermost term t of the form $f(t_1...t_k)$. The terms $t_1,...,t_k$ are composed only of constructors and variables. The normalized substitution σ^* replaces every variable by a ground irreducible subterm in GC (by (H1)) . The terms $(\sigma^* t_i)$ are hence in GC, for $1 \le i \le k$. By (HB), there is an equation $l=r$ of E and a substitution ζ such that $\zeta l = \sigma^* f(t_1...t_k)$. Hence l and t are unifiable with the m.g.u. $\lambda \cup \tau$, and there exists a substitution η such that $\sigma^* = \eta\tau$; the clause $(\tau C)^*$ is a clause N_q of Cs $(1 \le q \le p)$. Let G be the clause $(\sigma^* C)$. We have : $G = \eta\tau C$, hence $G^* = (\eta N_q)^*$.

Now U_q is true, so :

$(P_{qj}$ E-valid , for all $P_{qj} \in \Theta(N_q)$) ==> (N_q E-valid) , therefore

$(\eta\ P_{qj}$ E-valid , for all $P_{qj} \in \Theta(N_q)$) ==> ($\eta\ N_q$ E-valid)

Now ηN_q is E-valid iff $(\eta N_q)^*$ is E-valid ,i.e iff G^* E-valid. So ηN_q E-valid iff G is E-valid; so :

$(\eta P_{qj}$ E-valid , for all $P_{qj} \in \Theta(N_q)$) ==> (G E-valid) (I)

Now $\{P_{qj}\} = \{ \delta_{qj} C \ / \ \delta_{qj} \in Pd(\tau) \}$

For any $\delta_{qj} \in Pd(\tau)$, $\eta\delta_{qj}$ is a member of $Pd(\eta\tau)$, i.e. of $Pd(\sigma^*)$.

Therefore, $\eta\ P_{qj}$ is a clause $\rho_j C$, with $\rho_j \in Pd(\sigma^*)$. From (I), we have then

($\rho_j C$ E-valid , for all $\rho_j \in Pd(\sigma^*)$) ==> (G E-valid) , so

(ρ_j C E-valid , for all $\rho_j \in Pd(\sigma^*)$) ==> ($\sigma^* C$ E-valid)

From the structural induction principle, it follows that $\sigma^* C$ is E-valid; therefore σC is E-valid. Thus, whatever the ground substitution σ is, σC E-valid. It follows that C is E-valid. ▪

5.1.2. Hypothesis decomposition

Let C be a clause of the form $L_1 \vee .... \vee L_r$.

Let N be a narrowing of the form $N_1 \vee \dots \vee N_r$, and let $\Theta(N)$: $\{H_1,\dots,H_q\}$ be the hypothesis set of N. For $1 \le j \le q$, H_j is of the form $h_{j1} \vee \dots \vee h_{jr}$

The r^q functions γ of $\{1,\dots,q\}$ into $\{1,\dots, r\}$ define the so-called *sampling functions* of N. Given a sampling function γ, the collection H_γ of literals $\{ h_{j\gamma(j)} / j=1,\dots,q \}$ is called a *hypothesis sample* of N.

The set $\{ H_\gamma$ / for all sampling function $\gamma \}$ is called the *hypothesis decomposition* of N and is denoted $\Pi(N)$.

example

We consider the functions formp, eval, optimize defined by Boyer-Moore (see [BM], p.258). The definitions given in [BM] can be transformed into canonical sets of equations defining respectively formp on {true,false} and eval,optimize on {nil,cons} ; in particular, we have the following equations :

E1: (formp (cons w (cons x (cons y z)))) = (and (formp x) (formp y))
E2: (eval (cons w (cons x (cons y z)))) e) = (apply w (eval x e) (eval y e)
E3: (optimize (cons w (cons x (cons y z))))
= (cons w (cons (optimize x) (cons (optimize y) nil)))

Let us now consider the translated form of the OPTIMIZE.CORRECTNESS.THEOREM
C: (formp x) = false $\vee$ (eval (optimize x) e) = (eval x e)

The narrowing of E3 into C, using the m.g.u
σ : { x ← (cons w (cons x (cons y z)))) }, gives :
N: (and (formp x) (formp y)) = false $\vee$
(apply w (eval (optimize x) e)(eval (optimize y) e))
= (apply w (eval x e) (eval y e))

$\Theta(N)$ = { (formp x) = false $\vee$ (eval (optimize x) e) = (eval x e) ,
(formp y) = false $\vee$ (eval (optimize y) e) = (eval y e) }

The sets H_γ are :
H_1 = { (formp x) = false, (formp y) = false }
H_2 = { (formp x) = false, (eval (optimize y) e) = (eval y e) }
H_3 = { (eval (optimize x) e) = (eval x e), (formp y) = false }
H_4 = { (eval (optimize x) e) = (eval x e), (eval (optimize y) e) = (eval y e) } .
$\Pi(N) = H_1 \cup H_2 \cup H_3 \cup H_4$

From the properties of conjunction-disjunction, follows :

proposition 5.2

Let N_i be a narrowing of an equation of E into a clause C. Then
(P_{ij} E-valid , for all $P_{ij} \in \Theta(N_i)$ ==> (H_γ E-valid , for some sample H_γ of $\Pi(N_i)$)

Furthermore :

proposition 5.3

Let N_i be a narrowing of an equation of E into a clause C.
Let V_i be the following statement :
(H_γ E-valid ==> N_i E-valid) , for every $H_\gamma \in \Pi(N_i)$.
If V_i is true, then U_i is true.

proof :
Let us suppose that V_i is true (a), and that P_{ij} is E-valid, for all $P_{ij} \in \Theta(N_i)$ (b), and let us show that N_i is E-valid. From (b) and proposition 5.2, it follows that there exists an E-valid sample H_δ in $\Pi(N_i)$; so by (a), N_i is E-valid, q.e.d.

Let N be a narrowing, and H a sample of Π(N). In the following, H is regarded as a set of rewrite rules for N. All the variables of H appear in N ; these variables are thus not universally quantified but "frozen" as new constants.
In order to prevent infinite sequences of reduction, T is provided in a classical way, with a well-founded ordering $<_0$ [Pl,De1,JLR]. Equations of H will be used as rewrite rules only if their sides are $<_0$-comparable. Inequations of H, as for them, will be also considered as rewrite rules, so far as they can transform possible instanciated forms within N into tautologies (thus, with the inequation $t \neq u$, one would "rewrites" the instanciation $\sigma t \neq \sigma u$ into the trivial equation x=x).
The term $N\downarrow_H$ will denote an irreducible clause obtained by rewriting N with the subset of rewriting rules in H.

example

For the narrowing of the former example, we have :

$N\downarrow_{H1}$: (and false false) = false $\vee$
(apply w (eval (optimize x) e) (eval (optimize y) e))
= (apply w (eval x e) (eval y e))

$N\downarrow_{H2}$: (and false (formp y)) = false $\vee$
(apply w (eval (optimize x) e) (eval y e) = (apply w (eval x e) (eval y e))

$N\downarrow_{H3}$: (and (formp x) false) = false $\vee$
(apply w (eval x e) (eval (optimize y) e)) = (apply w (eval x e) (eval y e))

$N\downarrow_{H4}$: (and (formp x) (formp y)) = false $\vee$
(apply w (eval x e) (eval y e) = (apply w (eval x e) (eval y e))

Note that $N\downarrow_{H4}$ is subsumed by x=x, and that $N\downarrow_{Hi}$ (i=1,2,3) is also subsumed by x=x, after rewriting with { (and false x)=false, (and x false)=false) }.

In the following, given a clause D, R(D) denotes the set of selected resolvents of D against clauses of **C** $\cup$ {x=x}, and N_{inner}(D) denotes the set of selected narrowings for the chosen innermost class of D. V denotes the current set of clauses known to be valid.
Resolution against **C** $\cup$ {x=x} is now used as a means to split goals into subgoals. Indeed, we have :

proposition 5.4

Let D be an equational clause. Suppose the φ-selected literal is either an inequation over GCV, or an equation where each side is a variable or a term beginning with a constructor.
Then D is E-valid iff either R(D) is empty or every clause of R(D) is E-valid.

Let us now consider the following procedure PROC2.

PROC2 (C)

Initially V = $E \cup$ **C**
If C is subsumed by a clause of $\{x=x\} \cup$ **C**,
then stop "proof"
else TOP (C)

TOP (D)
If φ(D) is of the form : $x=y$,with $x,y \in V$
or $c(t_1...t_k)=d(u_1...u_l)$, with $c,d \in C$ and $t_i,u_j \in T$ $(1 \le i \le k, 1 \le j \le l)$
or $x=c(t_1...t_k)$ (resp. $c(t_1...t_k)=x$), with $c \in C$, $x \in V$ and $t_i \in T$ $(1 \le i \le k)$
or $t \ne u$, with $t,u \in GCV$
then RES (D)
otherwise NAR (D)

RES (D)
compute $R(D) = \{R_1,...,R_n\}$
if □ is a member of R(D), then stop "failure".
let $R' = \{R_{i1},...,R_{ip}\}$ be the set obtained from R(D) through elimination of clauses subsumed by a clause of $\{x=x\} \cup$ **C**,
if $R' = \phi$, then BACKTRACK (D)
else TOP (R_{i1})

NAR (D)
compute $N_{inner}(D) = \{N_1,...,N_n\}$
let $N' = \{N_{i1},...,N_{ip}\}$ be the set obtained from $N_{inner}(D)$ through elimination of clauses N_i such that $N_i{\downarrow}_{Hij}$ is subsumed by a clause of $\{x=x\} \cup$ **C**,
for all $Hij \in \Pi(N_i)$.
if $N' = \phi$, then BACKTRACK (D)
else TOP (R_{i1})

BACKTRACK (D)
Put D into V
If D has a righthand brother D_r , TOP (D_r)
else if D has a parent D_p , BACKTRACK (D_p)
else stop "proof"

5.1.3. Soundness of PROC2

The procedure PROC2 attempts to show the E-validity of clauses either directly or indirectly.
If C is subsumed by a clause of $\{x=x\} \cup$ **C**, then C is directly proved E-valid. Otherwise, either all the resolvents (case RES), or all the narrowings (case NAR) are computed. The procedure attempts then to establish the direct validity of all the children (by subsumption (case RES) or by rewriting with hypotheses then subsumption (case NAR)). In case it fails, the procedure is recursively applied to the leftmost child non directly validated.

The procedure backtracks to a node D, when all the children of D have been (directly or indirectly) validated. D is then put in the set V of validated clauses; actually, either :

- the children of D are resolvents; they have been directly subsumed, or indirectly validated on former backtracking. In all cases, the children are all valid, and hence D is valid (by proposition 5.4), or
- the children N_i of D are narrowings;then,either N_i is such that $N_i{\downarrow}_{Hij}$ is subsumed for all Hij $\in \Pi(N_i)$, or N_i has been indirectly validated on a former backtracking. In the first case, V_i is true, hence U_i is true (by prop. 5.3).
  In the second case, U_i is true again.
  So in any case, U_i is true. Hence D is valid (by prop. 5.1).

If a backtracking occurs towards a clause D without parents, then the validated clause D is the top clause C, hence C is E-valid.

Thus, we can state :

THEOREM 5.1 (soundness)
If PROC2 stops with "proof", then the input clause C is E-valid.

Remark
PROC2 can be modified without loss of soundness, in case certain function symbols being commutative and associative, by using Associative-Commutative unification and pattern matching [Sl,Pe,Fa], instead of ordinary ones.

Similar strategies to that of PROC2 apply, if (HB) holds besides (H0-H1),and if there is a procedure $\mathbf{P}_{val}$ of E-validity on *GCV*. The general strategy is expressed as follows : "Split the goal into subgoals by narrowing or $\mathbf{P}_{val}$, then try to prove each subgoal through hypothesis rewriting or $\mathbf{P}_{val}$. Split the unproved subgoals , and so on until all subgoals have been proved."

5.2. Rewriting with lemmas and goal assumptions

Rewriting lemmas [BM,Pa] are very useful in proving a goal. Usually, lemmas are considered as conditional rewriting rules of the form P & Q ==> $l=r$. Rewriting is performed if the left-hand side matches a subterm within the goal and if one can prove recursively that the corresponding instances of P and Q are satisfied. This strategy is risky because it may attempt to prove unsatisfiable conditions, or it may indefinitely appeal to lemmas to establish the conditions of other lemmas (infinite backwards chaining). Instead, we propose ***subsumptive rewriting*** , a much weaker form of rewriting but still always safe.

We say that a clause is ***oriented*** if its rigtmost literal is an oriented equation (from left to right).

definition
Let D' be an oriented clause $(D'_1, t_1 \to t_2)$
Let D be a clause $\mathbf{v}_{i=1,\dots,p}\ L_i$.
D is ***subsumptive reducible*** by D' if there is a subterm t, at occurrence o, in a literal L_j $(1 \le j \le p)$ such that :

- $t = \sigma t_1$, for some substitution σ

- $\mathbf{v}_{i=1,\dots,p\ \&\ i\neq j}\ L_i \subseteq \sigma D'_1$

The clause D'': $L_1\ \mathbf{v}\dots\mathbf{v}\ L_{j-1}\ \mathbf{v}\ L_j[o\leftarrow\sigma t_2]\ \mathbf{v}\ L_{j+1}\ \mathbf{v}\dots\mathbf{v}\ L_p$ is the ***subsumptive reduction*** of D by D'.

Remarks

D'' is a paramodulant of D' into D, and conversely D is a paramodulant of D' into D''.
Subsumptive reduction and ordinary reduction coincide when D' is the monoliteral clause $(t_1 \rightarrow t_2)$.

Furthermore, when trying to prove a goal and more precisely the goal φ-selected literal, we can rightfully make use of goal assumptions, by assuming the remaining literals are false (see [BM,Pa]).

definition

Let C : $\mathbf{v}_{k=1,\dots,p}\ t_k=t'_k\ \mathbf{v}_{l=1,\dots,q}\ u_l\neq u'_l$ be the goal to prove, and let $t_i=t'_i$ (resp. $u_j\neq u'_j$) be the φ-selected literal, with $1\leq i\leq p$ (resp. $1\leq j\leq q$). Then the ***goal assumption set*** is the following monoliteral clause set : $\{u_l=u'_l\}_{1\leq l\leq q}$ (resp. $\{u_l=u'_l\}_{1\leq l\leq q\ \&\ l\neq j}$).

Remarks

Variables of the goal assumption clauses are frozen as constants in the same way as those of hypothesis sample.
The goal assumption set can be extended with equations of the form t_k=false, when there is a (non φ-selected) equation t_k=true within C. More generally, it can be extended with inequations of the form $t_k\neq t'_k$, viewed as rewrite rules in the manner mentioned in § 5.1.2.

Throughout, we consider only subsumptive reduction performed at once with rewriting clauses D' known to be valid (lemmas or solved subgoals) and with goal assumptions. These clauses form the current set **W** of locally valid clauses. Clauses in **W** are oriented - therefore used as rewriting clauses - only if their rightmost equation sides are $<_0$-comparable. A clause which cannot be reduced by any clause of **W** is said to be subsumptive irreducible.
A ***normalized resolvent*** is a subsumptive irreducible clause obtained by s.b.r. followed by subsumptive rewriting with **W**. A ***normalized narrowing*** is a subsumtive irreducible clause obtained by s.n. followed by subsumptive rewriting with **W**.

For a clause D, R**(D) denotes the set of normalized resolvents of D, N**(D) denotes an innermost class of normalized narrowings of D, and D** denotes a normalized form of D under **W**.

definition

Let PROC2' be the procedure obtained from PROC2 by replacing
- "Initially V = $E \cup$ **C**" by "Initially **W** = $E \cup$ **C** $\cup$ {lemmas}"
- all the occurrences of "C","R(D)","N(D)" by "C**","R**(D)","N**(D)"
- "Put D into V" by 'Put D into **W**".

It can be shown without difficulty that PROC2' is sound

Remark

Subsumptive reduction with **W**, in PROC2', performs a weak role of forward subsumption. Suppose indeed that a clause D' in W subsumes the current clause D, then the reduced form of D by D' is subsumed by x=x and then discarded.

To illustrate the running of PROC2', we give two automated proof examples in appendix. The first one is the FLATTEN.MAC.FLATTEN.THEOREM (from [BM]). The proof is achieved through only two narrowings. The second one is the COMMUTATIVITY.OF.TIMES.THEOREM . The proof is given to illustrate how the procedure handles permutative equations.

6. Conclusion

The final narrowing procedure constitutes a method to prove equational clauses in the initial algebra defined by the set E of equations, when E forms a canonical set of rules satisfying a certain principle of definition. Its simplicity and relative efficiency are the main positive aspects of this method. In keeping with Dershowitz's interpretation of rewrite systems [De2], the procedure can be regarded as a prover of program properties.
Several strategies, such as generalization [BM,Au,KC] and conditional rewriting [Re,Ka], are still not embodied in the method. However, our work may hopefully provide logic programming with a basic tool for the incorporation of induction and equality.

ACKNOWLEDGEMENT

I am grateful to Gerard Huet for his helpful criticism and advice.

REFERENCES

[Au] Aubin R., "Mechanizing Structural induction", TCS 9 ,1979.
[Bi] Bidoit M, "Proofs by induction in "fairly" specified equational theories", Proc. 6th German Workshop on Artificial Intelligence, Sept. 1982, pp 154-166.,
[BM] Boyer R., Moore J.S., A computational logic, Academic press, 1979.
[Bu] Burstall R.M., "Proving properties of program by structural induction", Comput. J 12,1969.
[De1]Dershowitz N., "Ordering for term rewriting systems", TCS 17-3, 1982, pp 279-301.
[De2]Dershowitz N., "Computing with Rewrite Systems", Report No ATR-83 (8478)-1, The Aerospace Corporation, El Segundo, California, 1983.
[Fa] Fages F. "Associative-commutative Unification", Proc. CADE-7, 1984.
[Fr] Fribourg L., "A Superposition Oriented Theorem Prover", Proc. IJCAI-83, pp923-925.
[Go] Goguen J.A., "How to Prove Algebraic Inductive Hypotheses Without Induction, with Applications to the Correctness of Data Type Implementation", Proc. CADE 5, Les Arcs, July 1980.
[HH] Huet G., Hullot J.M., "Proofs by induction in equational theories with constructors", 21st FOCS, 1980, pp 96-107
[HO] Huet G., Oppen D., "Equations and Rewrite Rules: A Survey", Formal Languages Perspective and Open Problems, Ed. Book R, Academic Press, 1980, pp 349-406
[JLR]Jouannaud J.P., Lescanne P.,Reinig F., "Recursive Decomposition Ordering", Formal description of programming concepts 2, Ed. Bjorner, North-Holland, 1982.
[Jo] Joyner W.H., "Resolution Strategies as Decision Procedures" J.ACM 23:3, Jul. 1976.
[Ka] Kaplan S., "Conditional Rewrite Rule Systems and Termination", Report L.R.I, Orsay (to appear).
[KB] Knuth D.,Bendix P., "Simple Word Problems in Universal Algebras", Computational Problems in Abstract Algebras, Pergamon Press, 1970, pp 263-297.
[KC] Kodratoff Y., Castaing J., "Trivializing the proof of trivial theorems", Proc. IJCAI-83, pp 930-932.
[KK] Kowalski R., Kuehner D., "Linear Resolution with Selection Function", Artif. Intelligence 2, 1971, pp 227-260.
[La] Lankford D, "Canonical Inference", Report ATP-32, U. of Texas,1975.
[Lo] Loveland D.,"Automated Theorem Proving : A logical basis", Fundamental Studies in Computer Science,North Holland,1978.
[Mu] Musser D.L., "On Proving Inductive Properties of Abstract Data Types", Proc. 7th POPL, Las Vegas,1980.
[Pa] Paulson L., "A Higher-Order Implementation of Rewriting", Science of Computer Programming 3,1983, pp 119-149.
[Pl] Plaisted D.A., "A recursively defined ordering for proving termination of term rewriting systems", U. of Illinois, Report n° R-78-943, 1978.
[Re] Remy J.L. "Proving conditional identities by equational case reasoning rewriting and normalization", Report 82-R-085, C.R.I.N., Nancy, 1982.
[Sl] Slagle J.R., "Automated Theorem Proving for Theories with Simplifiers, Commutativity and associativity", J.ACM 21:4, Oct 1974, pp 622-642.
[St] Stickel M.E., "A unification algoritm for associative commutative functions" J.ACM 28:3,1981, pp 423-434.
[Th] Thiel J.J. "Un algorithme interactif pour l'obtention de definitions completes" Proc. 11th POPL, 1984.

APPENDIX

1) THE FLATTEN.MAC.FLATTEN.THEOREM

constructors nil, cons
axioms
E1 : (append nil y) = y
E2 : (append (cons x1 x2) y) = (cons x1 (append x2 y))
E3 : flatten nil = (cons nil nil)
E4 : flatten (cons x y) = (append (flatten x) (flatten y))
E5 : (macflatten nil y) = (cons nil y)
E6 : (macflatten (cons x1 x2) y) = (macflatten x1 (macflatten x2 y))

lemma
C0 : (append (append x y) z) = (append x (append y z))

theorem
C1 : (macflatten x y) = (append (flatten x) y)

the occurrence chosen for narrowing on C1 is the occurrence of macflatten in the left hand side

- narrowing of E5 into C1 (σ : {x←nil}, Pd = ϕ ; Θ = ϕ)
  cons nil y = cons nil y , subsumed by x=x

- narrowing of E6 into C1 (σ : {x←(cons x1 x2)}, Pd = {x1, x2};
  Θ = { macflatten x1 y = append (flatten x1) y
  macflatten x2 y = append (flatten x2) y})

macflatten x1 (macflatten x2 y) = append (append (flatten x1) (flatten x2 y))
which is rewritten, with $\Theta \cup$ C0, into :
append(flatten x1 (append(flatten x2) y))
= append(flatten x1 (append(flatten x2) y)) , which is subsumed by x=x.

This achieves the proof of C1.

2) THE COMMUTATIVITY.OF.TIMES.THEOREM

constructors : 0 , succ
AC operators : +
axioms :
E1 : x + 0 = x
E2 : x + succy = succ(x+y)
E3 : x × 0 = 0
E4 : x × succy = x × y + x

theorem
C1 : x × y = y × x

(Throughout the proof, the narrowing occurrence chosen is always the occurrence of '×' in the left-hand side)

- narrowing of E3 into C1 (σ : y←0, Pd = ϕ, Θ = ϕ)
  C2 : $0 \times x = 0$

  - narrowing of E3 into C2 (σ : x←0, Pd = ϕ, Θ = ϕ)
    $0 = 0$, which is subsumed by x=x

  - narrowing of E4 into C2 (σ : x←succ x', Pd = {x←x'}, Θ = { $0 \times x' = 0$ })
    $0 \times x' = 0$, which is rewritten by Θ
    into $0 = 0$, which is subsumed by x=x

This achieves the proof of C2, which is included into W.

- narrowing of E4 into C1 (σ : y←succ y', Pd = {y←y'}, Θ = { $x \times y' = y' \times x$ })
  C3 : succ $y' \times x = x \times y' + x$
  (C3 cannot be rewritten by Θ, because the equation in Θ cannot be oriented)

  - narrowing of E3 into C3 (σ : x←0, Pd = ϕ, Θ = ϕ)
    $0 \times y' = 0$, which is rewritten in 0=0 by C2 and then is subsumed by x=x

  - narrowing of E4 into C3 (σ : x←succ x', Pd = {x←x'},
    Θ = { succ $y' \times x' = x' \times y' + x'$ })

    succ (succ $x' \times y' + x'$) = succ (succ $y' \times x' + y'$)
    which is transformed into :
    succ (succ $x' \times y' + x'$) = succ ($x' \times y' + x' + y'$) (by reduction with Θ)
    then into : succ $x' \times y' + x' = x' \times y' + x' + y'$, so :

  C4 : succ $x \times y + x = x \times y + x + y$

  - narrowing of E3 into C4 (σ : y←0, Pd = ϕ, Θ = ϕ)
    x=x , which is subsumed by x=x

  - narrowing of E4 into C4 (σ : y←succ y', Pd = {y←y'},
    Θ = { succ $x \times y' + x = x \times y' + x + y'$ })
    succ(succ $x \times y' + x + x$) = succ ($x \times y' + x + x + y'$)
    which is transformed into:
    succ($x \times y' + x + y' + x$) = succ($x \times y' + x + x + y'$)
    (by rewriting with Θ), and then is subsumed by x=x .

This achieves the proof of C4, hence of C3 and consequently of C1.

A GENERAL INDUCTIVE COMPLETION ALGORITHM AND APPLICATION TO ABSTRACT DATA TYPES

Helene Kirchner*
Centre de Recherche en Informatique de Nancy
Campus Scientifique BP 239
54506 Vandoeuvre-les-Nancy Cedex FRANCE

ABSTRACT : This paper states the connection between hierarchical construction of equational specifications and completion of equational term rewriting systems. A general inductive completion algorithm is given, which turns out to be a well-suited tool to build up specifications by successive enrichments. Moreover, the same algorithm allows verifying consistency of a specification or proving theorems in its initial algebra without using explicit induction.

INTRODUCTION

In this paper, we address some general problems of equational logic from the point of view of abstract data type specifications and term rewriting systems. More precisely, we are dealing with three kinds of questions :

- How to prove theorems in the initial algebra of an equational theory, without using explicit induction ?

- How to enrich a basic specification with new operators in a hierarchical way ?

- How to prove correctness of some equational specification with respect to another one ?

This work extends previous results obtained in the following framework:

Let F be a set of function symbols split into a set of constructors F_0 with at least two elements, and a set of defined function symbols F_1. We assume that there is no relation between constructors and that the relations between constructors and defined function symbols are expressed via a set of axioms A, such that A provides a complete definition of operations of F_1. That implies

This research has been supported by GRECO de Programmation and by Agence pour le Developpement de l'Informatique, under contract no 82/767 that for any term, without variables, with at least one symbol of F_1, there exists an equivalent term modulo A expressed with symbols of F_0 only. If in addition axioms of A can be directed, giving a terminating and confluent term rewriting system R, the well-known Knuth and Bendix's completion procedure can be used to prove properties in the initial algebra of the equational theory A. After the pioneer work of Musser [MUS,80], the method has been clarified and improved by Goguen [GOG,80], Huet and Oppen [H&O,80], and Lankford [LAN,81]. In addition, Huet and Hullot [H&H,80] have shown how to modify the completion algorithm in order to take into account the existence of constructors, providing a powerfull technique. These works have given rise to various implementations : the systems AFFIRM , OBJ , FORMEL, REVE. Some of them accept associative commutative symbols in F_1, using for instance in FORMEL, the Peterson and Stickel's completion algorithm [P&S,81].

In order to allow relations between constructors, Remy [REM,82] introduced the concept of structured specification, in which equations on constructors can be directed into a confluent and noetherian term rewriting system.

On the other hand, Jouannaud [JOU,83] developed a general framework to decide equality in equational theories where some axioms (like commutativity) cannot be directed without loosing the termination property. An equational term rewriting system is then composed of a set of equations E and a set of rewrite rules R ; a suitable Chuch-Rosser property allows deciding (R U E)-equality of two terms by using rewritings only. In [J&K,84] and [KIR,83], we proved a general completion algorithm, roughly based on the same principles as the Knuth and Bendix's one, which allows completing an equational term rewriting system into another one which has the Church-Rosser property and defines the same congruence on terms. Our aim is to adapt this completion algorithm in order to deal with the problems of consistency or proofs in initial algebra, but for the wider class of equational theories which admit an equational term rewriting system. Moreover this inductive completion algorithm turns out to be a suitable tool for proofs in hierarchical specifications.

PRELIMINARIES

We assume that the reader is familiar with the basic notions of many-sorted algebras used in the algebraic approach of data type specification [GTW,78]. For more readibility we only use there the one-sorted case and assume that there exists at least one constant symbol. But all the results carry over to many-sorted algebras, assuming this condition for each sort [H&O,80].

I(F,X) denotes the free F-algebra over a set of variables X, whose domain is the set of terms T(F,X) constructed from operators of F and variables of X. For any term t, V(t) denotes the set of variables occurring in t, t/u the subterm of t at occurrence u and t[u<-t'] the term obtained by replacing t/u by t' in the term t. G(t) is the set of non variable occurrences in t.

I(F,Ø) is the initial F-algebra and T(F,Ø), denoted also T(F), is the set of ground terms, i.e. terms built up with symbols of F only.

Given a set A of axioms, the equational variety generated by A is the class of F-algebras which are models of axioms of A. The smallest congruence generated by A over T(F,X) is denoted by $=^A$. $|-|^A$ denotes one step of A-equality. The restriction of A-equality to T(F) is also denoted $=^A$. The initial algebra I(F,A) of the variety is the quotient $I(F)/=^A$.

An equation t=t' holds in I(F,A) iff $s(t) =^A s(t')$ for any substitution s from V(t)UV(t') to T(F), called a ground substitution. In this case we write $t =^{ind(A)} t'$ and $=^{ind(A)}$ is called inductive equality on T(F,X).

We use in what follows that, for two sets of axioms A and A', $=^{ind(A)}$ coincides with $=^{ind(A')}$ iff $=^A$ coincides with $=^{A'}$ on ground terms of T(F).

A specification (S,F,A) is an algebra of the variety generated by A and is composed of a set S of sorts, a finite set F of S-sorted function symbols and a set A of axioms .

Given a basic specification BSPEC=(S,BF,BA), we now want to add new operators DF defined by a set of new axioms DA. The specification SPEC=(S,BF U DF,BA U DA) is called an enrichment of BSPEC. SPEC is a protected enrichment of BSPEC if intuitively the new operations and equations do not modify I(BF,BA). A sufficient condition is that DA defines DF completely and consistently on BSPEC. More formally :

Definition 1 : * SPEC = (S,F,A) is complete w.r.t. BSPEC = (S,BF,BA) if for any term t in T(F), there exists some term t' in T(BF) such that $t =^{A} t'$.

* SPEC is consistent w.r.t. BSPEC if for any terms t,t' in T(BF), $t =^{A} t'$ implies $t =^{BA} t'$.
[]

Definition 2 : SPEC = (S,F,A) is a protected enrichment of BSPEC = (S,BF,BA) iff SPEC is consistent and complete w.r.t. BSPEC.
[]

The problem of deciding A-equality can be approached by the use of rewrite rules, that is directed equations, or more generally by the use of mixed sets of rules R and equations E defining an equational term rewriting system [JOU,83] :

Definitions 3 : * An equational term rewriting system (ETRS in short) is a pair (R,E) of a term rewriting system R and a set of equations E, satisfying :

- for any equation g=d in E, V(g)=V(d)
- for any rule l->r in R, V(r) is a subset of V(l).

* Let $\to^{E.R}$ the composed relation $\to^{R} \circ =^{E}$, which simulates the relation induced by $\to^{R}$ in the set of equivalence classes modulo $=^{E}$, and $\to^{R'}$ any rewrite relation satisfying $\to^{R} \sqsubseteq \to^{R'} \sqsubseteq \to^{E.R}$.

* R is said E-noetherian (or E-terminating) iff $\to^{E.R}$ is noetherian. $\to^{R'}$ is then also noetherian.

* $=^{(R \cup E)}$ is the reflexive, symmetric, transitive closure of $=^{E} \cup \to^{R}$, and $\overset{*}{\to}{}^{R'}$ denotes the reflexive transitive closure of $\to^{R'}$.

* R is R'-Church-Rosser modulo E on T(F) iff for any ground terms t1, t2 in T(F), $t1 =^{RUE} t2$ implies there exist ground terms t'1, t'2 s.t.

$$t1 \overset{*}{\to}{}^{R'} t'1 =^{E} t'2 \ {}^{R'}\overset{*}{\leftarrow} t2.$$
[]

The rewriting relation R', introduced by Jouannaud in [JOU,83], is used to compute in E-equivalence classes, even when these classes are infinite. Actually, if R is R'-Church-Rosser modulo E, then $\to^{E.R}$ is Church-Rosser, the converse being false.

In practice, we shall use for R' one of the rewriting relations $\to^{R}$, $\to^{R,E}$ (defined by Peterson and Stickel [P&S,81]) or $(\to^{L} \cup \to^{NL,E})$ (defined in [J&K,83]), where L are left-linear rules of R and NL the other ones. To use the two last relations, we need a matching algorithm for the theory $=^{E}$ and to prove the R'-Chuch-Rosser property, we need a complete unification algorithm for $=^{E}$.

Examples : (1) E = {x+y = y+x}   R = {x+0 -> x}

The term (x + (0 + y)) is not R-reducible but R,E-reduces to (x + y).

(2) E = {x+y = y+x}   L = {x+0 -> x}   NL = {x+(-x) -> 0}

The previous term is no more R'-reducible because no commutativity step is allowed before applying a rule of L.

But the term (-(x+y) + (x+y)) is R'-reducible to 0.

STRUCTURED SPECIFICATION AND EQUATIONAL TERM REWRITING SYSTEMS

Following Remy [REM,82], we introduce the notion of structured specification which allows us to translate in terms of rewriting systems, the properties of consistency and completeness and to give sufficient conditions to check them.

Definition 4 : Let BF U DF be a partition of a set of function symbols F.

We call * BF-equation g=d a pair of terms in T(BF,X) such that V(g) = V(d).

* BF-rule l->r a directed pair of terms in T(BF,X) such that V(r) is a subset of V(l)

* DF-equation g=d a pair of terms in T(F,X)\T(BF,X) s.t. V(g) = V(d)

* DF-rule l->r a directed pair of terms s.t. l is in T(F,X) \ T(BF,X) and V(r) is a subset of V(l).

[]

Notice that, according to this definition, an equation x=t with t in T(F,X) is not a DF-equation, and a rule x->t is not a DF-rule.

Definition 5 : A specification SPEC=(S,F,A) is a structured specification based on BSPEC=(S,BF,BA) iff

F=BF U DF, A=E U R, E=BE U DE, R=BR U DR, BA=BE U BR, such that

1) BE is a set of BF-equations, DE is a set of DF-equations

BR is a set of BF-rules, DR a set of DF-rules

2) BR is BR'-Church-Rosser modulo BE on T(BF), where BR' satisfies $\to^{BR} \subseteq \to^{BR'} \subseteq \to^{BE.BR}$.

3) R is E-noetherian

4) Any term t of T(F) has a R'-normal form $t!_{R'}$ in T(BF), where $\to^{R'}$ is a relation such that $\to^{R} \subseteq \to^{R'} \subseteq \to^{E.R}$ and $\to^{BR'} \subseteq \to^{R'}$.
[]

The notion of structured specification can be illustrated by a classical, but nevertheless significant example :

Example: Type integer   S = integer

BSPEC :	SPEC :
Function symbols : BF	DF
ZERO : --> integer SUCC : integer --> integer PRED : integer --> integer	OPP : integer --> integer PLUS : integer, integer --> integer
rules : BR	DR
SUCC(PRED(x)) -> x PRED(SUCC(x)) -> x	OPP(ZERO) ->ZERO OPP(SUCC(x)) -> PRED(OPP(x)) OPP(PRED(x)) -> SUCC(OPP(x)) PLUS(ZERO,y) -> y PLUS(SUCC(x),y) -> SUCC(PLUS(x,y)) PLUS(PRED(x),y) -> PRED(PLUS(x,y))
axioms : BE	DE
	PLUS(x,y) = PLUS(y,x) PLUS(PLUS(x,y),z) = PLUS(x,PLUS(y,z))

Some remarks are useful about definition 5 :

- In practice, we shall use for R'and BR' one of the rewriting relations previously mentioned, assuming implicitely that complete BE and E-unification algorithms are known if necessary.

- The condition 4) implies the completeness property of SPEC w.r.t. BSPEC. An equivalent formulation is : " any R'-normal form of a term t in T(F) is in T(BF)", since if a term t has some R'-normal form t_0 which is not in T(BF), applying condition 4) yields that t_0 would be R'-reducible to another term of T(BF), which is impossible. Thus t_0 is in T(BF).

The condition 4) is also equivalent to :

for any term $t=f(t_0,\ldots,t_n)$ such that f belongs to DF and $t_0,\ldots t_n$ are BR'-irreducible terms of T(BF), t is DR'-reducible.

An efficient decision algorithm for this last property can be obtained from [THI,84]. But an easy way to satisfy this condition is to define the new symbols of DF by structural induction on BR'-irreducible terms of T(BF).

Let us first point out some properties of a structured specification, resulting from the syntactical conditions of definition 4.

Lemma 1 : Let (S,F,A) a specification satisfying the condition 1 of definition 5. Then for any terms t and t' such that t is in T(BF,X) :

a) $t =^{BE} t' \Rightarrow t'$ is in T(BF,X)

b) $t \xrightarrow{*}^{BR'} t' \Rightarrow t'$ is in T(BF,X)

c) $t =^{E} t' \Rightarrow t =^{BE} t'$ and t' is in T(BF,X)

d) $t \xrightarrow{*}^{R'} t' \Rightarrow t \xrightarrow{*}^{BR'} t'$ and t' is in T(BF,X).

Proof : a) by induction on the length of the smallest proof that $t =^{BE} t'$. b) by induction on the length of the derivation.

c) and d) If t is in T(BF,X), neither equations of DE, nor rules of DR apply to t. An easy induction yields the result.
[]

The relation between structured specifications based on BSPEC and enrichments of BSPEC is expressed via term rewriting systems as follows :

Theorem 1 : Let SPEC=(S,F,A) a structured specification based on BSPEC such that A = (R U E) and R is R'-Church-Rosser modulo E on T(F), with R' as usual. Then SPEC is a protected enrichment of BSPEC.

Proof : 1) Since R is R'-Church-Rosser modulo E on T(F), $t =^{A} t'$ implies that there exist t1 and t'1 s.t. $t \xrightarrow{*}^{R'} t1 =^{E} t'1 \ {}^{R'}\!\!\xleftarrow{*} t'$. By lemma 1, since t and t' are in T(BF), $t \xrightarrow{*}^{BR'} t1 =^{BE} t'1 \ {}^{BR'}\!\!\xleftarrow{*} t'$.

2) Since SPEC satisfies the condition 4 of definition 5, let us define t' as the R'-normal form of t, which satisfies $t!_{R'}$ belongs to T(BF). t' clearly satisfies $t =^{A} t'$.
[]

The next lemma is the key point of our proofs. Its intuitive meaning is the following : to enrich a complete specification by equations which hold on ground terms does not modify the initial algebra and both specifications are consistent together.

Validity lemma :

Let SPEC=(S,F,A) be complete w.r.t. BSPEC=(S,BF,BA) and A' a set of axioms such that $=^{BA} \subseteq =^{A} \subseteq =^{A'}$. Then the three next conditions are equivalent :

(1) The specification SPEC'=(S,F,A') is consistent w.r.t. BSPEC

(2) a) SPEC is consistent w.r.t. BSPEC

b) any equation of A' holds in I(F,A).

(3) a) SPEC is consistent w.r.t. BSPEC

b) $=^{A}$ coincides with $=^{A'}$ on ground terms of T(F).

Proof : Let us first prove that (1) implies (2). a) is obvious.

b) If there exists some equation t=t' in A' which does not hold in I(F,A), there exists a ground substitution s such that $s(t) \neq^{A} s(t')$. As s(t) and s(t') are in T(F), since SPEC is complete w.r.t. BSPEC, there exist t_0 and t'_0 in T(BF) such that $s(t) =^{A} t_0$ and $s(t') =^{A} t'_0$. But $t_0 \neq^{A} t'_0$ and thus $t_0 \neq^{BA} t'_0$. On the other hand, $s(t) =^{A'} s(t')$ and since $=^{A} \subseteq =^{A'}$, $t_0 =^{A'} s(t) =^{A'} s(t') =^{A'} t'_0$.
Which yields a contradiction with the consistency of SPEC'=(S,F,A') w.r.t. BSPEC = (S,BF,BA).

The proof that (2) implies (3), or equivalently that if t and t' are two terms in T(F) s.t. $t =^{A'} t'$ then $t =^{A} t'$, is easily obtained by induction on the length of the smallest proof that $t =^{A'} t'$.

Finally (3) implies (1) is obvious.
[]

We now give a different version of this result, using equational term rewriting systems :

Validity theorem in equational term rewriting systems :

Assume that * SPEC = (S,F,R U E) is a structured specification based on BSPEC = (S,BF,BR U BE),

* SPEC' = (S, F, R U R" U E U E") is an enrichment of SPEC by a set E" of DF-equations and a set R" of DF-rules

* a complete (E U E")-unification algorithm is known.

If there exists a rewriting system R# such that

- R# is R#'-Church-Rosser modulo (E U E") on ground terms with R#' as usual.

- $=_{=}^{(R\# \cup E \cup E")} = {}_{=}^{(R \cup R" \cup E \cup E")}$ on ground terms of T(F)

- SPEC" = (S, F, R# U E U E") is a structured specification based on BSPEC, then

then SPEC is consistent w.r.t. BSPEC and all the equations of R" U E" hold in I(F, R U E).

Proof : Since R# is R#'-Church-Rosser modulo (E U E"), SPEC" is consistent and complete w.r.t. BSPEC by theorem 1 and thus also SPEC'=(S, F, R U R" U E U E"). The structured specification SPEC based on BSPEC is complete w.r.t. BSPEC, and since $=^{(BR \cup BE)}$ is included into $=^{(R \cup E)}$, the validity lemma then applies to BSPEC, SPEC and SPEC'. []

In order to schematize this theorem and to clarify what is done in the following, let us say that intuitively, from the first situation :

	BSPEC	SPEC	SPEC'
Function symbols :	BF	F = BF U DF	F = BF U DF
equations :	BE (BF-equations)	E = BE U DE	E U E" (DF-equations)
rules :	BR (BF-rules)	R = BR U DR	R U R" (DF-rules)
properties :	BR'-Church-Rosser modulo BE	structured spec. based on BSPEC	complete (EUE")-unification algorithm

we want a process that allows getting the second situation :

	BSPEC#	SPEC"
Function symbols :	BF	F = BF U DF
equations :	BE (BF-equations)	E U E"
rules :	BR# (BF-rules)	R# (DF-rules and BF-rules of BR#)
properties :	BR#'-Church-Rosser modulo BE	R#'-Church-Rosser modulo E U E" structured specif. based on BSPEC

with I(BF, BR# U BE) = I(BF, BR U BE) and I(F, R# U E U E") = I(F, R U E).

Notice that BR can be replaced by BR# during the process, assuming however that the initial algebra I(BF, BR U BE) is not modified.

This process is a general inductive completion algorithm which has a twofold interest :

First, it can be used to build up a hierarchy of enrichments or to add a level to such a construction. More precisely :

-- at level 0, R_0 and E_0 are respectively the sets of rules and equations on the set F_0 of constructors. By the inductive completion algorithm, the ETRS (R_0, E_0) is completed into (R_0#, E_0) such that R_0# is R_0#'-Churh-Rosser modulo E0 and =(R0# U E0) is equal to =(R0 U E0). We assume in addition that $=^{ind(R_0 \, U \, E_0)}$ is decidable (and thus also $=^{ind(R_0\# \, U \, E_0)}$)

-- at level 1, we add a complete definition of new operators of F_1 together with some properties that we want to prove. If, during the completion process, some equation on constructors is generated, it is proved or disproved in the basic specification $SPEC_0$, using the same mechanism. If the inductive completion process does not terminate with failure or disproof, the resulting specification (S, F_0 U F_1, E_0 U E_1 U E" U R#) is a protected enrichment of $SPEC_0$. In $SPEC_0$, some additional properties, valid in I(F_0, R_0 U E_0), have possibly been added.

-- at any level k, all the rules and axioms previously introduced make up the basic specification BSPEC=(S, BF, BR U BE) and again this specification can be enriched like at level 1.

On the other hand, at any level of this hierarchy, the validity theorem can be used in three particular ways : assume that BR is BR'-Church-Rosser modulo BE and BE-noetherian on T(BF).

- In order to prove the consistency of a specification SPEC=(S,BF U DF, BR U DR U BE U DE) w.r.t. BSPEC, the general inductive completion algorithm is initialized with DR as set of pairs, BR as set of rules and (BE U DE) as set of equations. It will generate, assuming it does not stop or abort, a term rewriting system R# which is R#'-Church-Rosser modulo (BE U DE) such that (S,F, R# U BE U DE) is a consistent and complete specification w.r.t.

BSPEC and $=^{(R\# \cup BE \cup DE)}$ is equal to $=^{(BR \cup DR \cup BE \cup DE)}$ on T(BF U DF). Thus SPEC is also consistent w.r.t. BSPEC.

- Assume now that we want to perform proofs in the initial algebra of a structured specification (S,F,R U E) based on BSPEC. For instance, our aim is to prove a set of equations split into a subset of DF-rules R" and a subset of DF-equations E", such that a complete (E U E")-unification algorithm is known. The general inductive completion algorithm initialized with R" as set of pairs, R as set of rules and (E U E") as set of equations, will generate a rewriting system R# satisfying the conditions of the validity theorem. The equations of R" U E" thus hold in the initial algebra I(F, E U R).

- Moreover it is possible to introduce at the same time the definitions of operators (expressed via DR U DE) together with some of their properties (expressed via R" U E") and to prove both the consistency of the new specification and the validity of the properties.

The second part of the paper is the proof of these points, after a more precise description of the general inductive completion algorithm. This algorithm attempts to check the hypotheses of the validity theorem and if they are not satisfied, tries to generate R#, by adding rules g->d such that g=d holds in I(F,RUE).

THE INDUCTIVE COMPLETION PROCEDURE

Let us assume that function symbols, rules and equations are organized in a hierarchical way. Function symbols are divided into F_0, $F_1, \ldots F_k, \ldots F_j$ where F_k is the set of operators defined at level k. The main feature of the algorithm is that it can be called at any level k, and then only works with rules and equations which do not contain any symbol of $F_{k+1} \cup \ldots \cup F_j$. For a given k, we need to distinguish rules and equations introduced at this level and rules and equations which make up the basic specification.

More formally, for all the considered sets of rules R and equations E, let us define the functions B_k and D_k in the following way :

* $B_k(F) = F_0 \cup \ldots \cup F_{k-1}$, $B_0(F)$ is empty. $D_k(F) = F_k$.

* $R = B_k(R) \cup D_k(R)$, where $D_k(R)$ is the subset of F_k-rules and $B_k(R)$ is the subset of $B_k(F)$-rules of R. $B_0(R)$ is empty.

* $E = B_k(E) \cup D_k(E)$, where $D_k(E)$ is the subset of F_k-equations and $B_k(E)$ is the subset of $B_k(F)$-equations of E. $B_0(E)$ is empty.

In each subset $D_k(R)$ or $D_k(E)$, elements are labelled using distinct sets of labels, isomorphic to positive integers. If the rule $l \to r$ is labelled by n, we denote either $n: l \to r$ or $l_n \to r_n$. The set of all the labels of rules is completely ordered in such a way that any rule of $D_k(R)$ has a label less than any rule of $D_{k+1}(R)$. We assume the same on the labels of all the equations.

The inductive completion procedure is obtained by modifying the E-completion procedure, described in [KIR,83] and [J&K,84], that generates from a set of axioms $R \cup E$, a term rewriting system R# which is R#'-Church-Rosser modulo E. Moreover $=^{(R\# \cup E)}$ is equal to $=^{(R \cup E)}$ on $T(F,X)$.

In the same way as the E-completion procedure, the inductive completion procedure described below works with three kinds of objects :

* a constant set of equations denoted SE. We assume given a SE-reduction ordering < on terms [J&K,84], used to prove the termination of the successive sets of rewrite rules.

* a current set of directed pairs denoted SP. The two terms of all the pairs have to be compared, according to the chosen SE-reduction ordering. Each pair comes either from an equation resulting from the reduction of some rule by another one, or from the computation of some critical pair.

* a current rewriting system denoted SR. For theoretical reasons (cf.[KIR,83] or [J&K,84]), a rule labelled n in $D_k(SR)$ can be :

- marked as soon as all the critical pairs of n with equations in SE and with rules of SR with a label n' less than n have been computed.

- protected for coherence of another rule n' if it is non left linear, if there is a critical pair (p,q) between the rule n' and an axiom (g=d) of SE and if the rule n is used to $\to^{SR'}$-reduce p on top. The left-hand side of a protected rule is not allowed to be reduced any more. Notice that a rule can be protected for coherence of several rules.

- an extension of a rule n':l->r, if there is a critical pair (p,q) between the non-left-linear rule n' and an axiom (g=d) of SE such that p is not $->^{SR'}$-reducible. The extension is built from the so called "extension pair" (g[u<-l], g[u<-r]!$_{SR'}$) which is directed from left to right. The extension of rule n' is only allowed to disappear when the rule n' disappears. Then, all the non protected extensions of n' previously introduced by the process disappear at the same time.

Whenever the process introduces a new rule l->r in the current rewriting system, the other rules are checked for simplifiability : a rule is simplifiable by l->r iff

- its left-hand side is reducible by the rule l->r,
- it is protected only for coherence of rules simplifiable by l->r,
- it may be an extension but only of rules simplifiable by l->r.

Let us point out that, at level k, each newly introduced rule is a D_k(F)-rule. Thus only other D_k(F)-rules need to be checked for simplifiability and B_k(SR) is not modified.

We need a fairness selection hypothesis, in order to ensure that any rule and equation will be selected after a finite time to compute its critical pairs, except if it has been deleted.

The main procedure is INDUCTIVE COMPLETION. The DIRECT procedure attempts to direct a pair (p,q) into a rule. The SIMPLIFICATION procedure reduces as much as possible rules in D_k(SR). Rules that are simplifiable by l->r must become new pairs because their orientation may change. The CRITICAL-PAIRS procedure computes overlappings between the rule or the equation n' given as argument and other equations and rules. This procedure also sets protections or builds extension-pairs if necessary, and normalize pairs before adding them to the set of new pairs.

```
INDUCTIVE COMPLETION (SP, SR, SE, <, n, k)

IF SP is not empty THEN choose a pair (p,q) in SP ;

    CASE * p=^SE q THEN SR := INDUCTIVE COMPLETION(SP\{(p,q)},SR,SE,<,n,k)          (1)

         * k ≠ 0 AND p, q ∈ T(B_k(F),X) THEN

          IF (k=1 AND p≠^ind(R_0 U E_0) q) THEN STOP and RETURN "disproof" END IF

          RES := INDUCTIVE COMPLETION((p,q),B_k(SR),B_k(SE),<,card(D_k-1(SR)),k-1)

          EXCEPT WHEN it returns "disproof" THEN STOP and RETURN "disproof"

                 WHEN it returns "failure" THEN STOP and RETURN "failure"

                 END EXCEPT

          SR := INDUCTIVE COMPLETION(SP\{(p,q)},RES U (SR\B_k(SR),SE,<,n,k)   (2)

         * ELSE 1->r := DIRECT(p,q)

          EXCEPT WHEN it returns "failure" THEN STOP and RETURN "failure"

                 END EXCEPT

          (SP,SR) := SIMPLIFICATION(SP, SR, 1->r)

          SR := INDUCTIVE COMPLETION(SP, SR U {n:1->r}, SE, <, n+1, k)        (3)

          END IF
    END CASE

ELSE IF all the elements of D_k(SR) and D_k(SE) are marked THEN STOP and RETURN SR

   ELSE Choose a non marked element  n'  in  D_k(SR) U D_k(SE)  according  to  the
        fairness selection hypothesis

        SP := CRITICAL-PAIRS(SR, SE, n')

        Mark the rule n'

        SR := INDUCTIVE COMPLETION(SP, SR, SE, <, n, k)                        (4)
   END IF
END IF
END INDUCTIVE COMPLETION


DIRECT(p, q)

CASE.(q∈T(F,X)\T(B_k(F),X) AND p∈T(B_k(F),X) AND q>p)THEN RETURN q->p

    .(p∈T(F,X)\T(B_k(F),X) AND q∈T(B_k(F),X) AND p>q)THEN RETURN p->q

    .(p, q ∈ T(F,X)\T(B_k(F),X) AND p>q)  THEN RETURN p->q

    .(p, q ∈ T(F,X)\T(B_k(F),X) AND q>p)  THEN RETURN q->p

    .ELSE STOP and RETURN "failure"
END CASE
END DIRECT
```

SIMPLIFICATION(SP,SR,l->r)

$K=\{m \in SR \backslash B_k(SR) \mid m$ is simplifiable by $l \to r\}$; $K1=\{m \in K \mid m$ is not an extension$\}$

$SP1 = SP \backslash \{(p,q)\} \cup \{(l'_m, r_m) \mid m$ in K1 and $l_m \to l'_m$ using $l \to r\}$

$SR1 = \{l_m \to r'_m \mid m$ not in K, $l_m \to r_m$ is in SR, r'_m is the

SR'-normal form of r_m using rules of SR and $l \to r\}$

RETURN(SP1,SR1)
END SIMPLIFICATION

Example : Applied to the specification of integers given as example, the algorithm effectively terminates, stating thus the consistency of the specification SPEC. In addition, we obtain that SPEC is a structured specification based on BSPEC.

Moreover, if we want to prove in this specification the assertions :

OPP(OPP(x)) = x

OPP(PLUS(x,y)) = PLUS(OPP(x),OPP(y)),

they have to be directed from left to right for giving R". Once more the general inductive completion algorithm terminates, thus proving the validity of these assertions.

In order to enrich SPEC by the new operator MULT, we add a set R_2 of rules defining MULT w.r.t. each operator of $F_0 \cup F_1$:

Function symbols : $F_0 \cup F_1$ F_2

```
ZERO : --> integer                          MULT : integer, integer --> integer
SUCC : integer --> integer
PRED : integer --> integer
OPP : integer --> integer
PLUS : integer, integer --> integer
```

Rules : $R_0 \cup R_1$ R_2

```
SUCC(PRED(x)) -> x                          MULT(ZERO,x) -> ZERO
PRED(SUCC(x)) -> x                          MULT(SUCC(x),y) -> PLUS(y,MULT(x,y))
OPP(ZERO) ->ZERO                            MULT(PRED(x),y) -> PLUS(OPP(y),MULT(x,y))
OPP(SUCC(x)) -> PRED(OPP(x))                MULT(PLUS(x,y),z) ->
OPP(PRED(x)) -> SUCC(OPP(x))                           PLUS(MULT(x,z),MULT(y,z))
PLUS(ZERO,y) -> y                           MULT(OPP(x),y) -> OPP(MULT(x,y))
PLUS(SUCC(x),y) -> SUCC(PLUS(x,y))
PLUS(PRED(x),y) -> PRED(PLUS(x,y))
```

Equations : $E_0 \cup E_1$ E_2

```
PLUS(x,y) = PLUS(y,x)                       MULT(MULT(x,y),z) = MULT(x,MULT(y,z))
PLUS(PLUS(x,y),z) = PLUS(x,PLUS(y,z))       MULT(x,y) = MULT(y,x)
```

The completion algorithm then generates and proves the new rules :

```
PLUS(x,OPP(x)) -> ZERO.
OPP(OPP(x)) -> x
OPP(PLUS(x,y)) -> PLUS(OPP(x),OPP(y)),
```

and terminates then with success. This example has been processed with the help of the system FORMEL due to G. Huet, G.Cousineau and their co-workers at INRIA.

PROOF OF THE INDUCTIVE COMPLETION ALGORITHM

For a given k, let SR_i be the current set of rewrite rules at the ith terminal recursive call (1), (2), (3) or (4) of the procedure. $R+ = \bigcup_i SR_i$ is the set of all the rules generated during the process. R# is the set of the rules which are never reduced neither on the right or the left. We are going to prove by successive lemmas the main theorem :

Theorem 2 : Let $F = F_0 \cup \ldots \cup F_{k-1} \cup F_k$ be a set of function symbols, SPEC = (S, F, R U E) a structured specification based on BSPEC = $(S, B_k(F), B_k(R) \cup B_k(E))$, R" a set of $D_k(F)$-rules and E" a set of $D_k(F)$-equations s.t. a complete (E U E")-unification algorithm is known.

Assume that $=^{ind(R_0 \cup E_0)}$ is decidable and that the general inductive completion procedure is initialized with R", R, EUE", <, card(D_k(R)) and k. Then

a) If it does not stop with failure nor disproof,

- R# is R#'-Church-Rosser modulo (E U E") on ground terms T(F) and $=^{ind(R\# \cup E \cup E")} = =^{ind(R \cup R" \cup E \cup E")}$.
- SPEC" = (S, F, R# U E U E") is a structured specification based on BSPEC# = $(S, B_k(F), B_k(R\#) \cup B_k(E))$.
- SPEC" is consistent and complete w.r.t. BSPEC.
- SPEC = (S, F, R U E) is consistent w.r.t. BSPEC,
- every equation of R" U E" holds in I(F,R U E).
- I(F, R U E) = I(F, R# U E U E") = I(F, R U R" U E U E").

b) If the procedure stops with disproof,

- either some equation in R" U E" does not hold in I(F, R U E)
- or SPEC is not consistent w.r.t. BSPEC.

c) Conversely, if some equation in E" U R" does not hold in I(F,R U E), or if SPEC is not consistent w.r.t. BSPEC, then the procedure stops with either disproof or failure.

Remark : The decidability of $=^{ind(R_0 \cup E_0)}$ is an open problem. Nevertheless, in some equational theories, the condition " $p =^{ind(R_0 \cup E_0)} q$ " is equivalent to "$p =^{(R_0 \cup E_0)} q$ ". Such theories are called "inductively complete" in [PAU,84]. An example is given by the theory $(R_0 \cup E_0)$ in the integers : if the equation t=t' is valid in $I(F_0, R_0 \cup E_0)$, t and t' both contain only one variable, say x. Then let us choose the substitution s defined by s(x)= ZERO. From the sequence of $(R_0 \cup E_0)$-equality steps :
$s(t) \mid\!-\!\mid^{(E_0 \cup R_0)} \ldots \mid\!-\!\mid^{(E_0 \cup R_0)} s(t')$, we can deduce a sequence of $(R_0 \cup E_0)$-equality steps from t to t' just by applying the inverse transformation of the substitution s. In such cases, the procedure stops with disproof as soon as p and q belong to $T(F_0,X)$ and are not E_0-equal.

The proof of theorem 2 is by induction on the level k in the hierarchy of specifications. At level 1, we need to prove that we can add to $SPEC_0$ some F_0-equations valid in $I(F_0, R_0 \cup E_0)$.

Lemma 2 : Let F_0 be the set of constructors, R_0 a set of F_0-rules, E_0 a set of F_0-equations s.t. a complete E_0-unification algorithm is known, and R" a set of F_0-equations which hold in $I(F_0 \cup R_0)$.

Then INDUCTIVE COMPLETION(R",R_0,E_0,<,card(R_0),0) either loops forever, or stops with failure, or returns $R_0\#$ s.t. $R_0\#$ is $R_0\#'$-Church-Rosser modulo E_0 and $I(F_0, R_0\# \cup E_0) = I(F_0, R_0 \cup E_0)$.

Proof : With k=0, we find again the usual general E-completion procedure and according to [KIR,83], we get a system $R_0\#$ such that $R_0\#$ is $R_0\#'$-Church-Rosser modulo E_0 and $=^{(R_0\# \cup E_0)} = =^{(R_0 \cup R" \cup E_0)}$. The equality of initial algebras then follows from the validity lemma applied with $A = BA = R_0 \cup E_0$ and $A' = R_0 \cup R" \cup E_0$.
[]

In order to prove a), let us assume that the procedure does not stop with failure nor disproof. The following preliminary lemma states a useful property about each terminal recursive call of the procedure :

Lemma 3 : On T(F), for any i, $=^{(SP_{i+1} \cup SR_{i+1} \cup SE)} = =^{(R \cup R'' \cup E \cup E'')}$ and on $T(B_k(F))$, for any i, $=^{(B_k(SRi) \cup B_k(SE))} = =^{(B_k(R) \cup B_k(E))}$

Proof : by introspecting each terminal recursive call in the procedure.
[]

Corollary : On T(F), $=^{(R+ \cup SE)} \subseteq =^{(R \cup R'' \cup E \cup E'')}$ and on $T(B_k(F))$, $=^{(B_k(R+) \cup B_k(SE))} = =^{(B_k(R) \cup B_k(E))}$.

We are now ready to state the successive points of Theorem 2 a) :

Lemma 4 : R# is R#'-Church-Rosser modulo (E U E") on ground terms of T(F) and $=_{ind}^{(R\# \cup E \cup E'')} = =_{ind}^{(R \cup R'' \cup E \cup E'')} = =_{ind}^{(R+ \cup E \cup E'')}$.

Proof : by the same technics as in [KIR,83], restricted to T(F).
[]

Lemma 5 : SPEC" is a structured specification.

Proof : By construction, SPEC" satisfies the condition 1 of definition 5

- On $T(B_k(F))$, $B_k(R\#)$ is $B_k(R\#)$'-Church-Rosser modulo $B_k(E)$ since if t and t' belong to $T(B_k(F))$, $t =^{(B_k(R\#) \cup B_k(E))} t' \Rightarrow t =^{(R\# \cup E \cup E'')} t'$. Then by lemma 4, $t \xrightarrow{*}{}^{R\#'} t_1 =^{(E \cup E'')} t'_1 \; {}^{R\#'}\!\xleftarrow{*} t'$ and by lemma 1 : $t \xrightarrow{*}{}^{B_k(R\#)'} t_1 =^{B_k(E)} t'_1 \; {}^{B_k(R\#)'}\!\xleftarrow{*} t'$.

- R# is E-noetherian.

We are left to prove that the condition 4 of definition 5 is also satisfied : let t be a term in T(F) and t_2 its R#'-normal form. Since SPEC is structured, let t_1 be the R'-normal form of t which is in $T(B_k(F))$. Since, on ground terms of T(F), $=^{(R \cup E)}$ is contained into $=^{(R \cup R'' \cup E \cup E'')} = =^{(R\# \cup E \cup E'')}$, $t_1 =^{(R\# \cup E \cup E'')} t_2$. Because R# is R#'-Church-Rosser modulo (E U E") on T(F), $t_1 \xrightarrow{*}{}^{R\#'} t'_1 =^{(E \cup E'')} t_2$. By lemma 1, $t'_1 =^{B_k(E)} t_2$ and t_2 is in $T(B_k(F))$.
[]

Lemma 6 : If the procedure does not stop with failure nor disproof, then

- SPEC" is consistent and complete w.r.t. BSPEC.
- SPEC = (S, F, R U E) is consistent w.r.t. BSPEC,
- every equation of R" U E" holds in I(F,R U E),
- I(F, R U E) = I(F, R# U E) = I(F, R U R" U E U E").

Proof : By theorem 1, SPEC" is then complete and consistent w.r.t. BSPEC#. It is easily proved that $=^{(B_k(R) \cup B_k(E))} = =^{(B_k(R\#) \cup B_k(E))}$ on ground terms of $T(B_k(F))$.

Thus BSPEC# = BSPEC and SPEC" is consistent w.r.t. BSPEC.

The theorem of validity in the ETRS then applies and yields that SPEC is consistent w.r.t. BSPEC# and any equation of (R" U E") holds in I(F, R U E). The identity of initial algebras are easily obtained from validity lemma (with BA = $(B_k(R) \cup B_k(E))$, A = (R U E) and A' = (R U R" U E U E")) and lemma 4.
[]

The point b) of theorem 2 is based on the following lemma :

Lemma 7 : If the procedure stops with disproof at some recursive call i,

- $SPEC_i$ = (S, F, $SP_i \cup SR_i \cup SE$) is not consistent w.r.t. BSPEC,
- SPEC' = (S, F, R U R" U E U E") is not consistent w.r.t. BSPEC.

Proof : If the procedure stops with disproof, there exists a pair (p,q) in SP_i such that p and q are both in $T(B_k(F))$ and not SE-equal. Then either k=1 and p=q does not hold in $I(F_0, R_0 \cup E_0)$, or INDUCTIVE COMPLETION((p,q), $B_k(SR_i)$, $B_k(SE)$, card($D_{k-1}(SR_i)$), k-1) stops with disproof and by induction hypothesis, p=q does not hold in $I(B_k(F), B_k(SR_i) \cup B_k(E))$. There exists then a substitution s from X to $T(B_k(F))$ such that $s(p) \neq^{(B_k(SR_i) \cup B_k(E))} s(q)$. Thus $s(p) \neq^{(B_k(R) \cup B_k(E))} s(q)$ and $SPEC_i$ is not consistent w.r.t. BSPEC.

Then, from lemma 3, on $T(B_k(F))$ $=^{(SP_i \cup SR_i \cup SE)}$ is included into $=^{(SP_0 \cup SR_0 \cup SE)} = =^{(R \cup R" \cup E \cup E")}$.

Thus if $SPEC_i$ is not consistent w.r.t. BSPEC, SPEC' is clearly not consistent w.r.t. BSPEC.
[]

In order to obtain the second point b) of the theorem 2, we apply the validity lemma to BSPEC, SPEC and SPEC'.

Let us now prove c) :

Lemma 8 : If there exists some assertion of R" U E" which does not hold in I(F,R U E) or if SPEC = (S, F, R U E) is not consistent w.r.t. BSPEC, the inductive completion procedure stops with either disproof or failure.

Proof : Applying the validity lemma to BSPEC, SPEC and SPEC', we obtain that SPEC' is not consistent w.r.t. BSPEC.

Thus there exist two terms t and t' in $T(B_k(F))$ such that : $t =^{(R \cup R'' \cup E \cup E'')} t'$ and $t \neq^{(B_k(R) \cup B_k(E))} t'$. If the procedure terminates or loops, (according to [KIR,83] or [J&K,84]), there exist terms t_i and t'_i s.t. $t \xrightarrow{*}{}^{SR_i'} t_i =^{(E \cup E'')} t'_i \; {}^{SR_i'}\xleftarrow{*} t'$.

But since t and t' belong to $T(B_k(F))$, we then deduce by lemma 1, that $t =^{(B_k(SR_i) \cup B_k(E))} t'$, thus $t =^{(B_k(R) \cup B_k(E))} t'$, which yields a contradiction. Thus the procedure stops with disproof or failure.
[]

Notice that, if the procedure stops with failure, nothing can be said. Perhaps consistency or validity properties are not satisfied, or the chosen SE-reduction ordering is not powerful enough, or the method is not applicable at all, or a valid equation in the basic specification is generated but it cannot be directed into a rule with the given ordering. In this last case, one can try again after adding the valid equation to the set of equations, assuming a complete unification algorithm is known for the whole set.

CONCLUSION

This procedure is being implemented in the system REVE [LES,83] which soon provides the classical mechanism of the inductive completion described in [H&H,80]. As a counterpart of an increased generality, our procedure needs complete unification algorithms, which are often runtime consuming. On the other hand, some problems are yet open, for example how to decide $=^{ind(R_0 \cup E_0)}$? Is it possible, in some cases, to terminate with disproof only by considering the top symbols of the two terms ?

BIBLIOGRAPHY

[EKP,78] EHRIG H., KREOWSKY H.J., PADAWITZ P. : "Stepwise specification and implementation of abstract data types" Proc. 5th Int. Colloquium on Automata, Languages and Programming, Udine (1978)

[GOG,80] GOGUEN J.A. : "How to prove algebraic inductive hypotheses without induction, with application to the correctness of data types implementation" Proc. 5th CADE, Les Arcs (1980)

[GTW,78] GOGUEN J.A., THATCHER J.W., WAGNER E.G. : "An initial algebra approach to the specification, correctness and implementation of abstract data types" in "Current trends in programming methodology", vol. 4, pp 80-149, Ed. Yeh R., Prentice-Hall (1978)

[H&H,80] HUET G., HULLOT J.M. : "Proofs by induction in equational theories with constructors" Proc. 21th FOCS (1980) and JCSS 25-2 (1982)

[H&O,80] HUET G., OPPEN D. : "Equations and rewrite rules : a survey" in "Formal languages : perspectives and open problems" Ed. Book R., Academic Press (1980)

[J&K,84] JOUANNAUD J.P., KIRCHNER H. : "Completion of a set of rules modulo a set of equations" Proc of POPL (1984).

[JOU,83] JOUANNAUD J.P. : "Confluent and coherent sets of reductions with equations. Application to proofs in data types." Proc. 8th Colloquium on Trees in Algebra and Programming (1983)

[KIR,83] KIRCHNER H. : "A general completion algorithm for equational term rewriting systems and its proof of correctness" Rep. CRIN, Nancy (1983)

[K&K,82] KIRCHNER C., KIRCHNER H. : "Unification dans les theories equationnelles" Proc. Journees GROSSEM-82, Marseille and RCP Algorithmique, Limoges (1982)

[LAN,81] LANKFORD D.S. : "A simple explanation of inductionless induction" Louisiana Tech. University, Math. Dep., Rep. MTP-14 (1981)

[LES,83] LESCANNE P. : "Computer experiments with the REVE term rewriting system generator" Proc. 10th POPL Conference (1983)

[MUS,80] MUSSER D.R. : "On proving inductive properties of abstract data types" Proc. 7th POPL Conference, Las Vegas (1980)

[PAU,84] PAUL E. : "Preuve par induction dans les theories equationnelles en presence de relations entre les constructeurs" Proc of 9th Colloquium on trees in Algebra and Programming (1984)

[P&S,81] PETERSON G.E., STICKEL M.E. : "Complete sets of reductions for equational theories with complete unification algorithms" J.ACM 28, no. 2, pp 233-264 (1981)

[REM,82] REMY J.L. : "Etude des systemes de reecriture conditionnels et application aux types abstraits algebriques" These d'Etat, Nancy (1982)

[THI,83] THIEL J.J. : "Un algorithme interactif pour l'obtention de definitions completes" Proc of POPL (1984).

ACKNOWLEDGMENTS : I thank J.P. Jouannaud, P. Lescanne, C. Kirchner and J.L.Remy for their helpfull comments about preliminary versions of this paper.

THE NEXT GENERATION OF INTERACTIVE THEOREM PROVERS

Patrick Suppes

Stanford University

1. History

Prior to discussing what I see as desirable and achievable features of the next generation of interactive theorem provers, I want to say something about the history of my own work and that of my colleagues, which forms the basis for the view of the future I sketch in the remainder of this paper. Simple uses of an interactive theorem prover for the teaching of elementary logic began more than twenty years ago. I remember well our first demonstrations with elementary-school children in 1963. For a number of years we concentrated on teaching elementary logic and algebra to bright elementary- and middle-school children. We felt at the time that this was the right level of difficulty to reach for in terms of computer capacity and resources that could be devoted to the endeavor. All of this early work was done on one of the low-serial-number PDP-1's, which John McCarthy and I jointly purchased from grants at Stanford in 1963. This early work has been described in Suppes and Binford (1965) and Suppes (1972).

By the late sixties it became clear that we could aim at something a step more advanced, and by 1972 I was able to convert the elementary-logic course at Stanford to a course taught entirely at computer terminals. By that time we had introduced a better and more powerful interactive theorem prover. That course is probably the longest-running show anywhere on earth having an interactive theorem prover used on a regular basis day in and day out by large numbers of persons. Approximately 100 students each term at Stanford enroll for this course and it is given every term. Access to the computer is pretty much around the clock seven days a week so that at almost any time of the day or night a routine use of our interactive theorem prover is taking place. The content of the course in elementary logic is comparable to that of my text (Suppes, 1957). It is obvious enough that the theorem-proving demands of such an elementary course are not very severe. To update also the computer framework, by the early seventies we had moved from a PDP-1 to a PDP-KA10.

The next natural move up was to a course in axiomatic set theory, roughly corresponding to my text in the subject (Suppes, 1960/1972). Here there were nontrivial theorems to be proved, above all the classical organization of a mathematical subject into a long sequence of theorems, with no hope of individual theorems' being proved from scratch directly from the axioms. Since 1974 this course also has been offered every term as a course in computer-assisted instruction, with students' getting all of the instruction at terminals. The enrollment is much smaller than that of the logic course. The average enrollment each term is eight or nine students, but the enrollment is greater than it was in the days before the course was computerized. By the time this course was introduced we had moved to a PDP-KI10, and a little later we were able to add a second KI10. We are still running both courses on the two KI10s running as a dual processor. We have ported the logic course to a number of other systems and it is running in 15 or 20 places around the world, but the set theory is a much more elaborate course. It is probably a matter of at least two man-years to convert the course to a portable framework. The details of the set-theory course are described in several articles in Suppes (1981) and more recently in McDonald and Suppes (1984). I shall not recapitulate more details than you will want to hear about this course. It is organized in terms of somewhat more than 600 theorems. Depending upon the grades students seek they prove somewhere between 30 and more than 50 theorems. Some of the theorems are too hard to require them to prove in a beginning course of this kind. After all, one of the theorems on transfinite induction is essentially the main content of von Neumann's dissertation. I also do not want to give the wrong impression about the finished character of the setup. I think that the interactive theorem prover we are using has many good features but it is still awkward to prove the hardest theorems in the sequence. Much work remains to be done to make the proofs of those theorems as natural and easy as they really should be.

Roughly speaking, I would describe the main features of the interactive theorem prover used in the set-theory course under three headings. First, elementary rules of inference, roughly the kind we associate with first-order logic, are available and can be used by the students. Second, students can call a resolution theorem prover that will run for a few seconds of machine time. What they give as input to the theorem prover are the definitions or previous theorems that are to be used to infer a desired formula. Third, a goal structure is provided for helping the student in determining the structure of his proof. A goal structure represents the kind of expert knowledge that is not yet perfected in all respects but that is used extensively by the students and provides a good deal of meaningful assistance.

There are several important remarks to be made about the way in which the students use the theorem prover. First, the most frequently used rule is the use of a previous theorem to make a fairly direct inference. Second, contrast must be sharply drawn between the highly interactive nature of the way the students use the theorem prover and the way proofs can be printed out under a review function at the end. The interactive phase of creating the proof looks from the outside world like a mess. It has the kind of highly interactive discourse structure that is not easy to follow at a glance. It is meant to be easy for the students to use and to provide considerable help to them. In contrast, the proofs that are printed out at the end under the heading of review are organized and systematic and put in a standard crisp form that gives a minimal insight into the interaction that took place in creating the proofs. As an object of study of how students are able to create proofs, the interactive proof is a much more important object than the cleaned-up version. Almost all mathematical study of the structure of proofs has been concerned with the latter rather than the former objects.

Let me give a couple of partial examples. Here is the beginning of an interactive proof of Cantor's theorem that the power of a set is strictly less than the power of its power set, taken from McDonald and Suppes (1984).

In the first three frames the student merely accepts the program's suggestion (note that ρA is the power set of A):

Goal G1: $(\forall A)\ A \prec \rho A$

[Reduce the current goal with a universal reduction]
(reduce) *! (Student input is underlined.)
Doing universal reduction

Goal G2: $A \prec \rho A$

[Reduce the current goal with an introduction reduction using the definition of less power]
(reduce) *!
Doing introduction reduction

Goal G3: $A \preccurlyeq \rho A$ and not $\rho A \preccurlyeq A$

[Reduce the current goal with a conjunction reduction]
(reduce) *!
Doing conjunction reduction

In the fourth frame the student begins to exert control by using the sufficiency condition given in theorem 4.2.1, which reads $(\forall A,B)(A \preccurlyeq B$ iff $(\exists F)$ F: A inj B), rather than expanding the definition of leq power.

Goal G4: $A \preccurlyeq \rho A$

[Reduce the current goal with an introduction reduction using the definition of leq power]
(reduce) *$reduce
Which proof procedure? (introduction) *theorem <Name> *4.2.1

The next goal shown will be labeled G6 since G5 was created at the same time as G4, when the conjunction G3 was reduced. In this frame, the student does a short proof using forward-chaining commands. Since the set-theory course uses a sorted language, it is necessary to prove in line 2 (see below) that the indicated term is actually a function before it can be existentially generalized in line 3. The command VERIFY invokes a resolution theorem prover used as a black box by the student. TM:1:1 denotes the first term on line 1, and is used to simplify typing. EXCHECK, the set-theory program, does partial recognition of input strings, so "injection" indicates that the student typed "inj$", the program extended that to "injecti" (which is ambiguous between injection and injective), and the student then typed "o$" to resolve the ambiguity. Remember that $ indicates the ESC key. Typing ? instead of "o$" would have displayed the ambiguity. In the command for line 3, "occurrences (1-1)" is a mildly awkward way of indicating all occurrences, in this case the first through the first.

```
Goal G6:      (∃ F) F: A inj ρA

[Use establish to infer the current goal]
 (establish) *theorem <Name> *4.1.10
    (∀ A) {z: (∃ x)(z = <x,{x}> & x ε A)}: A inj ρA

Do you want to specify? (yes) *$yes
Substitute for A? (A) *$A
theorem 4.1.10
        (1)  {z: (∃ x)(z = <x,{x}> & x ε A)}: A inj ρA
[Use establish to infer the current goal from line 1]
 (establish) *lverify  (2) *func(tm:1:1)
Using *definition <Name> *injection
Using *definition <Name> *map
Using *ok

[Use establish to infer the current goal from line 1, line 2]
 (establish) *leg
Replace term  *tm:1:1 Variable *F
F and {z: (∃ x)(z = <x,{x}> & x ε A)  aren't in the proper sort relation.
Line justifying a sort for {z: (∃ x)(z = <x,{x}> & x ε A)} *2
Occurrences  (1-1) *$1-1
1 eg    (3)  (∃ F) F: A inj ρA
Goal 6 fulfilled by line 3.
3 implies using theorem 4.2.1
        (4)  A ≺ ρA
Goal 4 fulfilled by line 4.
```

In contrast, here is a proof not at the interactive stage but at the finished stage when the student asks for a printout of the proof that he has created. The proof is for the theorem that if a set is finite then another set that is a subset of it must also be finite. The proof is so transparent that it requires little comment. Each of the steps were not taken by the student, but were taken automatically by the program, for example, the elementary conditional proof inference in line (16) and the routine application of universal generalization in line (17) to bind free variables. The abbreviations on the left for rules used are fairly self-explanatory.

Derive:

$(\forall A,B)[\text{finite}(A)\ \&\ B \subseteq A \text{ -> } \text{finite}(B)]$

assume (1) Finite(A) and $B \subseteq A$

assume (2) $B1 \neq 0$ and $B1 \subseteq \rho B$

definition finite

(3) Finite(A) iff $(\forall B)(B \neq 0\ \&\ B \subseteq \rho A \text{ -> } (\exists C)\ C \text{ min-elt } B)$

1 simp (4) Finite(A)

4 implies using 3

(5) $(\forall B)(B \neq 0\ \&\ B \subseteq \rho A \text{ -> } (\exists C)\ C \text{ min-elt } B)$

5 us (6) If $B1 \neq 0\ \&\ B1 \subseteq \rho A$ then $(\exists C)$ C min-elt B1

1 simp (7) $B \subseteq A$

7 theorem using theorem 2.12.8

(8) $\rho B \subseteq \rho A$

2 simp (9) $B1 \subseteq \rho B$

9, 8 theorem using theorem 2.4.2

(10) $B1 \subseteq \rho A$

2 simp (11) $B1 \neq 0$

11, 10 implies using 6

(12) $(\exists C)$ C min-elt B1

2, 12 cp

(13) If $B1 \neq 0\ \&\ B1 \subseteq \rho B$ then $(\exists C)$ C min-elt B1

13 ug (14) $(\forall B1)(B1 \neq 0\ \&\ B1 \subseteq \rho B \text{ -> } (\exists C)\ C \text{ min-elt } B1)$

14 introduction using definite finite

(15) Finite(B)

1, 15 cp

(16) If finite(A) & $B \subseteq A$ then finite(B)

16 ug (17) $(\forall B,A)[\text{finite}(A)\ \&\ B \subseteq A \text{ -> } \text{finite}(B)]$

QED

There is a great deal more I could say about the use of the interactive theorem prover in the set-theory course, but since this is meant to be a talk about not what has been done but what should be done in the future I will say no more about it.

2. Desirable Features for the User

In discussing the desirable features of the next generation of interactive theorem provers, it is natural to break the analysis into two parts. The most important is from the standpoint of the user but for reasons that I shall try to bring out it is almost equally important that the desirable features for authors creating courses be given serious and thoughtful consideration.

Flexibility--ease of interaction. First, above all, in the list of features is easy and flexible use of the theorem-proving machinery. There can be in the construction of only a moderately difficult theorem for a student what corresponds to 25 or 30 pages of interaction as long as it is the kind of interaction that is easy for the student. There is, of course, more than one criterion of ease. If the student has to go through an awkward path to construct a proof because of the severe limitations on the theorem prover, then it does not have the proper sense of flexibility. There is a general problem of human engineering of the proper interface between the student and the interactive theorem prover that is too easy to forget about. I suppose the point I would stress the most is that the interface must be such that it can be used by someone with no programming experience of any kind. This is one criterion we have had to meet as a strict test in having our interactive theorem provers be used by large numbers of students. No programming requirements are placed on the students and they come to the course with, in many cases, no prior background in programming. We emphasize that they will learn nothing about programming or about computers in the courses. These are courses teaching them a given subject matter. Computers are being used just as they are used in banks or in factories. Students are not in these courses to learn about computers or to gain any programming experience. The interface has got to be thought of in this fashion, it seems to me, in order for us to have a successful next generation of theorem provers. If we put in more power and generality we must be careful that this power and generality do not impose strains on the use of the system by relatively naive users.

Minimum input. The technical typing of mathematical formulas is an arcane and difficult art. It is something that we do not want to get in the way of students' giving proofs. This means that we want to think of the interactive theorem prover as offering as much as possible a control language to the students, not directly a language for writing mathematics. There is a tension here that will not go away and that will remain with us forever, for there will be, on the one hand, the desire to make the student input as natural as possible in terms of ordinary mathematical practice of writing informal proofs, and, on the other hand, even informal proofs require, in subjects with any development, fairly elaborate mathematical formulas that are painful and unpleasant to type. We will therefore have a tension between a relatively more arcane control and the use of mathematical

English in the giving of proofs. At the stage of development we should see in the next generation I think we should continue to concentrate, as we have in the set-theory course, on minimum input and the conception that we are offering to a student a control language rather than an informal mathematical language as his main vehicle for expressing the proof he wants to give.

Power. The theorem-proving machinery in the set-theory course is not as powerful as one would like, and natural inferences cannot be made directly and easily. The criterion here is the inferential leaps that are naturally and easily made by teachers and students in giving proofs. Now the leaps and jumps that will be made by different students and different instructors will vary widely, but I think there is a common understanding of when matters are too tedious and too much time is being spent on routine that should be swept under the rug. Power can be increased by having at the heart of matters more powerful resolution theorem provers, but I think that what I have to say under the inclusion of expert knowledge in the form of heuristic guidance and automatic "subject-matter" inference is probably as important an ingredient of power as direct computation. Do not misunderstand me. Increasing the power of the resolution theorem provers or related theorems that can be called by students is of importance and should by no means be neglected. The increasing cheapness of sheer computational power makes the prospects here rather bright.

Heuristic guidance. The incorporation of expert knowledge about a given subject matter and, in particular, a given course by specific heuristic guidance available to students in a variety of forms, especially under the form of goal structures and responses to calls for help is one of the most difficult, tedious, and time-consuming aspects of good interactive theorem provers. It is unfortunately an aspect that I see little prospect for being able to get right in a general way. Certainly it would seem for the next generation of theorem provers the best we can hope is to incorporate highly specific knowledge of a given subject matter, put together most likely by an experienced instructor in the subject. In fact, I guess I would express my skepticism that the kind of specific heuristic guidance and construction of goal structures required would ever become a matter of generalized routine. I do say something about the need for making it easier for authors to implement such guidance in the next section.

Graphics. As far as I know, there is no standard regular use of a theorem prover anywhere in the world that extensively and directly interacts with graphic displays related to that which is being proved. It is obvious that already at the level of high-school geometry a powerful use of graphics is called for. There is every reason to be hopeful about the kind of hardware that will be available to us. We are still, as far as I can see, a long way from having all the tools needed to create a really first-rate course even in high-school geometry. The uses in a variety of other courses should be obvious. I will say no more about this but take it as understood that the extensive interaction with graphic displays should be a high-priority feature.

3. Desirable Features for Authors

It is too easy to concentrate on the kind of end product that should be available to users. It is obvious to me when I look back on the agony of effort that has gone into creating the set-theory course that if we are to have the kind of widespread use of interactive theorem provers that can be extremely useful in meeting teaching needs in mathematics and science, we must also worry about helping the authors who will actually create the courses using the tools I am calling for. Let me mention here four desirable features.

Nonprogramming environment. The first and most essential requirement is that a sufficiently rich author language be built up that authors can create a new course without having to do any programming, or, ideally, even having the assistance available of a programming staff. We are a very great distance from achieving this objective at the present time, but I see no reason in principle why it is not even a feasible objective for the next generation of interactive theorem provers. I cannot stress too much its importance if we are to see widespread use of theorem provers in both high school and college instruction in mathematics and science.

Easy to use author language. It would be possible to create a nonprogramming environment but one that is so awkward and tedious to use that only authors of the hardiest nature would be willing to tangle with it. It is important that the author language that is created be one that authors like and feel at home with. There is a great deal to be said on this subject. I would just emphasize that once again as much as possible we would like to minimize input on the part of authors. We would like to give them as much as possible a control language for creating a course. As far as I can tell we are very far from having such facilities anywhere in the world at the present time.

Flexible course structure. It is also an important requirement that authors have available a clearly formulated and flexible course structure that they can use without programming assistance. The course structures in the elementary logic and set-theory courses with which I have been associated closely myself are not, I think, impressive at all. We did not concentrate on what I generally call the course driver in these cases but more on the theorem-proving apparatus. But good courses using interactive theorem provers need to have the possibility for instructors to not be locked into a single course structure but to satisfy their own particular teaching plans and to fit the course into the curriculum of their particular institution. Again, it is easy to underestimate the importance of this kind of flexibility in terms of making the use of interactive theorem provers a success.

Easy ways to add expert knowledge. Above all, we need to make it easy for instructors without programming experience or programming assistance to add expert knowledge to give the course the full-bodied character it should have. I do not want to underestimate either my ambitions in this area or the difficulties. What is probably most important is not to think in terms of encoding a fixed body of knowledge but to build dynamic procedures that interact with what the student is doing in powerful ways to give pertinent and cogent guidance to the student. Let me give the simplest kind of example, but also one of the most important, that arises in any use of interactive theorem provers. The student begins to construct a proof. He is, let us say, a certain distance into the proof and although he had a reasonable idea to begin with he is now at a loss as to how to continue. He asks for help. A dumb expert system will make him start over and give advice in terms of some preset ideas of what a proof of a given theorem should look like. A smart system of expert knowledge will work in a very different way. It will look at the proof as developed thus far by the student and be able, if he has a reasonable idea, to give him help on continuing and completing the proof he has already begun. Now we all know it is easy to say this but either as instructors ourselves or as ones creating such expert systems, it is no mean feat to come up with such a system. We have had in the last decade a great deal of discussion of such expert systems of help in such trivial subjects as elementary arithmetic. I have myself devoted some time to these matters and so when I say 'trivial' I do not mean to denigrate the work that has been done but just to put it in proper perspective. The kind of work associated with BUGGY created by John Seely Brown and others simply has no obvious and easy extension to a subject at the level, let us say, even of the first course in axiomatic set theory, not to speak of more advanced subjects. The difficulties of creating really good systems of expert knowledge of the kind I am calling for cannot be underestimated. It is, I think, in many respects almost the first item on the agenda for the next generation of interactive theorem provers.

4. Next Round of Courses

Let me just conclude by listing some of the courses that I think are just right in difficulty for attack by the next generation of interactive theorem provers. None of the courses reaches above the undergraduate level. I think it is going to take one more generation beyond the next one before we can have interactive theorem provers that can be seriously used in graduate courses of instruction in mathematics or science. To emphasize the generality of the framework that needs to be created, let me briefly describe seven standard courses that would be of considerable significance to have available in a computer-based framework and with good interactive theorem-proving facilities. The first three courses could as well be offered to able high school students as to college undergraduates.

Elementary geometry. This course is in fact a high school course. It is a bit of a scandal that we do not yet have a production version of a good elementary geometry course with a good interactive theorem prover available anywhere in the country as far as I know. There are some formidable problems to be solved in creating the theorem-proving facilities required for such a course, especially in having the proper interaction between proofs and graphs of the figures being constructed, but the problems are not of great fundamental difficulty. A standard criticism of many axiomatic theorem-proving elementary geometry courses is that there is too much emphasis on theorems whose geometrical content is limited. I think that a computer-based course can avoid this problem in the way that we have avoided it in the set-theory course. Students could be given individual lists of theorems and they could be led to expect to have to use previous theorems that they themselves have not proved in giving their own proofs. In this way it would be possible carefully to select theorems of geometrical interest that are still sufficiently elementary to let students try them and to deal with the problem of giving adequate proofs. I have some slightly idiosyncratic ideas about this course that I shall not go into here. I think there is a place for a quantifier-free formulation of elementary geometry that has a highly constructive formulation and that could be a basis of a course that would avoid some of the logical intricacies inherent in the quantified formulas that are so much a part of a standard geometry course.

Linear algebra. As has been emphasized by many mathematicians in the past several decades, an elementary course in linear algebra might well replace the elementary geometry course in high school. In any case, a course in linear algebra is now standard fare in every undergraduate mathematics curriculum. There are many nice things about linear algebra from the standpoint of being a computer-based course. Much of the course, for example, requires only a small number of types of variable. One of the standard bookkeeping problems in computer-based courses is that of embodying in a natural and easy way the standard informal use of typed variables. Moreover, most of the proofs in linear algebra are relatively easy and rather computational in spirit. Of the courses I mention in this list I think it would be the easiest to implement. As we all know, it is possible to continue development of the course so that it becomes relatively difficult, but even then I do not see the proofs as being as difficult as the harder proofs in the first course in axiomatic set theory described above.

Differential and integral calculus. The undergraduate teaching staffs are so familiar with this course and it is such a standard service offering, it might be wondered why it should be considered for development as a computer-based course with an interactive theorem prover. I think the main argument for this is that there is a definite place for it in the more than 20,000 high schools in this country. For many of these high schools it is really not economical to staff a small course in the calculus for the small number of students interested in taking it. From a broad national standpoint, however, it is important that such courses be offered to the willing and able students. We know from a great deal of experience that very bright sixteen- and seventeen-year-olds, for example, can do just as well in such a course as students who are a couple of years older. I can see offering an excellent course with theorem-proving facilities but also offering some additional graphical and symbolic facilities as desired of the kind that have been developed in recent years. In fact, one of the problems of the more powerful facilities for elementary integration, for example, is that of knowing exactly what should be available to the student at a given point in the course. It would also be possible to offer such a course in the calculus with a new viewpoint, for example, the viewpoint of nonstandard analysis. That would be a difficult decision because many of the schools at which a computer-based course would be directed would not have instructors who would feel at home with a nonstandard approach to the calculus. In any case, the desirability of such a course seems clear.

Differential equations. Again, this course might just as well be one that would be offered to the very best students in some of the high schools. I must confess that I know of no one who has yet tackled even the first course in differential equations as a computer-based course using an interactive theorem prover. It seems to me, however, that there is nothing that stands in the way of such a course. It is true that many instructors would, and so I would myself, emphasize concepts and applications perhaps more than proofs in the first course, but there is no reason that a computer-based framework could not offer a good approach to these matters as well. Also, here again is a case where the use of sophisticated graphics could be highly desirable.

Introduction to analysis. In this list I am developing, by now the student will be ready for a first course in analysis. Here a theorem prover would really get a proper workout but again I find that the proofs in most books that are billed as a first course in analysis are not especially more difficult or complex than the proofs in set theory. Also in such proofs there is a fairly restricted typing of variables. By careful and judicious arrangements of theorems I see no difficulties in principle, just the difficulties of actually working out all the details in a way that will give a smooth-running course with the kinds of facilities that could be offered in the near future.

Introduction to probability. The deductive organization of the first course in probability that assumes a background in the calculus is a natural subject to put within a computer-based framework. Also, it is a course that some faculties are not particularly interested in teaching. It could also be made available again for the very best students in high school. Most of the introductory courses in probability at the level I am talking about do not require very elaborate proof procedures. The machinery, I think, would not be too difficult to implement.

Theory of automata. An undergraduate course in the theory of automata is again at the right level of difficulty. There are some problems in this course. The notation is harder than the proofs themselves. The structures being considered are complicated but most of the proofs are not of a comparable difficulty. At the present time this is something of a problem in computer-based frameworks but it is something that I am sure we will see solved in a reasonable and intuitive fashion in the near future. I mention this only as one theoretical course in computer science. It is clear that there are other undergraduate courses in computer science that will, on occasion, be thinly staffed in many colleges and universities. The opportunities for computer-based courses using sophisticated interactive theorem provers in this domain are perhaps among the best of any of the areas I have touched on.

In closing I want to return to my point that what we need is a general facility for creating such courses. The logic and set-theory courses with which I have been closely associated myself have an inevitably parochial character in their organization and conception. This is because they were created from scratch and were focused on solving the immediate problems at hand for the subject at hand and not on creating a general framework usable by many different people for many different courses. At the time we were creating these courses it would have been premature to aim at such generality. It is not now. It is what we need in the near future in order to fulfill the promise of the role that theorem provers should be playing in the teaching of mathematics and science.

REFERENCES

McDonald, J., and Suppes, P. Student use of an interactive theorem prover. In W. W. Bledsoe and D. W. Loveland (Eds.), _Automated theorem proving: After 25 years_. Providence, R.I.: American Mathematical Society, 1984.

Suppes, P. _Introduction to logic_. New York: Van Nostrand, 1957.

Suppes, P. _Axiomatic set theory_. New York: Van Nostrand, 1960. Slightly revised edition published by Dover, New York, 1972.

Suppes, P. Computer-assisted instruction at Stanford. In _Man and computer_. Basel: Karger, 1972.

Suppes, P. (Ed.), _University-level computer-assisted instruction at Stanford: 1968-1980_. Stanford, Calif.: Stanford University, Institute for Mathematical Studies in the Social Sciences, 1981.

Suppes, P., and Binford, F. Experimental teaching of mathematical logic in the elementary school. _The Arithmetic Teacher_, 1965, _12_, 187-195.

The Linked Inference Principle, II: The User's Viewpoint*

L. Wos
Mathematics and Computer Science Division
Argonne National Laboratory
9700 South Cass Avenue
Argonne, IL 60439

R. Veroff
Department of Computer Science
University of New Mexico
Albuquerque, NM 87131

B. Smith
Mathematics and Computer Science Division
Argonne National Laboratory
9700 South Cass Avenue
Argonne, IL 60439

and

W. McCune
Department of Electrical Engineering and Computer Science
Northwestern University
Evanston, IL 60201

1. Introduction

In the field of automated reasoning, the search continues for useful representations of information, for powerful inference rules, for effective canonicalization procedures, and for intelligent strategies. The practical objective of this search is, of course, to produce ever more powerful automated reasoning programs. In this paper, we show how the power of such programs can be sharply increased by employing inference rules called linked inference rules. In particular, we focus on linked UR-resolution, a generalization of standard UR-resolution [2], and discuss ongoing experiments that permit comparison of the two inference rules. The intention is to present the results of those experiments at the Seventh Conference on Automated Deduction. Much of the treatment of linked inference rules given in this paper is from the user's viewpoint, with certain abstract considerations reserved for Section 3.

*This work was supported in part by the Applied Mathematical Sciences Research Program (KC-04-02) of the Office of Energy Research of the U.S. Department of Energy under Contract W-31-109-Eng-38 (Argonne National Laboratory, Argonne, IL 60439).

Employment of linked inference rules enables an automated reasoning program to draw conclusions in one step that typically require many steps when standard (unlinked) inference rules are used. Each of the linked inference rules is obtained by applying the "linked inference principle", a principle of reasoning that is fully developed in [8]. Application of the linked inference principle yields generalizations of a number of well-known inference rules--for example, binary resolution, UR-resolution, hyperresolution, and paramodulation. The larger steps permitted by the use of various linked inference rules are made possible by substituting semantic criteria for syntactic criteria. In particular, the usual clause boundaries that define certain well-known inference rules are replaced by criteria defined in terms of the meaning of the predicates and functions being employed. For example, consider the case in which the natural representation of a problem produces two nonunit clauses, each containing negative literals. Further, assume that some reasoning step that you would like a reasoning program to take requires the simultaneous consideration of the two nonunit clauses. Neither standard hyperresolution nor standard UR-resolution suffice on syntactic grounds. However, linked UR-resolution would not be so restricted for the criteria that are employed governing application of the rule are semantic and not syntactic.

Although the discussion focuses chiefly on the inferential process, we shall touch on some consequences and benefits of a strategic nature that are present when employing linked inference. (We often discuss various strategic and inferential approaches as if they are integrally connected when the connection is in fact primarily historical. The study of linked inference has led to the discovery of strategies [8] that can be used outside of linking, strategies such as the target strategy and the extension of the set of support strategy [7], for example.) Notwithstanding our emphasis on the user, we shall briefly discuss the abstract notion of linking in order to show how it unifies a number of concepts. In addition to furthering the objective of producing more effective and more powerful reasoning programs, linking contributes to two other objectives. First, the user is provided greater control over the performance of an automated reasoning program when attacking some specific problem. In particular, the user has control over the size of the steps that occur in a deduction, and also can instruct an automated reasoning program to restrict its deductions to those directly relevant to some chosen concept or goal. Second, the user is permitted more freedom in the use of a natural presentation of a specific problem. The choice of notation is not dictated by the syntactic flavor of the clauses to be deduced. In contrast, because of such syntactic considerations, many (unlinked) inference rules place limitations on the choice of notation.

The problems we solve with linked inference are taken from the world of puzzles, circuit design, and program verification. By selecting from the first of these three areas, we can immediately give a sample of how linked inference works. We choose two fragments of a puzzle called the "jobs puzzle". The jobs puzzle concerns four people who, among them, hold eight jobs. Each of the four people--Roberta, Thelma, Steve, and Pete--holds exactly two jobs. Each of the eight jobs--actor, boxer, chef, guard, nurse, police officer, teacher, and telephone operator--is held by one person. You are asked to determine which jobs are held by which people. Among the clues, you are told that Roberta is not the chef, and that the husband of the chef is the telephone operator.

Perhaps you have jumped to the conclusion that Thelma is the chef. To see if you are right, the following clauses can be used to represent this puzzle fragment, and UR-resolution can be used to draw conclusions.

Roberta is not the chef.

(1) -HASAJOB(Roberta,chef)

If a job is held by a female, then the female is Roberta or Thelma.

(2) -FEMALE(jobholder(y)) HASAJOB(Roberta,y) HASAJOB(Thelma,y)

If a person is a wife, then that person is female.

(3) -HUSBAND(x,y) FEMALE(y)

The person who holds the job of telephone operator is the husband of the person who holds the job of chef.

(4) -HASAJOB(x,telop) HUSBAND(x,jobholder(chef))

(A second clause

-HUSBAND(x,jobholder(chef)) HASAJOB(x,telop)

is needed to complete the representation of this fact, but it does not participate in the illustration.)

For every job, there is a person holding that job.

(5) HASAJOB(jobholder(y),y)

UR-resolution suffices to yield the desired conclusion in three steps. From 5 as satellite and 4 as nucleus,

(6) HUSBAND(jobholder(telop),jobholder(chef))

is obtained. From 6 as satellite and 3 as nucleus,

(7) FEMALE(jobholder(chef))

is obtained. And finally, from 7 and 1 as satellites with 2 as nucleus,

(8) HASAJOB(Thelma,chef)

is deduced as the desired result.

A person solving this puzzle fragment would simply and naturally conclude clause 8 by (simultaneously) considering the information contained in clauses 1 through 5. The information contained in clauses 6 and 7 would not exist, at least explicitly. One variant of linked UR-resolution would also immediately deduce clause 8 by simultaneously considering clauses 1 through 5 without deducing clauses 6 and 7. In particular, in the terminology to be illustrated in this paper, the user could instruct a reasoning program to choose clause 1 as the "initiating satellite" and the predicate HASAJOB as the "target". With those choices, clause 2 is the "nucleus", and clauses 3 and 4 are "linking clauses" that "link" clause 2 to the "satellite", clause 5.

In the example just discussed, the choice of the "target" is motivated by the natural interest in who holds which jobs. The intent is to produce a unit clause as output, since the goal is to establish a (simple) fact rather than a choice of possibilities. The conclusion, HASAJOB(Thelma,chef), is a descendant of a literal in clause 2. Thus, in this variation of linked UR-resolution, the nucleus is the "target clause". In the following variation, however, the nucleus is not the target clause, for the conclusion is not a descendant of a literal in the nucleus.

For this variant, we select another fragment from the "jobs puzzle". Another clue in the puzzle is that the job of nurse is held by a male. From this clue, you quickly deduce that Roberta is not the nurse. (Incidentally, this fragment is the example that led to the first application of the principle of linked inference and, in fact, to the first variant of linked UR-resolution.) The deduction can be obtained with the following clauses.

(9) FEMALE(Roberta)

(10) -FEMALE(x) -MALE(x)

(11) -HASAJOB(x,nurse) MALE(x)

Here again the user chooses the natural target and the inference rule of linked UR-resolution, the two choices motivated as discussed earlier. Clause 9 is the initiating satellite, clause 10 the nucleus, and clause 11 the target clause. The literal MALE(x) of clause 11 links to the literal -MALE(x) of clause 10. The other literal of clause 11, -HASAJOB(x,nurse), is the target literal, and is the parent of the deduced clause (as a literal)

(12) -HASAJOB(Roberta,nurse)

and we have an illustration of a variation on the preceding example of linked UR-resolution.

The two examples illustrate two variants of linked UR-resolution. They show how the user can rely on a natural representation for the problem, without regard to syntactic tricks required by the wish for using a particular inference rule. They show how the step size need not be limited by the obvious clause boundaries that occur when using this natural formulation of the puzzle. Finally, they suggest the possible increase in program control and possible decrease in user effort that can occur when employing linked inference. The increase in control is derived in part from avoiding the generation of certain classes of intermediate clauses and in part from keying on semantically chosen targets. The decrease in effort is derived in part from the ability to rely on a natural representation without being so concerned with the need to generate (intermediate) clauses that are required to be unit clauses.

2. Overview

In this section, we briefly review certain material covered in [8]. In particular, we discuss the need for linking, some of its properties, and the motivation for its existence. At the most general level, the motivating forces are the desire to rely on semantic considerations in place of syntactic, the intention of increasing the power and efficiency of automated reasoning programs, and the desire to provide the user with more control over the actions taken by a reasoning program while simultaneously placing less burden on the user. At the more specific level, linking addresses a number of problems commonly faced when using a reasoning program. Of course the problems are extremely interconnected, but let us discuss them as if they are somewhat separate.

The first problem focuses on the size and nature of the steps that ordinarily occur in a deduction. Because many inference rules are constrained and defined in terms of syntactic criteria alone, and because clause boundaries currently prohibit certain combinations of facts from being simultaneously considered, the steps taken by a reasoning program are often smaller than necessary or desired. The termination of a deduction step is often given in terms of syntactic criteria such as the signs of the literals and the number of the literals. Users of an automated reasoning program might well prefer the termination condition to rely on the significance and the meaning of the conclusion. In standard hyperresolution, for example, a reasoning program might be forced to accept a conclusion containing two positive literals, while linked hyperresolution might produce a conclusion with but one positive literal, the second literal being removed with a negative unit clause.

The second problem concerns representation and its interconnection with the choice of inference rule(s) and strategy. Too often the user wishes to use one choice of predicates and functions dictated by a desire for convenience, readability, and naturalness, and finds that, for example, binary resolution must then be included as an inference rule. If binary resolution is to be avoided--after all, typically its use results in too many clauses of too small a step size--then more clauses must be added to the set of support. Thus, in these situations, the user is forced to choose between using an inference rule that may be too prolific and weakening the power of the chosen strategy.

The third problem addresses user control of the actions taken by an automated reasoning program. In many cases, the user does not wish to be forced to read through and examine a myriad of conclusions, but instead wishes to be presented only with important conclusions. The intent of using linked inference is to provide each user with a means for telling a reasoning program which concepts are interesting, in turn permitting the program to present only conclusions consistent with the given instruction. By judicious choices, the program can be prohibited from generating intermediate clauses by in effect classifying them as relatively insignificant. Many such clauses are merely links between one significant statement and another significant one. Depending of course on the price paid (measured in terms of time) for achieving this reduced clause space, a sharp increase in efficiency results.

The fourth and final problem is that of extending the power and flavor of the set of support strategy. The strategy, as currently defined and employed, partitions the input clauses into those with support and those without. The strategy prohibits application of an inference rule to a set of clauses when none of its members has support. By doing so, many clauses that would have been generated are not generated. This action is far more efficient than actions that generate clauses and then purge them subject to various criteria. However, as pointed out by Luckham [1], this partitioning of the input clauses is not recursively present--is not present at higher levels of the clause space. Of course, levels 2, 3, 4, and so on are smaller than they would have been because of the clauses not present at level 1 when using the strategy. But the recursive power, were it present, would further partition the retained level 1 clauses into two sets, one with support and one without. With this action, the level 2 clause space would be smaller than with the standard definition of the set of support strategy. The problem thus is to provide a means for partitioning each level of retained clauses--as currently occurs for level 0--and, as a result, to continually reduce the size of levels greater than 1 comparable to the way that level 1 is reduced. Of course, the object is to extend the set of support strategy in this fashion without (operationally) losing refutation completeness. (Questions of refutation completeness are not addressed in this paper.)

Application of the linked inference principle to produce various linked inference rules addresses these various problems. Intuitively, linked inference rules enable an automated reasoning program to "link" together as many clauses as are required to skip the less significant intermediate results. The object is to draw a conclusion, or term the deduction step complete, only when a significant result has been obtained, where significance is defined by the user. Rather than totally abandoning syntactic notions such as sign of literals and number of literals and clause boundaries, linked inference rules permit the user to combine such criteria with certain semantic notions. Thus, for example, linked UR-resolution requires that the conclusion be a unit clause but, when the appropriate strategy is employed, broadens the definition of standard UR-resolution to require that the unit clause satisfy additional specified criteria. This extension enables a reasoning program to avoid terminating the deduction step merely because of having produced some unit clause. The extension also allows the use of clauses that would normally not be considered satellites. Linked UR-resolution is to standard UR-resolution as standard UR-resolution is to unit resolution.

3. The Linked Inference Principle

Note that, when referring to a clause, here we mean an occurrence of a clause. Thus the mention of two clauses admits the possibility that the clauses are identical--are merely copies of the same clause. Similarly, the reference to a literal means an occurrence of a literal. Thus the mention of two literals does not imply that the two literals are distinct, but rather that they are different occurrences of (possibly the same) literals. Although we could define the linked inference principle to cover literal merging and to cover factoring [6], such a definition interferes slightly with an understanding of the principle and is not consistent with the current implementation of the corresponding inference rules. Employment of the principle, and hence of various inference rules derived from it, requires using factoring as an additional inference rule. Finally, we choose here to define the principle at the literal level, presenting its extension to the term level in another paper. Thus, equality-oriented inference rules such as paramodulation are not covered in this treatment.

3.1 Definition

The linked inference principle is a principle of reasoning that considers a finite set S of two or more (not necessarily distinct) clauses with the objective of deducing a single clause that follows logically from the clauses in S. The principle applies if there exists an appropriate function f and an appropriate unifier u that depends on f. (The formal discussion of appropriate functions and unifiers is found in [8].) Such a function--that required for the linked inference rule to apply--pairs literals in the same sense that certain standard inference rules do. For example, standard binary resolution considers two clauses and pairs a literal in one with a literal of opposite sign in the other with the intent of the two literals unifying. As a different example, standard UR-resolution considers a nucleus and a set of satellites, and pairs all but one of the literals in the nucleus with the satellites. Again the intent is to find an appropriate unifier that depends on the pairings, and that is required to simultaneously unify the chosen pairs.

In the same spirit, linked UR-resolution considers a nucleus, a set of satellites, and various linking clauses. It pairs literals in the nucleus with satellites and possibly with literals from linking clauses with the intent of employing a unifier that simultaneously unifies the chosen pairs. The object is to "cancel", with one exception, all literals in all clauses--the exception being the linked UR-resolvent that results if the application is successful. With this perspective, standard UR-resolution can be viewed as an instance of linked UR-resolution. Just as standard UR-resolution and standard hyperresolution can be implemented as a sequence of the appropriate binary resolutions, so also can various linked inference rules. However, as expected, linked inference rules are not implemented in this fashion, but instead are implemented to avoid the generation of intermediate clauses.

3.2 Inference Rules

Before briefly discussing other concepts covered by the linked inference principle, we focus on inference rules. For example, binary resolution is captured by the linked inference principle, even when a clause is resolved with itself. In that case, S consists of two copies of the same clause, and each of the literals not involved in the unification is mapped to itself. Factoring, on the other hand, is not captured by the principle as presented here. Hyperresolution, negative hyperresolution, and UR-resolution are also captured. Equally, linked UR-resolution and linked hyperresolution are captured by the linked inference principle.

We can now give the following formal definition of linked UR-resolution.

Definition. Linked UR-resolution is that inference rule that requires the selection of a unit clause, called the initiating satellite, and a nonunit clause, called the nucleus, such that the literal of the initiating satellite unifies with a literal of opposite sign in the nucleus. In addition, with at most one exception, for each of the remaining literals of the nucleus, there must exist a (unit or nonunit) clause containing a literal of opposite sign that unifies with that literal. The unit clauses are called immediate satellites or satellites of ancestor-depth 1, and the nonunit clauses are called links of ancestor-depth 1. Further, for each of the literals in the ancestor-depth 1 links that are not paired with a literal in the nucleus, with at most one exception, there must exist a satellite of ancestor-depth 2 or a link of ancestor-depth 2 that provides a literal of opposite sign that unifies with that literal. ... There must be an n such that no satellites or links of ancestor-depth greater than n participate. Next, the number of so-called exception literals in the set consisting of the nucleus and the links must be exactly one. Finally, there must exist a unifier that simultaneously unifies pairwise all pairs designated by the given conditions. The linked UR-resolvent is obtained by applying the unifier to the unpaired literal.

In the unification requirement, the reason for allowing the possibility of no exception literals in the nucleus but instead allowing an exception in one of the links is that the deduced unit clause may be descended from a literal contained in one of the links. Accidental deduction of a unit clause, as can occur with merging, is not permitted. As in standard UR-resolution, we are interested in a definition that operationally predicts, if all conditions are satisfied, that a unit clause will be deduced. Allowing in the definition the unit clause that is accidentally produced by merging leads to an implementation of linked UR-resolution, as well as of standard UR-resolution, that is less effective. The broader definition would force exploration of many paths that in fact would not produce a unit clause.

The definitions, from the abstract viewpoint of the linked inference principle or from the user viewpoint, of linked hyperresolution can be obtained by focusing on the objective of deducing a positive clause rather than a unit clause. The corresponding definitions for linked negative hyperresolution and for linked binary resolution can be obtained by focusing on the corresponding objectives. The linked inference principle is thus seen to capture a number of inference rules.

3.3 Other Applications of the Principle

The linked inference principle also captures other concepts in addition to capturing various inference rules. For example, it captures, with one exception, all classes of proof by contradiction that are signaled by the deduction of the empty clause. The exception is illustrated with the two clauses

P(x) P(y)
-P(x) -P(y)

which, taken together, are an unsatisfiable set. The proof of unsatisfiability requires the use of factoring. The linked inference principle as given here can be extended to capture this type of proof as well by replacing the requirement of the one-to-one property of the required function f. In the extension, rather than simply considering pairs of literals l,f(l), the function is allowed to admit paired sets T,f(T). All literals in any such T must be from a single clause C in S, all literals in f(T) must be from a single clause D, and of course the literals of T must simultaneously unify, so must the literals of f(T), and the two resulting literals must unify and be of opposite sign. This extension is similar to the first definition published for binary resolution [3].

The linked inference principle also captures, with one exception, the notion of the deduction of a clause D from a given set S of clauses. As expected, again the exception is any deduction that requires factoring.

Finally, the linked inference principle can be viewed as capturing the deductive aspects of Prolog [4]. The procedural aspects can be captured by an appropriate strategy. The execution of a Prolog program can be achieved with a single linked hyperresolution.

The point of noting the various concepts captured under the linked inference principle is merely to observe that the principle provides a unifying framework for a number of rather distinct concepts. No claim is being made that, for example, when seeking a proof by contradiction, the user should instruct a reasoning program to search for the appropriate single linked inference. In fact, unless the user has a well-tailored algorithm for solving the problem under consideration, searching for such one-step proofs by contradiction is essentially a waste of time.

4. Experiments

We cannot overstate the importance of proper experimentation in evaluating new ideas. Because our implementation of the linked inference rules is still in a very early stage, we have not yet been able to do the extensive testing that is required to properly assess the value of the concept. In this section we briefly summarize the few experiments that we are running, and include other examples, obtained by hand, to further illustrate the potential of linking.

The problems are selected from three areas: solving puzzles, designing circuits, and proving properties of programs. The experiments focus on comparisons between standard and linked UR-resolution. Because of the obvious need for brevity, we include here only sketches of solutions to problems. A more detailed discussion of the problems is found in [8].

4.1 Solving Puzzles

The first experiment focuses on the "jobs puzzle", a fragment of which was described in the introduction. There are four people: Roberta, Thelma, Steve, and Pete. Among them, they hold eight different jobs. Each holds exactly two jobs. The jobs are: actor, boxer, chef, guard, nurse, police officer (gender not implied), teacher, and telephone operator. The job of nurse is held by a male. The husband of the chef is the telephone operator. Roberta is not a boxer. Pete has no education past the ninth grade. Roberta, the chef, and the police officer went golfing together. Who holds which jobs?

The second experiment concerns the "fruit puzzle". A merchant wishes to sell you some fruit. He places three boxes of it on a table. Each box contains only one kind of fruit: apples, bananas, or oranges. Each box contains a different type of fruit. Each box is mislabeled. Box a is labeled apples, box b oranges, and box c bananas. Box b contains apples. What do the other boxes contain?

4.2 Designing Circuits

Circuit design is an application of automated reasoning that has generated much interest [5]. The basic approach is to describe with axioms the available components and the way they interact, and then deny that a circuit with the desired properties can be constructed. Examination of a proof of the corresponding theorem contains all of the information necessary to specify the design of the desired circuit. The experiments are with multiple-valued logic circuits employing T-gates.

The following clause defines a T-gate.

```
(1)  -CKT(x) -CKT(y) -CKT(z) CKT(tgate(x,y,z,w))
```

The following clauses define the circuit problem to be solved.

```
(2)  CKT(0)

(3)  CKT(1)

(4)  CKT(2)

(5)  -CKT(tgate(tgate(0,1,0,i1),tgate(1,2,1,i1),
                tgate(0,1,0,i1),i2))
```

With standard UR-resolution, the choice of clause 5 as set of support does not yield a proof that the desired circuit can be designed. In fact, no clauses are generated with that choice. And yet, clause 5 is an obvious choice, while the inclusion of any of clauses 2 through 4 is less justified. Thus, you must choose between modifying your choice of inference rule and modifying your choice of set of support. With linked UR-resolution, on the other hand, you are not required to modify the natural choice of having clause 5 as the only input clause in the set of support.

4.3 Proving Properties of Programs

Automated reasoning is applicable to a wide variety of formal verification problems. One area of particular interest is that of proving that a given computer program has certain properties it is claimed to have. For example, you might wish to prove that the program correctly implements a given algorithm.

Many of the programs that we study involve some use of integers, vectors, and matrices. The corresponding axiom sets have the property that very closely related facts are deducible from them when standard inference rules are employed. For example, if the reasoning program deduces that a is less than b, then it soon deduces that b is not less than a. The presence of such closely related facts can seriously impair a reasoning program's performance. Since employment of linked inference rules, by relying heavily on semantic criteria, enables a reasoning program to avoid drawing such closely related conclusions, efficiency is enhanced. Correspondingly, verification problems for computer programs with such axiom sets appear to be very amenable to attack with a linked inference rule.

We include here a portion of a proof (obtained by hand) that a given program correctly finds the maximum element of an array. The predicates, functions, and constants of this problem have the following interpretations.

LT(x,y)	-	x is less than y
IB(x,y)	-	index y is in bounds in array x
eval(x,y)	-	the value in position y of array x
s(x),pd(x)	-	successor and predecessor functions
cc,numl,cn	-	array cc is dimensioned from numl through cn
cj, cmax	-	constants representing program variables
a,b	-	other Skolem constants that come from the statement of the theorem

The relevant portion of the problem is the following.

Special hypothesis:

(1) LT(a,cj)
(2) -LT(cj,numl)
(3) -LT(cn,cj)
(4) -LT(s(cn),cj)
(5) -LT(a,numl)
(6) -IB(cc,a) EQUAL(cmax,eval(cc,a))
(7) LT(x1,numl) -LT(x1,cj) -IB(cc,x1) -LT(cmax,eval(cc,x1))
(8) -IB(cc,cj) -LT(cmax,eval(cc,cj))

Denial of the conclusion:

(9) -LT(b,numl) -IB(cc,cj) LT(s(cj),numl) LT(s(cn),s(cj))
LT(a,numl) -LT(a,s(cj)) -IB(cc,a)
-EQUAL(cmax,eval(cc,a))

(10) LT(b,s(cj)) -IB(cc,cj) LT(s(cj),numl) LT(s(cn),s(cj))
LT(a,numl) -LT(a,s(cj)) -IB(cc,a)
-EQUAL(cmax,eval(cc,a))

(11) LT(cmax,eval(cc,b)) -IB(cc,cj) LT(s(cj),numl)
LT(s(cn),s(cj)) LT(a,numl) -LT(a,s(cj)) -IB(cc,a)
-EQUAL(cmax,eval(cc,a))

In addition to these clauses, there is a set of general axioms that gives properties of the integers, of ordering relations, of arrays, and other relevant information.

Note that there are seven literals that are common to each of the clauses of the denial, clauses 9, 10, and 11. A single application of linked UR-resolution to each of the three clauses removes the seven common literals and yields the following three unit clauses.

(12) -LT(b,numl)
(13) LT(b,s(cj))
(14) LT(cmax,eval(cc,b))

Before completing the proof, which requires two additional linked UR steps, we show how linking removes each of the seven literals. The process is described by associating each literal with the appropriate axioms and set of support clauses that are required to remove it from the nucleus. Simply for illustrative purposes, each of the seven literals is viewed as a unit clause that is used to deduce the empty clause.

Note the difference between linked UR-resolution and standard UR-resolution in the treatment of this proof fragment. In standard UR, it would be necessary to deduce in separate steps an appropriate unit clause to remove each of the seven literals from clauses 9, 10, and 11. To see precisely what suffices--to derive the standard UR proof from the linked UR proof we are about to give--merely consider the linking clauses associated with each of the seven literals, but in reverse order.

```
a. -IB(cc,cj)

        LT(x1,num1) LT(cn,x1) IB(cc,x1)     (axiom)
        -LT(cj,num1)                        (clause 2)
        -LT(cn,cj)                          (3)

b. LT(s(cj),num1)

        -LT(x,y) -LT(y,z) LT(x,z)           (axiom)
        LT(x,s(x))                          (axiom)
        -LT(cj,num1)                        (2)

c. LT(s(cn),s(cj))

        -LT(x,y) -LT(pd(x),pd(y))           (axiom)
        EQUAL(pd(s(x)),x)                   (demodulator)
        -LT(cn,cj)                          (3)

d. LT(a,num1)

        -LT(a,num1)                         (5)

e. -LT(a,s(cj))

        -LT(x,y) -LT(y,z) LT(x,z)           (axiom)
        LT(x,s(x))                          (axiom)
        LT(a,cj)                            (1)

f. -IB(cc,a)

        LT(x1,num1) LT(cn,x1) IB(cc,x1)     (axiom)
        -LT(a,num1)                         (5)

     ---> LT(cn,a) (intermediate result added for clarity)

        LT(a,cj)                            (1)
        -LT(x,y) -LT(y,z) LT(x,z)           (axiom)
        -LT(cn,cj)                          (3)

g. -EQUAL(cmax,eval(cc,a))

        -IB(cc,a)  EQUAL(cmax,eval(cc,a))  (6)
        (see solution of literal "f" to complete)
```

Having shown how the seven literals can be removed with linked UR-resolution to deduce the unit clauses

```
(12)  -LT(b,num1)
(13)  LT(b,s(cj))
(14)  LT(cmax,eval(cc,b))
```

we show how the proof can be completed. With clause 12 as the initiating unit and clause 7

```
(7)  LT(x1,num1) -LT(x1,cj)  -IB(cc,x1)  -LT(cmax,eval(cc,x1))
```

as the nucleus, the intermediate clause

```
-LT(b,cj) -IB(cc,b) -LT(cmax,eval(cc,b))
```

can be deduced by binary resolution. The first literal can be transformed into a (different) unit clause, and simultaneously the second and third literals can be removed by linking. The output of this linked UR step is the following unit clause.

```
(15)  EQUAL(b,cj)
```

This application of linked UR can be described as we described the action of linking in removing the seven literals.

```
h.  -LT(b,cj)

        LT(x,y)  LT(y,x)  EQUAL(x,y)        (axiom)
        -LT(x1,s(x2))  -LT(x2,x1)           (axiom)
        LT(b,s(cj))                         (13)

    --->   EQUAL(b,cj)      (result of inference)

i.  -IB(cc,b)

        LT(x1,num1) LT(cn,x1) IB(cc,x1)     (axiom)
        -LT(b,num1)                         (clause 12)

     ---> LT(cn,b)  (intermediate result for clarity)

        -LT(x,y)  LT(z,y)  LT(x,z)          (axiom)
        -LT(cn,cj)                          (3)

     ---> LT(cj,b)  (intermediate result for clarity)

        -LT(x1,s(x2))  -LT(x2,x1)           (axiom)
        LT(b,s(cj))                         (13)
```

j. -LT(cmax,eval(cc,b))

```
        LT(cmax,eval(cc,b))                (14)
```

Note that the new equality (clause 15, which was derived from literal "h") back demodulates clause 14 to produce the following unit clause.

(16) LT(cmax,eval(cc,cj))

To complete the proof, it is sufficient to show how clause 16 can be used to deduce a unit clause that conflicts with an existing unit.

k. LT(cmax,eval(cc,cj))

```
        -IB(cc,cj)  -LT(cmax,eval(cc,cj))   (8)
         LT(x1,num1) LT(cn,x1) IB(cc,x1)    (axiom)
         -LT(cj,num1)                       (clause 2)
```

The result of this linked UR-resolution step is the unit clause

(17) LT(cn,cj)

which conflicts with clause 3. Note that the intermediate result -IB(cc,cj) is itself a linked UR-resolvent that would be generated with this step. The entire problem is solved with 5 linked UR steps.

5. Conclusions

We have discussed the linked inference principle and, specifically, the application that yields linked UR-resolution. Linked UR-resolution enables a reasoning program to take much larger steps than standard UR-resolution does. By employing this inference rule, the user is usually free to choose a natural representation for presenting a problem in clause form. In particular, while standard inference rules often force you to choose between expanding the set of support and adding binary resolution to the inference rules being used, linked inference rules often avoid both undesirable choices.

We have described certain experiments to test the position that employment of linked inference increases the effectiveness of automated reasoning programs. In addition, we have included other illustrations (carried out by hand) to aid in understanding how the linked inference principle works. The experiments and illustrations focus on problems from the world of puzzles, circuit design, and proving properties of computer programs. We intend to present the results of those experiments at the Seventh Conference on Automated Deduction.

Although we have not given a detailed account of precisely how the user gives the reasoning program the instructions that in turn result in controlling linked inference, we have illustrated how a strategy, the target strategy, can be used to employ semantic considerations rather than the usual syntactic ones that define standard inference rules. The motivation for formulating the linked inference principle is to provide greater control over a reasoning program's attack on a problem, and also to increase the role of semantic criteria. Questions of soundness and refutation completeness as well as a detailed discussion of the fine points of the linked inference principle are found in [8]. Our main objective here is to demonstrate that the use of linked inference rules may increase the potential of relying on a reasoning program to function as an automated reasoning assistant.

References

[1] Luckham, D. C., "Some Tree-paring Strategies for Theorem Proving," Machine Intelligence 3 (ed. Michie, D.), Edinburgh University Press 1968, pp. 95-112.

[2] McCharen, J., Overbeek, R. and Wos, L., "Problems and experiments for and with automated theorem proving programs," IEEE Transactions on Computers, Vol. C-25(1976), pp. 773-782.

[3] Robinson, J., "A machine-oriented logic based on the resolution principle," J. ACM, Vol. 12(1965), pp. 23-41.

[4] Warren, D. H. D., Implementing Prolog - compiling predicate logic programs, DAI Research Reports 39 & 40, University of Edinburgh, May 1977.

[5] Wojciechowski, W. and Wojcik, A., "Automated design of multiple-valued logic circuits by automatic theorem proving techniques," to appear in IEEE Transactions on Computers.

[6] Wos, L., Carson, D. and Robinson, G., "The unit preference strategy in theorem proving," Proc. AFIPS 1964 Fall Joint Computer Conference, Vol. 26, Part II, pp. 615-621 (Spartan Books, Washington, D.C.).

[7] Wos, L., Carson, D. and Robinson, G., "Efficiency and completeness of the set of support strategy in theorem proving," J. ACM, Vol. 12(1965), pp. 536-541.

[8] Wos, L., Smith, B. and Veroff, R., "The Linked Inference Principle, I: The Formal Treatment," in preparation.

A NEW INTERPRETATION OF THE RESOLUTION PRINCIPLE

Etienne Paul

Centre national d'etudes des Telecommunications
38/40 Rue du General Leclerc
92131 Issy les Moulineaux, France

Abstract

We show in this paper that the application of the resolution principle to a set of clauses can be regarded as the construction of a term rewriting system confluent on valid formulas. This result allows the extension of standard properties and methods of equational theories (such as Birkhoff's theorem and Knuth and Bendix completion algorithm) to quantifier-free first order predicate calculus.

These results are extended to first order predicate calculus in an equational theory, as studied by Plotkin [15], Slagle [17] and Lankford [12].

This paper is a continuation of the work of Hsiang [5], who has already shown that rewrite methods can be used in first order predicate calculus. The main difference is that Hsiang applies rewrite methods only as a refutational proof technique, trying to generate the equation TRUE=FALSE. We generalize these methods to satisfiable theories; in particular, we show that the concept of confluent rewriting system, which is the main tool for studying equational theories, can be extended to any quantifier-free first order theory. Furthermore, we show that rewrite methods can be used even if formulas are kept in clausal form.

1:INTRODUCTION

We show in this paper that the resolution algorithm applied to a set of clauses has the same goal as the Knuth and Bendix algorithm applied to a set of equations, namely the construction of a confluent rewriting system, but with the restriction that the system obtained is confluent only on valid formulas.

More specifically, let S be a set of clauses. We show that the application of the resolution principle to S produces a rewriting system R such that any quantifier-free first order formula F which is a valid consequence of S reduces to TRUE, using reductions from R in any order. Conversely, only valid consequences of S reduce to TRUE. But the system R is not confluent in general, for F may possess several normal forms if it is not a valid consequence of S.

Such partly confluent rewriting systems have already been studied in equational theories: for example, the Greendlinger-Bucken algorithm [2] for finitely presented groups constructs a rewriting system which, under certain conditions, is confluent on the relators (i.e. words equal to the identity in the group). The word problem can then be decided as follows: two words a and b are equal iff the normal form of the word a+(-b) is the identity.

In the same way, the equivalence between two quantifier-free first order formulas F and G in the theory defined by the set of clauses S can be decided by computing the normal form of the formula F<=>G: F and G are equivalent iff this normal form is TRUE.

From this first result, we deduce that standard properties and methods of equational theories (such as Birkhoff's theorem and Knuth and Bendix completion algorithm) can be extended to quantifier-free first order predicate calculus.

The organization of the paper is as follows:

In section 2, we introduce a rewriting system in propositional calculus confluent on valid formulas (which are merely tautologies). Since we do not need general confluence, we will not use the system discovered by Hsiang [5], because this system is not practical for manipulating clauses. Our system will be constructed from the usual boolean connectors "and", "or", "not".

In section 3, we extend this result to first order predicate calculus: we describe a completion algorithm based on the resolution principle which generates, from any initial set of clauses, a rewriting system confluent on valid formulas.

In section 4, we prove Birkhoff's theorem for first order predicate calculus. This theorem states that equational-type reasoning (i.e. reasoning by instantiation and replacement of equivalents by equivalents) is complete for quantifier-free first order predicate calculus. From this theorem, we deduce a new completion algorithm, based on the Knuth and Bendix completion algorithm and the Hsiang system for propositional calculus.

In section 5, the previous results are extended to first order predicate calculus in an equational theory.

2:CONFLUENCE ON VALID FORMULAS IN PROPOSITIONAL CALCULUS

2-1: review of equational term rewriting systems

It is assumed that the reader is familiar with the literature on equational theories and term rewriting systems. See [7] for a full description.

We start with a vocabulary of variables and function symbols; we define terms, occurrences, substitutions, unification, most general unifier (m.g.u). If M is a term and u an occurrence of M, M/u denotes the subterm of M at occurrence u, and M[u<-N] the term M in which this subterm is replaced by N. If s is a substitution, s(M) is denoted Ms. The set of terms is denoted T.

An equational system E is a set of pairs M=N, where M and N are two terms. <->E denotes the one step equality. The equality relation =E or <*>E generated by E is the reflexive-transitive closure of <->E, i.e. the smallest congruence over T containing all pairs <Ms,Ns> for M=N in E and s an arbitrary substitution.

A term rewriting system R is a set of directed pairs l->r such that each variable in r occurs in l. A term t1 R-reduces at occurrence u to a term t2 using the rule l->r iff there exists a substitution s such that t1/u = ls and t2 = t1[u<-rs]. We write t1 ->R t2.

Given E and R as above, a term t1 E,R-reduces to t2 iff there exists a term t3 such that t1 =E t3 and t3 ->R t2. We write t1 ->E,R t2. The relation ->E,R can also be regarded as the relation induced by R in the equivalence classes of terms modulo =E. *>E,R denotes the reflexive-transitive closure of ->E,R, and <*>E,R or =E,R denotes the reflexive-symmetric-transitive closure of ->E,R.

The pair (E,R) is called an equational term rewriting system (ETRS). Such systems are a generalization of usual rewriting systems for handling non-terminating equations such as commutativity. They are studied in detail in [10],[14].

(E,R) is terminating if there is no infinite sequence of E,R-reductions from any term.

(E,R) is inter-reduced if for each rule l->r in R, r is (E,R)-irreducible and l is ((E,R)-(l->r))-irreducible.

(E,R) is confluent if for any terms t, t1 and t2, t *>E,R t1 and t *>E,R t2 implies: there exist t3 and t4 such that:
t1 *>E,R t3, t2 *>E,R t4 and t3 =E t4.

(E,R) is canonical if it is both terminating and confluent.

A term t1 is a normal form of t2 iff t2 *>E,R t1 and t1 is (E,R)-irreducible. It is easy to see that if (E,R) is canonical, then every term has a unique irreducible normal form (up to =E).

2-2: Propositional calculus

Formulas of propositional calculus are constructed from the following vocabulary:
two constants TRUE and FALSE.
boolean connectors:

```
v : or          <=> : equivalence
& : and          => : implication
¬ : not           ! : exclusive or
```

a denumerable set of boolean variables, each variable can have two values: TRUE or FALSE.

A formula is VALID or TAUTOLOGICAL iff its value is TRUE for each assignment of values to its variables, according to the rules of boolean calculus. We embody these rules into the following ETRS:

-Equational system E:

```
      XvY    =   YvX
  (XvY)vZ    =   Xv(YvZ)
      X&Y    =   Y&X
  (X&Y)&Z    =   X&(Y&Z)
```

-Rewriting system R:

```
(1)    X<=>Y      ->   (Xv(¬Y))&(Yv(¬X))
(2)     X=>Y      ->   (¬X)vY
(3)      X!Y      ->   (XvY)&((¬X)v(¬Y))

(4)   ¬(TRUE)     ->   FALSE
(5)   ¬(FALSE)    ->   TRUE
```

```
(6)     ¬(XvY)     ->    (¬X)&(¬Y)
(7)     ¬(X&Y)     ->    (¬X)v(¬Y)
(8)      ¬(¬X)     ->    X

(9)     Xv(Y&Z)    ->    (XvY)&(XvZ)

(10)    XvTRUE     ->   TRUE
(11)    XvFALSE    ->    X
(12)      XvX      ->    X
(13)     Xv(¬X)    ->    TRUE

(14)    X&TRUE     ->    X
(15)    X&FALSE    ->   FALSE
(16)      X&X      ->    X
(17)     X&(¬X)    ->    FALSE
```

Rules (1) to (3) eliminate the connectors <=>, =>, !. Rules (4) to (8) eliminate all the connectors ¬ which are not directly applied to variables. Rules (10),(11),(14),(15) eliminate the constants TRUE and FALSE. Rule (9) converts the formula into conjunctive normal form C1&C2&...&Cm, each Ci being in the form:

Ci = X1 v X2 v ... v Xp v (¬Xp+1) v (¬Xp+2) v ... v (¬Xn)

For a given Ci, the Xj's are distinct variables from rules (12) and (13).

Therefore, in this ETRS which we denote BOOL and which is obviously terminating, the normal forms of terms are:
-The constants TRUE and FALSE
-The formulas C1&C2&...&Cm, each Ci being a disjunction of variables and of negations of variables which are all distinct.

This system is not confluent on all formulas. Moreover, it is well known that it is impossible to construct a finite confluent rewriting system in propositional calculus based on connectors v,&,¬, due to the fact that the prime implicant representation of Boolean terms is not unique.

But this system is confluent on valid formulas. For if F were a valid formula and had the normal form C1&C2&...&Cm, with variables in each Ci distinct, it would be easy to assign values to the variables of F such that, for example, C1=FALSE and hence F=FALSE. Therefore, the unique normal form of F is TRUE.

Given two formulas F1 and F2, we can decide whether F1 and F2 are equivalent by computing the normal form of F1<=>F2: F1 and F2 are equivalent iff this normal form is TRUE. For example, the absorption law (i.e. X&(XvY) = X) can be proved by reducing the formula (X&(XvY))<=>X:

```
(X&(XvY))<=>X  ->  ((X&(XvY))v(¬X))&(Xv(¬(X&(XvY))))  by rule (1)
               ->... ->  TRUE  by rules (4) to (17)
```

Note that both terms X&(XvY) and X are irreducible.

Of course, this method is less efficient than using Hsiang's system, in which two formulas are equivalent iff they have the same normal form. But we shall retain this system for the moment, for it is easier to manipulate clauses with it than with

Hsiang's system (we will introduce Hsiang's system in section 4).

Note that if we replace rule (9) by the other distributivity rule: X&(YvZ) -> (X&Y)v(X&Z), we obtain a system confluent on unsatisfiable formulas (FALSE being the unique normal form of these formulas).

In this new system, we can decide the equivalence of two formulas F1 and F2 by computing the normal form of F1!F2: F1 and F2 are equivalent iff this normal form is FALSE.

3:CONFLUENCE ON VALID FORMULAS IN PREDICATE CALCULUS

3-1: Review of first order predicate calculus and resolution

We start with a vocabulary of variables, function symbols, and predicate symbols. Terms, atoms and literals are introduced next (see for example [3] for a full description).

Throughout this paper, we only deal with quantifier-free first order predicate calculus; that means that all formulas manipulated are in prenex form with all their variables implicitly universally quantified.

A clause is a finite disjunction of zero or more literals; the order and parenthesis of the literals are irrelevant; in other words, a clause is assimilated to its equivalence class modulo =E. The empty clause is assimilated to the constant FALSE. A tautological clause is a clause which contains a complementary pair of literals (i.e. L and ¬L).

A clause with repeated literals is considered as different from the same clause without repeated literals, because both clauses are not in the same equivalence class modulo =E (note that it would be the case if the rule XvX -> X of BOOL were placed into E, i.e. regarded as a non oriented equation).

The resolution operation is decomposed into binary resolution and binary factoring, which are defined in the usual way. Deletion of repeated literals in a clause is considered as a particular case of factoring.

The clause C1 subsumes the clause C2 if there is a substitution s such that C1s $\subseteq$ C2; in this inclusion, repeated literals in C1s or C2 are taken into account only once.

A set S of clauses is regarded as equivalent to the conjunction of these clauses. The first order formula F is a valid consequence of S, or is valid in the theory specified by S, iff F is true in all models of S. In particular, TRUE is a valid consequence of any set S, and FALSE is a valid consequence of S iff S is unsatisfiable.

We will use the following result obtained by R.C.Lee:

<u>Theorem 3.1</u> (completeness of resolution for consequence finding):

Let S be a set of clauses.
Let C be a non tautological clause.
C is a valid consequence of S iff there is a clause I deduced from S by resolution which subsumes C.

This theorem links the semantic concept of truth and the syntactic concept of provability (by resolution). It is proved in [13].

3-2: <u>confluence on valid formulas</u>

If S is a set of clauses, we associate with S an equational term rewriting system which is defined as the union of the two following systems:
-The system BOOL defined in section 2.2.
-The set of rules {C -> TRUE} with C in S.

We denote also S this ETRS .

We say that S is confluent on valid formulas iff for any formula F which is a valid consequence of S, F *>S TRUE.

<u>Theorem 3.2</u>:

Let S be a set of clauses which do not contain the empty clause. S is confluent on valid formulas iff the following conditions are met:
(i) for each binary factor F of a clause in S, F *>S TRUE
(ii) for each binary resolvent R of two clauses in S, R *>S TRUE

This theorem corresponds to the Knuth and Bendix critical pairs theorem [11]. In fact, the computation of binary factors and binary resolvents can be regarded as a computation of critical pairs between rules of S restricted to certain critical pairs (the details of this computation are given in annex). We do not need to compute all critical pairs because we do not require the general confluence, but only the confluence on valid formulas.

Proof: the conditions (i) and (ii) are obviously necessary. Conversely, let us suppose that (i) and (ii) are true. To prove that S is confluent on valid formulas, we need the following lemmas:

<u>Lemma A</u>: if C is a non tautological clause such that C *>S TRUE, there is a clause D in S, a sub-clause B of C and a substitution s such that B = Ds.

Proof: obvious.

<u>Lemma B</u>: if C is a clause such that C *>S TRUE, and F(C) is a binary factor of C, F(C) *>S TRUE.

Proof: if C is tautological, F(C) is also tautological.
Hence F(C) *>BOOL TRUE and F(C) *>S TRUE since BOOL $\subseteq$ S.

If C is not tautological, by using lemma A, C can be written C = Ds v Cl, with D in S.

If the factoring of C is done between two literals of Cl, or between a literal in Cl and a literal in Ds, it is easy to check that F(C) *>S TRUE by application of the rule D -> TRUE.

If the factoring is done between two literals L1s and L2s of Ds, we have: C = L1s v L2s v D1s v Cl, with D = L1 v L2 v D1.

Let t be the m.g.u of L1s and L2s. We have:

F(C) = L1st v D1st v Clt

L1 and L2 are unifiable, since L1s and L2s are unifiable. Let u be the m.g.u of L1 and L2: F(D) = L1u v D1u is a binary factor of D. By hypothesis (i), F(D) *>S TRUE. From the definition of the m.g.u, there is a substitution w such that st = uw. Hence F(C) = F(D)w v C1t and F(C) *>S TRUE.

Lemma C: if C and D are two clauses such that C *>S TRUE and D *>S TRUE, and R(C,D) is a binary resolvent of C and D, R(C,D) *>S TRUE.

Proof: similar to the proof of lemma B.

We can now prove that S is confluent on valid formulas: let F be a formula which is a valid consequence of S.

If F is a tautology, F *>BOOL TRUE and hence: F *>S TRUE.

If F is not a tautology, F *>BOOL C1&...&Cp, each Ci being a non tautological clause which is also a valid consequence of S.

Let us choose for example the clause C1: by theorem 3.1, there is a clause I deduced from S by resolution which subsumes C1. Therefore, there is a sub-clause D of C1 and a substitution s such that Is and D contain exactly the same literals (possibly repeated for Is, but not for D because D is reduced by the the system BOOL).

I is deduced from S by resolution, i.e. by a sequence of binary resolution and binary factoring. Hence, from lemmas B and C: I *>S TRUE. Therefore Is *>S TRUE.

D is obtained from Is by eventual deletion of repeated literals, which is a particular case of factoring. Hence, by lemma B, D *>S TRUE. Since S does not contain the empty clause, D is not the empty clause.

Since D is a non-empty sub-clause of C1, C1 *>S TRUE.

We prove in the same way that Ci *>S TRUE for all i. Therefore F *>S TRUE, which ends the proof of theorem 3.2.

Note that we have also proved that S is satisfiable, since we do not have FALSE *>S TRUE; hence FALSE is not a valid consequence of S.

3-3: Completion algorithm

From theorem 3.2, we can derive a completion algorithm similar to the Knuth and Bendix algorithm, which will generate from any set of clauses an equivalent (i.e. having the same models) rewriting system confluent on valid formulas.

In the following, Ei and Ri are two finite set of clauses. The current rewriting system is the system associated with Ri, i.e. the system: BOOL u {C -> TRUE, C in Ri}. This rewriting system is also denoted Ri.

Each clause in Ri has a label, which is a unique integer. We denote by k:C the clause C with label k. Finally, each clause in Ri is marked or unmarked.

Initial data: a (finite) set of clauses S.

1-Initialization: Let E0 = S, R0 = BOOL, i = 0, p = 0.

2-If Ei # emptyset, go to 4.

3-Compute binary resolvents and binary factors:

If all clauses in Ri are marked, stop with success. Otherwise, select an unmarked C clause in Ri, say with label k. Do:
Ei+1 = {binary factors of C} u {binary resolvents of (C,D) for any clause D of Ri of label not greater than k}
Ri+1 = Ri with clause C marked
i = i+1, go to 2.

4-Introduction of new rules:
Select clause C in Ei. Let C! a Ri-normal form of C.
-If C! = FALSE, stop with answer: S unsatisfiable.
-If C! = TRUE, do: Ei+1 = Ei-C, Ri+1 = Ri, i = i+1 and go to 2.
-If C! # FALSE and C! # TRUE:
Let K be the set of labels of clauses in Ri reducible by the new rule C! -> TRUE. Do:
Ei+1 = Ei-C u {j:D with j in K}
p = p+1
Ri+1 = {j:D with D in Ri and j not in K} u {p:C!}
i = i+1, go to 2
In Ri+1, the clauses coming from Ri are marked or unmarked as they were in Ri, the new clause p:C! is unmarked.

To ensure the completeness of this algorithm, we need an assumption concerning the selection of clause at step 3: for every clause label k, there is an iteration i such that either the clause of label k is deleted from Ri (i.e. k is in K at iteration i), or the clause of label k is selected at step 3. This assumption ensures that no clause will be ignored indefinitely by the selection process.

The deletion at step 4 of clauses which are reduced to TRUE corresponds to the deletion, in resolution algorithm, of clauses which are tautologies or which are subsumed by other clauses.

The only difference with the Knuth and Bendix completion algorithm, as presented for instance in [6], is the replacement of the computation of critical pairs by the computation of binary factors and binary resolvents, i.e. computation of only certain critical pairs. Consequently, there is no case of stop with failure, because the right-hand side of the added rules is always TRUE. The algorithm may:
-either stop with the conclusion: S is unsatisfiable,
-either stop with success,
-or run forever.

In the last two cases, let R_{∞} be the final rewriting system constructed by the algorithm. R_{∞} is the set of all the rules which belong to some Ri and to all Rj's for j>i; i.e. which are never reduced by other rules. In the first case, we define R_{∞} as the single rule X -> TRUE. Note that this rule is obtained by superposing the rule FALSE -> TRUE, generated by the algorithm, and the rule XvFALSE -> X of BOOL; furthermore, this rule X -> TRUE entails the deletion of all previous rules by redundancy (including rules of BOOL).

<u>Theorem 3.3</u>:

The rewriting system R_{∞} has the following properties:
(i) R_{∞} is inter-reduced.
(ii) R_{∞} is equivalent to the initial set of clauses S.

(iii) $R\infty$ is confluent on valid formulas.
Furthermore, for a given S, $R\infty$ is the only rewriting system associated with a set of clauses which has these properties.

Proof: if S is unsatisfiable, $R\infty$ is reduced to the only rule X -> TRUE, which means that any formula can be considered as a valid consequence of S. Properties (i) to (iii) are obvious.

If S is satisfiable, (i) comes from the structure of the algorithm. (ii) comes from the fact that at each iteration i of the algorithm, S is equivalent to Ei u Ri. Finally, (iii) comes from theorem 3.2.

To prove the unicity, let us suppose that there exists another rewriting system Q, associated with a set of clauses, such that Q has the properties (i) to (iii). As Q is equivalent to S by (ii), Q is equivalent to $R\infty$.

Let C1 be a clause in Q. C1 is a valid consequence of $R\infty$, and since $R\infty$ is confluent on valid formulas, C1 $\overset{*}{\to}_{R\infty}$ TRUE.

C1 is not tautological (otherwise, it could be reduced by rules of BOOL, and the system Q would not be interreduced). Therefore, by lemma A, there is a clause C2 in $R\infty$ and a substitution s such that C2s $\subseteq$ C1. This clause C2 is a valid consequence of Q, and we can prove in the same way that there is a clause D1 in Q and a substitution t such that D1t $\subseteq$ C2.

Hence D1ts $\subseteq$ C1. Since Q is interreduced, that entails D1=C1, hence ts = identity and C1 and C2 are identical (up to the names of variables). Therefore Q $\subseteq$ $R\infty$. We prove in the same way that $R\infty$ $\subseteq$ Q, therefore Q = $R\infty$.

Note the difference with the Knuth and Bendix algorithm, in which it is possible to generate several different confluent rewriting systems, depending on the orientation chosen for rewrite rules.

4: EQUATIONAL METHODS IN FIRST ORDER PREDICATE CALCULUS

4-1: Theorem of Birkhoff

Let S = {E1,...,En} be a set of quantifier-free formulas. E1,...,En are not necessarily in clausal form.

We associate with S an equational system which is defined as the union of the following systems:
-The system BOOL defined in section 2.2 (regarded as a set of equations).
-The set of equations {Ei = TRUE} with Ei in S.

We denote also S this equational system.

Note that the symbol =, linking two boolean formulas, corresponds here to the boolean connector <=> and is different from the equality predicate, linking two non boolean terms, which can also exist in S.

This equational system defines an equality relation on the set of quantifier-free formulas. We denote this relation =S or <*>S.

Theorem 4.1:

Let F and G be two quantifier-free formulas.

F and G have simultaneously the same values (TRUE or FALSE) for every assignment of their variables in all the models of S iff F =S G.
An equivalent statement is:
The formula F<=>G is a valid consequence of S iff F =S G.

Theorem 4.1 states that equational-type reasoning (i.e. reasoning by instantiation and replacement of equivalent by equivalent) is complete for quantifier-free first order predicate calculus. The boolean connector <=> plays exactly the same role as the equality predicate in the usual Birkhoff theorem for equational theories. A comparison between the scope of application of these two theorems is performed in section 5.4.
To prove theorem 4.1, we need the following lemmas:

Lemma D: for every set of formulas S, there is an equivalent set of clauses S1 such that: <*>S = <*>S1.

Proof: by using BOOL, each Ei in S can be transformed into a conjunction of clauses. Let S1 be the corresponding set of clauses. For each Ei in S, we have the following relation:

Ei <*>BOOL C1&C2&...&Cp with Cj in S1.

Since BOOL $\subseteq$ S1, we have Ei <*>S1 TRUE. Hence <*>S $\subseteq$ <*>S1.
Conversely, the following sequence holds for each Cj:

Cj <->BOOL Cj&TRUE <->S Cj&Ei <*>BOOL Cj&(C1&...&Cp)
<*>BOOL C1&...&Cp <*>BOOL Ei <->S TRUE

Therefore Cj <*>S TRUE. Hence <*>S1 $\subseteq$ <*>S.

Lemma E: let S be a set of clauses, and R∞ the rewriting system built from S by the completion algorithm. We have <*>S = <*>R∞.

Proof: if C is a clause in S, C *>R∞ TRUE. Then <*>S $\subseteq$ <*>R∞.
Conversely, if C is a clause in R∞, the equation C = TRUE is obtained from the clauses in S by a finite sequence of binary factoring and binary resolution. These operations are particular cases of computation of critical pairs (see annex). Consequently, the equation C = TRUE is an equational consequence of S, and therefore <*>R∞ $\subseteq$ <*>S.

Lemma F: the boolean connector <=> has the following properties:

Commutativity:	X<=>Y	<*>BOOL	Y<=>X
Associativity:	(X<=>Y)<=>Z	<*>BOOL	X<=>(Y<=>Z)
Identity :	X<=>TRUE	<*>BOOL	X
Nilpotence :	X<=>X	<*>BOOL	TRUE

Proof: straightforward, by reducing each expression to its normal form in the rewriting system BOOL.

We can now prove theorem 4.1:

Let F and G be two quantifier-free formulas. If F =S G, it

is obvious that F<=>G is a valid consequence of S.

Conversely, let us suppose that F<=>G is a valid consequence of S. From lemma D, we can suppose that S is clausal. Let R_∞ be the rewriting system obtained from S by the completion algorithm. We have:

```
F<=>G  *>R∞  TRUE
```

Hence F<=>G <*>S TRUE by lemma E.
Then, using lemma F, we can write the following sequence:

```
F  <*>BOOL  F<=>TRUE  <*>S  F<=>(F<=>G)  <*>BOOL  (F<=>F)<=>G
   <*>BOOL  TRUE<=>G  <*>BOOL  G
```

Hence F =S G, which ends the proof.

Remark 1: theorem 4.1 has the following particular cases:
-With G = TRUE: F is a valid consequence of S iff F =S TRUE.
-With F = FALSE and G = TRUE: FALSE is a valid consequence of S (i.e S is unsatisfiable) iff FALSE =S TRUE. That is a well known definition of the inconsistency of a specification.

Remark 2: theorem 4.1 remains true if the system BOOL is replaced by any specification BOOL1 of propositional calculus such that <*>BOOL = <*>BOOL1. In particular, theorem 4.1 is true if BOOL1 is the Hsiang system.

4-2: <u>general confluence in first order predicate calculus</u>

Theorem 4.1 states that equational reasoning is complete for quantifier-free first order predicate calculus. Consequently, we can try to build from any specification S a rewriting system confluent on all formulas and not only on valid formulas, by using the Knuth and Bendix algorithm instead of the resolution algorithm (i.e. by computing all critical pairs and not only certain critical pairs). But we will not succeed if we start with the rewriting system BOOL, because even if we restrict ourselves to propositional calculus, it is impossible to build from BOOL a finite complete rewriting system.

Therefore, we must use for this purpose Hsiang's system.

4-2-1: <u>The Hsiang system and the dual system</u>

The Hsiang system is the following:

```
(1)        X&X  ->  X
(2)     X&TRUE  ->  X
(3)    X&FALSE  ->  FALSE
(4)    X!FALSE  ->  X
(5)        X!X  ->  FALSE
(6)    X&(Y!Z)  ->  (X&Y)!(X&Z)

(7)      X<=>Y  ->  X!Y!TRUE
(8)         ¬X  ->  X!TRUE
(9)        XvY  ->  (X&Y)!X!Y
(10)      X=>Y  ->  (X&Y)!X!TRUE
```

This system is based on the connectors & and !. Rules (7) to (10) eliminate other connectors. This system is confluent modulo the associativity/commutativity of & and !.

As noted by Hsiang, there exists also a dual system. To formalize the construction of this system, we define the dual d(f) of any boolean fonction f with arity n by:

```
d(f)(X1,X2,...,Xn) = ¬(f(¬X1,¬X2,...,¬Xn))
```

The following relations are straightforward:

```
        d(TRUE)   =  FALSE      d(FALSE) =  TRUE
        d( v )    =  &          d( & )   =  v
        d( <=> )  =  !          d( ! )   =  <=>
        d( ¬ )    =  ¬
        (in general, d(d(f)) = f)
        d(X)      =  X    (X being a boolean variable)
 d(f(P1,...,Pn)) = d(f)(d(P1),...,d(Pn))  for any connector f
```

These relations allow us to compute the dual of any expression constructed from boolean variables and connectors, by induction over the structure of this expression.

If P1 and P2 are two equivalent expressions, d(P1) and d(P2) are also equivalent. This property allows us to apply a transformation by duality to Hsiang's system. We obtain the following system:

```
(1)          XvX  ->  X
(2)      XvFALSE  ->  X
(3)       XvTRUE  ->  TRUE
(4)     X<=>TRUE  ->  X
(5)        X<=>X  ->  TRUE
(6)  Xv(Y<=>Z)    ->  (XvY)<=>(XvZ)

(7)          X!Y  ->  X<=>Y<=>FALSE
(8)           ¬X  ->  X<=>FALSE
(9)          X&Y  ->  (XvY)<=>X<=>Y
(10)        X=>Y  ->  (XvY)<=>Y
```

(The rule (10) is directly computed, since the dual of => is not a usual connector)

This system is built from v and <=> instead of & and !. It is confluent modulo the associativity/commutativity of these two connectors.

4-2-2: <u>completion algorithm</u>

Let S = {E1,...,En} be a set of quantifier-free formulas. To complete S, we run the Knuth and Bendix algorithm in its associative/ commutative version, initializing the set of rewrite rules with the Hsiang system (or the dual system) and the set of equations with {Ei = TRUE}. If a formula Ei is already in the form Fi<=>Gi, we can initialize the set of equations with the equation Fi = Gi instead of (Fi<=>Gi) = TRUE.

From now on, we suppose that the completion algorithm does

not stop with failure because of the generation of an incomparable critical pair. Let R_∞ be the (finite or infinite) rewriting system built by the algorithm.

<u>Theorem 4.2</u>:

Let F and G be two quantifier-free formulas.
The formula F<=>G is a valid consequence of S iff F and G have the same R_∞ normal form. In particular, F is a valid consequence of S iff F $\overset{*}{\to}_{R_\infty}$ TRUE.

Proof: from theorem 4.1 and the general properties of the Knuth and Bendix algorithm, as explained in [6].

Remark: if the algorithm stops with failure, we can in certain cases run it again after putting the incomparable critical pair into the set of equations, with the associativity/commutativity of ! and & (or <=> and v if we use the dual system). For example, commutative predicates can be handled in this way.

Example: let us consider the following clausal specification:
S = {P(f(x)) v P(x) , ¬P(f(x)) v ¬P(x)}

If we apply to this specification the completion algorithm of section 3.3, based on resolution, we generate an infinite number of rewrite rules:

```
(system BOOL +)
 P(f(x)) v P(x)  ->  TRUE
¬P(f(x)) v ¬P(x)  ->  TRUE
¬P(f(f(x))) v P(x)  ->  TRUE
 P(f(f(f(x)))) v P(x)  ->  TRUE
¬P(f(f(f(f(x))))) v P(x)  ->  TRUE
...
```

But if we apply to this specification the Knuth and Bendix completion algorithm, this algorithm will stop with only a finite number of rules, namely:

```
(Hsiang's system +)
P(f(x))  ->  P(x) ! TRUE
```

Let us consider another example: let S be the following specification for a program to test if an element u is a member of a sequence z:

(1) x=x
(2) ¬Elem(u,NIL)
(3) Elem(u,w.x) <=> (u=w) v Elem(u,x)

We have added (1), simple reflexivity, the only equality property needed here. Using the dual Hsiang system and the Knuth and Bendix algorithm, we obtained at once the following complete system:

(dual Hsiang's system +)

```
x=x -> TRUE
Elem(u,NIL) -> FALSE
Elem(u,w.x) -> (u=w) v Elem(u,x)
```

If we use the resolution algorithm, we must split the equivalence into three clauses. We obtain an infinite system:

```
(system BOOL +)
x=x -> TRUE
¬Elem(u,NIL) -> TRUE
¬Elem(u,w.x) v (u=w) v Elem(u,x) -> TRUE
¬(u=w) v Elem(u,w.x) -> TRUE
¬Elem(u,x) v Elem(u,w.x) -> TRUE
¬Elem(u,w.NIL) v (u=w) -> TRUE
Elem(u,u.x) -> TRUE
¬(u=w) v Elem(u,wl.(w.x)) -> TRUE
Elem(u,wl.(u.x)) -> TRUE
...
```

In both examples, the Knuth and Bendix algorithm is preferable to the resolution algorithm. That is due to the fact that we can use equivalence relations (such as (3)) as simplifiers.

In other cases, both algorithms will run in parallel. For example, if S is {¬P(x) v P(f(x))}, each algorithm generates infinitely many rules. Each rule:

¬P(x) v P(f(f(f(...(x))))) -> TRUE

produced by the resolution algorithm is associated with the rule:

P(x) & P(f(f(f(...(x))))) -> P(x)

produced by the Knuth and Bendix algorithm.

Actually, the major problem with the Knuth and Bendix algorithm is the orientation of rules; for example if we run the Knuth and Bendix algorithm with S = axiomatization of equality, we generate at once non-orientable critical pairs such as: <(x=y)&(x=z) , (x=y)&(y=z)>. Furthermore, with the techniques available now, we cannot put this rule into the set of equations, because we do not have a unification algorithm for this kind of equation.

4-2-3: <u>Knuth-Bendix algorithm as a refutational proof technique</u>

Theorem 4.2 can be particularized for unsatisfiable specifications as follows:

<u>Theorem 4.3</u>:

Let S be a set of quantifier-free formulas.
S is unsatisfiable iff the Knuth and Bendix completion algorithm generates one of the following rules (X being a boolean variable):

```
     X -> TRUE         X -> FALSE
 FALSE -> TRUE      TRUE -> FALSE
```

Proof: taking F=FALSE and G=TRUE in theorem 4.2, we obtain: FALSE is a valid consequence of S (i.e. S is unsatisfiable) iff FALSE and TRUE have the same normal form in R^{∞}. Therefore, either FALSE or TRUE must be reducible by R^{∞}. Hence, one of the above listed rules has been generated by the algorithm.

Theorem 4.3 is close to results of Hsiang [5] and Fages [4]. However, there are two differences:
-We do not suppose in theorem 4.3 that formulas in S are initially in clausal form.
-In exchange for removing this restriction, we have to compute all critical pairs; consequently, the algorithm can stop with failure if it generates an incomparable critical pair.

5-FIRST ORDER PREDICATE CALCULUS IN AN EQUATIONAL THEORY

5-1: preliminaries

We are now in first order predicate calculus with equality. Furthermore, we suppose that the set of clauses that we consider is divided into two parts:
-A set T of unit clauses of the form {M=N} which define an equational theory.
-A set S of clauses which do not contain any equality predicates.

We suppose that the equational theory T can be compiled into a canonical term rewriting system R. We introduce a new inference rule, called "narrowing" defined as follows:
Given a clause C, if there is a non variable occurrence u in C such that C/u is unifiable with the left-hand side of a rule (l->r) in R with m.g.u s, the clause N(C) = C(u<-r)s is a narrowing of C.
This is Hullot's definition, and not Lankford's or Slagle's, for we do not normalize N(C) by R. Note that the pair <N(C),TRUE> is a critical pair between the two rules l->r and C->TRUE.
A resolution of the clauses C1 and C2 producing the clause C3 is said to be a blocked resolution if the three clauses C1,C2,C3 are in R-normal form. We similarly define a blocked binary resolution and a blocked binary factoring.
We need the following result, which is analogous to theorem 3.1 (Lee's theorem):

Theorem 5.1:

Let C be a non tautological clause in R-normal form, which does not contain any equality predicates.
C is a valid consequence of S u T in predicate calculus with equality iff there is a clause I deduced from S by narrowing and blocked resolution which subsumes C.

Proof: we first consider the case where all clauses in S are ground. If S! is the set of R-normal forms of clauses in S, S! is deduced from S by reduction by R, which is a particular case of narrowing; and C is a valid consequence of S! u T in predicate calculus with equality. Therefore S! u T u ¬C is equality unsatisfiable.
Since S! and ¬C are in R-normal form, by using methods of

Lankford [12] we obtain that S! u ¬C is unsatisfiable.
Therefore C is a valid consequence of S!.

From Lee's theorem, there is a clause I deduced from S! by resolution which subsumes C. Moreover, this deduction is blocked.

This proof is easily lifted to the general case by using methods of Lee [13], the usual lifting lemma for resolution, and the following lifting lemma for narrowing:

Lemma G: let C be a clause. Let s be a R-normalized substitution (i.e Xs is R-irreducible for each variable X). Let D be the R-normal form of Cs.
There exist a clause C1 derived from C by a finite sequence of narrowings and a substitution t such that C1t = D.

Proof: see Hullot [8].

5-2: confluence on valid formulas

We associate with S u T an equational term rewriting system which is defined as the union of the three following systems:
-The system BOOL defined in section 2.2.
-The rewriting system R associated with the equational theory T.
-The set of rules {C -> TRUE} with C in S.

We denote RS this equational term rewriting system.

We say that RS is confluent on valid formulas iff for each formula F without equality predicate which is a valid consequence of S u T in predicate calculus wih equality, F *>RS TRUE.

Theorem 5.2:

Let S u T be defined as above. We suppose that S does not contain the empty clause.

The ETRS RS is confluent on valid formulas iff the following conditions are met:
(i) for each binary factor F of a clause in S, F *>RS TRUE
(ii) for each binary resolvent R of two clauses in S, R *>RS TRUE
(iii) for each narrowing N of a clause in S, N *>RS TRUE

Proof: the run of the proof follows closely the proof of theorem 3.2. Lemmas A,B,C have to be proved only for clauses in R-normal form (that is sufficient because we use only blocked resolution in theorem 5.1). This fact allows us to ignore the rewriting system R; consequently, the proofs of these lemmas are exactly the same as for theorem 3.2.

We need the following additional lemma, which extends lemmas B and C to the narrowing operation:

Lemma H: if C is a clause such that C *>RS TRUE, and N(C) is a narrowing of C, N(C) *>RS TRUE.

Proof: let U be the equational term rewriting system obtained from RS by using only the following rules of BOOL:

¬(TRUE) -> FALSE

$\neg$(FALSE) -> TRUE
XvTRUE -> TRUE
XvFALSE -> X
XvX -> X
Xv($\neg$X) -> TRUE

Let C be a clause such that C *>RS TRUE and N(C) a narrowing of C. We have C *>U TRUE because the rules of BOOL which are not in U cannot be applied to a clause.

We are going to prove that U is confluent w.r.t the associativity/ commutativity of v, by using the confluence criterion of Peterson and Stickel. This criterion consists in checking the confluence of all AC-critical pairs between rules of U and their extensions. AC-critical pairs means that we use AC-unification instead of ordinary unification (see [14] for more details).

It is easy to check that all critical pairs are confluent:
-critical pairs between rules of R are reduced because R is confluent.
-critical pairs between rules of R and rule D -> TRUE, D in S, are confluent from hypothesis (iii), because such critical pairs correspond to a narrowing of D.
-critical pairs between rule D -> TRUE, D in S, and rule XvX -> X of BOOL (or its extension YvXvX -> YvX) are confluent from hypothesis (i), because such critical pairs correspond to a binary factoring of D.
-critical pairs between rule D1 -> TRUE and rule D2 -> TRUE, D1 and D2 in S, with:
D1 = D3 v $\neg$L1
D2 = L2
L1 and L2 being two positive unifiable literals, are confluent from hypothesis (ii), because such critical pairs correspond to a binary resolution between D1 and D2.
-other critical pairs are obviously confluent (in particular, the sub-system of BOOL which we use is confluent).

Since C *>U TRUE, we have C =U TRUE. Hence N(C) =U TRUE from the definition of narrowing. Hence N(C) *>U TRUE by confluence of U. Therefore N(C) *>RS TRUE since U $\subseteq$ RS, which ends the proof of lemma H.

We can now prove that the system RS is confluent on valid formulas by using theorem 5.1 which caracterizes the valid formulas. The proof is exactly the same as the end of the proof of theorem 3.2 and so is omitted.

From theorem 5.2, we deduce a completion algorithm which is similar to the completion algorithm of section 3.3. The differences are:
-We initialize the set of rules R0 to BOOL u R.
-We add at step 3 the computation of narrowings.

If $R\infty$ is the final rewriting system produced by the algorithm, we have the theorem:

Theorem 5.3:

The rewriting system R^∞ has the following properties:
(i) R^∞ is interreduced
(ii) R^∞ is equivalent to $S \cup T$
(iii) R^∞ is confluent on valid formulas
Furthermore, for a given rewriting system R associated with the equational theory T, R^∞ is the only rewriting system associated with a set of clauses which has these properties.

Proof: this proof follows closely the proof of theorem 3.3 and is left to the reader.

5-3: extension to equational term rewriting system

We suppose now that the equational theory T can be compiled into a canonical equational term rewriting system (P,R), P being a set of equations and R being a set of rewrite rules, as described in section 2.1.

We suppose that there is a finite and complete algorithm of P-unification. We define P-binary resolution, P-binary factoring, P-narrowing as binary resolution,... in which P-unification is used instead of ordinary unification.

To extend the previous results, we need an additional property of the ETRS (P,R). This property has been introduced by Jouannaud and is called P-coherence. Roughly speaking, P-coherence allows to replace the reduction relation ->P,R defined in section 2.1 by a weaker relation ->R,P defined as follows:

The term t1 R,P-reduces at occurence u to a term t2 using the rule l->r in R iff there exists a substitution s such that t1/u =P ls and t2 = t1[u<-rs]. Note that R,P-reduction is the same as R-reduction except that we use P-matching instead of matching.

Jouannaud gives sufficient conditions for testing simultaneously confluence and P-coherence of an ETRS. See [10] for details. These results extend the previous results of Peterson and Stickel.

For our purpose, P-coherence allows us to generalize lemma G (lifting lemma for narrowing) if we use P-narrowing instead of narrowing. This generalization is done by Jouannaud and Kirchner [9], where it is used for the construction of unification algorithms.

All other results of sections 5.1 and 5.2 are carried over without difficulty. In particular, the confluence criterion of Jouannaud can be applied to prove the confluence of the system U used in proof of lemma H.

Note that in this framework, the set of non oriented equations is the union of two systems:
-The equational system P.
-The set of associativity/commutativity equations for the boolean connectors v and &.

If R is empty, these results are close to some results of Plotkin [15].

5-4: Theorem of Birkhoff

In this section, we extend the results of section 4 to first order predicate calculus in an equational theory.

Let T be an equational theory, and S = {E1,...,En} a set of quantifier-free formulas which do not contain any equality predicates. E1,...,En are not necessarily in clausal form.

We suppose that we can build from T a canonical term rewriting system (or a canonical and coherent equational term rewriting system as in section 5.3).

We associate with T u S an equational system which is defined as the union of the three following systems:

-The system BOOL defined in section 2.2 (regarded as a set of equations).
-The equational system T.
-The set of equations {Ei = TRUE} with Ei in S.

Note that the symbol = corresponds in this system either to the boolean connector <=> or to the equality predicate.

This equational system defines an equality relation on the set of quantifier-free formulas without equality predicates. We denote this relation =TS or <*>TS.

<u>Theoreme 5.4</u>

Let F and G be two quantifier-free formulas without equality predicates.
F and G have simultaneously the same values (TRUE or FALSE) for every assignment of their variables in all the models of T u S iff F =TS G.
Another equivalent statement is:
The formula F<=>G is a valid consequence of T u S in predicate calculus with equality iff F =TS G.

Proof: the proof follows closely the proof of theorem 4.1 and is left to the reader. Although this proof is valid only if we can build from T a canonical rewriting system, we conjecture that this theorem is true for any equational theory T.

Note the difference between theorem 4.1 and theorem 5.4: in theorem 5.4, we only deal with formulas which do not contain the equality predicate. In exchange for this restriction, we do not use explicitly the axiomatization of equality (reflexivity, symmetry, transitivity, substitution). In fact, this axiomatization is used implicitly when we use an equation M=N in T for replacing in a formula an instance of M by the corresponding instance of N (if we use theorem 4.1, we are only allowed to replace an instance of (M=N) by TRUE).

Let us now compare theorem 5.4 with usual Birkhoff's theorem in equational theories: for this purpose, let us consider a many-sorted equational theory Q with a sort Bool, defined by the union of the three equational systems given above. Usual Birkhoff's theorem states that two expressions are equal in all the models of the theory Q iff they can be proved equal by pure equational reasoning.

This result is identical to the result of theorem 5.4, but there is a major <u>semantic</u> difference: we need to consider all the models of the equational theory Q. In some of these models, the carrier of the sort Bool can be different from the standard two-values model {TRUE,FALSE}. Actually, the carrier of sort Bool can be any boolean algebra.

Theorem 5.4 is much stronger, for it proves that equational reasoning is still complete if we restrict ourselves to the models whose carrier of the sort Bool is the standard model; for these models are the only one to be considered in first order predicate calculus. These models are indeed the most interesting.

That is due to the fact that we use a complete specification BOOL for propositional calculus; this completeness means that two expressions F and G of propositional calculus are equal for all assignments of their variables in the standard model of Bool (i.e. F<=>G is valid) iff they can be proved equal by pure equational reasoning. In other words, inductive reasoning is useless for propositional calculus.

Theorem 5.4 (and also theorem 4.1) are the extensions of this property of completeness from the propositional calculus to the (quantifier-free) predicate calculus.

Note also that theorem 5-4 can be extended by the usual Birkhoff theorem to the case where F and G are two non-boolean terms, if we replace the boolean connector <=> by the equality predicate.

5-5: <u>general confluence in first order predicate calculus</u>

From theorem 5.4, we obtain a method for building rewriting systems confluent on formulas without equality predicate, based on the Knuth and Bendix completion algorithm and the Hsiang system for propositional calculus. Theorems 4.2 (for general confluence) and 4.3 (for refutational proof) are easily extended in this framework.

Note that if the canonical rewriting system for the equational theory T is not provided at the beginning, we can run the completion algorithm simultaneously on T and S.

Example: let us consider the following specification:

Equational theory T: X+0 = X, X+Y = Y+X, X+(Y+Z) = (X+Y)+Z
Set of formulas S: ¬(0>X), X+1>0, X+1>Y+1 <=> X>Y

The standard model of this specification is the set of natural numbers, with the usual meaning of 0,1,+,>.

Applying the Knuth and Bendix completion algorithm, we obtain the following system, complete w.r.t the associativity/commutativity of +, !, &:

```
(Hsiang's system + )
X+0 -> X
0>X -> FALSE
1>1 -> FALSE
1>0 -> TRUE
X+1>Y+1 -> X>Y
X+1>0 -> TRUE
1>X+1 -> FALSE
```

```
X+1>1 -> X>0
```

If we use the (resolution + narrowing) algorithm of theorem 5.3,, we must split the equivalence into two clauses. We obtain an infinite rewriting system:

```
(system BOOL +)
X+0 -> X
¬(0>X) -> TRUE
X+1>0 -> TRUE
¬(X+1>Y+1) v (X>Y) -> TRUE
¬(X>Y) v (X+1>Y+1) -> TRUE
¬(X+1+1>Y+1+1) v (X>Y) -> TRUE
¬(X+1+1+1>Y+1+1+1) v (X>Y) -> TRUE
...
```

ANNEX

It is proved in this annex that the computation of binary factors and binary resolvents can be considered as a computation of critical pairs restricted to certain critical pairs. To take into account the associativity/ commutativity of v, we consider for each rule its extension rule, as defined by Peterson and Stickel [14].

1-Binary factor

C being a clause and F(C) a binary factor of C,the rule F(C) -> TRUE is obviously a critical pair between the rule C -> TRUE and the the rule XvX -> X (or its extension YvXvX -> YvX) of BOOL.

2-Binary resolvent

Let C = Cl v L and D = Dl v ¬P be two clauses, L and P being two positive unifiable literals. Let s be the m.g.u of L and P. A binary resolvent of C and D is R(C,D) = Cls v Dls.
The rewrite rule associated with C and D are:

```
 Cl v L  ->  TRUE
Dl v ¬P  ->  TRUE
```

We suppose that Cl and Dl are not the empty clause (i.e. C and D are not unit clauses). The proof is easily extended if Cl and/or Dl are the empty clause.

The rule R(C,D) -> TRUE can be obtained by a computation of critical pairs as follows:

The rules of BOOL:

```
X v (Y & Z)  ->  (X v Y)&(X v Z)
     X & ¬X  ->  FALSE
```

can be superposed, generating the rule:

```
(X v Y)&(X v ¬Y)  ->  X      (1)
```

Rule (1) can be superposed with the rule Cl v L -> TRUE, generating the rule:

Cl v ¬L -> Cl (2)

The rule (2) can be considered as a kind of "extension" of the rule Cl v L -> TRUE. Such additional rules are also used by Hsiang [5] and Fages [4] for computing critical pairs.

The two rules:

X v Cl v ¬L -> X v Cl (extension of rule (2))
Y v Dl v ¬P -> TRUE (extension of rule Dl v ¬P -> TRUE)

can be superposed since L and P are unifiable. The rule generated is: Cls v Dls -> TRUE, i.e. R(C,D) -> TRUE.

Note that we retain only the rule R(C,D) -> TRUE to build the system R^{∞} confluent on valid formulas. It is not necessary to retain the intermediate rules (1) and (2).

REFERENCES

[1] BIRKHOFF G. : On the structure of abstract algebras. Proc.Cambridge Phil.Soc.31, pp 433-454 (1935).

[2] BUCKEN H. : Reduction systems and small cancellation theory. Proc. Fourth Workshop on Automated Deduction, 53-59 (1979)

[3] CHANG C.L. and LEE R.C. : Symbolic logic and mechanical theorem proving. Academic Press, New-York (1973)

[4] FAGES F. : Formes canoniques dans les algebres booleennes, et application a la demonstration automatique en logique du premier ordre. These de 3me cycle, Universite Pierre et Marie Curie, Juin 1983.

[5] HSIANG J. and DERSHOWITZ N. : Rewrite methods for clausal and non-clausal theorem proving. ICALP 83, Spain.(1983).

[6] HUET G. : A complete proof of correctness of the KNUTH-BENDIX completion algorithm. INRIA, Rapport de recherche No 25, Juillet 1980.

[7] HUET G., OPPEN D.C., Equations and rewrite rules: a survey , Technical Report CSL-11,SRI International, Jan.1980.

[8] HULLOT J.M., Canonical form and unification. Proc. of the Fifth Conference on Automated Deduction, Les Arcs. (July 1980).

[9] JOUANNAUD J.P., KIRCHNER C. and KIRCHNER H. : Incremental unification in equational theories. Proc. of the 21th Allerton Conference (1982).

[10] JOUANNAUD J.P. : Confluent and coherent sets of reduction with equations. Application to proofs in data types. Proc. 8th

Colloquium on Trees in Algebra and Programming (1983).

[11] KNUTH D. and BENDIX P., Simple word problems in universal algebra Computational problems in abstract algebra, Ed. Leech J., Pergamon Press, 1970, 263-297.

[12] LANKFORD D.S. : Canonical inference, Report ATP-32, Department of Mathematics and Computer Science, University of Texas at Austin, Dec.1975.

[13] LEE R.C. : A completeness theorem and a computer program for finding theorems derivable for given axioms. Ph.D diss.in eng., U. of California, Berkeley, Calif.,1967.

[14] PETERSON G. and STICKEL M. : Complete sets of reductions for some equationnal theories. JACM, Vol.28, No2, Avril 1981, pp 233-264.

[15] PLOTKIN G.: Building-in Equational Theories. Machine Intelligence , pp 73-90. (1972)

[16] ROBINSON J.A : A machine-oriented logic based on the resolution principle. JACM, Vol.12, Nol, Janvier 1965, pp 23-41

[17] SLAGLE J.R : Automatic theorem proving for theories with simplifiers, commutativity and associativity. JACM 21, pp.622-642. (1974)

Using Examples, Case Analysis, and Dependency Graphs in Theorem Proving

David A. Plaisted
Department of Computer Science
University of Illinois
1304 West Springfield Avenue
Urbana, Illinois 61801

This work was partially supported by the National Science Foundation under grant MCS 81-09831.

1. Introduction

The use of examples seems to be fundamental to human methods of proving and understanding theorems. Whether the examples are drawn on paper or simply visualized, they seem to be more common in theorem proving and understanding by humans than in textbook proofs using the syntactic transformations of formal logic. What is the significance of this use of examples, and how can it be exploited to get better theorem provers and better interaction of theorem provers with human users? We present a theorem proving strategy which seems to mimic the human tendency to use examples, and has other features in common with human theorem proving methods. This strategy may be useful in itself, as well as giving insight into human thought processes. This strategy proceeds by finding relevant facts, connecting them together by causal relations, and abstracting the causal dependencies to obtain a proof. The strategy can benefit by examining several examples to observe common features in their causal dependencies before abstracting to obtain a general proof. Also, the strategy often needs to perform a case analysis to obtain a proof, with different examples being used for each case, and a systematic method of linking the proofs of the cases to obtain a general proof. The method distinguishes between positive and negative literals in a nontrivial way, similar to the different perceptions people have of the logically equivalent statements $A \supset B$ and $(\neg B) \supset (\neg A)$. This work builds on earlier work of the author on abstraction strategies [17] and problem reduction methods [18], and also on recent artificial intelligence work on annotating facts with explanatory information [6,7,9]. This method differs from the abstraction strategy in that it is possible to choose a different abstraction for each case in a case analysis proof; there are other differences as well. For other recent work concerning the use of examples in theorem proving see [1] and [2].

1.1 Comparison with previous work

Several methods have been proposed for using examples or semantics in theorem provers. Gelernter [10] developed a geometry theorem prover which used back chaining and expressed semantic information in the form of diagrams; this enabled unachievable subgoals to be deleted. Reiter [19] proposed an incomplete natural deduction system which could represent arbitrary interpretations and use them as counterexamples to delete unachievable subgoals. His method also could use models to suggest instantiations of free variables. Slagle [20] presented a generalization of hyper-resolution to arbitrary models; his system gives a semantic criterion for restricting which resolutions are performed. Ballantyne and Bledsoe [1] give techniques for generating counterexamples in topology and analysis, and also show how examples can help in a positive sense in finding a proof. This idea is extended by Bledsoe [2] who gives methods for instantiating set variables to help prove theorems with existential quantifiers.

Our method differs in the following ways: a) We actually apply a transformation to the clauses themselves and do a search on the transformed clauses. This transformation is based on semantic information. b) We split the set of input clauses to obtain Horn clauses and use the splits to structure the case analysis in the proof. c) We construct a dependency graph representing the assertions that can conceivably contribute to a proof, and then restrict the search to these assertions. Our method permits different models to be used for each case in a case analysis proof. Also, our methods of generating examples and counterexamples are not nearly as sophisticated as those in [1] and [2]; we concentrate on methods that are simple and general and easily mechanized. However, eventually more complex, domain-dependent approaches such as those in [1] and [2] will undoubtedly be necessary.

1.2 An example

We informally discuss an example of how the method works, before giving a formal presentation. This example should make the main features of the method clear. The theorem is the following: for all natural numbers x, x*(x+1) is even. Assume we have the axioms even(x) $\vee$ odd(x), even(x) $\supset$ odd(x+1), odd(x) $\supset$ even(x+1), even(x) $\supset$ even(x*y), and some axioms about arithmetic. The theorem, when negated and Skolemized, becomes $\neg$even(c * (c + 1)), which can be viewed as a goal of even(c*(c+1)). We first split the non-Horn clause even(x) $\vee$ odd(x) to obtain the two clauses even(x) and odd(x) which must be dealt with separately. For semantics we use the standard model of the integers; this will be the "initial model" introduced below. Now, c may be interpreted to an arbitrary integer. Intuitively, when dealing with the case even(x), we would like to interpret c as an even integer, and when dealing with the case odd(x), we would like to interpret c as an odd integer. However, in general, there are technical problems so that when doing a case analysis, it is not always possible to find an interpretation making the case true. Thus when doing the case even(x), it is conceivable that we might be forced to look at an interpretation in which x is odd. For now suppose this does not happen. Suppose we do the case even(x) and interpret c to be 2. Suppose we also interpret x to be 2. Then we construct causal relations between such interpreted facts; we say even(2) causes even(2*3). From these causal relations we construct a proof for the case even(x). Similarly, for the case odd(x), we interpret x and c as 3, say. We say odd(3) causes even(4) and even(4) causes even(4*3). From these causal relations we construct a dependency graph relating odd(3), even(4), even(4*3), and even(3*4) (making use of x*y = y*x). From this graph we construct a general proof for the case odd(x); these proofs are then combined to obtain a complete proof.

2. The Horn Case

First we consider the case in which the input clauses are all Horn clauses. We assume the reader is familiar with the usual concepts of a term, a literal, a clause, a substitution, and so on. For an introduction to such theorem proving terminology see [5] or [15]. An *atom* is a positive literal. A *Horn clause* is a clause in which at most one literal is positive; thus $P \wedge Q \supset R$, considered as the clause $\neg P \vee \neg Q \vee R$, is a Horn clause. Consider the task of finding a proof of a positive literal M (called the *goal literal)* from a set S of Horn clauses. Assume for now that none of the clauses of S are all-negative clauses. This implies that S is consistent. Therefore there is a *Herbrand model* of S, which for our purposes is a set of ground atoms which are assumed to be true. Also, this model must make S true, in the sense that all clauses in S are true in the interpretation in which literals in the model are true and all other literals are false. Since S is a Horn set and has no all-negative clauses, there is in fact a *minimal* Herbrand model of S, which consists of all positive ground literals that may be derived from S by hyper-resolution (equivalently, all positive ground literals that are logical consequences of S). We denote this model by Cl(S), the *closure* of S. The first step in the proposed theorem proving method is to interpret the function symbols of S and interpret the terms of Cl(S) in a corresponding way to obtain a more "concrete" model that will serve as an example for our theorem prover. Let I be an interpretation of the function and constant symbols of S. If t is a ground term composed of function and constant symbols of S, let t^I be the value of t in I. Assume for simplicity that S is untyped, so that I has one domain D_I and each function symbol f is interpreted as a function from D_I^n to D_I for n the arity of f. Then I can be extended to a model of S by properly interpreting the predicate symbols of S; let $M_I(S)$ be the minimal such extension of I. We define an *interpreted literal* to be a literal whose arguments are elements of D_I. Thus an interpreted literal is of the form $P(d_1, \cdots, d_k)$ where P is a predicate symbol of S and the d_i are elements of D_I, or is of the form $\neg P(d_1, \cdots, d_k)$. An *interpreted clause* is a clause which is the disjunction of interpreted literals. If L is a ground literal and I is an interpretation, let L^I be the interpreted literal in which each term of L is replaced by its value in I. Thus if L is $P(t_1, \cdots, t_k)$ and $t_i^I = d_i$ then L^I is $P(d_1, \cdots, d_k)$. If C is a ground clause and I is an intepretation, let C^I be the interpreted clause in which each literal L of C is replaced by L^I. For example, if C is $a \leq b \vee b < a$ and $a^I = 2$ and $b^I = 3$ then C^I is $2 \leq 3 \vee 3 < 2$. Let S^I be $\{ C^I$: C is a ground instance of a clause in S $\}$. Let $Cl^I(S)$ be $\{$

L^I : L $\in$ Cl(S) }. One can show that $Cl^I(S) \subset Cl(S^I)$. If D_I is finite, then S^I is a finite set of ground clauses, and its closure is finite and can be computed in a finite number of steps. Now $Cl(S^I)$ is a model of S^I. Also, $Cl(S^I)$ is a model of S, in the sense that if I is extended by making $P(d_1, \cdots, d_k)$ true iff $P(d_1, \cdots, d_k)$ is in $Cl(S^I)$, then this extension of I is a model of S. In fact, this extension of I is just $M_I(S)$. Recall that $M_I(S)$ is *minimal,* in the sense that removing any literals would not yield a model of S. Thus, given any interpretation I of the function symbols of S such that D_I is finite, a minimal model $M_I(S)$ of S having domain D^I can be computed in a completely mechanical way. For certain standard sets S of clauses and interpretations I, $M_I(S)$ will be known in advance and need not be derived. For example, for the integers, we can quickly compute that $2 < 5$, $\neg(3<1)$, et cetera. We will also be interested in models whose domains are infinite. We consider such a model to be an example, and the elements of $Cl(S^I)$ as facts which are to be connected by causal relations, as explained below. Suppose I is specified. If D_I is infinite, then $Cl(S^I)$ may be infinite, so in practice we need to restrict attention to a relevant subset J of $Cl(S^I)$. Even if D_I is finite, it still may be useful for efficiency reasons to restrict attention to a relevant subset J of $Cl(S^I)$. We can think of the literals in J as "relevant facts" for the proof.

The next step (step 2) is to attach *causal relations* to the literals of J. Literals of J which are (positive) unit clauses of S^I will be called *assumption* literals. Also, if L1, L2, ..., Lk and L are literals of J such that the clause $L1 \wedge L2 \wedge \cdots \wedge Lk \supset L$ is in S^I, we say L is *caused by* L1 and L2 and ... and Lk. We call this a causal relation between {L1, L2, ..., Lk} and L, and use the notation {L1 , ..., Lk} $\rightarrow$ L. For convenience in remembering the correspondence between literals in clauses and literals in causal relations, we consider {L1, ..., Lk} to be an ordered sequence, so that the order of the literals is significant. Similarly, the order of literals in a clause is significant. The causal relation {L1 , ..., Lk} $\rightarrow$ L. is annotated with the clauses C of S having ground instances C1 such that $C1^I$ is is $L1 \wedge L2 \wedge \cdots \wedge Lk \supset L$. Note that there may be more than one such clause C. Also, the set {L1, L2, ..., Lk} may be empty, in which case L is an assumption literal. In this way we see that an assumption literal is a special case of a causal relation. The collection of annotated causal relations will be called the *dependency graph* for J. This graph has nodes labeled with the literals of J, and relations between the nodes corresponding to the causal relations of J. These relations are annotated with clauses of S, as above. The particular literals labeling nodes in the graph are not as important as the dependency relations between them. Therefore we say that two dependency graphs are *isomorphic* if they are the same except for the replacement of some nodes and labels by others. A literal L of J which is an instance of the goal literal M is called a *goal literal* of J, and the associated node of the dependency graph is called a *goal node.* If $\{L_1, \cdots, L_k\} \rightarrow$ L is a causal relation of a dependency graph G, and $L_1, \cdots, L_k, L$ are the labels of nodes $N_1, \cdots, N_k, N$ of G, respectively, then we also consider $\{N_1, \cdots, N_k\} \rightarrow$ N to be a causal relation of G, for convenience.

From the dependency graph, we obtain a *causal chain* relating some goal node to assumption nodes of the dependency graph; this is step 3 of the method. This causal chain is a tree, the root of the tree being labeled with a goal literal, and for each node N of the tree, there is a causal relation between the labels {L1, L2, ..., Lk} of the sons of N and the label L of N. Thus the leaves of the tree must be labeled with assumption literals of J. It is permissible for a literal to label more than one node of the tree; in fact, a literal may even label both a node and one of its sons or more distant descendents. Hence for a given dependency graph there may be infinitely many possible causal chains. Each such tree represents a combination of causal relations by which some goal literal may be related to assumption literals. A causal chain may be considered as an attempt to obtain a general proof by looking first at the causal relations in this particular example.

Step 4 of the proof method is to consider several examples, and for these examples to find causal chains which have the same structure (the labels of the nodes may be diferent but the trees must have the same shape and the same clauses of S annotating the causal relations at corresponding nodes). Step 4 poses an interesting algorithmic problem, and there are many possible algorithms that could perform this step efficiently. The idea is to look at all the examples at the same time and gradually extract a common causal chain from all of them, working backwards from the goal. Later (in section 4.1) we will give an algorithmic solution to step 4. Step 5 is to obtain a general proof by lifting this common causal chain to a proof of M from S. It is not difficult to see that this method is complete (subject to the proper choice of J), since positive unit resolution for Horn sets is complete[11]. The intent of the method is that the number of causal chains will be reduced by the examples under consideration, so that proofs will not even be attempted which do not correspond to causal chains from an example. This reduction in search space should be even more pronounced if several examples are used. The causal chains give the prover some idea of where it is trying to go in a global sense, as opposed to the common approach of resolving any two clauses which satisfy some local property specified by a resolution strategy. Note that if all domains of interpretations are finite, step 5 is the only part of the method that involves general resolution and undecidability; every other part of the method deals only with ground clauses and ground resolution. Therefore the first four parts should be reasonably fast, and hopefully the search space for part 5 will be reduced by the preprocessing of steps 1 through 4. If some of the interpretations have infinite domains, then constructing $Cl(S^I)$ involves a potentially infinite search, but this can be limited by choosing a finite "relevant" subset J of $Cl(S^I)$. Also, we note that the method is still complete if interpretations with infinite domains are never chosen, but such infinite interpretations may be useful for intuitive reasons and possibly may reduce the search space in step 5 in some cases, as we shall indicate in section 5.

Some models I are much more useful than others in guiding the search for a proof. The function of the model is to eliminate as many causal chains as possible, so that it is only necessary to search among a small subset of the causal chains to find one leading to a proof. We say a model $M_I(S)$ is *sparse* if it makes each predicate symbol true on a small number of sets of arguments. Clearly if $M_I(S)$ makes each predicate symbol true everywhere, then many possible causal chains can be generated from the dependency graph, and the search for a proof is more difficult. It is an interesting question to determine for which interpretations I does $M_I(S)$ have good sparseness properties. Note that the method is still complete if we choose *any* model of S, not just models of the form $M_I(S)$. However, one can show that for any model M1 of S there exists I such that $M_I(S) \subset M1$, so the models $M_I(S)$ are the best ones for purposes of reducing the search space.

2.1 Example 2

Let S be {Pa, Px $\supset$ Pfx} and let the goal literal be Pfffa. We are omitting parentheses for simplicity. Then Cl(S) is {Pa, Pfa, Pffa, ... }. Let I have domain D_I = {b, c, d, e}, a interpreted as b, and f interpreted so that fb = c, fc = d, and fd = b. Then S^I is {Pb, Pb $\supset$ Pc, Pc $\supset$ Pd, Pd $\supset$ Pb} and $Cl(S^I)$ is {Pb, Pc, Pd}. Thus $M_I(S)$ is the model of S which interprets a and f as does I and interprets P so that Pb, Pc, and Pd are true but Pe is false. Let J be {Pb, Pc, Pd}. The dependency graph has nodes N1 labeled Pb, N2 labeled Pc, and N3 labeled Pd. Also, N1 is an assumption node and N2 is a goal node. There is a causal relation on N1 annotated with the clause Pa of S, a causal relation {N1} → N2 annotated with the clause Px $\supset$ Pfx, a causal relation {N2} → N3 annotated with the clause Px $\supset$ Pfx, and a causal relation {N3} → N1 annotated with the clause Px $\supset$ Pfx. The graph is thus as follows:

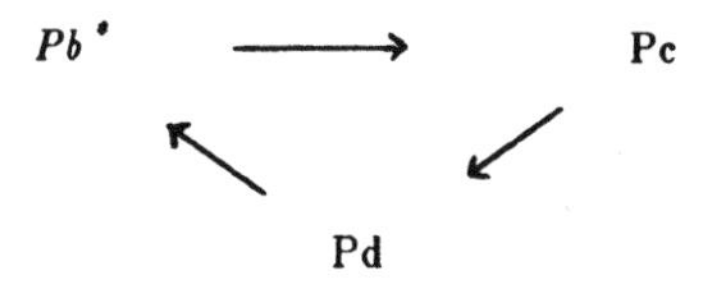

The superscript * indicates an assumption node. There are an infinite number of causal chains connecting the goal literal N2 to the assumption literal N1; one of them is the tree containing nodes M1 and M2 labeled with Pb and Pc, respectively, and the other is the tree containing nodes M1, M2, M3, M4, and M5 labeled with Pb, Pc, Pd, Pb, and Pc, respectively. In both trees, M1 has a causal relation annotated with the clause Pa, and there is a causal relation between {M1} → M2 annotated with the clause Px $\supset$ Pfx. In the second causal chain, {M2} → M3, {M3} → M4, and {M4} → M5 are causal relations annotated with the clause Px $\supset$ Pfx. The first causal chain cannot be lifted to obtain a proof, but the second causal chain can be lifted to obtain the following proof: Pa (assumption), Pfa (using Px $\supset$ Pfx), Pffa, Pfffa,

Pfffa (all using Px $\supset$ Pfx).

2.2 Example 3

This is from Chang and Lee [5]. The problem is to show that in a group, if the square of every element is the identity, then the group is commutative. When phrased in relational form (so that P(x y z) means x*y = z), this yields the following set S of clauses:

Pexx Pxex

Pxxe Pabc

$$Pxyu \wedge Pyzv \wedge Pxvw \supset Puzw$$

$$Pxyu \wedge Pyzv \wedge Puzw \supset Pxvw$$

Also, the goal literal is Pbac. Here x, y, u, z, v, and w are variables and e, a, b, and c are constants. In this example, the closure Cl(S) of S consists of the sixteen literals of the form P(x y z) where x and y are all possible combinations of e, a, b, and c, and z is as required by the "multiplication table" for this group. Thus we have P(x e x) for all x, P(e x x) for all x, P(x x e) for all x, P(a b c), P(b a c), P(a c b), P(c a b), P(b c a), and P(c b a). Let I have domain {a b c e} and interpret a as a, b as b, c as c, and e as e for simplicity. Then S^I consists of the ground instances of clauses in S; that is, the ground instances obtained by replacing variables by a, b, c, and e. Also, $Cl(S^I)=Cl(S)$ in this case, and $M_I(S)$ is the interpretation in which a, b, c, and e are interpreted as themselves and P(x y z) is true for the sixteen combinations of a, b, c, and e given above and false otherwise. Let J (the relevant set of literals) be $Cl(S^I)$, which is the above sixteen literals. Now, Pbac is the goal node of the dependency graph, all ground instances of P(x e x), P(e x x) and P(x x e) are the assumption literals of this graph, and for each ground instance $L1 \wedge L2 \wedge L3 \supset L$ of the last two clauses such that L1, L2, L3, and L are all true in $M_I(S)$ there is a causal connection between {L1, L2, L3} and L in the dependency graph. These causal connections are annotated with the appropriate clauses of S. From this dependency graph, causal chains may be extracted and lifted to obtain a proof of the goal literal from S.

2.3 Example 4

This example is intended to show the reduction in search space that may be obtained by using examples. Theoretically, there may be many literals L, perhaps infinitely many, corresponding to a given interpreted literal L^I. When using interpretation I as an example, all these literals will be considered at once. This makes it possible to learn about the structure of the search space as a whole while examining a comparitively small set of literals. The information gained in this way may be used to guide the search in the full search space.

Consider the following axioms:

Pf(c)	$x \geq y \supset fx \geq fy$
$Pfx \supset Pffx$	$x \geq y \supset gx \geq gy$
$Pfx \supset Pgfx$	$x \geq x$
$Pgx \supset Pggx$	$x \geq y \wedge y \geq z \supset x \geq z$
$gfx \geq fgx$	

Theorem:
$(\exists x)P(x) \wedge x \geq ffgfggc$

The first four axioms permit derivation of $Pg^k f^j c$. The remaining axioms permit g and f to interchange, decreasing the value in the partial ordering. We consider two interpretations, I_a and I_b.

First Interpretation

We interpret c as 0, f(x) as x, g(x) as x+1, P(x) as x > 0, and x ≥ y as x = y. This model basically counts occurrences of g.

Second Interpretation

We interpret c as 0, f(x) as x+1, g(x) as x, P(x) as x ≥ 0, and x ≥ y as x = y. This model basically counts occurrences of f.

The theorem, negated, becomes $\neg P(x) \vee \neg(x \geq ffgfggc)$. This is converted to $P(x) \wedge x \geq ffgfggc \supset GOAL$ where GOAL is the goal literal. For the first interpretation, assuming J is chosen large enough, we obtain the following dependency graph, where literals have been omitted that cannot contribute to a proof of the goal:

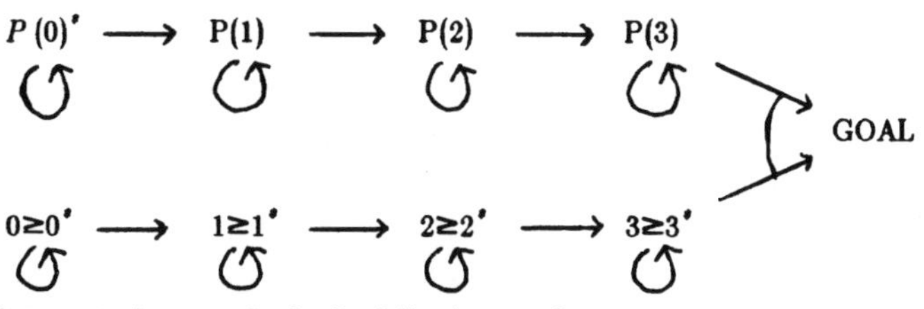

$P(0)' \longrightarrow P(1) \longrightarrow P(2) \longrightarrow P(3)$

GOAL

$0 \geq 0' \longrightarrow 1 \geq 1' \longrightarrow 2 \geq 2' \longrightarrow 3 \geq 3'$

For the second interpretation, we obtain the following graph:

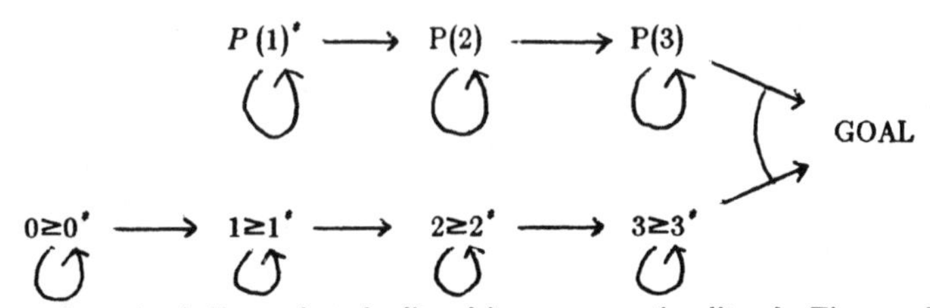

As before, superscripts of * indicate that the literal is an assumption literal. The graph for the first interpretation will prevent the theorem prover from examining literals with too many g's, and the graph for the second interpretation will prevent the prover from examining literals with too many f's. These graphs can be matched, as will be explained in section 4.1. This essentially results in graphs whose literals are ordered pairs of literals from the above two graphs, and results in a graph which keeps track of numbers of f and g symbols at the same time, further reducing the search space. Note that in the matched graph, the single pair (P(3), P(3)) represents literals having 3 f symbols and 3 g symbols; there are 20 such literals, but only one entry in the graph. This illustrates how these graphs form a compact representation of the full search space. Although this example may be artificial, it should illustrate the possibilities for controlling search using examples in this way. The reduction in search space is similar to that achieved by the abstraction strategy [17].

2.4 Comments

As mentioned above, this method of using examples is similar to the "abstraction" strategies presented in [17]. However, in this method, it is not necessary to use multiclauses (in which a literal may appear more than once); ordinary clauses suffice. The method may easily be modified to find refutations (proofs of FALSE) rather than proofs of a goal literal. For this, it is permissible for S to contain all-negative Horn clauses. In this case, S is modified to obtain a new set S1 of clauses by replacing each all-negative clause $\neg L1 \vee \cdots \vee \neg Lk$ by $L1 \wedge L2 \wedge \cdots \wedge Lk \supset GOAL$, where GOAL is a new predicate symbol interpreted as TRUE. The method is then applied to S1 with GOAL as the goal literal.

2.5 Equality

The method also works for problems containing equality axioms. However, certain choices of models and representations are better than others. It may be that Cl(S) contains literals of the form $s = t$ for ground terms s and t. If the interpretation is chosen so that s and t have the same value for all such pairs s and t, then reasoning involving the equality axioms is difficult to extract from the example. For example, if I is an interpretation such that for terms s and t, $s^I = t^I$ if $S \models s = t$, and L is a literal in $Cl(S^I)$ of the form $q = r$, then q and r will be identical, which gives no information about how the literal $q = r$ can be derived. One solution is to choose interpretations I in which equality is a *congruence* relation, not necessarily the equality relation on D_I. This permits s and t to have different values in I even if $s = t$ is a logical consequence of S. Another solution is to redefine the concept of an interpreted literal. Suppose an interpretation I with domain D_I is specified. Then we can define an interpreted literal to be a ground literal in which certain subterms are replaced by their values in I. For example, if literal L is S(S(S(0))) > S(S(0)), and I is the usual interpretation of arithmetic with S the successor relation, then the following are all interpreted literals corresponding to L: 3 > 2, 3 > S(1), S(2) > S(1), S(2) > S(S(0)), et cetera. Then L^I may be chosen to be some such literal corresponding to L. It is necessary to choose L^I consistently, that is, if L and M are complementary ground literals, then L^I and M^I must also be complementary. The advantage of this method of defining L^I is that more information can be retained when using equality. For example, it is possible to have a causal relation between even(4*3) and even(3*4) using the commutativity of multiplication. This gives more information than having a causal relation between even(12) and even(12), which is all that would be permitted by the previous definition of interpreted literal. Also, since the equality axioms are cumbersome to use explicitly, it seems better to use some representation such as Brand's "modification method"[3] to eliminate the need for most of the equality axioms. With these choices of representation and interpretations, the method should work reasonably well for problems involving equality. It should be possible to incorporate some term-rewriting ideas[13] to only construct proofs in which "complex" terms are replaced by "simpler" terms, reducing the search space. Such a method was shown complete by Lankford [14].

2.5.1 Equational systems

If S is of the form $S1 \cup E$ where E is an equational system having an efficient unification algorithm, the method can be used together with E unification as follows: The interpretation I must be chosen so that all equations in E are true on D_I. Then steps 1 through 4 of the method can be applied exactly as before, using the set S1 instead of S. Thus the equations in E are ignored for steps 1 through 4. In step 5, to lift the proof, we use E unification instead of unification. With this modification, the method is complete and can take advantage of the many efficient unification algorithms known for various sets of equations, such as the associative-commutative unification algorithms of [21] and [12] and many other algorithms [8].

3. The Non-Horn Case

The idea for extending the method to the non-Horn case is to use splitting[4] to obtain Horn clauses and to prove the theorem for each set of Horn clauses obtained by splitting. Suppose S is a set of clauses, possibly including non-Horn clauses. Let I be some interpretation of the function and constant symbols of S, as before. Then the domain D_I, the interpreted literals L^I, the interpreted clauses C^I, and the set S^I of interpreted clauses are defined as before. Let S_H be S with non-Horn clauses replaced by Horn clauses in the following way: Suppose $L_1 \wedge \cdots \wedge L_k \supset M_1 \vee \cdots \vee M_n$ is a clause of S, in which the L_i and M_j are all positive literals and n > 1. This non-Horn clause is replaced by the Horn clause $L_1 \wedge \cdots \wedge L_k \supset M_1$ with the literals $M_2 \vee \cdots \vee M_n$ considered as *alternatives* to this clause. We call this new Horn clause $L_1 \wedge \cdots \wedge L_k \supset M_1$ a *split clause*. Then S_H is S with all non-Horn clauses replaced by split clauses this way. We can now apply the method as usual to S_H using interpretation I. The difference is that the split clauses have alternatives associated with them, so that the annotations of the dependency graph also include the alternatives of split clauses. The causal chains similarly have alternatives of split clauses included. Then we lift to get a general proof of the goal literal GOAL from S_H as before. This proof can be converted in a simple way to a proof of $GOAL \vee M_1 \vee \cdots \vee M_p$ from S, where all the alternatives of split clauses used in the proof have instances among the M_i. It could be that no split clauses were used in the proof; in this case we are done. If not, the next task is to find a proof of the goal literal from $S \cup \{ M_i \}$ for each M_i in turn, by the same procedure applied recursively. The procedure must be modified so that only one instance of each M_i is used in the proof; this is easy to do but the details are tedious so we omit them here. Also, after finding a proof of the goal literal from S $\cup \{ M_i \}$, it is necessary to find which instance of M_i was actually used in the proof, and to apply the same substitution to the remaining M_j. Again, we omit details for simplicity; the general idea should be clear. It is most convenient for this method if the goal literal is a ground literal. Otherwise, it is possible to replace all free variables in the goal literal by new Skolem constants. If one desires to find all instances of the goal literal that are derivable, there could be a complication, because it may be that P(a) $\vee$ P(b) is derivable even if neither P(a) nor P(b) is derivable separately. Again, the method for the non-Horn case can be modified in a rather direct way to take care of this possibility and in general to return instances of the goal literal that are derivable, if such instances exist. This method of dealing with non-Horn clauses is intuitively appealing because it corresponds naturally to the human method of performing case analysis; the alternatives M_i may be thought of as other cases that need to be examined later, and cases that arise in the proof of the goal literal from S $\cup \{ M_i \}$ may be thought of as subcases.

3.1 Example 5

We give an example to illustrate the use of splitting and case analysis for non-Horn clauses. Suppose S consists of the following clauses:

$$x \geq y \supset max(x, y) = x$$
$$y > x \supset max(x, y) = y$$

$x \geq y \quad \vee \quad y > x$

$x = y \quad \supset \quad x \geq y$

$x \geq y \quad \wedge \quad y > z \quad \supset \quad x \geq z$

Suppose the goal is to show that $max(x, \ y) \geq x$. We first introduce new Skolem constants c and d and transform the goal to $max(c, \ d) \geq c$. Let I be the interpretation in which the domain D_I is the natural numbers {0, 1, 2, ... } and max(x, y) gives the maximum of the numbers x and y. Suppose I interprets c as 2 and d as 4. The non-Horn clause $x \geq y \ \vee \ y > x$ is replaced by the split clause $x \geq y$ with alternative $y > x$. The Horn set S_H is obtained by replacing $x \geq y \ \vee \ y > x$ in this way. Now S_H^I has infinitely many clauses, including $2>3 \ \supset \ 3=2$ and $4>1 \ \supset \ 4=4$ (from the first clause), $4>2$ and $1>5$ etc. (from the split clause), and in general we get all clauses resulting from replacing the variables of clauses in S_H with natural numbers and evaluating max in the standard way. The closure $Cl(S_H^I)$, is all ground literals derivable from S_H^I by hyper-resolution. This includes all literals of the form $x \geq y$ (from the split clause) and all literals of the form $y = x$ (from the first clause, using literals of the form $x \geq y$). Let J be all such literals whose arguments x and y are in the range {0, 1, 2, 3, 4, 5}, say. From J it is possible to construct a dependency graph, a causal chain, and a proof of $max(c, d) \geq c \ \vee \ d > c$. There are also many other proofs that could be constructed. We then attempt a proof of the goal literal from $S \cup \{d > c\}$. This proof can be found without using the split clause, so we are done. This example illustrates the use of splitting. However, it is not entirely satisfactory because the closure $Cl(S_H^I)$ includes all possible literals, which means that in $M_I(S)$, both "=" and "≥" are interpreted as identically true. Thus $M_I(S)$ is "dense." This does not seem intuitively appealing, and in addition results in a dependency graph which does not help much in reducing the search space. We will show how to overcome these problems below. Despite the problems, this example should give an idea of how splitting works, and should help to motivate the discussion of section 5.

4. Matching and Searching Dependency Graphs

We now give a more detailed description of how the dependency graphs are actually used to guide the search for a proof, and how several dependency graphs may be matched to obtain a graph which contains the common features of all the separate graphs. It is not best to extract causal chains and then search separately for a proof using each causal chain, because there may be many subproofs in common among the various causal chains and these will be found repeatedly. It is better to work directly with the dependency graphs. The number of vertices in the dependency graph is bounded by the number of literals in J; however, there may be infinitely many causal chains, even for a finite number of literals. Even if the depth of the causal chains is limited to d, the number of such chains may still be double exponential in d (an exponential number of possible nodes, and for each node more than one possible literal). We show how to work directly with the dependency graphs, and thereby avoid this problem. Also, we show how to attach depth information to the nodes of the dependency graph to make it easy to search for proofs at restricted depths.

4.1 Matching Dependency Graphs

Suppose dependency graphs $G_1, \cdots, G_n$ are given and we want to find a graph G which contains the common features of all the G_i. More precisely, we want a graph G such that the set of causal chains of G is the intersection of the sets of causal chains of the G_i. To obtain G, we first define a procedure "match" which matches two graphs F_1 and F_2 to yield a graph F whose set of causal chains is the intersection of the causal chains of F_1 and F_2. This procedure is then applied to G_1 and G_2 to produce a new dependency graph H_1; then H_1 and G_3 are matched to produce H_2; then H_2 and G_4 are matched to produce H_3; and so on, until all graphs are matched to produce H_{n-1}, which is the desired graph G.

The graph G = match(G_1, G_2) is defined as follows, for arbitrary dependency graphs G_1 and G_2: Let Nodes(G) be the nodes of G and Rel(G) be the set of causal relations of G. The nodes of G are ordered pairs $<N^1, N^2>$ for $N^i \in Nodes(G_i)$. To obtain G, first let Nodes(G) be $\emptyset$. Then add $<N^1, N^2>$ to Nodes(G) for all $N^i \in Nodes(G_i)$ such that the N^i are assumption nodes of G_i and both N^1 and N^2 are annotated with the same clause {M}. The ordered pair $<N^1, N^2>$ is an assumption node of G and is added to Rel(G) as a causal relation, annotated with all such clauses {M}, which will be positive unit clauses. Then successively add to Nodes(G) all nodes as follows, and add to Rel(G) all causal relations as follows, until no more can be added: If N^i and $N_1^i, \cdots, N_k^i$ are in $Nodes(G^i)$ for i = 1, 2, and the causal relations $\{N_1^i, \cdots, N_k^i\} \rightarrow N^i \in$ Rel(G^i) for i = 1, 2, both causal relations annotated with the same clause C, and $<N_j^1, N_j^2> \in$ Nodes(G) for $1 \leq j \leq k$, then add $<N^1, N^2>$ to Nodes(G) and add the causal relation $\{<N_1^1, N_1^2>, \cdots, <N_k^1, N_k^2>\} \rightarrow <N^1, N^2>$ to Rel(G), this causal relation annotated with all such clauses C.

4.2 Assigning Depths to the Dependency Graph

After matching the dependency graphs as above, depth information is added to each node telling at what depths the node can possibly contribute to a proof of the goal node. Each node N of the dependency graph G has a *forward depth* $d_f(N)$ and a *backward depth* $d_b(N)$ assigned to it. If no matching on graphs has been done, then the forward depth of N tells the smallest depth at which the literal labeling N can possibly be derived by hyper-resolution; the backward depth of N tells the smallest possible depth of a proof of the goal node such that the label of N occurs in the proof. To be precise, $d_f(N)$ is the depth of the smallest causal chain whose root node is labeled with the label of N. Also, $d_b(N)$ is the depth of the smallest causal chain whose root node is a goal node, and which also contains some node labeled with the literal labeling N. Therefore, when searching for proofs of depth d or less, it is only necessary to consider the subgraph G_d of G consisting of nodes N such that $d_b(N) \le d$, together with all associated causal relations. The forward depths are assigned as follows: First assign $d_f(N)=\infty$ for all N which are not assumption nodes; if N is an assumption node, assign $d_f(N)=0$. Then iterate the following equation on all causal relations $\{N_1, \cdots, N_k\} \rightarrow$ N until no more change occurs:

$$d_f(N)=\min(d_f(N),1+\max(d_f(N_1), \cdots, d_f(N_k)))$$

To assign backward depths, first assign $d_b(N)=\infty$ for all N except the goal node N_G, for which $d_b(N_G)=d_f(N_G)$. Then iterate the following equation on all causal relations $\{N_1, \cdots, N_k\} \rightarrow$ N until no more change occurs:

$$d_b(N_i)=\min(d_b(N_i),\max(d_b(N),1+\max(d_f(N_1), \cdots, d_f(N_k))))$$

4.3 Searching for Proofs

To search for proofs at depth d or less, we consider a subgraph G_d of the dependency graph G, where G_d consists of all nodes N of G such that $d_b(N) \le d$, together with their associated causal relations. With each such node N, we associate a set clauses(N) of (clause, depth) pairs as follows: Initially clauses(N) = Ø for all nodes N of G_d. Then add <{M}, 0> to clauses(N) for assumption nodes N, where {M} is a positive unit clause annotating N. Thereafter, do the following as often as possible until no more clauses can be generated or until a proof is found: Suppose $\{N_1, \cdots, N_k\} \rightarrow$ N is a causal link of G_d, annotated with the clause $L_1, \cdots, L_k \supset L$, and suppose $<M_i, d_i> \in$ clauses(N_i) and $d_i < d$ for $1 \le i \le k$. Let M be the hyper-resolvent of $L_1, \cdots, L_k \supset L$ and $M_1, \cdots, M_k$, if such a hyper-resolvent exists. Add $<M, 1+max(d_1, \cdots, d_k)>$ to clauses(N). When doing splitting it may be necessary to continue even after some element of clauses(N_G) has been derived (where N_G is the goal node), since different proofs may use different instances of the alternatives.

5. Splitting Refinements

We present semantic refinements of splitting that are intuitively appealing and also reduce the search space. The motivation is that it is desirable to reduce the number of split clauses used in a proof, since each such clause used introduces alternatives which must also be considered. The refinements discussed here are intended to deal with problems such as those mentioned in example 5. Suppose S is a set of clauses; let H(S) be the set of Horn clauses of S containing exactly one positive literal. (We assume all-negative clauses C1 of S have been replaced by $C1 \vee GOAL$ as indicated earlier.) Let C be a non-Horn clause of S which can be expressed as $C_1 \vee \cdots \vee C_k$ where C_1 is a Horn clause and the remaining C_i are the remaining positive literals of C. Let D be a ground instance of C, and let D_i be the corresponding ground instances of the C_i. Now, we want to minimize the use of such ground instances D in proofs, as mentioned above. If $H(S) \models D_i$ for some i, then there is a proof which does not use D, since some D_i can be derived entirely from Horn clauses. Therefore it is sufficient to use only D such that no D_i is a logical consequence of $H(S)$. Furthermore, there are certain models of $H(S)$ from which it is easy to determine which literals are logical consequences of $H(S)$. For an arbitrary set S of Horn clauses, we say an interpretation I of the function and constant symbols of S is *S-minimal* if for all ground literals L, $S \models L$ is equivalent to $S^I \models L^I$. Note that if I is S-minimal, then $S \models L$ iff $L^I \in Cl(S^I)$. Therefore by examining the model $Cl(S^I)$, it is possible to determine which literals are logical consequences of S, which may be useful for restricting splitting, as indicated above. We say an interpretation I of the function and constant symbols of S is *S-initial* if I is S-minimal and furthermore for all terms s and t composed of the function and constant symbols of S, $s^I = t^I$ iff $S \models s = t$. Initial interpretations are particularly intuitive because they only identify terms which are equal in all models. Often, initial interpretations correspond to standard models, such as the standard model of the integers. According to the above remarks, it is not necessary to use ground instances D such that $D_i^I \in Cl(S^I)$ if I is S-minimal. For example, we would not want to use D $= 2 \geq 3 \vee 3 > 2$ in the usual axioms for inequalities and natural numbers, since $3 > 2$ is itself a consequence of the axioms.

There is another approach to restricting the use of non-Horn clauses. It seems counter-intuitive to use some "false" literal such as $2 \geq 3$ as a split literal. When we do a proof by case analysis, if we assume some case D_i is true, we generally have in mind a model in which D_i is true. To force consideration of models in which D_i is false is unnatural and also increases the search space (because such models are less sparse). Frequently the goal literal is of the form $\forall x_1 \cdots \forall x_m A(x_1, \cdots, x_m)$ for some formula A. To prove this, we convert the x_i into new Skolem constants which may appear in the ground instances used in the proof. Often all function symbols have standard interpretations except for these Skolem constants. Then, if the D_i do not contain any Skolem constants, they will be true or false in the standard interpretation. If some D_i is true, then it can be derived from the axioms and so splitting is not needed. Furthermore, not all D_i can be false since D is true in the standard interpretation. Now, if the D_i do contain Skolem constants, we are free to choose I to interpret them in any way desired. We would like to say that for each D_i such a choice can be made so that D_i will be true in $M_I(S)$. If this is not possible, it must be that $S \cup D_i$ is inconsistent. Therefore it seems reasonable for each D_i to restrict I to interpret the Skolem constants so that D_i is true. (It is permissible and natural to choose a different I for each case in the case analysis.) If a proof cannot be found, then attempt to show that $S \cup D_i$ is inconsistent. Thus if the non-Horn clause were $c \leq d \quad \vee \quad d < c$, where c and d are such Skolem constants, when considering the case $c \leq d$ we would at first only consider I interpreting c to be some natural number less than or equal to d. In this step, the dependency graphs would be less dense than if I is unrestricted, since we do not have to consider the possibility that literals such as 3≤2 are true. If this step fails, then we can look for proofs that $S \cup c \leq d$ is inconsistent. In the special case in which there are only two D_i, instead of attempting to prove that $S \cup D_1$ is inconsistent, we can attempt to prove that $S \models D_2$ (since presumably $D_1 \quad \vee \quad D_2$ is true in the standard interpretation), and this may be much easier since a more natural semantics should be available.

5.1 Initial versus first-order semantics

The use of semantics and splitting introduces some subtleties concerning models of H(S). Suppose S is consistent, and D_i are as above. Thus D = $\vee_i D_i$ and D is a ground instance of some clause C of S. Suppose the goal literal is of the form $\forall x_1 \quad \cdots \quad \forall x_m A(x_1, \quad \cdots, \quad x_m)$, as above. Suppose some of the D_i contain Skolem constants c_j obtained from x_j in the theorem. Let I be some S-initial interpretation of H(S). It is conceivable that no matter how the Skolem constants are interpreted as elements of the domain of I, all the D_i will still be false in $M_I(H(S))$, even though D is consistent with S. The reason is that restricting the Skolem constants to be elements of the domain of I is like restricting attention to the initial model semantics of H(S) instead of the usual first-order semantics. It appears possible at first glance that D may be false in the initial model but true in some other models. This would make it necessary in some cases to use split literals all of which are inconsistent with the initial interpretation, contrary to intuition. However, the following result shows that this need never happen.

Theorem. Suppose S is consistent, and D is a ground instance of a clause C of S. Suppose D contains new Skolem constants c_j in addition to the function symbols in S. Let I be an S-initial interpretation of the function and constant symbols in S. Then no matter how I is extended to interpret the Skolem constants c_j as elements of the domain D_I of I, D^I is consistent with S^I.

Proof. Since S is consistent and I is S-initial, S^I is consistent. Let D_1 be D with the Skolem constants c_j replaced uniformly by variables x_j. Then D_1 is also an instance of C since the c_j do not occur in C. Since S is consistent, D_1 is consistent with S. Let D_2 be an arbitrary ground instance of D_1, such that D_2 contains none of the symbols c_j. Then $S \cup D_2$ is consistent, so $S^I \cup D_2^I$ is consistent. Therefore, no matter how we assign elements of D_I as values of the x_j, D_1 is consistent with S^I. Therefore, no matter how we assign elements of D_I as values of the c_j, D is consistent with S^I. Thus values can be assigned to the c_j so that at least one of the literals of D is not contradictory with S^I.

This guarantees that if S is consistent, then one of the split literals D_j will be consistent with S^I, when the Skolem constants are interpreted as elements of the domain of the initial model of S. For example, if D is $c \leq d \quad \vee \quad d < c$ then no matter how we interpret c and d as integers, one of the literals of D will be consistent with the standard model of the integers. Thus if the order of considering split literals is flexible, we can at least guarantee that the first case considered preserves consistency. After the first split, usually more will be known about the instance D of C since the variables of C will be restricted by the proof found for the first case. Still, after the first split, we may not know that $S \cup D_i$ is consistent. Possibly when proving theorems using the initial semantics, other strategies can be used.

6. Similarities with Human Theorem Proving Methods

There are a number of similarities between this general approach and the methods used by humans to prove theorems. This method makes heavy use of examples and diagrams, as people do. It considers causal connections between known facts, using general axioms. The method separates the proof search into two steps: 1. Finding a *relevant* set of facts and 2. Connecting these facts together to get a proof. Frequently for humans, one of the most important steps to finding a proof is to find a relevant set of facts. This method has an explicit case analysis structure, which is often buried in conventional theorem provers. Also, in section 5 we considered ways in which semantics can be integrated naturally into the case analysis structure. The method emphasizes relevant *facts* instead of relevant axioms as in [16]. This makes the strategy somewhat more independent of the particular axiom system chosen. Finally, the method is sensitive to the difference between positive and negative literals, as people are; to replace P by ¬P everywhere would significantly alter the behavior of the method.

7. Extensions and Modifications

It would be useful to have heuristics for adding relevant facts to a given set of facts; these heuristics would depend on the problem domain considered. It might be interesting to consider causal relations that correspond not to a single clause but perhaps to short proof steps; possibly specialized decision procedures could be used to incorporate specialized axioms into the causal links so these axioms would not have to be represented explicitly. It seems useful to consider only *minimal* causal relations; that is, relations $\{L_1, \cdots, L_k\} \rightarrow L$ such that L cannot be derived in one step from any proper subset of the L_i. It might be possible to search for proofs of theorems of the form $\forall x P(x)$ by individually proving P(a), P(b), P(c), ... and somehow matching up these proofs, using methods similar to those of section 4.

8. References

1. Ballantyne, A., and Bledsoe, W., On generating and using examples in proof discovery, Machine Intelligence 10 (Harwood, Chichester, 1982) 3-39.

2. Bledsoe, W., Using examples to generate instantiations for set variables, IJCAI(1983)892-901.

3. Brand, D., Proving theorems with the modification method, SIAM J. Comput. 4 (1975)412-430.

4. Chang, C., The decomposition principle for theorem proving systems, Proc. Tenth Annual Allerton Conference on Circuit and System Theory, University of Illinois(1972)20-28.

5. Chang, C. and Lee, R., Symbolic Logic and Mechanical Theorem Proving (Academic Press, New York, 1973).

6. Charniak, E., Riesbeck, C., and McDermott, D., Data dependencies, Artificial Intelligence Programming (Lawrence Erlbaum Associates, Hillsdale, N.J., 1980) 193-226

7. Doyle, J., A truth maintenance system, Artificial Intelligence 12 (1979) 231-272.

8. Fay, M., First-order unification in an equational theory, Proceedings 4th Workshop on Automated Deduction, Austin, Texas (1979)161-167.

9. Fikes, R., Deductive retrieval mechanisms for state description models, Proceedings of the Fourth International Joint Conference on Artificial Intelligence, Tbilisi, Georgia, USSR (1975) 99-106.

10. Gelernter, H., Realization of a geometry theorem-proving machine, Proc. IFIP Congr. (1959)273-282.

11. Henschen, L. and Wos, L., Unit refutations and Horn sets, J. ACM 21(1974)590-605.

12. Huet, G., An algorithm to generate the basis of solutions to homogeneous linear diophantine equations, Information Processing Letters 17 (1978)144-147.

13. Huet, G. and Oppen, D., Equations and rewrite rules: a survey, in Formal Languages: Perspectives and Open Problems (R. Book, ed.), Academic Press, New York, 1980.

14. Lankford, D., Canonical algebraic simplification in computational logic, Memo ATP-25, Automatic Theorem Proving Project, University of Texas, Austin, TX, 1975.

15. Loveland, D., Automated Theorem Proving: A Logical Basis (North-Holland, New York, 1978).

16. Plaisted, D., An efficient relevance criterion for mechanical theorem proving, Proceedings of the First Annual National Conference on Artificial Intelligence, Stanford University, August, 1980.

17. Plaisted, D., Theorem proving with abstraction, Artificial Intelligence 16 (1981) 47 - 108.

18. Plaisted, D., A simplified problem reduction format, Artificial Intelligence 18 (1982)227 - 261.

19. Reiter, R., A semantically guided deductive system for automatic theorem proving, Proc. 3rd IJCAI (1973) 41-46.

20. Slagle, J., Automatic theorem proving with renamable and semantic resolution, J. ACM 14 (1967) 687-697.

21. Stickel, M., A unification algorithm for associative-commutative functions, J. ACM 28 (1981)423-434.

EXPANSION TREE PROOFS AND THEIR CONVERSION TO NATURAL DEDUCTION PROOFS*

Dale A. Miller
Department of Computer and Information Science
University of Pennsylvania
Philadelphia, PA 19104

Abstract: We present a new form of Herbrand's theorem which is centered around structures called expansion trees. Such trees contains substitution formulas and selected (critical) variables at various non-terminal nodes. These trees encode a shallow formula and a deep formula — the latter containing the formulas which label the terminal nodes of the expansion tree. If a certain relation among the selected variables of an expansion tree is acyclic and if the deep formula of the tree is tautologous, then we say that the expansion tree is a special kind of proof, called an ET-proof, of its shallow formula. Because ET-proofs are sufficiently simple and general (expansion trees are, in a sense, generalized formulas), they can be used in the context of not only first-order logic but also a version of higher-order logic which properly contains first-order logic. Since the computational logic literature has seldomly dealt with the nature of proofs in higher-order logic, our investigation of ET-proofs will be done entirely in this setting. It can be shown that a formula has an ET-proof if and only if that formula is a theorem of higher-order logic. Expansion trees have several pleasing practical and theoretical properties. To demonstrate this fact, we use ET-proofs to extend and complete Andrews' procedure [4] for automatically constructing natural deductions proofs. We shall also show how to use a mating for an ET-proof's tautologous, deep formula to provide this procedure with the "look ahead" needed to determine if certain lines are unnecessary to prove other lines and when and how backchaining can be done. The resulting natural deduction proofs are generally much shorter and more readable than proofs build without using this mating information. This conversion process works without needing any search. Details omitted in this paper can be found in the author's dissertation [16].

Key Words: Higher-order Logic, Expansion Trees, ET-proofs, Natural Deduction, Matings.

1. Introduction

Problem solving in mathematics involves many different kinds of reasoning processes: about propositional connectives, about individual objects in a given domain, about equality and order relations, about sets and functions, and, among a host of others, the more exotic reasoning by example, analogy, etc. Approaches to theorem proving have generally focused on studying the first three of these reasoning processes. Reasoning of the more exotic kinds have also been studied by various artificial intelligence researchers. Although logics based on the ability to reason about sets and functions (higher-order logics) have been studied (see [1, 2, 8, 11, 12, 14, 18, 19, 20, 22]), until very recently few implementations of theorem provers in such logics have been described in the literature.

* This work was supported by NSF grant MCS81-02870.

The importance of doing theorem proving in higher-order logic has been argued by several people, including Andrews in [15] and Robinson in [20]. In fact, Robinson concludes [20] with:

> It is important to recognize that it is higher-order logic, and not first-order logic, which is the natural technical framework for the 'mechanization of mathematics'. We have in fact attempted ... to pursuade those engaged in mechanical theorem-proving research, and those proposing to start such research, to focus their attention henceforth on mechanizing higher-order logic.

The computer system, TPS, described in [15], is the result of an ongoing project in which many automatic and interactive approaches to theorem proving in higher-order logic have been developed. Initially, the logical language described in the next section was implemented, along with Huet's higher-order unification algorithm [12]. Using this unification algorithm, a mating enumeration theorem prover, described in [5], was then implemented. The mating enumeration strategy, being conceived for first-order logic theorem proving, provided TPS with an automatic theorem prover which was complete for first-order logic. The use of Huet's algorithm also permitted genuinely higher-order theorems to be proved. In particular, TPS found a proof of Cantor's theorem, *i.e.* there is no one-to-one mapping from a power set of a set back to that set. The classical diagonal argument was found by discovering a non-trivial higher-order substitution term [3]. However, the collection of theorems for which TPS could (theoretically) find a proof was only a very modest extension of first-order logic. Currently, TPS is still very incomplete in the general higher-order setting, since substitution terms containing quantifiers and binary, logical connectives are not discovered by a straightforward use of Huet's unification algorithm.

One way to describe this inadequacy is to say that TPS was not searching through the proper "search space" of proof structures for higher-order theorems. What we shall present in this paper is a useful characterization of the search space for a theorem prover in higher-order logic. This characterization is based on a structure called an expansion tree. Among the set of expansion trees for a given theorem, certain ones will be considered proofs, called ET-proofs. In this paper we shall define and present several important properties of expansion trees, but we shall not discuss the many issues surrounding how to automate the search for ET-proofs. Approaches to this problem are currently being studied and implemented in TPS.

In the next section, we shall describe the higher-order logic $\mathcal{T}$ on which we base the rest of this paper. In Section 3, we present the definition of expansion trees and ET-proofs. As it turns out, expansion trees and ET-proofs are useful structures for the study of proofs in both first-order and higher-order logic. All the results in this and following sections will work equally well in both logics. (See [17] for examples of how expansion trees can be used in first-order logic metatheory.) The fact that this one kind of structure can actually be used in both setting is clearly one of its strengths, since most definitions of search spaces for first-order logic do not work in the higher-order case. In Section 4, we present a list representation for expansion trees which are succinct and easily implemented. We warn the reader that these 3 sections may prove to be difficult to read, especially for a reader not familiar with higher-order logic and λ-conversion.

In the remaining sections of this paper, we show another strength of expansion trees: Given an ET-proof, it is easily converted to a natural deduction proof. Such a feature is also very important to the TPS project since a considerable amount of effort has gone into providing the TPS user with not only automatic tools for proving theorems but also interactive tools (see [4] and [15]). For example, TPS provides a interactive editor for constructing natural deduction proofs in a top-down and bottom-up fashion (much as is also done in [6]). At any point in editing such a proof, the user can have an unfinished portion of the proof given to the automatic theorem prover. If the theorem prover finds a proof — that is, an ET-proof — the methods described in the last sections can be used to finish the unfinished portion of the natural deduction proof which the user started. In this way, the theorem prover can explain the proof it found in a readable fashion. This capability is clearly valuable not only to researcher using TPS but also to beginning logic students who use the interactive proof editor to learn the process of building proofs. If the student gets stuck, the automatic theorem prover could provide hints or complete instructions on how to complete the proof. See [17] for more discussion of this feature, along with the description of how it is possible to convert natural deduction proofs to ET-proofs. In that paper, Pfenning also describes an algorithm which will convert a resolution-style refutation of a theorem into an ET-proof of that theorem. Thus the results about expansion trees mentioned in this paper and Pfenning's can be made available to those systems which are based on resolution theorem proving.

All the results described below are contained in [16], and the reader is referred to this dissertation for details of proofs which are omitted below.

2. Logical Preliminaries

It has often be observed that first-order logic is inadequate for formulating mathematics. For example, consider Tarski's lattice-theoretical fixpoint theorem [23]:

> If $\langle L, \leq\rangle$ is a complete lattice and if f is an increasing function on L, then f has a fixpoint, *i.e.* there is an $x \in L$ such that $f(x) = x$.

One difficult in representing this theorem and its proof in first-order logic is the need to quantify over a set variable in the axiom concerning a *complete* lattice. If we let $\langle L, \leq\rangle$ be a lattice and let B be a set variable (a higher-order variable), an informal mathematical representation of the completeness axiom would be:

$$\begin{aligned}&\forall B\,[\forall x\,[x \in B \supset x \in L] \supset\\ &\quad \exists z\,[z \in L \wedge \forall x\,[x \in B \supset x \leq z]\\ &\qquad \wedge \forall y\,[[y \in L \wedge \forall x\,[x \in B \supset x \leq y]] \supset z \leq y]]]\end{aligned}$$

In the proof of the fixpoint theorem, this axiom is used by applying it to the set $\{x | x \in L \wedge x \leq f(x)\}$. Informally this is done referring to the property used to define B wherever $x \in B$ appears in this instance of the axiom. In other words, we actually replace $x \in B$ with $x \in L \wedge x \leq f(x)$. Here, again first-order logic is inadequate to represent this kind of substitution — an atomic formula $x \in B$ becomes the non-atomic formula $x \in L \wedge x \leq f(x)$. We now present a higher-order logical system which solves these two problems: explicit quantification of set (and function) variables and the possible change of atomic subformula occurrences into non-atomic subformulas under substitution.

The higher-order logic, $\mathcal{T}$, which we shall consider here is essentially the simple theory of types given by Church in [10], except that we do not use the axioms of extensionality, choice, descriptions, or infinity. $\mathcal{T}$ contains two base types, o for boolean and ι for individuals. All other types are functional types, *i.e.* the type $(\beta\alpha)$ is the type of a function with domain type α and codomain type β. In particular, the type $(o\alpha)$, being the type of a function from type α to a boolean, *i.e.* a characteristic function, is used in $\mathcal{T}$ to represent the type for sets (predicates) of elements of type α. For example, if the lattice L mentioned earlier is a set of elements of type α, then we say that L has type $(o\alpha)$. Formulas are built up from logical constants, variables, and parameters (non-logical constants) by λ-abstraction and function application. Hence, the type of $[\lambda x_\alpha A_\beta]$ is $(\beta\alpha)$ while the type for $[A_{(\beta\alpha)} B_\alpha]$ is β. (We shall seldom adorn formulas with type symbols, but rather, when the type of a formula, say A, cannot be determined from context, we will add the phrase "where A is a formula$_\alpha$" to indicate that A has type α.) For the convenience of making definitions in the next section, the formulas of $\mathcal{T}$ which we shall consider contain only the logical constants $\sim_{oo}$ (negation), $\vee_{(oo)o}$ (disjunction), and $\Pi_{o(o\alpha)}$ (the "universal α-type set recognizer"). Other logical constants will be considered abbreviations, *i.e.* $A \wedge B$ stands for $\sim . \sim A \vee \sim B$, $A \supset B$ stands for $\sim A \vee B$, $\forall x\ P$ stands for $\Pi[\lambda x P]$, and $\exists x\ P$ stands for $\sim\Pi[\lambda x. \sim P]$. In particular, we write $L_{o\alpha} x_\alpha$ to denote the expression $x \in L$. This definition of the universal and existential quantifier may look rather peculiar, but it is very simple to explain. The meaning of the logical constant Π is such that $\Pi_{o(o\alpha)} B_{o\alpha}$ is true if and only if $B_{o\alpha}$ is the "universal" set of type $o\alpha$. Hence, $\Pi[\lambda x_\alpha P_o]$ is true if and only if $\lambda x_\alpha P_o$ is the universal set of type $(o\alpha)$, *i.e.* P_o is true for all x_α. We shall take as axioms of $\mathcal{T}$ the following formulas (p, q, and r are formulas$_o$):

$$p \vee p \supset p$$
$$p \supset p \vee q$$
$$p \vee q \supset . q \vee p$$
$$p \supset q \supset . r \vee p \supset . r \vee q$$
$$\Pi_{o(o\alpha)} f_{o\alpha} \supset f_{o\alpha} x_\alpha$$
$$\forall x_\alpha\ [p \vee f_{o\alpha} x_\alpha] \supset p \vee \Pi_{o(o\alpha)} f_{o\alpha}$$

Here, α is a type variable, and the last two axioms represent axiom schemes. The rules of inference are substitution, modus ponens, universal generalization, and λ-conversion. We shall write $\vdash_{\mathcal{T}} A$ to denote that A has a Hilbert-style proof using these axioms and inference rules. The deduction theorem holds for $\mathcal{T}$.

At first glance $\mathcal{T}$ may look rather esoteric, but it can be described as being simply first-order logic in which we permit unrestricted comprehension via the use of λ-terms. The type structure is necessary here in order to avoid the paradoxes (like Russell's paradox) which arise from unrestricted comprehension. The use of λ-terms in substitutions can make the nature of deductions in $\mathcal{T}$ more complex than in first-order logic. In the fixpoint example, the result of substituting B with the term $[\lambda x. Lx \wedge x \leq f(x)]$ in the completeness axiom will change the subformulas of the form Bx to $[\lambda x.Lx \wedge x \leq f(x)]x$ which λ-converts to $Lx \wedge x \leq f(x)$. For another example, if we have the formula (where Y is a variable$_{o\iota}$ and D and T are variables$_{o(o\iota)}$)

$$\forall D\,[DY \supset TY]$$

and we wished to do a universal instantiation (a derived rule of inference) of this formula with the term $\lambda Z[TZ \wedge \forall x\,.Zx \supset Ax]$, *i.e.* the set of all sets of individuals which are members of T and are subsets of A, we would then have

$$[\lambda Z.\,TZ \wedge \forall x\,.Zx \supset Ax]Y \supset TY.$$

We can now apply the λ-conversion inference rule to this formula to deduce

$$[TY \wedge \forall x\,.Yx \supset Ax] \supset TY.$$

Notice how the structure of this last formula is much more complex than that of the formula it was deduced from. This last formula contains occurrences of logical connectives and quantifiers which are not present in the original formula. Notice also that Y now has the role of a predicate where this was not the case in the first formula. None of these structural changes can occur in first-order logic. The discovery of such substitution terms as the one used to instantiate D is a much more complex problem than can be achieved by simply applying unification. TPS, for example, cannot currently discover terms of this kind. Radical new heuristics for finding substitutions must be developed, and we hope that expansion trees will provide a vehicle for formalizing such attempts. Bledsoe in [8] and [9] has made some exciting progress in the development of just such heuristics.

3. Expansion Trees and ET-Proofs

All references to trees below will actually refer to finite, ordered, rooted trees in which the nodes and arcs may or may not be labeled, and that labels, if present, are formulas. In particular, nodes may be labeled with simply the logical connectives $\sim$ and $\vee$. We shall picture our trees with their roots at the top and their leaves (terminal nodes) at the bottom. In this setting, we say that one node *dominates* another node if it they are on a common branch and the first node is *higher* in the tree than the second. This dominance relation shall be considered reflexive. All nodes except the root node will have *in-arcs* while all nodes except the leaves will have *out-arcs*. A node labeled with $\sim$ will always have one out-arc, while a node labeled with $\vee$ will always have two out-arcs. We shall also say that an arc dominates a node if the node which terminates the arc dominates the given node. In particular, an arc dominates the node in which it terminates. Also, we say that an arc dominates another arc if their respective terminal nodes dominate each other in the same order.

3.1. Definition. Let A be a formula$_o$. An occurrence of a subformula B in A is a *boolean subformula occurrence* if it is in the scope of only $\sim$ and $\vee$, or if A is B. A formula$_o$ A is an *atom* if its leftmost non-bracket symbol is a variable or a parameter.

A formula B is a ***boolean atom*** (***b-atom***, for short) if its leftmost non-bracket symbol is a variable, parameter or Π. A ***signed atom*** (***b-atom***) is a formula which is either an atom (b-atom) or the negation of an atom (b-atom). ∎

3.2. Definition. Formulas$_o$ of $\mathcal{T}$ can be considered as trees in which the non-terminal nodes are labeled with $\sim$ or $\vee$, and the terminal nodes are labeled with b-atoms. Given a formula$_o$, A, we shall refer to this tree as the ***tree representation of A***. ∎

3.3. Example. Figure 1 is the tree representation of $\sim[\Pi B \vee Ax] \vee \sim\sim\Pi[\lambda x.Ax \vee Bx]$. This formula is equivalent to $\sim[\forall y\, By \vee Ax] \vee \sim\sim \forall x\, .Ax \vee Bx$. ∎

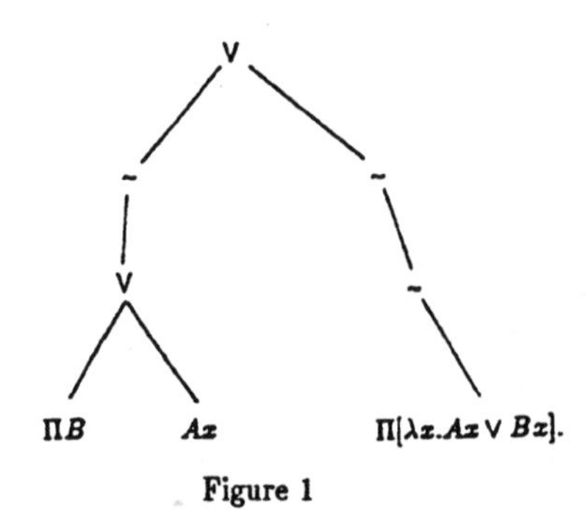

Figure 1

We shall adopt the following linear representation for trees. If the root of the tree Q is labeled with $\sim$, we write $Q = \sim Q'$, where Q' is the proper subtree dominated by Q's root. Likewise, if the root of Q is labeled with $\vee$, we write $Q = Q' \vee Q''$, where Q' and Q'' are the left and right subtree of Q. The expression $Q' \wedge Q''$ is an abbreviation for the tree $\sim[\sim Q' \vee \sim Q'']$.

3.4. Definition. Let Q, Q' be two trees. Let N be a node in Q and let l be a label. We shall denote by $Q +_N^l Q'$ the tree which results from adding to N an arc, labeled l, which joins N to the root of the tree Q'. This new arc on N comes after the other arcs from N (if there are any). In the case that the tree Q is a one-node tree, N must be the root of Q, and we write $A +^l Q'$ instead of $Q +_N^l Q'$, where A is the formula which labels N. ∎

3.5. Example. Figure 2 contains three trees, Q, Q' and $Q +_N^c Q'$, where N is a node of Q and c is some label. The nodes and arcs of Q and Q' may or may not have their own labels.

∎

3.6. Definition. Let Q be a tree, and let N be a node in Q. We say that N occurs positively (negatively) if the path from the root of Q to N contains an even (odd) number of nodes labeled with $\sim$. We shall agree that the root of Q occurs positively in Q. If a node N in Q is labeled with a formula of the form ΠB, then we say that N is ***universal*** (***existential***) if it occurs positively (negatively) in Q. A terminal node which is not labeled with a formula of the form ΠB is called a ***neutral*** node. A universal

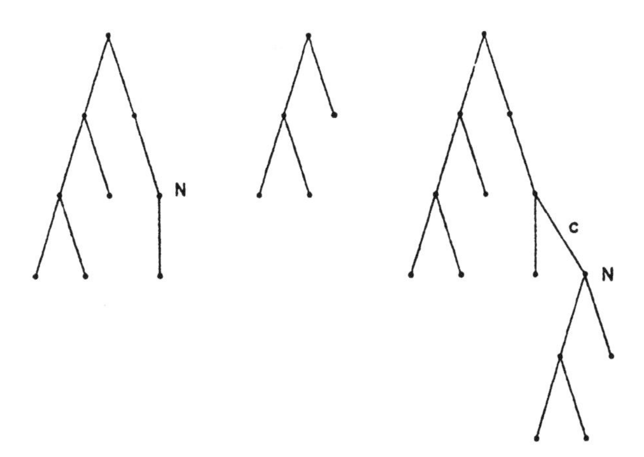

Figure 2: The trees Q, Q', and $Q +^c_N Q'$.

(existential) node which is not dominated by any universal or existential node is called a *top-level* universal (existential) node. A labeled arc is a *top-level labeled arc* if it is not dominated by any other labeled arc. ∎

3.7. Definition. Let Q be a tree with a terminal node N labeled with the formula ΠB, for some formula$_{o\alpha}$ B. If N is existential, then an *expansion* of Q at N with respect to the list of formulas$_\alpha$, $\langle t_1, \ldots, t_n \rangle$, is the tree $Q +^{t_1}_N Q_1 +^{t_2}_N \cdots +^{t_n}_N Q_n$ (associating to the left), where, for $1 \leq i \leq n$, Q_i is the tree representation for some λ-normal form of Bt_i. The formulas $t_1, \ldots, t_n$ are called *expansion terms* of the resulting tree. We say that each of these terms are used to *expand N*.

If N is universal, then a *selection* of Q at N with respect to the variable$_\alpha$ y, is the tree $Q +^y_N Q'$, where Q' is the tree representation of some λ-normal form of By, and y does not label an out-arc of any universal node in Q. We say that the node N *is selected by y*.

The set of all *expansion trees* is the smallest set of trees which contains the tree representations of all λ-normal formulas$_o$ and which is closed under expansions and selections. ∎

Expansion trees are, in a sense, generalized formulas. The main difference is that expansion trees can contain labeled arcs. An expansion tree which contains no labeled arcs can easily be interpreted as a formula.

3.8. Definition. Assume that Q is an expansion tree. Let $\mathbf{S}_Q$ be the set of all variable occurrences which label the out-arcs from (non-terminal) universal nodes in Q, and let Θ_Q be the set of all occurrences of expansion terms in Q. ∎

Expansion trees, a generalization of Herbrand instances, do not use Skolem functions as is customary in Herbrand instances. Skolem functions can be used in this setting, but their occurrences in substitution terms must be restricted in ways that are not apparent from the first-order use of Skolem functions. The reader is referred to [16] for details. In order to do without Skolem functions, we need to place a restriction on selected variables which models the way in which Skolem terms would imbed themselves in other Skolem terms. This restriction amounts to requiring that the following binary relation on $\mathbf{S}_Q$ be acyclic.

3.9. Definition. Let Q be an expansion tree and let $\prec^0_Q$ be the binary relation on $\mathbf{S}_Q$ such that $z \prec^0_Q y$ if there exists an expansion term occurrence $t \in \Theta_Q$ such that z is free in t and y is selected for a node dominated by the arc labeled by t. Let $\prec_Q$ be the transitive closure of $\prec^0_Q$. $\prec_Q$ is called the *imbedding relation* because it reflects how Skolem terms, represented by the variables in $\mathbf{S}_Q$, are imbedded in one another. ∎

We next define two formulas which are encoded in an expansion tree. The "deep" formula $Dp(Q)$ of the expansion tree Q is a formula whose b-atoms correspond to the leaves of Q. The "shallow" formula $Sh(Q)$ of Q is a formula whose b-atoms correspond to the top-level universal, existential, and neutral nodes in Q.

3.10. Definition. Let Q be a tree such that either Q or $\sim Q$ is an expansion tree. We define $Dp(Q)$ by induction on the structure of Q.

(1) If Q is a one-node tree, then $Dp(Q) = A$, where A is the formula which labels that one-node.

(2) If $Q = \sim Q'$ then $Dp(Q) := \sim Dp(Q')$.

(3) If $Q = Q' \vee Q''$ then $Dp(Q) := Dp(Q') \vee Dp(Q'')$.

(4) If $Q = \Pi B +^{t_1} Q_1 + \ldots +^{t_n} Q_n$ then $Dp(Q) := Dp(Q_1) \wedge \ldots \wedge Dp(Q_n)$. ∎

3.11. Definition. Let Q be a tree such that either Q or $\sim Q$ is expansion tree. We define $Sh(Q)$ by induction on the top-level boolean structure of Q.

(1) If Q is a one-node tree, then $Sh(Q) = A$, where A is the formula which labels that one-node.

(2) If $Q = \sim Q'$ then $Sh(Q) := \sim Sh(Q')$.

(3) If $Q = Q' \vee Q''$ then $Sh(Q) := Sh(Q') \vee Sh(Q'')$.

(4) If $Q = \Pi B +^{t_1} Q_1 + \ldots +^{t_n} Q_n$ then $Sh(Q) := \Pi B$. ∎

Notice, that if A is a formula$_o$, and Q is the tree representation of A, then $Dp(Q) = A = Sh(Q)$.

3.12. Definition. Let Q be an expansion tree. Q is *sound* if no variable in $\mathbf{S}_Q$ is free in $Sh(Q)$. Q is an *ET-proof* if Q is sound, $Dp(Q)$ is tautologous, and $\prec_Q$ is acyclic. Q is an expansion tree *for A* if Q is sound and $Sh(Q)$ is a λ-normal form of A. Q is an ET-proof *for A* if Q is an ET-proof and Q is an expansion tree for A. ∎

3.13. Example. Let A be the theorem $\exists y \,\forall x \,.Px \supset Py$. An ET-proof for A would then be the tree Q given as:

$$\sim [[\Pi\lambda y. \sim \Pi\lambda x. \sim Px \vee Py] +^u \sim [[\Pi\lambda x. \sim Px \vee Pu] +^v [\sim Pv \vee Pu]] \\ +^v \sim [[\Pi\lambda x. \sim Px \vee Pv] +^w [\sim Pw \vee Pv]]].$$

Here, $Dp(Q) = \sim[\sim[\sim Pv \vee Pu] \wedge \sim[\sim Pw \vee Pv]]$. The imbedding relation is the pair $v \prec_Q w$. Notice, that if we had used u instead of w, $\prec_Q$ would have been cyclic. In Example 4.2 we give a more readable representation of this expansion tree. ∎

3.14. Soundness and Relative Completeness for ET-Proofs. *Let A be a formula$_o$. $\vdash_{\mathcal{T}} A$ if and only if A has an ET-proof.*

This theorem is what we shall consider our higher-order version of Herbrand's theorem. The reader is referred to [16] for the details of this proof. The relative completeness result, *i.e.* if $\vdash_{\mathcal{T}} A$ then A has an ET-proof, is proven by using the Abstract Consistency Property in [1]. The central result concerning Abstract Consistency Properties is based on Takahashi's proof of the cut-elimination theorem for higher-order logic [22]. Since $\mathcal{T}$ is non-extensional, Henkin-style general models do not correctly characterize derivability in $\mathcal{T}$. Hence, the completeness result is stated relative to the notion of derivability and is not based on a notion of validity.

3.15. Definition. An expansion tree is *grounded* if none of its terminal nodes are labeled with formulas of the form ΠB. An ET-proof is a grounded ET-proof if it is also a grounded expansion tree. ∎

A formula has an ET-proof if and only if it has a grounded ET-proof.

4. List Representations of Expansion Trees

We shall now present a representation of expansion trees which is more succinct and more suitable for direct implementation on computer systems. We shall no longer consider the logic connectives $\wedge$ and $\supset$ and the quantifiers $\forall$ and $\exists$ to be abbreviations. This will help make list representations of expansion trees more compact.

The set of all list structures over a given set, Ξ, is defined to be the smallest set which contains Ξ and is closed under building finite tuples.

Since expansion and selection nodes in an expansion tree must occur under an odd and even number of occurrences of negations respectively, we need to be careful how we imbed expansion trees under negations when we attempt to build up larger expansion trees from smaller ones. This explains why we need to consider so many cases in the following definition.

4.1. Definition. Let Ξ be the set which contains the labels SEL and EXP and all formulas of $\mathcal{T}$. Let $\mathcal{E}$ be the smallest set of pairs $\langle R, A\rangle$, where R is a list structure over Ξ and A is a formula$_o$, which satisfies the conditions below. We say that a variable y is *selected* in the list structure R if it occurs in a sublist of the form (SEL y R').

(1) If A is a boolean atom and R is a λ-normal form of A, then $\langle R, A\rangle \in \mathcal{E}$ and $\langle \sim R, \sim A\rangle \in \mathcal{E}$. Here, $\sim R$ is shorthand for the two element list $(\sim R)$.

(2) If $\langle R, A\rangle \in \mathcal{E}$ then $\langle R, B\rangle \in \mathcal{E}$ where A conv B.

(3) If $\langle R, A\rangle \in \mathcal{E}$ then $\langle \sim\sim R, \sim\sim A\rangle \in \mathcal{E}$.

In cases (4), (5), and (6), we assume that R_1 and R_2 share no selected variables in common and that A_1 (A_2) has no free variable selected in R_2 (R_1).

(4) If $\langle R_1, A_1\rangle \in \mathcal{E}$ and $\langle R_2, A_2\rangle \in \mathcal{E}$ then $\langle(\vee\ R_1\ R_2), A_1 \vee A_2\rangle \in \mathcal{E}$ and $\langle(\wedge\ R_1\ R_2), A_1 \wedge A_2\rangle \in \mathcal{E}$.

(5) If $\langle\sim R_1, \sim A_1\rangle \in \mathcal{E}$ and $\langle\sim R_2, \sim A_2\rangle \in \mathcal{E}$ then $\langle\sim(\vee\ R_1\ R_2), \sim.A_1 \vee A_2\rangle \in \mathcal{E}$ and $\langle\sim(\wedge\ R_1\ R_2), \sim.A_1 \wedge A_2\rangle \in \mathcal{E}$.

(6) If $\langle\sim R_1, \sim A_1\rangle \in \mathcal{E}$ and $\langle R_2, A_2\rangle \in \mathcal{E}$ then $\langle(\supset\ R_1\ R_2), A_1 \supset A_2\rangle \in \mathcal{E}$ and $\langle\sim(\supset\ R_2\ R_1), \sim.A_2 \supset A_1\rangle \in \mathcal{E}$.

In cases (7), (8), and (9), we assume that y is not selected in R and that y is not free in $[\lambda x P]$ or in B.

(7) If $\langle R, [\lambda x P]y\rangle \in \mathcal{E}$ then $\langle(\text{SEL}\ \ y\ \ R), \forall x\, P\rangle \in \mathcal{E}$.

(8) If $\langle\sim R, \sim[\lambda x P]y\rangle \in \mathcal{E}$ then $\langle(\sim(\text{SEL}\ \ y\ \ R)), \sim\exists x\, P\rangle \in \mathcal{E}$.

(9) If $\langle R, By\rangle \in \mathcal{E}$ then $\langle(\text{SEL}\ \ y\ \ R), \Pi B\rangle \in \mathcal{E}$.

In cases (10), (11), and (12), we must assume that for distinct i, j such that $1 \leq i, j \leq n$, R_i and R_j share no selected variables and that no variable free in $[\lambda x P]t_i$ is free in R_j.

(10) If for $i = 1, \ldots, n$, $\langle R_i, [\lambda x P]t_i\rangle \in \mathcal{E}$ then $\langle(\text{EXP}\ \ (t_1\ R_1)\ldots(t_n\ R_n)), \exists x\, P\rangle \in \mathcal{E}$.

(11) If for $i = 1, \ldots, n$, $\langle\sim R_i, \sim[\lambda x P]t_i\rangle \in \mathcal{E}$ then $\langle\sim(\text{EXP}\ \ (t_1\ R_1)\ldots(t_n\ R_n)), \sim\forall x\, P\rangle \in \mathcal{E}$.

(12) If for $i = 1, \ldots, n$, $\langle\sim R_i, \sim Bt_i\rangle \in \mathcal{E}$ then $\langle\sim(\text{EXP}\ \ (t_1\ R_1)\ldots(t_n\ R_n)), \Pi B\rangle \in \mathcal{E}$. ∎

The pair $\langle R, A\rangle \in \mathcal{E}$ represents — in a succinct fashion — an expansion tree. Notice that the only formulas stored in the list structure R are those used for expansions and selections and those which are the leaves of the expansion tree. Expansion trees as defined in §2 contain additional formulas which are used as "shallow formulas" to label expansion and selection nodes. These formulas, however, can be determined up to λ-convertibility if we know what the expansion tree is an "expansion" for. Notice, that one list structure alone may represent several expansion trees. For example, (EXP $(a\ Paa)$) could represent an expansion tree for $\exists x\ Pxx$, $\exists x\ Pax$, and $\exists x\ Paa$. If we keep this complication in mind, we can informally considered list structures as expansion trees.

4.2. Example. The expansion tree in Example 3.13 can be written as the list structure:

$$(\text{EXP}\ \ (u\ (\text{SEL}\ \ v\ (\supset Pv\ Pu)\ \ ))\ (v\ (\text{SEL}\ \ w\ (\supset Pw\ Pv)\ \ ))).$$

5. Natural Deductions

Beyond the fact that ET-proofs are sound and (relatively) complete for $\mathcal{T}$, they also have several other pleasing properties, for both theoretical and practical concerns. We shall illustrate this claim by showing how ET-proofs can be converted to natural deduction-style proofs. This investigation is an immediate extension of the work described by Andrews in [4]. In that paper, Andrews showed how natural deduction proofs could be constructed by processing incomplete proofs, called *outlines*, in both a top-down and bottom-up fashion. In these outlines, certain lines, called *sponsoring* lines, were not justified. To each sponsoring line is associated a (possibly empty) list of justified lines which appear earlier in the proof and which might be required for completing the proof of the sponsoring line. These lines are called ***supporting*** lines. Proof lines which are either supporting or sponsoring are called ***active***. Incomplete proofs built in this fashion are such that their assertions are subformulas of the original theorem. (Notice that in higher-order logic, this is stretching the usual meaning of subformulas.) Using this fact, we shall be able to attach to each active line an expansion tree (actually a list representation) ***for*** the assertion in that line. These expansion trees, which are essentially sub-trees of the ET-proof of the original theorem, provide the information necessary to determine how an active line should be "processed."

Beyond the fact that the conversion process describe below works for higher-order logic, this process differs in two other important ways from the process described in [4]. First, Andrews used a structure called a ***plan*** to provide the information which would indicate how to process active lines. ET-proofs, when restricted to first-order logic, contain the same kind of information as plans. Plans, however, are defined with respect to several global properties of formulas. This makes it awkward (in theory and practice) to construct new plans for new subproofs. Since subtrees or the negation of subtrees of expansion trees are themselves expansion trees, it is much easier to build new ET-proofs for new subproofs. Secondly, Andrews actually considered subproofs to be based on a sponsoring line and its hypotheses while we consider subproofs to be based on sponsoring lines and their supports. These differences allow us to give a complete analysis of this transformation process.

Below we provide formal definitions for the concepts informally discussed above. In the rest of this paper, all ET-proofs will be assumed to be grounded.

5.1. Definition. By a natural deduction proof we mean a Suppes-style proof structures [21]. Such systems emphasize reasoning from hypotheses instead of axioms. An ***incomplete natural deduction proof*** is a list of proof lines some of which are justified by NJ — the non-justification label. Such lines represent subproofs which must be completed. The rules of inference in this system are those listed in [4] along with a rule for λ-conversion. The rules of existential generalization and universal instantiation are examples of two rules of inference. ∎

5.2. Example. The following is an example of an incomplete natural deduction proof.

(1)	1	$\vdash$	$\exists c\, \forall p\, .[\exists u\, .pu] \supset .p.cp$	Hyp
(2)	2	$\vdash$	$\forall x\, \exists y\, .Pxy$	Hyp
(3)	3	$\vdash$	$\forall p\, .[\exists u\, .pu] \supset .p.cp$	Hyp
(16)	2, 3	$\vdash$	$\exists f\, \forall z\, .Pz.fz$	NJ
(17)	1, 2	$\vdash$	$\exists f\, \forall z\, .Pz.fz$	$RuleC$: 1, 16
(18)	1	$\vdash$	$[\forall x\, \exists y\, .Pxy] \supset .\exists f\, \forall z\, .Pz.fz$	$Deduct$: 17
(19)		$\vdash$	$[\exists c\, \forall p\, .[\exists u\, .pu] \supset .p.cp] \supset$ $[\forall x\, \exists y\, .Pxy] \supset \exists f\, \forall z\, .Pz.fz$	$Deduct$: 18

Here c is a variable$_{\iota(o\iota)}$, p is a variable$_{o\iota}$, P is a variable$_{o\iota\iota}$, f is a variable$_{\iota\iota}$, and x, y, z, u are variables$_{\iota}$. ∎

In what follows, we shall use $\perp$ to represent a false statement. It can be treated as an abbreviation for $p \wedge \sim p$. We shall also let $\perp$ stand for both the expansion tree for $\perp$ and for the list representation for this expansion tree. If $\perp$ occurs as one of the disjuncts of a formula, we shall assume that that formula is an abbreviation for the formula which results from removing $\perp$ as a disjunct.

5.3. Definition. A *proof outline*, $\mathcal{O}$, is the triple, $\langle L, \rho, \{R_l\}\rangle$, where:

(1) L is a list of proof lines which forms an incomplete natural deduction proof. A line with the justification NJ corresponds to a subproof which must be completed. Let L_0 be the set of all lines labels in L which have this justification. These are called the *sponsoring* lines of $\mathcal{O}$.

(2) ρ is a function defined on L_0 such that whenever $z \in L_0$, $\rho(z) \subset L \setminus L_0$ and all the lines in $\rho(z)$ precede z in the list L. Whenever $l \in \rho(z)$, we say that z *sponsors* l, l *supports* z, z is a *sponsoring* line, and l is a *supporting* line. A line is *active* if it is either a supporting line or a sponsoring line which does not assert $\perp$. (In the outlines we shall consider, only sponsoring lines may assert $\perp$.)

(3) $\{R_l\}$ represents a set of list structures, one for each active line, such that if l is a supporting line, then $\langle \sim R_l, \sim l\rangle \in \mathcal{E}$ and if l is a sponsoring line, then $\langle R_l, l\rangle \in \mathcal{E}$.

(4) If line a supports line z then the hypotheses of a are a subset of the hypotheses of z.

If L_0 is not empty, we define the following formulas and expansion trees. For each $z \in L_0$ set $A_z := [\bigvee_{l\in\rho(z)} \sim l] \vee z$ (where line labels stand for their assertions) and let Q_z be the expansion tree for A_z represented by the list structure $(\vee\ (\bigvee_{l\in\rho(z)} \sim R_l)\ R_z)$. The following condition must also be satisfied by an outline.

(5) If L_0 is not empty, then Q_z is a (grounded) ET-proof for A_z for each $z \in L_0$.

It is easy to show that $\mathcal{O}$ has an active line if and only if L_0 is not empty. We say that $\mathcal{O}$ is an outline *for* A if the last line in $\mathcal{O}$ has no hypotheses and asserts A. ∎

The ET-proof Q_z roughly corresponds to a plan for the sponsoring line z as described in [4].

5.4. Definition. Let A be a formula and R a list representation of an ET-proof for A. Let z be the label for the proof line

$$(z) \qquad \vdash \ A \qquad\qquad\qquad NJ,$$

and set $L := \langle z\rangle$, $\rho(z) = \emptyset$ and $R_z := R$. Then $\mathcal{O}_0 := \langle L, \rho, \{R_l\}\rangle$ is clearly an outline. We call this outline the *trivial outline for A based on R*. ∎

5.5. Example. An example of a proof outline is given by setting $L = \langle 1, 2, 3, 16, 17, 18, 19\rangle$, $\rho(16) = \{2, 3\}$ and

$$R_2 = (\text{EXP}\ \ (z\ (\text{SEL}\ \ y\ \ Pzy)))$$
$$R_3 = (\text{EXP}\ \ (Pz\ (\supset (\text{EXP}\ \ (y\ Pzy))\ Pz.c.Pz)))$$
$$R_{16} = (\text{EXP}\ \ ([\lambda v.c.Pv]\ \ \ (\text{SEL}\ \ z\ \ Pz.c.Pz)))$$

where the lines in L are those listed in Example 5.2. It is easy to verify that $(\vee \sim R_2\ (\vee \sim R_3\ R_{16}))$ represents an ET-proof of $\sim 2 \vee \sim 3 \vee 16$ and that $\langle L, \rho, \{R_2, R_3, R_{16}\}\rangle$ is an outline. ∎

5.6. Definition. A formula t is *admissible in* $\mathcal{O}$ if no free variable in t is selected in R_l for any active line l. ∎

The D- and P- (deducing and planning) transformations described in [4] can now be used in this setting if we describe how each such transformation attributes expansion trees to each new active line. We illustrate how this is done with the P-Conj and D-All transformations.

If some sponsoring line z in an outline $\mathcal{O} = \langle L, \rho, \{R_l\}\rangle$ is of the form

$$(z) \quad \mathcal{H} \quad \vdash \quad A_1 \wedge A_2 \qquad\qquad NJ$$

then R_z is of the form $(\wedge\ R_1\ R_2)$. Applying P-Conj to line z will result in an outline $\mathcal{O}' = \langle L', \rho', \{R'_l\}\rangle$, where L' contains the new sponsoring lines

$$(x) \quad \mathcal{H} \quad \vdash \quad A_1 \qquad\qquad NJ$$

$$(y) \quad \mathcal{H} \quad \vdash \quad A_2 \qquad\qquad NJ$$

and line z has its justification changed to RuleP: x, y. Also, $\rho'(x)$ and $\rho'(y)$ are set equal to $\rho(z)$, and $R'_x := R_1$, $R'_y := R_2$. ρ' agrees on all other sponsoring lines of $\mathcal{O}'$, and $R'_l := R_l$ for all active lines of $\mathcal{O}'$ other than x and y. This application of P-Conj has reduced the subproof based on line z to the two subproofs based on lines x and y.

If the outline $\mathcal{O}$ contains a supporting line a of the form

$$(a) \quad \mathcal{H} \quad \vdash \quad \forall x\, P \qquad\qquad RuleX$$

for some justification RuleX (other than NJ), then R_a has the form (EXP $(t_1\ R_1)\ldots(t_n\ R_n)$). If any one of the terms $t_1, \ldots, t_n$ is admissible within $\mathcal{O}$, say t_i, then D-All can be applied to line a by doing a universal instantiation of it with t_i. L' is then equal to L with the line b, shown below, inserted after line a.

$$(b) \quad \mathcal{H} \quad \vdash \quad \mathbf{S}^x_{t_i} P \qquad\qquad \forall I : a.$$

Here it is assumed that in this substitution, bound variables are systematically renamed to avoid variable capture. Also, $R'_b := R_i$. If $n \geq 2$ then line a must remain active, so

$$R'_a := (\text{EXP}\ \ (t_1\ R_1)\ldots(t_{i-1}\ R_{i-1})\ (t_{i+1}\ R_{i+1})\ldots(t_n\ R_n))$$

and for each sponsoring line z such that $a \in \rho(z)$, set $\rho'(z) := \rho(z) \cup \{b\}$ (*i.e.* b is a cosupport with a). If $n = 1$, then line a is no longer active so b replaces a as a support — that is, for each sponsoring line z such that $a \in \rho(z)$, set $\rho'(z) := \rho(z) \setminus \{a\} \cup \{b\}$. In either case, $\rho'(z) := \rho(z)$ for all other sponsoring lines of $\mathcal{O}$ and $R'_l := R_l$ for all active lines $l \neq a$ of $\mathcal{O}$. It is straightforward to verify that $\mathcal{O}' = \langle L', \rho', \{R'_l\}\rangle$ is an outline.

It is possible to show that at least one expansion term associated with such active lines in $\mathcal{O}$ must be admissible, so requiring that the terms introduced in a universal instantiation (or introduced in a bottom-up fashion by P-Exists) be admissible is always possible to meet. This restriction to admissible terms is necessary to guarantee that when variables are selected in the P-All and P-Choose transformations, they do not already have a free occurrence in the current proof outline.

A simple, naive process of transforming an ET-proof, represented by the list structure R, for the theorem A, would then start by successively applying either D- or P-transformations to the trivial outline for A based on R and finish when all the subproofs generated can be recognized as instances of the RuleP transformation.

6. Focused Construction of Proof Outlines

The proof outlines produced by the naive method described above will often turn out to be very inelegant for at last two reasons, which we will examine here. An implementation of this naive algorithm was made in the computer program TPS (see [15]) and it was frequently found that many of the supporting lines for a given sponsoring line were not really needed to prove that sponsoring line. The naive algorithm contained no way of checking for this since it was provided with no ability to "look ahead." Hence, may applications of D- and P- rules were not necessary and the resulting, completed natural deduction proofs were much longer and redundant than necessary. The naive algorithm was also not equipped to recognize when it could backchain on a supporting line which asserted an implication, since backchaining also requires looking ahead to see it if can actually be applied. Hence, the naive procedure always treated such implicational lines in the most general possible way — by using its equivalent disjunctive form in the form of an argument from cases. Implicational support lines were always used in a very unnatural fashion.

The information which would supply a transformation process with the necessary ability to look ahead is contained in a *mating* which is present in the tautology encoded in the ET-proofs of each subproof of a given outline. We now need several definitions.

6.1. Definition. If $\mathcal{A}_1$ and $\mathcal{A}_2$ are sets, define $\mathcal{A}_1 \uplus \mathcal{A}_2 := \{\xi_1 \cup \xi_2 \mid \xi_1 \in \mathcal{A}_1, \xi_2 \in \mathcal{A}_2\}$. Let D be a λ-normal formula$_o$. We shall define two sets, $\mathcal{C}_D$ and $\mathcal{V}_D$, which are both sets of sets of b-atom subformula occurrences in D, by joint induction on the boolean structure of D. $\mathcal{C}_D$ is the set of clauses in D while $\mathcal{V}_D$ is the set of "dual" clauses in D. Dual clauses have been called vertical paths by Andrews (see [5]).

(1) If D is a b-atom, then $\mathcal{C}_D := \{\{D\}\}$ and $\mathcal{V}_D := \{\{D\}\}$.

(2) If $D = \sim D_1$ then $\mathcal{C}_D := \mathcal{V}_{D_1}$ and $\mathcal{V}_D := \mathcal{C}_{D_1}$.

(3) If $D = D_1 \vee D_2$ then $\mathcal{C}_D := \mathcal{C}_{D_1} \uplus \mathcal{C}_{D_2}$ and $\mathcal{V}_D := \mathcal{V}_{D_1} \cup \mathcal{V}_{D_2}$.

(4) If $D = D_1 \wedge D_2$ then $\mathcal{C}_D := \mathcal{C}_{D_1} \cup \mathcal{C}_{D_2}$ and $\mathcal{V}_D := \mathcal{V}_{D_1} \uplus \mathcal{V}_{D_2}$.

(5) If $D = D_1 \supset D_2$ then $\mathcal{C}_D := \mathcal{V}_{D_1} \uplus \mathcal{C}_{D_2}$ and $\mathcal{V}_D := \mathcal{C}_{D_1} \cup \mathcal{V}_{D_2}$. ∎

6.2. Definition. Let D be a λ-normal formula$_o$. Let $\mathcal{M}$ be a set of unordered pairs, such that if $\{H, K\} \in \mathcal{M}$ and H and K are b-atom subformula occurrences in D, then H and K are contained in a common clause in D, H conv-I K, and either H occurs positively and K occurs negatively in D, or H occurs negatively and K occurs positively in D. Such a set $\mathcal{M}$ is called a *mating* for D. If $\{H, K\} \in \mathcal{M}$ we say that H and K are $\mathcal{M}$-*mated*, or simply *mated* if the mating can be determined from context. If it is also the case that for all $\xi \in \mathcal{C}_D$ there is a $\{H, K\} \in \mathcal{M}$ such that $\{H, K\} \subset \xi$, then we say that $\mathcal{M}$ is a *clause-spanning mating* (cs-mating, for short) for D. In this case, we shall also say that $\mathcal{M}$ *spans* D. If $\mathcal{D}$ is a set of λ-normal formulas$_o$, we say that $\mathcal{M}$ is a mating (cs-mating) for $\mathcal{D}$ if $\mathcal{M}$ is a mating (cs-mating) for $\vee\mathcal{D}$. Here, the order by which the disjunction $\vee\mathcal{D}$ is constructed is taken to be arbitrary but fixed. ∎

The notion of a mating used by Andrews in [5] is a bit more general than the one we have defined here. In that paper, a mating, $\mathcal{M}$, is a set of ordered pairs, $\langle H, K\rangle$, such that there is a substitution θ which makes all such pairs complementary, *i.e.* $\theta K = \sim\theta H$. Except for this difference, the notion of a cs-mating corresponds very closely to his notion of a p-acceptable proof*-mating. Bibel in [7] also exploits matings for various theorem proving and metatheoretical application.

6.3. Proposition. *Let D be in λ-normal form. D is tautologous if and only if D has a cs-mating.*

6.4. Definition. Let $\mathcal{D}$ be a finite, nonempty set of formulas$_o$, and let $\mathcal{M}$ be a mating for $\mathcal{D}$. With respect to $\mathcal{D}$ and $\mathcal{M}$, define $\approx^0$ to be the binary relation on $\mathcal{D}$ such that when $D_1, D_2 \in \mathcal{D}$, $D_1 \approx^0 D_2$ if D_1 contains a b-atom subformula occurrence H and D_2 contains a b-atom subformula occurrence K such that $\{H, K\} \in \mathcal{M}$. Let $\approx$ be the reflexive, transitive closure of $\approx^0$. Clearly $\approx$ is an equivalence relation on $\mathcal{D}$. If $D \in \mathcal{D}$, we shall write $[D]_\approx$ to denote the equivalence class (partition) of $\mathcal{D}$ which contains D. The following proposition is easily proved. ∎

6.5. Proposition. *Let $\mathcal{D}$ be a finite, nonempty set of formulas$_o$. If $\mathcal{M}$ is a cs-mating for $\mathcal{D}$ then $\mathcal{M}$ spans at least one of the $\approx$-partitions of $\mathcal{D}$. The converse is trivially true.*

6.6. Definition. Let $\mathcal{O} = \langle L, z, \{R_l\}\rangle$ be an outline. Let D_l be the formula $Dp(Q_l)$ if l is a sponsoring line or $Dp(\sim Q_l)$ if l is a supporting line. Now define $\mathcal{D}_z :=$ $\{D_z\} \cup \{D_l \mid l \in \rho(z)\}$ if z does not assert $\perp$ and $\mathcal{D}_z := \{D_l \mid l \in \rho(z)\}$, otherwise. Notice that for each $z \in L_0$, $Dp(Q_z) = \bigvee \mathcal{D}_z$. Now let $\mathcal{M}_z$ be a cs-mating for $Dp(Q_z)$ for each $z \in L_0$ and set $\mathcal{M} := \bigcup_{z\in L_0} \mathcal{M}_z$. $\mathcal{M}$ is called a *cs-mating for* $\mathcal{O}$. (Notice that $\mathcal{M}$ is also a cs-mating for each $Dp(Q_z)$.) We say that $\mathcal{O}$ is $\mathcal{M}$*-focused* if for each $z \in L_0$, $\mathcal{D}_z$ is composed of exactly one $\approx$-partition. ∎

6.7. Example. If $\mathcal{O}$ is the outline in Example 5.5, then

$$D_2 = \sim Pzy$$
$$D_3 = \sim[Pzy \supset Pz.c.Pz]$$
$$D_{16} = Pz.c.Pz$$
$$\vee\mathcal{D}_{16} = \sim Pzy \vee \sim[Pzy \supset Pz.c.Pz] \vee Pz.c.Pz]$$

Notice that $\mathcal{D}_{16}$ is tautologous. If we let $A1, A2, A3, A4$ represent the four b-atom occurrences in $\mathcal{D}_{16}$ then a cs-mating for $\mathcal{D}_{16}$ would be $\{(A1, A2), (A3, A4)\}$. ∎

Let $\mathcal{O} = \langle L, z, \{R_l\}\rangle$ be an outline and let $\mathcal{M}$ be a cs-mating for $\mathcal{O}$. If $\mathcal{O}$ is not $\mathcal{M}$-focused, then there must be a $z \in L_0$ such that $\mathcal{D}_z$ has too many members, *i.e.* there are at least two $\approx$-partitions of $\mathcal{D}_z$. What we need is a thinning outline transformation which will permit us to deactivate lines in $\mathcal{O}$, there by removing elements from $\mathcal{D}_z$. As long as the resulting $\mathcal{D}_z$ is still spanned by $\mathcal{M}$, the result of the thinning transformation will satisify the requirements of being an outline.

The thinning transformation works as follows. Let outline $\mathcal{O}$ and a cs-mating $\mathcal{M}$ for $\mathcal{O}$ be such that $\mathcal{O}$ is not $\mathcal{M}$-focused. Let z be sponsoring line such that $\mathcal{D}_z$ contains more than one $\approx$-partition. By Proposition 6.5, there is at least one $\approx$-partition $\mathcal{P} \subset \mathcal{D}_z$ such that $\mathcal{M}$ spans $\mathcal{P}$. Set $\mathcal{P}' := \mathcal{P} \setminus \mathcal{D}_z$. For each supporting line l of z such that $l \in \mathcal{P}'$, the thinning transformation modifies the value of $\rho(z)$ by removing l from it. If it is the case that $D_z \in \mathcal{P}'$, then the supporting lines in $\mathcal{P}$ are strong enough to prove $\bot$, from which the assertion in line z follows immediately. In this case, the thinning transformation must add the new sponsoring line

$$(y) \quad \mathcal{H} \quad \vdash \quad \bot \qquad\qquad NJ,$$

where $\mathcal{H}$ is the set of hypotheses for line z. The justification for line z is changed to RuleP: y. The supports for line y are those lines which were supporting line z and were not thinned out as described above.

7. Backchaining

Using the mating information contained in the Dp-values of the expansion trees associated with each active line of an outline provides the outline transformation process with enough information to look ahead and identify unnecessary supporting and sponsoring lines. This same look ahead will help us determine when we should backchain on an implicational support line.

Consider the outline fragment

$$\begin{array}{llll} (a) & \mathcal{H} \vdash & \sim A_1 \vee A_2 & RuleX \\ (z) & \mathcal{H} \vdash & B & NJ \end{array} \qquad (\sigma)$$

where we have already determined that line a is a necessary support of line z, and RuleX is the justification for line a. One way to use line a in proving line z is to apply P-Cases (see [4]) to the lines in (σ), which would then yield the following lines.

$$\begin{array}{lllll} (a) & \mathcal{H} & \vdash & \sim A_1 \vee A_2 & RuleX \\ (b) & b & \vdash & \sim A_1 & Hyp \\ (m) & \mathcal{H}, b & \vdash & B & NJ \\ (n) & n & \vdash & A_2 & Hyp \\ (y) & \mathcal{H}, n & \vdash & B & NJ \\ (z) & \mathcal{H} & \vdash & B & Cases: a, m, y \end{array} \qquad (\tau_1)$$

It may turn out that this new outline is no longer focused for at least two reasons. First, line m may be proved indirectly from its sponsors, which now includes line b. In other words, $\mathcal{D}_m$ may contain a partition $\mathcal{P}$ such that $D_b \in \mathcal{P}$ but $D_m \notin \mathcal{P}$. Hence, $\sim A_1$ is used to prove $\bot$. The proof could, therefore, be reorganized so that we instead try to prove A_1 directly. In this case, we should apply the new D-ModusPonens transformation to the lines in (σ) to yield the following lines.

$$\begin{array}{lllll} (a) & \mathcal{H} & \vdash & \sim A_1 \vee A_2 & RuleX \\ (m) & \mathcal{H} & \vdash & A_1 & NJ \\ (n) & \mathcal{H} & \vdash & A_2 & RuleP: a, m \\ (y) & \mathcal{H} & \vdash & A_2 \supset B & NJ \\ (z) & \mathcal{H} & \vdash & B & RuleP: y, n \end{array} \qquad (\tau_2)$$

Lines m and y are new sponsoring lines and they share the supports which z had, less line a. Notice that R_a has the form $(\vee \sim R_1\ R_2)$ for some list structures R_1 and R_2. In the new outline, we set $R'_m := R_1$ and $R'_y := (\supset R_2\ R_z)$. The new outline will be focused.

Another way the outline containing the lines in (τ_1) may not be focused is that line y is proved indirectly from its supports. In this case, we need to backchain on the contrapositive form of line a, *i.e.* we should apply the new D-ModusTollens on the lines in (σ) to yield the following lines.

$$\begin{array}{llll} (a) & \mathcal{H} \vdash & \sim A_1 \vee A_2 & RuleX \\ (m) & \mathcal{H} \vdash & \sim A_2 & NJ \\ (n) & \mathcal{H} \vdash & \sim A_1 & RuleP : a, m \\ (y) & \mathcal{H} \vdash & \sim A_1 \supset B & NJ \\ (z) & \mathcal{H} \vdash & B & RuleP : y, n \end{array} \qquad (\tau_3)$$

If in fact the outline containing the lines in (τ_1) was focused, then neither D-ModusPonens or D-ModusTollens could not be used on line a, and we actually needed to treat line a as a disjunction by applying P-Cases. Of course, all these comments apply equally well when line a asserts a formula of the form $A_1 \supset A_2$.

8. Other Forms of Natural Deduction

There are several different formats of proofs which have been called natural deduction, and, at first glance, the problems encountered in converting ET-proofs to these other proof formats might appear to be quite different than the problems encountered in building the Suppes-style proofs of the previous sections. This is generally not the case. For example, the transformation process already described produces, in a sense, proofs in Gentzen's LK format [13]. For each sponsoring line z in a given outline, consider the sequent, $\rho(z) \rightarrow z$, where line labels are used to refer to their assertions. Hence, to each outline there corresponds a set of sequents which represent the unfinished subproofs of that outline. The D- and P- transformations can then be seen as ways of taking the sequents of one outline and replacing some of them with logicially simpler sequents. These simpler sequents can then be joined using derived rules of the LK-calculus to yield the sequents they replace. In this fashion, an entire LK derivation can be built. Of course, for this to work in higher-order logic, we would need to add an inference rule for λ-conversion, but this is the only essential addition needed for this accommodation. LK derivations built in this fashion will contain no instances of the cut inference rule. Thus, by using our relative completeness result for ET-proofs, if A is a theorem of $\mathcal{T}$, A has an ET-proof which can be converted to a cut-free LK derivation. Via the transformation process, our version of Herbrand's theorem can thus be used to prove Gentzen's *Hauptsatz*. See [16] for a complete account of how ET-proofs can be converted to LK deriviations.

9. Acknowledgements

I would like to thank Peter Andrews and Frank Pfenning for many valuable comments concerning this paper and the work reported in it.

10. Bibliography

[1] Peter B. Andrews, "Resolution in Type Theory," Journal of Symbolic Logic **36** (1971), 414–432.

[2] Peter B. Andrews, "Provability in Elementary Type Theory," Zeitschrift für Mathematische Logik und Grundlagen der Mathematik **20** (1974), 411–418.

[3] Peter B. Andrews and Eve Longini Cohen, "Theorem Proving in Type Theory," *Proceedings of the Fifth International Joint Conference on Artificial Intelligence* 1977, 566.

[4] Peter B. Andrews, "Transforming Matings into Natural Deduction Proofs," *Fifth Conference on Automated Deduction, Les Arcs, France*, edited by W. Bibel and R. Kowalski, Lecture Notes in Computer Science, No. 87, Springer-Verlag, 1980, 281–292.

[5] Peter B. Andrews, "Theorem Proving Via General Matings," Journal of the Association for Computing Machinery **28** (1981), 193–214.

[6] Maria Virginia Aponte, José Alberto Fernández, and Philippe Roussel, "Editing First-order Proofs: Programmed Rules vs. Derived Rules," *Proceedings of the 1984 International Symposium on Logic Programming*, 92–97.

[7] Wolfgang Bibel, "Matrices with Connections," Journal of the Association of Computing Machinery **28** (1981), 633–645.

[8] W. W. Bledsoe, "A Maximal Method for Set Variables in Automatic Theorem-proving," in *Machine Intelligence 9*, edited by J. E. Hayes, Donald Michie, and L. I. Mikulich, Ellis Horwood Ltd., 1979, 53–100.

[9] W. W. Bledsoe, "Using Examples to Generate Instantiations for Set Variables," University of Texas at Austin Technical Report ATP-67, July 1982.

[10] Alonzo Church, "A Formulation of the Simple Theory of Types," Journal of Symbolic Logic **5** (1940), 56–68.

[11] Gérard P. Huet, "A Mechanization of Type Theory," *Proceedings of the Third International Joint Conference on Artificial Intelligence* 1973, 139–146.

[12] Gérard P. Huet, "A Unification Algorithm for Typed λ-calculus," Theoretical Computer Science **1** (1975), 27–57.

[13] Gerhard Gentzen, "Investigations into Logical Deductions," in *The Collected Papers of Gerhard Gentzen*, edited by M. E. Szabo, North-Holland Publishing Co., Amsterdam, 1969, 68–131.

[14] D. C. Jensen and T. Pietrzykowski, "Mechanizing ω-Order Type Theory Through Unification," Theoretical Computer Science **3** (1976), 123–171.

[15] Dale A. Miller, Eve Longini Cohen, and Peter B. Andrews, "A Look at TPS," *6th Conference on Automated Deduction, New York*, edited by Donald W. Loveland, Lecture Notes in Computere and Science, No. 138, Springer-Verlag, 1982, 50–69.

[16] Dale A. Miller, "Proofs in Higher-order Logic," Ph. D. Dissertation, Carnegie-Mellon University, August 1983. Available as Technical Report MS-CIS-83-37

from the Department of Computer and Information Science, University of Pennsylvania.

[17] Frank Pfenning, "Analytic and Non-analytic Proofs," elsewhere in these proceedings.

[18] T. Pietrzykowski and D. C. Jensen, "A complete mechanization of ω-order type theory," *Proceedings of the ACM Annual Conference, Volume I*, 1972, 82–92.

[19] Tomasz Pietrzykowski, "A Complete Mechanization of Second-Order Type Theory," Journal of the Association for Computing Machinery **20** (1973), 333–364.

[20] J. A. Robinson, "Mechanizing Higher-Order Logic," *Machine Intelligence 4*, Edinburgh University Press, 1969, 151–170.

[21] Patrick Suppes, *Introduction to Logic*, D. Van Nostrand Company Ltd., Princeton, 1957.

[22] Moto-o-Takahashi, "A proof of cut-elimination theorem in simple type-theory," Journal of the Mathematical Society of Japan **19** (1967), 399–410.

[23] Alfred Tarski, "A Lattice-theoretical Fixpoint Theorem and Its Applications," Pacific Journal of Mathematics **5** (1955), 285–309.

Analytic and Non-analytic Proofs

Frank Pfenning
Department of Mathematics
Carnegie-Mellon University
Pittsburgh, PA 15213

0. Abstract

In automated theorem proving different kinds of proof systems have been used. Traditional proof systems, such as Hilbert-style proofs or natural deduction we call *non-analytic*, while resolution or mating proof systems we call *analytic*. There are many good reasons to study the connections between analytic and non-analytic proofs. We would like a theorem prover to make efficient use of both analytic and non-analytic methods to get the best of both worlds.

In this paper we present an algorithm for translating from a particular non-analytic proof system to analytic proofs. Moreover, some results about the translation in the other direction are reformulated and known algorithms improved. Implementation of the algorithms presented for use in research and teaching logic is under way at Carnegie-Mellon University in the framework of TPS and its educational counterpart ETPS.

Finally we show how to obtain non-analytic proofs from resolution refutations. As an application, resolution refutations can be translated into comprehensible natural deduction proofs.

1. Introduction

In automated theorem proving different kinds of proof systems have been used. Traditional proof systems, such as Hilbert-style proofs or natural deduction we call *non-analytic*, while resolution or mating proof systems we call *analytic*. There are many good reasons to study the connections between analytic and non-analytic proofs. We would like a theorem prover to make efficient use of both analytic and non-analytic methods to get the best of both worlds.

The advantages of analytic proofs are well known. One of the most important advantage is that they seem to be ideally suited for an efficient automatic search for a proof on the computer.

On the other hand there is much to gain from the use of non-analytic proof systems in addition to analytic methods. Non-analytic proofs can be presented in a comprehensible and pleasing format. If we can translate, say, resolution refutations into legible non-analytic proofs, we can help the mathematician understand the automatically generated proof. Valuable work here has been done by Miller [10]. The natural deduction proofs obtained from mating refutations are often elegant and easy to understand and use such mathematically common concepts as proof by contradiction and case-analysis, and make use of intuitive operations such as backchaining. Better translations which are the object of current research would make this even more useful for a wider class of theorems.

The ability to freely translate between analytic and non-analytic proofs also gives us a tool for creating a more elegant natural deduction style proof from a given one. We would translate a given proof into an analytic proof, possibly transform this analytic proof into a shorter one, and then build a new natural deduction style proof from it in a canonical fashion.

Good translation procedures can also serve as a valuable research tool. Heuristics and lemmas of use to a theorem prover can often be discovered and formulated naturally in some non-analytic proof style. The ability to translate these into an analytic format may help to incorporate them into a theorem prover. Moreover, if we can translate automatic proofs obtained with and without a certain heuristic, we may gain deeper insight into the nature and performance of the heuristics.

Another perhaps more immediately important application is in the use of these procedures in computer-aided instruction in logic. The student will attempt his proof in a deductive format, e.g. in a natural deduction style, on the computer. The analytic proof of the exercise can be found beforehand by an automated theorem prover employing resolution or a mating procedure, or even constructed from a sample natural deduction proof given by the teacher. This analytic proof can then be used to guide the student through his own attempts to prove the theorem by suggesting which inference rules may be appropriate when the student asks for help. Moreover, when the student is done, a "normalizing" procedure like the one described above can demonstrate to the student how he might have proven the theorem more elegantly or efficiently. A system called ETPS, which will contain all these features, is currently under development at Carnegie-Mellon University.

There is also a very good complexity-theoretic reason why a theorem prover may want to make use of non-analytic as well as analytic methods. A result by Statman [14] shows that there are theorems which have "short" non-analytic proofs, but no "short" analytic proofs whatsoever. He exhibits a sequence of theorems (from the theory of combinators) whose shortest possible analytic proof is $\left.2^{2^{\cdot^{\cdot^{\cdot^{2^l}}}}}\right\}d$. ($d$ is the number of connectives and quantifiers of a theorem X, and l the length of a non-analytic proof for X.) This lower bound is not Kalmar-elementary, and there are therefore theorems which cannot be practically proven by purely analytic methods which have short non-analytic proofs.

Let us now try to make more precise the distinction between analytic and non-analytic proof systems. The term "analytic" was introduced by Smullyan in [13] and conveys the idea that the proof (or refutation) procedure analyzes the given formula. An analytic proof has a very strong *subformula property*: Only subformulas of the theorem and their instances will appear in an analytic proof.

In the field of automated deduction the discovery of analytic proof systems such as resolution [12] went hand in hand with the beginning of research. The mating approach [3] and a similar method by Bibel [4] are other examples of analytic proof systems.

Examples of non-analytic methods in automated theorem proving can be found in Bledsoe's survey [6] of non-resolution theorem proving. This includes approaches like term-rewriting, built-in inequalities, forward-chaining, models, and even counterexamples. Some of these approaches may be called *non-analytic*, since they sometimes consider formulas not part of the proposed theorem. Many of the stimuli here come from mathematics rather than pure logic. Hilbert-style, Gentzen-style [7], or natural deduction style systems are all examples of traditional non-analytic proof systems. In general they do not obey the subformula property. Usually Cut or Modus Ponens is used to eliminate the helpful formulas, which are not part of the theorem, but substitutivity of equivalence or equality may be used as well. The use of Cut itself does not characterize non-analytic proof systems, as can be seen from the case of resolution, where the cut formulas are all subformulas of the given theorem.

Andrews has shown in [2] how to convert matings into natural deduction proofs. Miller [9] took this work further by generalizing it to higher-order logic and also addressing questions of style in these proofs. Some related work was also done by Bibel in [5]. An algorithm translating in the other direction is the main contribution of this paper. The ability to readily translate in either direction between analytic and non-analytic proofs (in the case of the implementation in TPS between expansion proofs and natural deduction style proofs) gives us all the aforementioned advantages.

As a representative of non-analytic proof systems we pick $\mathcal{I}^*$, mainly for its conceptual clarity and simplicity of cut-elimination. $\mathcal{I}^*$ which is described in section 2 is closely related to the system LK of Gentzen [7] and a related system of Smullyan [13].

Following Miller in [9], who works in the setting of higher order logic, we define a purely analytic proof system in section 3. Expansion proofs, as they are called, are very natural and convenient and very concisely represent the information contained in an analytic proof.

In section 4 we give a new exposition of part of Miller's work in terms of our analytic and non-analytic first-order proof systems. This exposition provides the reader with a self-contained and unified treatment of the translations between the various proof styles. We also handle conjunction in a new way, thus creating stylistically different proofs.

As the main part of this paper, we give an explicit algorithm which translates $\mathcal{I}^*$-proofs into expansion proofs in sections 5, 6, and 7. Expansion proofs are very much different from the kind of analytic proofs Gentzen or Smullyan considered, though some of their ideas, in particular for cut-elimination, are used. Our *merge* algorithm which deals with the inference rule *Contraction* is a significantly improved version of Miller's [9] MERGE, which generally produces much larger expansion trees.

Andrews in [1] has given an algorithm which computes a mating from a resolution refutation. In section 8 we state and prove the correctness of a different algorithm which translates resolution refutations into expansion proofs, which do not make use of Skolem-functions or conjunctive normal forms and satisfy a quite different acceptability criterion from Andrews'. We thus give a two-step procedure by which resolution refutations can be translated into $\mathcal{I}^*$-proofs, or, in one more step, into natural deduction proofs.

Space does not permit to include here non-trivial examples illustrating the various algorithms. Detailed examples for all the translation procedures presented here are given by the author in [11].

2. The Systems $\mathcal{I}$ and $\mathcal{I}^*$

Our non-analytic proof system is $\mathcal{I}^*$, which builds upon similar systems of Gentzen [7] and Smullyan [13]. $\mathcal{I}^*$ is particularly well suited for the description of our algorithms. Notice, for instance, that any theorem derived in $\mathcal{I}^*$ is automatically in negation normal form. The work done here can easily be generalized to other superficially richer systems of first-order logic. To simplify some of our exposition we introduce a system $\mathcal{I}$ which is identical to $\mathcal{I}^*$ but does not contain the rule of *Mix* (a variant of *Cut*).

Our formulation of first-order logic includes the propositional connectives $\vee$, $\wedge$, $\neg$, the quantifiers $\exists$ and $\forall$ and an infinite number of individual variables and constants. Function constants of arbitrary finite arity are also permitted. An atomic formula is of the form $Pt_1 \ldots t_n$ for an n-ary predicate P and terms $t_1, \ldots, t_n$. A literal is of the form A or $\neg A$ for an atomic formula A. A formula is in negation normal form if the scope of each negation is atomic. Each first-order formula has a classically equivalent formula in negation normal form, and we generally assume our formulas to be in negation normal form. $X[v/a]$ is our notation for the result of substituting a for the free occurrences of v in X. We write **nnformula** for a formula in negation normal form. We do not assume that formulas are alphabetically normal, except in section 8 where we talk about resolution refutations. Sometimes we write ⋈ to indicate that an equation is valid for both conjunction and disjunction.

Nodes in a proof-tree in I we call **lines**. A line in I is a multi-set of formulas. This formulation is halfway between Gentzen's (sequents) and Smullyan's (sets). The reason for choosing this particular representation lies in the fact that contraction is an extremely powerful inference rule of our system. When we try to analyze how the effect of a contraction induces a change in an associated expansion tree, we will see that the transformation is really quite complicated. Thus we cannot leave contraction implicit, like Smullyan did, when he introduced sets of formulas as objects in the proof. Structural rules like exchange, however, have no impact on the logical contents of the formula or proof line. We therefore leave them implicit in the multi-set notation.

In general we let U and V stand for multi-sets of formulas, i.e. sets where we allow the same formula to appear more than once as a member. We often write U, X to mean $U \cup \{X\}$ if U is a multi-set.

The axioms of I are of the form

$$U, A, \neg A$$

where A is an atomic formula.

The inference rules can be divided into *structural rules*, *propositional rules*, and *quantificational rules*. The only structural rule in I is *contraction* (C). There is one propositional rule for each propositional connective: $\vee$-*introduction* $(\vee I)$ and $\wedge$-*introduction* $(\wedge I)$. There is also exactly one rule for the quantifiers: $\exists$-*introduction* $(\exists I)$ and $\forall$-*introduction* $(\forall I)$.

Structural rules

$$\text{Contraction:} \qquad \frac{U, X, X}{U, X}\, C$$

Propositional rules

$$\frac{U, X, Y}{U, X \vee Y}\, \vee I \qquad\qquad \frac{U, X \qquad V, Y}{U, V, X \wedge Y}\, \wedge I$$

Quantificational rules

$$\frac{U, X[v/t]}{U, \exists v X}\, \exists I\ , \quad t \text{ a term free for } v \text{ in } X.$$

$$\frac{U, X[v/a]}{U, \forall v X}\, \forall I\ , \quad a \text{ not free in } U, \forall v X.$$

U, V contain the side-formulas of an inference rule. They may be empty. The propositional and quantificational inference rules correspond to Smullyan's [13] rules α, β, γ, δ.

System I is complete in the sense that we can derive the negation normal form of every valid formula in classical first order logic. This follows almost immediatedly from Smullyan's form of the completeness result for Gentzen systems and we will not repeat the argument here.

We shall also use the system I^* which contains the rule of *Mix*:

$$\frac{U, X, \ldots, X \qquad V, \overline{X}, \ldots, \overline{X}}{U, V}\ Mix\,, \qquad X \notin U,\ \overline{X} \notin V$$

$\overline{X}$ is the negation normal form of $\neg X$. There must be at least one occurrence of X, the **mix formula**, in the left premise and at least one occurrence of $\overline{X}$ in the right premise. *Mix* was introduced by Gentzen and is a variant of the rule of *Cut*, and the two are easily shown to be equivalent.

3. Expansion Trees

Analytic proofs in this paper are presented as expansion trees. Expansion trees very concisely and naturally represent the information contained in an analytic proof, as we hope to show. They were first introduced by Miller [9] and are somewhat similar to Herbrand expansions [8]. Some redundancies can easily be eliminated for an actual implementation as done by Miller in the context of higher order logic. The shallow formula of an expansion tree will correspond to the theorem; the deep formula is akin to a Herbrand-expansion proving the theorem. Our formulation of expansion trees differs only trivially from Miller's in [10], if restricted to first-order logic. At several places it is convenient to allow n-ary conjunction and disjunction instead of treating them as binary operations.

3.1. Definition. We define **Expansion Trees** inductively. Simultaneously, we also define Q^D, the **deep formula** of an expansion tree, which is always quantifier-free, and Q^S, the **shallow formula** of an expansion tree. We furthermore place the restriction that no variable in an expansion tree may be selected more than once.

(i) A literal l (signed atom) is an expansion tree. $Q^D(l) = Q^S(l) = l$. Literals form the leaves of expansion trees.

(ii) If $Q_1, \ldots, Q_n$, $n \geq 2$, are expansion trees, so is

$Q =$

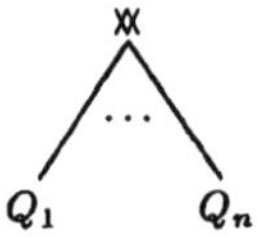

Then $Q^D = Q_1^D \wedge \cdots \wedge Q_n^D$,
and $Q^S = Q_1^S \wedge \cdots \wedge Q_n^S$.

(iii) If $Q_1, \ldots, Q_n$ are expansion trees such that If $Q_1^S = S[v/t_1], \ldots, Q_n^S = S[v/t_n]$, t_i a term free for v in S for $1 \leq i \leq n, n \geq 1$, then

$Q =$

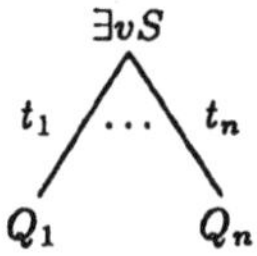

is an expansion tree. Then $Q^D = Q_1^D \vee \cdots \vee Q_n^D$,
and $Q^S = \exists vS$.

$\exists vS$ is called an **expansion node**; v is the **expanded variable**; $t_1, \ldots, t_n$ are the **expansion terms**.

(iv) If Q_0 is an expansion tree such that $Q_0^S = S[v/a]$ for a variable a, so is

$$Q = \begin{array}{c} \forall v S \\ \Big| a \\ Q_0 \end{array} \qquad \begin{array}{l} \text{Then } Q^D = Q_0^D, \\ \text{and } Q^S = \forall v S. \end{array}$$

$\forall v S$ is called a **selection node**; a is the **variable selected** for this occurrence of v.

To improve legibility of our diagrams we will frequently draw $\triangle X$ for an expansion tree with $Q^S = X$.

Since traditional proof systems do not contain Skolem-functions, we need a different mechanism to insure the soundness of our proofs. Following an idea of Bibel [4], which was picked up by Miller [9], we introduce a relation $<_Q$ on occurrences of expansion terms. The condition that $<_Q$ be acyclic replaces Skolemization in our analytic proof system. The reason for this definition will become clear in section 4. $<_Q$ is dual to $\prec$ defined in [10], and it is shown in [9] that they are equivalent. Later in section 8 we shall see how this relates to Skolemization.

3.2. Definition. Let Q be an expansion tree. $<_Q^0$ is a relation on occurrences of expansion terms such that $t <_Q^0 s$ iff there is a variable selected for a node below t in Q which is free in s. $<_Q$, the **dependency relation**, is the transitive closure of $<_Q^0$.

We define clauses only for quantifier-free nnformulas, since this is the only case we will need.

3.3. Definition. Let X be a quantifier-free nnformula. A **clause** in X is a list of literal occurrences defined inductively by

(i) $X = l$, a literal. Then $C = (l)$ is the only clause in X.

(ii) $X = A \vee B$. Then for all clauses $(a_1, \ldots, a_n)$ in A and $(b_1, \ldots, b_m)$ in B, $C = (a_1, \ldots, a_n, b_1, \ldots, b_m)$ is a clause in $A \vee B$.

(iii) $X = A \wedge B$. Then all clauses in A and all clauses in B are clauses in $A \wedge B$.

3.4. Definition. A relation on literal occurrences in a quantifier-free nnformula X is a **mating** $\mathcal{M}$ if $\neg l = k$ for every pair $(l, k) \in \mathcal{M}$ and there is at least one clause in X containing both l and k. If $(l, k) \in \mathcal{M}$, l and k are said to be $\mathcal{M}$-**mated**.

3.5. Definition. A mating $\mathcal{M}$ is said to **span a clause** C if there are literals $l, k \in C$ such that $(l, k) \in \mathcal{M}$. A mating $\mathcal{M}$ is said to be **clause-spanning** on a quantifier-free nnformula X if every clause in X is spanned by $\mathcal{M}$.

The significance of this definition is of course that a quantifier-free nnformula X is tautologous iff there is a mating clause-spanning on X (see Andrews [3], [1], and Miller [9]).

3.6. Definition. A pair $(Q, \mathcal{M})$ is called an **expansion tree proof** for a nnformula X if

(i) $Q^S = X$.

(ii) No selected variable is free in Q^S.

(iii) $<_Q$ is acyclic.

(iv) $\mathcal{M}$, a **mating on** Q^D, is clause-spanning on Q^D.

Our translations establish soundness and completeness of expansion tree proofs with respect to nnformulas. We rely on the soundness and completeness of $\mathcal{I}^*$, which is a simple consequence of results by Smullyan [13]. A similar, but necessarily less constructive argument was carried out by Miller [9] for expansion tree proofs in higher-order logic.

4. Building $\mathcal{I}$-Proofs from Expansion Tree Proofs

The algorithm follows ideas of Miller [9], but we provide a different treatment of conjunction. Our algorithm results in shorter proofs than the more naive algorithm that always applies case (vii) below for a conjunction, but we do not achieve the full power of Miller's *focusing* method. In return, our method is computationally faster.

In the exposition below we sometimes assume that there is a unique correspondence between the formulas in a line and an associated expansion tree, even though we like to think of the line as a multi-set where several identical members are indistinguishable. In general it is sufficient to pick any correspondence between those multiple occurrences of a formula in a line and the unique subtrees of the associated expansion tree.

4.1. Definition. A pair $(Q, \mathcal{M})$ is an expansion tree proof for a line $L = X_1, \ldots, X_n$ in an $\mathcal{I}$-proof iff $(Q, \mathcal{M})$ is an expansion tree proof for $X_1 \vee \cdots \vee X_n$.

4.2. Definition. Let $(Q, \mathcal{M})$ be an expansion tree proof for a line L in an $\mathcal{I}$-proof, and let X be a subformula of an element in L. Then $Q|_X$ is the part of the expansion tree Q representing X $(Q|_X^S = X)$, and $\mathcal{M}|_X$ is the restriction of $\mathcal{M}$ to pairs both of whose elements lie in $Q|_X^D$. We will sometimes talk about X^D instead of $Q|_X^D$, if the expansion tree Q is clear from the context.

We shall describe an algorithm which constructs an $\mathcal{I}$-proof from an expansion tree proof, starting with the nnformula to be proven and working upwards until every branch in the proof tree begins with an axiom. The cases given below can in principle be applied in any order. The ordering below will often, but not in general, result in the shortest proof that can be constructed with this algorithm.

If an $X \in L$ is such that $Q|_X^D$ has no literal in a pair in $\mathcal{M}$, then X is to be ignored and can only be part of a side-formula in an inference above L.

Now assume L is a given line in an $\mathcal{I}$-proof, and $(Q, \mathcal{M})$ is an expansion tree proof for L.

(i) $L = U, A, \neg A$. Then L is an axiom.

(ii) $L = U, X \vee Y$. Infer L by $\dfrac{U, X, Y}{U, X \vee Y} \vee\mathcal{I}$. $(Q, \mathcal{M})$ is an expansion tree proof for U, X, Y.

(iii) $L = U, \forall v S$. Infer L by $\dfrac{U, S[v/a]}{U, \forall v S} \forall\mathcal{I}$,

where a is the variable selected for this occurrence of $S[v/a]$.

In Q we replace the corresponding subtree $\forall v S$ —a— $S[v/a]$ by $S[v/a]$.

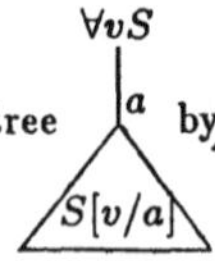

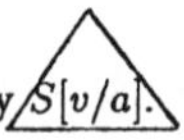

By definition 3.6 and the inductive assumption that $(Q, \mathcal{M})$ forms an expansion tree proof for $U, \forall v S$, a cannot be free in U or $\forall v S$, since a is a selected variable in Q.

(iv) $L = U, \exists v S$ and $\exists v S$ has n, $n \geq 2$ successors in Q.

Infer L by $\dfrac{U, \exists v S, \ldots, \exists v S}{\dfrac{\cdots}{U, \exists v S}} (n-1) \times C$.

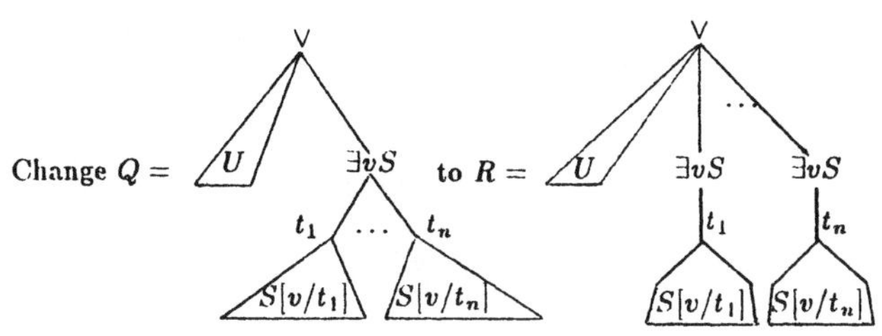

Since $Q^D = R^D$, $(R, \mathcal{M})$ is again an expansion tree proof for $U, \exists vS, \ldots, \exists vS$.

(v) $L = U, \exists vS$, and $\exists vS$ has exactly one successor $S[v/t]$, and no free variable in t is a variable to be selected in Q.

Infer L by $\dfrac{U, S[v/t]}{U, \exists vS}\ \exists I$, and replace $\exists vS$ $\overset{t}{|}$ $S[v/t]$ in Q by $S[v/t]$.

From the restriction on t it is clear that no variable to be selected will be free in $S[v/t]$, and therefore by inductive hypothesis in $U, S[v/t]$.

(vi) $L = U, V, X \wedge Y$ such that $\mathcal{M} = \mathcal{M}|_{U,X} \cup \mathcal{M}|_{V,Y}$, i.e. no literal in U^D or X^D is $\mathcal{M}$-mated to any literal in V^D or Y^D.

Here we have to consider three subcases.

(a) $\mathcal{M}|_U$ is clause-spanning for U^D. Then restrict the mating to $\mathcal{N} = \mathcal{M}|_U$. Then no literal in $V, X \wedge Y$ is involved in the mating and they will only appear as side formulas in any inference above L.

(b) $\mathcal{M}|_V$ is clause-spanning on V^D. This case is symmetric to case (a): Let $\mathcal{N} = \mathcal{M}|_V$.

(c) Neither case (a) nor case (b) apply. Then infer L by $\dfrac{U, X \qquad V, Y}{U, V, X \wedge Y}\ \wedge I$.

Since the problem is symmetric, we will simply show that $(Q|_{U,X}, \mathcal{M}|_{U,X})$ is an expansion tree proof for U, X. It then follows analogously that $(Q|_{V,Y}, \mathcal{M}|_{V,Y})$ is an expansion tree proof for V, Y. The only condition we have to test is whether $\mathcal{M}|_{U,X}$ is clause-spanning on $Q|^D_{U,X}$. Let P be a clause in $Q|^D_{U,X}$. Since neither case (a) nor case (b) applies, there is a clause O in V^D not spanned by $\mathcal{M}$. Let P' be the extension of P to a clause in Q^D such that $P'|_V = O$ and $P'|_{U,X} = P$. By inductive assumption, P' is spanned by $(l, k) \in \mathcal{M}$. Not both l and k are in V^D, since $\mathcal{M}$ does not span O. We also assumed $\mathcal{M} = \mathcal{M}|_{U,X} \cup \mathcal{M}|_{V,Y}$ and hence $(l, k) \in \mathcal{M}|_{U,X}$.

(vii) $L = U, X \wedge Y$ and case (vi) does not apply.

Then infer L by $\dfrac{U, U, X \wedge Y}{U, X \wedge Y}\ C$.

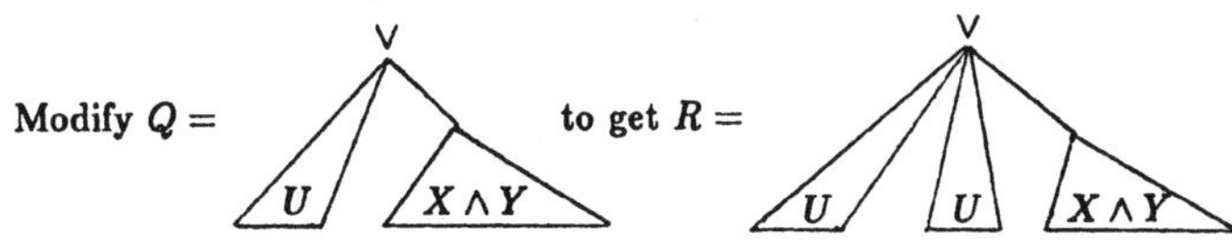

For every occurrence of a literal l in U, there are two occurrences of l in U, U. Call these l^1 and l^2 for the occurrences in the left and right copies of U, respectively. Let

$\mathcal{M}|^1_{U,X}$ [$\mathcal{M}|^2_{V,Y}$] be the result of replacing every occurrence of a literal l from U^D in $\mathcal{M}|_{U,X}$ [$\mathcal{M}|_{V,Y}$] by l^1 [l^2]. Then $\mathcal{N} = \mathcal{M}|^1_{U,X} \cup \mathcal{M}|^2_{V,Y}$ spans every clause in R^D. To see this, let P be a clause in R^D. Then P contains literals from either X or Y, but not both. Without loss of generality, assume P contains literals in X, and let O be the clause in Q^D which agrees with P on X and contains a literal l in U^D iff l^1 is in P. By inductive assumption, O is closed by a pair $(k,m) \in \mathcal{M}$. But then also $(k^1,m) \in \mathcal{M}|^1_{U,X} \subset \mathcal{N}$ (if m is in $Q|^D_X$), or $(k^1,m^1) \in \mathcal{M}|^1_{U,X} \subset \mathcal{N}$ (if m is in $Q|^D_U$). Thus P is spanned by $\mathcal{N}$. Since P was arbitrary, $\mathcal{N}$ spans every clause in R^D.

Now the case (vi) can be applied immediatedly, thus reducing the complexity of $L = U, X \wedge Y$ to the complexities of the lines U, X and U, Y.

Since the size of connected subformulas of the unjustified lines in the $\mathcal{I}$-proof is diminished in each step, all we need to show to prove correctness is that at least one of the cases always applies. One can see that only one problem may arise: all top-level nnformulas are existentially quantified, each of them has just one substitution term, and all of the substitution terms contain a free variable which is still to be selected. Since $<_Q$ has no cycles, there is a term t such that for no s, $s <_Q t$. If t contained a free variable a, which were still to be selected, then the node where a is selected has to lie below one of the top-level existential quantifiers in Q. But if s is the substitution term for this node, then by definition 3.2, $s <_Q t$. This is a contradiction, since $<_Q$ is acyclic and therefore case (v) must apply for at least one of the quantifiers.

5. Building Expansion Tree Proofs from $\mathcal{I}$-proofs

In this section we show how to construct an expansion tree proof from a proof in $\mathcal{I}$. This translation plays an important role in giving a translation procedure from $\mathcal{I}^*$ into expansion tree proofs. Some ideas of Miller [9] are used, but we proceed entirely constructively. Also, the procedure for *merge* presented in case (vi) below results in much smaller expansion trees than the ones obtained by Miller's MERGE algorithm. Moreover, because of the way we set up $\mathcal{I}^*$, a merge is necessary only for contraction and not inherently tied to any quantifier or logical connective. This allows a clearer exposition of the ideas which underly the translation from $\mathcal{I}$-proofs into expansion tree proofs.

The construction proceeds by induction on the $\mathcal{I}$-proof tree. Note that all cases except for *Contraction* are very simple. This supports our claim that the expansion tree proof induced by an $\mathcal{I}$-proof corresponds to the $\mathcal{I}$-proof "in a natural way". The basic "idea" underlying the original proof is retained.

We now assume we are given an inference (or axiom) in $\mathcal{I}$, and we have already constructed expansion tree proofs for the premise. We shall call this expansion tree proof $(Q, \mathcal{M})$ ($(Q_1, \mathcal{M}_1)$ and $(Q_2, \mathcal{M}_2)$ in the case of $\wedge I$). The expansion tree proof for the conclusion will be $(R, \mathcal{N})$.

(i) We have an axiom $U, A, \neg A$.

Then $\mathcal{N} = \{(\neg A, A)\}$ and $R =$

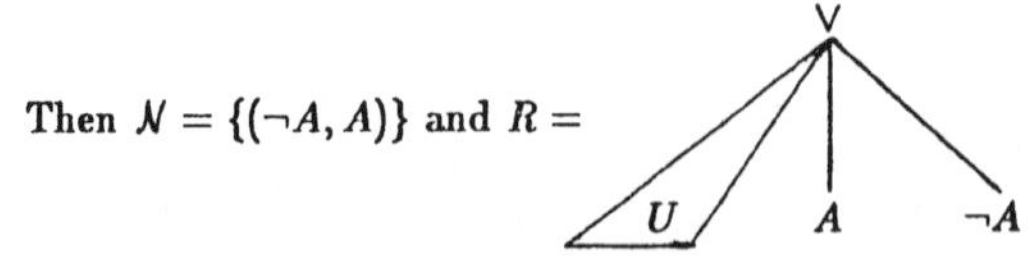

In $Q|_U$, let each existentially quantified variable expand to itself, and select a new unique variable for each universally quantified variable.

(ii) $\vee I$: $\dfrac{U, X, Y}{U, X \vee Y}$. Here $(R, \mathcal{N}) = (Q, \mathcal{M})$.

(iii) $\wedge I$: $\dfrac{U, X \qquad V, Y}{U, V, X \wedge Y}$. Here $\mathcal{N} = \mathcal{M}_1 \cup \mathcal{M}_2$ and

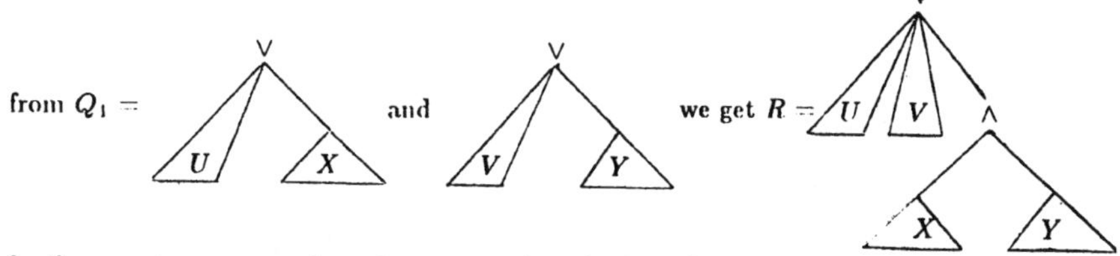

In the new tree we may have to rename the selections for some universal variables, to make sure that no free or selected variable from one branch of the I-proof tree is selected in the other branch.

(iv) $\exists I$: $\dfrac{U, S[v/t]}{U, \exists v S}$, t free for v in S.

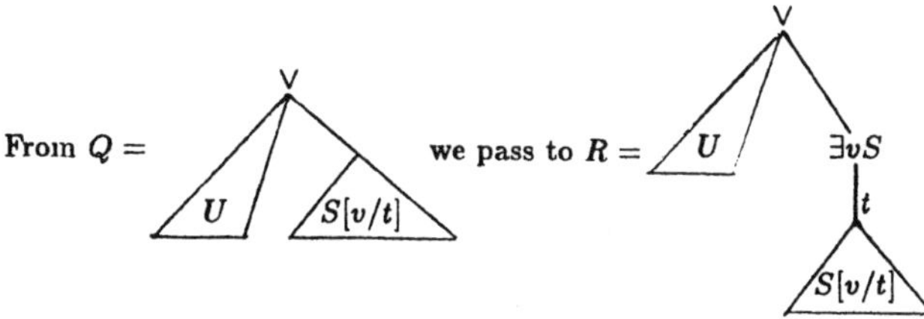

If v does not appear in S, we pick a new variable a to be t, a not selected in Q and not free in U, S.

Since $R^D = Q^D$, we can take $\mathcal{N} = \mathcal{M}$. What remains to be shown in this case is that $<_R$ is acyclic. Let a be a variable selected below $\exists v S$ in R. There may be expansion terms s_i, $t <_R^0 s_i$, but there is no term s such that $s <_R^0 t$. If $s <_R^0 t$ would hold, there had to be a variable b selected in R, and b free in t. But then also b free in $S[v/t]$ (otherwise t was selected to be a new variable), and hence b free in Q^S which contradicts the assumption that $(Q, \mathcal{M})$ is an expansion tree proof for $U, S[v/t]$.

(v) $\forall I$: $\dfrac{U, S[v/a]}{U, \forall v S}$, a a variable not free in U or $\forall v S$.

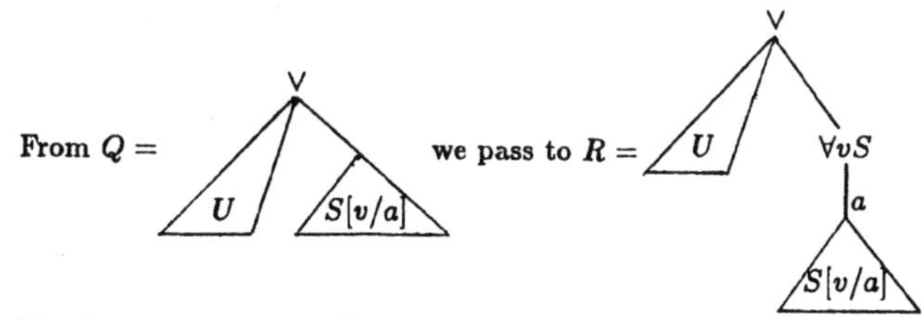

If v does not appear in S, we pick a new variable a not free in U or S or selected in Q.

Since $R^D = Q^D$, we can take $\mathcal{N} = \mathcal{M}$. Moreover, since a is not free in $U, \forall v S$, a is a valid selection. Moreover, a could not have been selected in Q, since a occurs free in $S[v/a]$ or had been chosen not to be selected in Q. Thus a is selected in R only once.

(vi) C: $\dfrac{U, X, X}{U, X}$

Let Q_1, Q_2 be the subtrees of Q with the root node being the left and right occurrences of X in the premise, respectively. We apply a recursive merging algorithm to obtain an expansion tree $Q_1 \oplus Q_2$ for the single occurrence of X in the conclusion. We will pass from

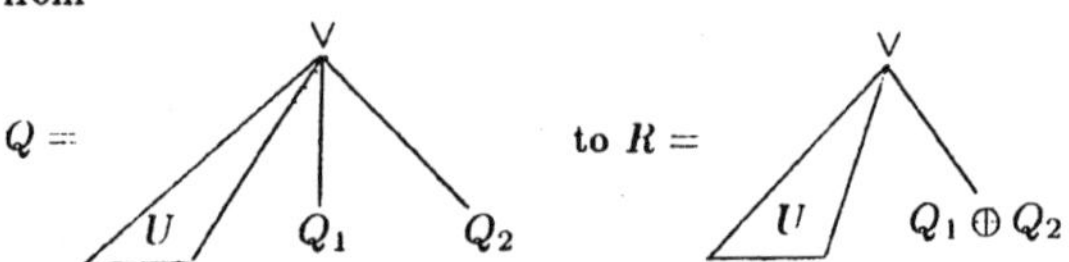

In order to apply $\oplus$ to two expansion trees P_1, P_2, we require $P_1^S = P_2^S$, which is certainly true of Q_1 and Q_2.

(a) $P_1 = l_1 = l = l_2 = P_2$. Then $P_1 \oplus P_2 = l$. We say we identify the distinct occurrences of the literal l.

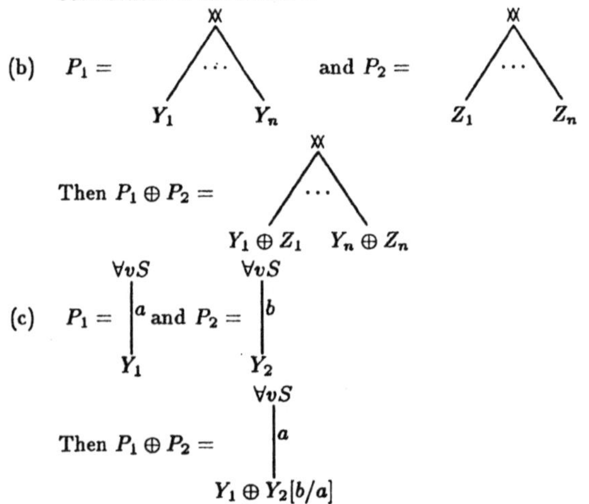

$Y_2[b/a]$ is the result of replacing every occurrence of b in the expansion tree Y_2 by a. But not only do we have to apply this change of names in Y_2, but in the whole expansion tree in which our merge takes place.

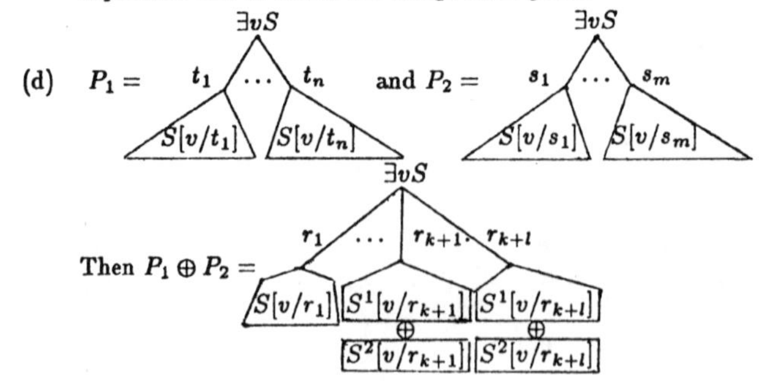

Here $r_1, \ldots, r_k$ are the expansion terms which appear only in one of $t_1, \ldots, t_n$ and $s_1, \ldots, s_m$; $r_{k+1}, \ldots, r_{k+l}$ are the expansion terms which appear in both. S^1 [S^2] stands for the occurrence of a subtree in P_1 [P_2]. If $r_{k+h} = t_i = s_j$ we say that r_{k+h} is the result of **identifying** the distinct occurrences of the expansion terms t_i and s_j.

We now show by induction on the number of identifications of expansion terms in $Q_1 \oplus Q_2$ that $<_R$ is acyclic. We define a sequence of relations, $<_Q = <^0$, $<^1, \ldots,$ $<^n = <_R$ such that each $<^i$, $1 \leq i \leq n$, is acyclic.

Note first that $<^0$ is acyclic, since $<_Q$ is acyclic. If no two literal occurrences were identified during the merge, $<^0 = <_R$ and we are done. Otherwise let $p_1, p_2, \ldots, p_n$ be all the literal occurrences in R which result from identifying expansion terms in Q_1 and Q_2 ordered in such a way that $i < j$ whenever p_i is above p_j in R. Now assume we have already defined $<^i$. Let q_1 in Q_1 and q_2 in Q_2 be the expansion terms which were identified to form p_i. We define $t <^{i+1} s$ iff $t <^i s$ or $t = p_i$ and $q_2 <^i s$ or $q_1 <^i s$ bearing in mind that each variable selected below q_1 is also selected below q_2 after merging, since q_1 and q_2 are identfied. This can only introduce a cycle into $<^{i+1}$ if $p_i <^{i+1} p_i$ which in turn can only happen if $q_1 <^i q_2$ or $q_2 <^i q_1$. But if for some s, $s <^i q_1$, then also $s <^i q_2$, since q_1 and q_2 have the same free variables. Thus this would mean $q_1 <^i q_1$ or $q_2 <^i q_2$, which is a contradiction to the inductive hypothesis that $<^i$ has no cycles.

One can finally see that $<_R = <^n$, since $t <_R s$ either since $t <_Q s$ or because of one of the identifications of distinct expansion term occurrences. The case where selected variables are being renamed and identified does not contribute any new pairs to $<_R$, since a selection is below a given expansion before identifying the selections iff it is below that expansion after identifying the selections.

To obtain $\mathcal{N}$ on R from $\mathcal{M}$ on Q, we simply identify in $\mathcal{M}$ all literal occurrences which were identified to form one literal occurrence. Then $\mathcal{N}$ spans every clause on R: Let $l^\oplus$ be defined as $l \oplus k$, if l and k are literal occurrences which were identifed using case (a) above when forming $Q_1 \oplus Q_2$, otherwise $l^\oplus = l$. Then $\mathcal{N} = \{(l^\oplus, k^\oplus) : (l, k) \in \mathcal{M}\}$. Now let C be a clause in R^D. Then there is a corresponding clause D in Q^D such that $l \in D$ iff $l^\oplus \in C$. D is spanned by a pair $(l, k) \in \mathcal{M}$. But then $(l^\oplus, k^\oplus) \in \mathcal{N}$ and consequently $\mathcal{N}$ spans C.

6. Cut Elimination in $\mathcal{I}^*$

Our cut elimination algorithm is based on similar algorithms of Gentzen [7] and Smullyan [13]. We reformulate these algorithms in terms of the system $\mathcal{I}^*$ in order to give a completely self-contained and unified treatment to all the translations between analytic and non-analytic proofs. If one wanted to write out the details of a procedure which computes an expansion tree proof for a formula B, given those for A and $\neg A \vee B$ directly in terms of expansion tree proofs, one could use the cases below in an inductive proof to show that such a direct procedure will result in the same expansion tree proof for B as the less direct procedure described in section 7.

The proof of termination relies on a double induction argument: At each step we transform one mix (which has no other mixes above it) into one or several mixes with lower degree, or, if the degree stays the same, with smaller rank. The degree of a mix is the number of quantifiers and connectives in the mix formula (the formula being eliminated). The left [right] rank of a mix is the number of lines in the left [right] premise of a mix which contain the mix formulas. The rank of a mix is the sum of left and right rank.

For many of the following cases there is an obvious symmetric case which can be treated completely analogously. It is to be understood that there could be more occurrences of the mix formula in the premises of a mix, but we do not write this out to keep the diagrams as simple as possible. First we consider the case that one of the premises of the mix is an axiom.

(i) The mix formula is the side-formula of the axiom. Then we eliminate the mix immediatedly:

$$\dfrac{U, A, \neg A, X \qquad V, \overline{X}}{U, V, A, \neg A}\,Mix \qquad \Rightarrow \qquad U, V, A, \neg A$$

(ii) The mix formula is not the side-formula of the axiom. Then we also eliminate the mix:

$$\dfrac{U, A \qquad V, A, \neg A}{U, V, A}\,Mix \qquad \Rightarrow \qquad \dfrac{\text{Add } V \text{ as a side-formula to every inference above } U, A}{U, V, A}$$

We will now treat the case that the rank of the mix (which contains no other mix above it) is 2.

(i) The mix formula is a literal A. Since the rank of the mix is 2, one of the previous two cases must apply.

(ii) $C = X \vee Y, \overline{C} = \overline{X} \wedge \overline{Y}$.

$$\dfrac{\dfrac{U, X, Y}{U, X \vee Y}\,\vee I \qquad \dfrac{V_1, \overline{X} \qquad V_2, \overline{Y}}{V_1, V_2, \overline{X} \wedge \overline{Y}}\,\wedge I}{U, V_1, V_2}\,Mix \qquad \Rightarrow \qquad \dfrac{\dfrac{U, X, Y \qquad V_1, \overline{X}}{U, V_1, Y}\,Mix \qquad V_2, \overline{Y}}{U, V_1, V_2}\,Mix$$

Each of the two new mixes has smaller degree.

(iii) $C = \forall v S, \overline{C} = \exists v \overline{S}$.

$$\dfrac{\dfrac{U, S[v/a]}{U, \forall v S}\,\forall I \qquad \dfrac{V, \overline{S}[v/t]}{V, \exists v \overline{S}}\,\exists I}{U, V}\,Mix \qquad \Rightarrow \qquad \dfrac{\dfrac{\vdots}{\text{replace } a \text{ by } t} \quad U, S[v/t] \qquad V, \overline{S}[v/t]}{U, V}\,Mix$$

Note that t is free for v in $\overline{X}$, hence in X, and therefore replacing a by t is a legal operation, transforming one $\mathcal{I}$-proof into another if we also rename some variables b which are free in t.

Now we consider the case where the rank is greater than 2. We treat the case where the left rank is greater than 1. The case where the right rank is greater than 1 can be treated analogously.

This case again breaks up into two subcases. The new formula on the left hand side of the premise may or may not be the same as the mix formula. First we show how to reduce a mix in case the new formula is not the same as the mix formula. Here we generally reduce the mix to a mix with the same degree but lower rank.

(i)

$$\dfrac{\dfrac{U, A, B, X}{U, A \vee B, X}\,\vee I \qquad V, \overline{X}}{U, V, A \vee B}\,Mix \qquad \Rightarrow \qquad \dfrac{\dfrac{U, A, B, X \qquad V, \overline{X}}{U, V, A, B}\,Mix}{U, V, A \vee B}\,\vee I$$

(ii) $$\dfrac{\dfrac{U_1,A,X \qquad U_2,B,X}{U_1,U_2,A\wedge B,X,X}\wedge I \qquad V,\overline{X}}{U_1,U_2,V,A\wedge B}\,Mix \quad\Rightarrow\quad \dfrac{\dfrac{\dfrac{U_1,A,X \qquad V,\overline{X}}{U_1,V,A}\,Mix \qquad \dfrac{U_2,B,X \qquad V,\overline{X}}{U_2,V,B}\,Mix}{U_1,U_2,V,V,A\wedge B}\wedge I}{U_1,U_2,V,A\wedge B}\,C$$

If X appears in only one premise of the $\wedge I$, this case simplifies in the obvious way.

(iii) $$\dfrac{\dfrac{U,A[v/t],X}{U,\exists vA,X}\exists I \qquad V,X}{U,V,\exists vA}\,Mix \quad\Rightarrow\quad \dfrac{\dfrac{U,A[v/t],X \qquad V,\overline{X}}{U,V,A[v/t]}\,Mix}{U,V,\exists vA}\exists I$$

(iv) $$\dfrac{\dfrac{U,A[v/a],X}{U,\forall vA,X}\forall I \qquad V,\overline{X}}{U,V,\forall vA}\,Mix \quad\Rightarrow\quad \dfrac{\dfrac{U,A[v/a],X \qquad V,\overline{X}}{U,V,A[v/a]}\,Mix}{U,V,\forall vA}\forall I$$

If a happens to be free in V, replace a by a new variable b everywhere above $V,\overline{X}$.

(v) $$\dfrac{\dfrac{U,A,A,X}{U,A,X}\,C \qquad V,\overline{X}}{U,V,A}\,Mix \quad\Rightarrow\quad \dfrac{\dfrac{U,A,A,X \qquad V,\overline{X}}{U,V,A,A}\,Mix}{U,V,A}\,C$$

The last case remaining occurs when the mix formula is also the formula introduced by the last inference rule on the left-hand side. The cases are analogous to the previous ones, except that one mix is now reduced to one mix of lower rank and another mix of left rank 1.

(i) $$\dfrac{\dfrac{U,A,B,A\vee B}{U,A\vee B,A\vee B}\vee I \qquad V,\overline{A}\wedge\overline{B}}{U,V}\,Mix \quad\Rightarrow\quad \dfrac{\dfrac{\dfrac{\dfrac{U,A,B,A\vee B \qquad V,\overline{A}\wedge\overline{B}}{U,V,A,B}\,Mix}{U,V,A\vee B}\vee I \qquad V,\overline{A}\wedge\overline{B}}{U,V,V}\,Mix}{U,V}\,C$$

(ii) $$\dfrac{\dfrac{U_1,A,A\wedge B \qquad U_2,B,A\wedge B}{U_1,U_2,A\wedge B,A\wedge B,A\wedge B}\wedge I \qquad V,\overline{A}\vee\overline{B}}{U_1,U_2,V}\,Mix$$

$$\Rightarrow\quad \dfrac{\dfrac{\dfrac{\dfrac{U_1,A,A\wedge B \qquad V,\overline{A}\vee\overline{B}}{U_1,V,A}\,Mix \qquad \dfrac{U_2,B,A\wedge B \qquad V,\overline{A}\vee\overline{B}}{U_2,V,B}\,Mix}{U_1,U_2,V,V,A\wedge B}\wedge I \qquad V,\overline{A}\vee\overline{B}}{U_1,U_2,V,V,V}\,Mix}{U_1,U_2,V}\,2\times C$$

This case simplifies if the mix formula does not appear in both premises of the $\wedge I$.

(iii) $$\dfrac{\dfrac{U,\exists vS,S[v/t]}{U,\exists vS,\exists vS}\exists I \qquad V,\forall v\overline{S}}{U,V}\,Mix \quad\Rightarrow\quad \dfrac{\dfrac{\dfrac{\dfrac{U,\exists vS,S[v/t] \qquad V,\forall v\overline{S}}{U,V,S[v/t]}\,Mix}{U,V,\exists vS}\exists I \qquad V,\forall v\overline{S}}{U,V,V}\,Mix}{U,V}\,C$$

(iv) $$\dfrac{\dfrac{U,\forall vS,S[v/a]}{U,\forall vS,\forall vS}\forall I \qquad V,\forall v\overline{S}}{U,V}\,Mix \quad\Rightarrow\quad \dfrac{\dfrac{\dfrac{\dfrac{U,\forall vS,S[v/a] \qquad V,\forall v\overline{S}}{U,V,S[v/a]}\,Mix}{U,V,\forall vS}\forall I \qquad V,\forall v\overline{S}}{U,V,V}\,Mix}{U,V}\,C$$

(v) $$\dfrac{\dfrac{U,X,X,X}{U,X,X}\,C \qquad V,\overline{X}}{U,V}\,Mix \quad\Rightarrow\quad \dfrac{U,X,X,X \qquad V,\overline{X}}{U,V}\,Mix$$

7. Building Expansion Tree Proofs from $\mathcal{I}^*$-proofs

Since we already showed how to construct expansion tree proofs from $\mathcal{I}$-proofs we have only to show how to construct an expansion tree proof, given expansion tree proofs for the two premises of a mix. We emphasize the constructiveness of our approach. Of course we could simply use any theorem proving procedure and arrive at a proof, since we already know we are dealing with a theorem. Our goal, however, is to construct an expansion tree proof which most closely reflects the structure of the two given original proofs, and moreover can be explicitly obtained from them.

Here is our procedure: If we do not already have mix-free $\mathcal{I}$-proofs for both premises, construct them with the algorithm described in section 4. Eliminate the mix from the resulting proof in $\mathcal{I}^*$ to obtain a proof in $\mathcal{I}$ using the algorithm in section 6. Finally, construct an expansion tree proof from this $\mathcal{I}$-proof using the procedure given in section 5.

In practice we do not have to explicitly contruct these $\mathcal{I}$-proofs. The procedure may be reformulated in terms of the expansion tree proofs themselves, but space does not permit to write out the rather laborious details here.

By looking at one of the critical cases, case (i) where a mix of rank 1 is eliminated, one can see the following: If d is the number of quantifiers and connectives in the mix formula (degree of the mix), l is the length of the proof (say, above the leftv premise), and $f(d,l)$ is a worst case lower bound of the length of the resulting mix-free proof, the following relation must hold: $f(d,l) \geq f(\frac{d}{2}, f(\frac{d}{2}, l))$. Thus we get $f(d,l) \geq \left. 2^{2^{\cdot^{\cdot^{\cdot^{2^l}}}}} \right\} d$.

Since an $\mathcal{I}$-proof is at most exponentially bigger than a corresponding expansion tree proof, the lower bound remains non-Kalmar-elementary when the resulting $\mathcal{I}$-proof is translated into an expansion tree proof. A result by Statman [14] mentioned in the introduction tells us that this can not be significantly improved. There cannot be a Kalmar-elementary translation from $\mathcal{I}^*$-proofs into $\mathcal{I}$-proofs.

In practice, however, the translation is often feasible and it is not clear which class of theorems will actually blow up the size of the proof by as much as $f(d,l)$.

8. Building Expansion Tree Proofs from Resolution Refutations

When describing the translation procedure from resolution refutations into expansion tree proofs care must be taken to avoid confusion between the different nnformulas and the clauses in them. Resolution refutations are stated for the negation of a theorem; expansion tree proofs are defined for the theorem itself. In both cases clauses play a central role. Thus we will call clauses in an expansion tree **paths**, while clauses in a resolution refutation will be called **clauses**. We say a path **intersects** a clause if they have a literal occurrence in common. Notice that our definition of a clause is slightly different from the customary definition as a set. Since matings are relations on literal occurrences, we cannot afford to regard different occurrences of the same literal as identical. During a resolution of two clauses we delete all occurrences of the literal resolved upon. Generally in this section we will assume nnformulas also to be $\alpha\beta$-normal, i.e. no variable occurs both free and bound and each variable is bound at most once.

Andrews [1] described an algorithm which translates resolution refutation into matings, but the setting here is essentially different. We do not work with conjunctive normal forms or Skolem-terms in expansion tree proofs and the condition that matings in expansion tree proofs must be clause-spanning is also quite different from Andrews' condition that every cycle in a mating must have a merge.

With the aid of this algorithm a resolution refutation can be translated into a non-analytic proof by first translating it into an expansion tree proof and then into a proof in $\mathcal{I}^*$ using the algorithm in section 5. This can be carried even further by translating the $\mathcal{I}^*$-proof into a proof in natural deduction style. A procedure for this translation is given by Miller in [10]. This can help a mathematician understand a proof by a resolution theorem prover since he can study it in a familiar format. It may also be a valuable research tool as indicated in the introduction.

8.1. Definition. Let X be an $\alpha\beta$-normal nnformula. Then X^*, the **Skolem-form** of X, is the result of replacing every subformula of the form $\exists vS$ by $S[v/f_v(w_1,\ldots,w_n)]$, where $w_1,\ldots,w_n$ are all the universally quantified variables in whose scope $\exists vS$ lies, and then deleting all the universal quantifiers. $f_v(w_1,\ldots,w_n)$ and instances thereof are called **Skolem-terms**, f_v the **Skolem-function** for v.

8.2. Definition. Let X be an $\alpha\beta$-normal nnformula. A **resolution refutation** of X is a list of clauses $c_1,\ldots,c_n$ such that

(i) $\exists m$ such that $\{c_j : 1 \leq j \leq m\}$ is a subset of the set of clauses of $\overline{X}^*$,

(ii) for each $j > m$ either

(a) c_j is a substitution instance ϕc_i for some $i \leq j$,

(b) c_j is the resolvent of c_{a_j} and c_{b_j}, where $a_j, b_j \leq j$, and c_j is formed by appending the results of deleting all occurrences of a literal l_j from c_{a_j} and $\neg l_j$ from c_{b_j}.

(c) $c_n = \Box$ (the empty clause).

In our translation we will have to select unique variables for Skolem-functions and their arguments. In general, if $f(w_1,\ldots,w_n)$ is a Skolem-term for arbitrary terms $w_1,\ldots,w_n$, then $\overline{f(w_1,\ldots,w_n)}$ is a unique corresponding variable. Note that this is just a notational convenience in our metalanguage. We must also occasionally model the effect of a substitution into a Skolem-term on the corresponding variables.

8.3. Definition. Let $\overline{f(w_1,\ldots,w_n)}$ be a variable, ϕ a substitution for variables which do not come from Skolem-terms. We extend ϕ to terms and formulas in the usual way, but also extend it to act on variables which come from Skolem-terms. Recursively define $\phi\overline{f(w_1,\ldots,w_n)} := \overline{f(\phi w_1,\ldots,\phi w_n)}$.

We are now ready to define what it means to apply a substitution to an expansion tree. Note that $(\phi Q)^S = \phi(Q^S)$.

8.4. Definition. Let Q be an expansion tree. Then we define ϕQ inductively.

(i) Q is a literal l. Then $\phi Q = \phi l$.

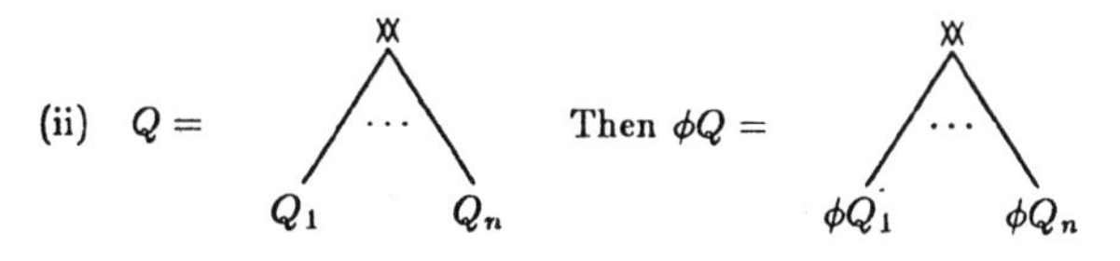

(iii) $Q =$

$$\begin{array}{ccc} & \exists v S & \\ t_1 \swarrow & \cdots & \searrow t_n \\ Q_1 & & Q_n \end{array}$$

We leave the original expansions intact, and add all terms which change under the substitution as new expansion terms. Let $t_{i_1}, \ldots, t_{i_m}$ be all the expansions terms t_i such that $\phi t_i \neq t_i$. Then

$\phi Q =$

$$\begin{array}{ccccc} & & \exists v S & & \\ t_1 \swarrow & \cdots & \mid \phi t_{i_1} & \cdots & \searrow \phi t_{i_m} \\ Q_1 & & \phi Q_{i_1} & & \phi Q_{i_m} \end{array}$$

(iv) $Q =$

$$\begin{array}{l} \forall v S \\ \mid \overline{f(w_1, \ldots, w_n)} \\ Q_0 \end{array}$$

Then $\phi Q =$

$$\begin{array}{l} \forall v S \\ \mid \overline{\phi f(w_1, \ldots, w_n)} \\ \phi Q_0 \end{array}$$

During the translation from resolution refutations to expansion tree proofs we associate an expansion tree and a mating with each line in the resolution refutation. These expansion trees have to satisfy all of the conditions of expansion tree proofs except that the mating does not have to be clause-spanning. We therefore define:

8.5. Definition. A **partial expansion tree proof** $(Q, \mathcal{M})$ for a nnformula X is an ordered pair consisting of an expansion tree Q and a mating $\mathcal{M}$ on Q^D such that

(i) $Q^S = X$.

(ii) No selected variable is free in Q^S.

(iii) $<_Q$ is acyclic.

A particular partial expansion tree will correspond to the part of the resolution proof which is constructed solely from the clauses in the negated and Skolemized theorem.

8.6. Definition. Let X be an $\alpha\beta$-normal nnformula. The **initial expansion tree** $\mathcal{Q}(X)$ for X is inductively defined for parts Y of X by

(i) $Y = l$ for a literal l. Then $\mathcal{Q}(Y) = l$.

(ii) $Y = Y_1 \bowtie \cdots \bowtie Y_n$. Then $\mathcal{Q}(Y) =$

$$\begin{array}{ccc} & \bowtie & \\ \swarrow & \cdots & \searrow \\ \mathcal{Q}(Y_1) & & \mathcal{Q}(Y_n) \end{array}$$

(iii) $Y = \exists v S$. Then $\mathcal{Q}(Y) =$

$$\begin{array}{l} \exists v S \\ \mid v \\ \mathcal{Q}(S) \end{array}$$

(iv) $Y = \forall v S$. Then $\mathcal{Q}(Y) = $

$$\begin{array}{c}\forall v S \\ \Big|\;\overline{f_v(w_1,\ldots,w_n)} \\ \mathcal{Q}(S[v/\overline{f_v(w_1,\ldots,w_n)}])\end{array}$$

where $f_v(w_1,\ldots,w_n)$ is the Skolem-term for v in $\overline{X}$.

Now we construct an expansion tree proof from a resolution refutation. Let a resolution refutation $c_1,\ldots,c_m,c_{m+1},\ldots,c_n = \square$ be given. For each clause c_j, $j \geq m$ we will recursively construct a partial expansion tree proof $(Q_j, \mathcal{M}_j)$ with the following property:

$(*)_j$ Let c_i, $i \leq j$ be a clause in the resolution refutation. Then every path through Q_j^D which does not intersect c_i contains a pair of $\mathcal{M}_j$-mated literals.

If we can show that $(*)_j$ holds for all $m \leq j \leq n$, the correctness of our translation is proven, since $c_n = \square$ and therefore no path through Q_n^D intersects c_n by $(*)_n$. Hence every path through Q_n^D must be spanned by $\mathcal{M}_n$ and $(Q_n^D, \mathcal{M}_n)$ is an expansion tree proof for X.

Now we come to the construction of $(Q_j, \mathcal{M}_j)$.

Let $(Q_m, \mathcal{M}_m) = (\mathcal{Q}(X), \{\})$. Since every path in $\mathcal{Q}(X)^D$ intersects every clause in $\overline{X}^*$, $(Q_m, \mathcal{M}_m)$ is a partial expansion tree proof for X and satisfies $(*)_m$.

Now assume $(Q_m, \mathcal{M}_m),\ldots,(Q_{j-1}, \mathcal{M}_{j-1})$ are partial expansion tree proofs for X and $(*)_i$ is satisfied for $m \leq i \leq j-1$. We have to distinguish cases, since c_j could either be a substitution instance or a resolvent of earlier clauses.

(i) Assume c_j is a substitution instance ϕc_i for some $1 \leq i \leq j-1$, ϕ a substitution for the free variables in c_i. If a variable is free in c_i it must be existentially quantified in X. Now we pass to a substitution θ such that θ agrees with ϕ if the substituent is not a Skolem-term, and $\theta v = \overline{f(w_1,\ldots,w_n)}$ if $\phi v = f(w_1,\ldots,w_n)$.

Let $Q_j = \theta Q_{j-1}$. $(Q_j, \mathcal{M}_j)$ is a partial expansion tree proof for X ($\mathcal{M}_j$ to be contructed later):

(a) $Q_j^S = Q_{j-1}^S = X$ by inductive assumption.

(b) From the way selections for universal variables in X are chosen and from the fact that X was $\alpha\beta$-normal, it is clear that every variable is selected at most once and that no selected variable is free in Q_j^S.

(c) $<_{Q_j}$ is acyclic. Assume, to the contrary, that there is a cycle

$$t_1 <^0_{Q_j} t_2 <^0_{Q_j} \cdots <^0_{Q_j} t_n = t_1.$$

The first relation means that there is a variable selected below t_1 which is free in t_2. Since the variable is selected below t_1 in the expansion tree, it has the form of a variable corresponding to a Skolem-term which contains t_1. Thus t_2 contains a term of the form $\overline{f_1(\ldots,t_1,\ldots)}$. Hence in the Skolem-form ϕ of the substitution, t_1 is free in t_2 The next relation would say that there is a variable selected below t_2 which is free in t_3. Thus a term of the form $\overline{f_2(\ldots,t_2,\ldots)}$ is free in t_3. Combined with the previous conclusion this gives us that t_1 is free in t_3. Iterating this process we finally arrive at the conclusion that t_1 is free in $t_n = t_1$. But this would mean that the original substitution ϕ was not legal, which is a contradiction. Therefore $<_{Q_j}$ must be acyclic.

Now we show how to construct $\mathcal{M}_j$. First note that because of definition 8.4 any literal occurrence in Q^D_{j-1} is still present in Q^D_j. Each *new* literal occurrence in Q^D_j is of the form θl for some l in Q^D_{j-1}. Then we simply let $\mathcal{M}_j = \mathcal{M}_{j-1} \cup \{(\theta l, \theta k) : (l,k) \in \mathcal{M}_{j-1}\}$.

(a) Consider c_h, $h < j$, P a path through Q^D_j not intersecting c_h. Since paths in Q^D_j can only be longer than paths in Q^D_{j-1}, there is a projection P' of P in Q^D_{j-1}. P' may be obtained by deleting all the new literals from P. Then P' is spanned by M_{j-1} by inductive hypothesis and hence P by $M_j \supset M_{j-1}$.

(b) Consider c_j, P a path through Q^D_j not intersecting c_j. Construct a path P' through Q^D_{j-1} as follows: Every literal occurrence l in Q^D_{j-1} such that there is a *new* literal occurrence $\theta l \in P$ is included. Furthermore all literal occurrences such that there is no *new* literal occurrence θl in Q^D_j, but $l \in P$ are also included. Then P' does not intersect c_i and is therefore spanned by a pair $(l,k) \in \mathcal{M}_{j-1}$. But then $\theta l, \theta k \in P$ (neither necessarily new) and $(\theta l, \theta k) \in \mathcal{M}_j$. Hence P is spanned by $\mathcal{M}_j$.

(ii) Assume c_j is the resolvent of c_{a_j} and c_{b_j} upon the literal $l_j \in c_{a_j}$, $\neg l_j \in c_{b_j}$, where $a_j, b_j < j$. Define $Q_j = Q_{j-1}$ and let $\mathcal{M}_j = \mathcal{M}_{j-1} \cup \{(l,k) : l$ an occurrence of l_j in c_{a_j}, k an occurrence of $\neg l_j$ in $c_{b_j}\}$.

Since $Q_j = Q_{j-1}$, Q_j is a partial expansion tree proof for X. What remains to be shown is that $\mathcal{M}_j$ spans every path through Q^D_j which does not intersect c_i, for all $i \leq j$. For $i < j$ this is obvious by the inductive hypothesis and the fact that $M_j \supset M_{j-1}$.

Now consider a path P through Q_j not intersecting c_j. There are three cases:

(a) P does not intersect c_{a_j}. By inductive hypothesis $\mathcal{M}_{j-1} \subset \mathcal{M}_j$ spans P.

(b) P does not intersect c_{b_j}. By inductive hypothesis $\mathcal{M}_{j-1} \subset \mathcal{M}_j$ spans P.

(c) P intersects both c_{a_j} and c_{b_j}. Since P does not intersect c_j, P must intersect c_{a_j} in one of the literal occurrences l_j resolved upon, and c_{b_j} in one of the literal occurrences $\neg l_j$. But then $\mathcal{M}_j$ spans P since $(l_j, \neg l_j) \in \mathcal{M}_j$.

9. References

[1] Peter B. Andrews, *Refutations by Matings*, IEEE Transactions on Computers C-25 (1976), 801-807.

[2] Peter B. Andrews, Transforming Matings into Natural Deduction Proofs, in *5th Conference on Automated Deduction, Les Arcs, France*, edited by W. Bibel and R. Kowalski, Lecture Notes in Computer Science 87, Springer-Verlag, 1980, 281-292.

[3] Peter B. Andrews, *Theorem Proving via General Matings*, Journal of the Association for Computing Machinery 28 (1981), 193-214.

[4] Wolfgang Bibel, *Automatic Theorem Proving*, Vieweg, Braunschweig, 1982.

[5] W. Bibel and J. Schreiber, *Proof search in a Gentzen-like system of first-order logic*. Proceedings of the International Computing Symposium, 1975, pp. 205-212.

[6] W. W. Bledsoe, *Non-resolution Theorem Proving*, Artificial Intelligence 9 (1977), 1-35.

[7] G. Gentzen, Investigations into Logical Deductions. In *The Collected Papers of Gerhard Gentzen*, M. E. Szabo, Ed.,North-Holland Publishing Co., Amsterdam, 1969, pp. 68-131.

[8] J. Herbrand, *Logical Writings*, Harvard University Press, 1972.

[9] Dale A. Miller, *Proofs in Higher Order Logic*, Ph.D. Th., Carnegie-Mellon University, August 1983.

[10] Dale A. Miller, *Expansion Tree Proofs and Their Conversion to Natural Deduction Proofs*. 7th Conference on Automated Deduction, Napa, May 1984.

[11] Frank Pfenning. Conversions between Analytic and Non-analytic Proofs. Tech. Report, Carnegie-Mellon University, 1984. (to appear)

[12] J. A. Robinson, *A machine-oriented logic based on the resolution principle*, Journal of the Association for Computing Machinery 12 (1965), 23-41.

[13] R. M. Smullyan, *First-Order Logic*, Springer-Verlag, Berlin, 1968.

[14] R. Statman, *Lower Bounds on Herbrand's Theorem*, Proceedings of the American Mathematical Society 75 (1979), 104-107.

Applications of Protected Circumscription

Jack Minker and Donald Perlis

Computer Science Department

University of Maryland

College Park, MD 20742

Abstract

We examine applications of an extension of circumscription that allows protection of certain objects against being included in the circumscription process. We show that this allows a clean handling of incomplete information in problems from artificial intelligence and databases.

1. Introduction

This paper amplifies on results proven elsewhere (see Minker & Perlis [1984]), in which we extended the idea of circumscription to allow prescription of what objects are or are not to be included in the circumscription process, broadening the applicability of the technique. A way to view circumscription is that it characterizes what it means for a set to be specified by means of various assertions. We review briefly the idea of circumscription, before discussing the extended version. We begin with a suggestive example.

Suppose a precious red sapphire, s, is purchased in India and brought to Denver, only to be lost. Then years later a youngster is found living alone in the Rocky Mountain wilderness wearing a red sapphire ring, r. The reader of the mystery is supposed to immediately think, Aha! That's the red sapphire that disappeared earlier! In fact, only one red sapphire exists, one presumes, at least as far as we need consider.

Yet such has not been stated, and to state it is to go further than we wish. Somehow we have great use for jumping to conclusions of this sort, although we realize they need not be true. Still, in order to get ideas to begin reasoning at all, we need to do some such associating, and often it is useful to use these associations as conclusions for immediate acceptance (at least until forced to alter them by weight of later evidence). How then are we to do this? It clearly is a kind of default problem, and one addressed recently by several workers in artificial intelligence (McDermott & Doyle [1980], McCarthy [1980], Reiter [1980]). The approach of McCarthy, predicate circumscription, applies particularly well to the above problem. In another paper (Minker & Perlis [1984]) we have extended McCarthy's formalism; here we are concerned with specific applications of the extension.

McCarthy's approach then is as follows: Given a predicate symbol P and a formula A[P] containing P, the circumscription of P by A[P] can be thought of as saying that the P-things consist of certain ones as needed to satisfy A[P] and no more, in the sense that any P-things Z satisfying A[Z] already include ALL P-things:

$$C_A^P[Z]: \quad [A[Z] \ \& \ (x)(Z(x) \rightarrow P(x))] \ \rightarrow \ (x)(P(x) \rightarrow Z(x))$$

To see how this 'solves' the sapphire problem, let P(x) say x is a red sapphire. We decide to circumscribe on P since red sapphires are, as far as we can judge, quite unusual and unlikely to be present without being recognized and well-known. Once mentioned, the gem becomes 'the' red sapphire s of the story until futher notice. So, the property of being a red sapphire becomes the only contextual information needed: A[P] is P(s). As long as it remains our judgement that red-sapphired-ness is appropriate to circumscibe, we will conclude that this red sapphire is also the one and only red sapphire, namely, the lost one. Thus we will be able to prove that $r = s$.

In detail, circumscription of P by P(s) (as the only information A[P] that initially pertains) can be applied by taking the predicate Z(x) to be $x = s$. Then A[Z] will be Z(s), i.e., $s = s$, resulting from replacing P by Z in A[P]. It follows by the above circumscription schema that P(x) --> Z(x), i.e., that the only red sapphire is s. This is seen as follows: first, Z(s) is obvious; and Z(x) --> P(x) follows from P(s). So the schema yields P(x) --> Z(x).

If we retain this conclusion on hearing about the sapphire r, then of course we must conclude that $r = s$. Of course, we have made two significant judgements here, neither of which is automatic: that red sapphires are things to circumscribe on, and that new data of the sort presented (the existence of g) does not alter the first judgement. We are not tackling this issue here, but simply the one of how to formally represent such reasoning.

2. Circumscription with Protected Terms

Here we discuss a simple syntactic device from Minker & Perlis [1984]. There we suggested that once A has been selected as appropriate for circumscribing P, and if (perhaps later) it is desired to protect S-things from this process so that circumscription will not be used to show S-things are not P-things, we can keep the same criteria A, but alter the form of the schema itself. Starting with P(x) & -S(x), which we write P/S(x) (and more generally T/U(x) for T(x) & -U(x)), we alter the circumscription schema to read as follows:

$$C^{P/S}_{A}[Z]: [A[Z] \ \& \ (x)(Z/S(x) \text{-->} P(x))] \ \text{-->} \ (x)(P/S(x) \text{-->} Z(x))$$

for all formulas Z. Intuitively, we are saying that conclusions are drawn only about non-S-things, as far as ruling out possible P-things goes. We refer to this schema as 'protected circumscription'; unless so indicated, circumscription will refer to McCarthy's schema. We write C[Z] when context makes clear what the A, P, and S (if protected) are.

It may appear that by circumscribing on the formula P(x)&-S(x) the same effect is achieved. Indeed intuitively this should be the case. However, circumscription, as defined by McCarthy, applies only for single predicate letters. It is not obvious how to extend it to general formulas. John McCarthy has communicated to us that he is currently pursuing this extension.

To return to our sapphire example, suppose in addition to the red sapphire that is lost, another precious stone has been brought from India by another Denver resident, but its precise gemology has not been revealed. In fact, we may suppose for the sake of story-line, that the two gem buyers are in fact obtaining gifts for their (one and the same) admiree, a third Denver resident whose birthday anniversary is to be celebrated soon. The reader may already feel a tingling sense of worry that the two gems may be identical in type and bound to produce embarrassment.

How then can we represent the reasoning that there are one and possibly two red sapphires, but no more, and that s is one, and the other stone, say g, may or may not be, in such a way that we still can conclude later that r = s (supposing g not to be lost)? Our schema will do this if we again let P(x) say x is a red sapphire, P(s) being the only information that is needed to circumscribe that very property (i.e., the axiom A[P] is simply P(s) itself) except that now we also state S(g) to protect g from being squeezed out of possible red-sapphired-ness. Again we let Z(x) be x=s, and further simply take S(g) as an axiom. S(x) will have no special meaning other than that x is 'selected' for protection from circumscription.

Then much as before we can conclude $P(x) \rightarrow Z(x) \vee S(x)$, i.e., any red sapphire either is the first one (s) or is the new untyped stone (g). Then on learning of the red sapphire ring r, it follows that either $r = s$ or $r = g$. If further it is known that g is not lost, indeed is in the firm possession of its owner, then we know $r \neq g$, hence $r = s$.

Notice the apparent non-monotonicity present in such a line of reasoning. Before we have heard of the second stone g, we conclude $r = s$; later with further information but (apparently) no loss of what was previously known, we no longer can make such a strong conclusion but instead have only $(r = s) \vee (r = g)$. In fact, of course, information has been retracted, namely our original unprotected treatment of red sapphires: now A[P] is {P(s),S(g)} whereas before it was just {P(s)}, so the previous schema has been replaced by a new one that in fact is not logically stronger.

3. Using Model-Theory

In McCarthy [1980] the concept of minimal model was discussed in the context of circumscription. In Minker & Perlis [1984] we re-defined minimal model in an manner appropriate to the new version of circumscription as follows: Let M and N be models of A[P]. We say $M <P/S\ N$ if the atomic truths of M are contained in those of N, if those atomic truths of M not using P are precisely those of N, and if the extension of P&S in M is also that in N, i.e., if {x | P(x) and S(x) holds in M} = {x | P(x) and S(x) hold in N}. Then M is a P/S-minimal model of A[P] if M is a model of A[P] minimal with respect to the relation $<P/S$.

As an example, suppose P(a)&P(b)&P(c)&-P(d)&Q(d) is the sentence A[P], and we wish to protect the constant c: S(c). Then the only model is {P(a) P(b) P(c) Q(d) S(c)}. (Here we indicate a model by writing the positive ground clauses that hold in it.) This model is the only minimal model. In this case protection is superfluous since P(c) is required to hold.

Now consider the sentence P(a)&P(b) where c is still a protected constant--S(c)--and d is an unprotected constant. Here we obtain four models:

M1 = {P(a) P(b) P(c) P(d) S(c)}

M2 = {P(a) P(b) P(c) S(c)}

M3 = {P(a) P(b) P(d) S(c)}

M4 = {P(a) P(b) S(c)}.

Of these only M2 and M4 are minimal, M2 being a P/S-minimal submodel of M1, and M4 of M3.

Finally, consider P(a) v P(b) v P(c) with S(a) and S(c). Then the models are

M1 = {P(a) P(b) P(c) S(a) S(c)}

M2 = {P(a) P(b) S(a) S(c)}

M3 = {P(a) P(c) S(a) S(c)}

M4 = {P(b) P(c) S(a) S(c)}

M5 = {P(a) S(a) S(c)}

M6 = {P(b) S(a) S(c)}

M7 = {P(c) S(a) S(c)}.

The minimal ones are M3, M5, M6, M7.

Using models to draw conclusions about derivablility relies on having appropriate soundness and completeness theorems tying model-theoretic truth to syntactic proof. McCarthy [1980] provides the soundness half of such a result for circumscription, but not the completeness part. As noted by Davis [1980] the fully general completeness result would be false. Nonetheless, Minker & Perlis [1984] have a soundness and completeness result that applies to cases of 'ground' theories (among others), i.e., ones with no variables, such as we are considering here: for such theories A[P], and for any ground formula B, we have

A[P] |P/S== B iff A[P] |P/S– B.

It is instructive to consider the following example: Let A[P] consist of the data P(a), -P(b) v -P(c). Then there are three models of A[P]:

1. {P(a)}
2. {P(a), P(b)}
3. {P(a), P(c)}

Of these, only 1 is minimal, and so the formulas true in 1 are the circumscriptive theorems of A[P], for all choices of Z at once! Notice that the theory A'[P] having ONLY P(a) as axiom also has these three models as well as: 4. {P(a), P(b), P(c)} which still is not minimal. So A and A' have the same minimal models and hence the same circumscriptive theorems. In fact in both theories we have the theorems -P(b) and -P(c), so that the axiom -P(b) v -P(c) in A is circumscriptively redundant.

Now suppose we wish to protect b and c in A so that ALL we know about P(b) and P(c) is that they are not jointly true, i.e., -P(b) v -P(c) represents real uncertainty. Then we find that 1, 2, and 3 are the only models and all are minimal. Furthermore, although -P(b) v -P(c) holds in each, neither -P(b) nor -P(c) does, so that the protection has really worked. But now if we pass to A' and protect b and c, we find still all four models as before and all are minimal, so that not even -P(b) v -P(c) holds.

Although the completeness result has shown us what the ground theorems of these four theories are, we see from this example that negative data (-P(b) v -P(c)) can have a non-redundant effect when there are protected constants. This shows a strong distinction from the situation for ordinary circumscription.

4. Applications to Databases

We believe that protected circumscription is applicable to belief systems, databases, and many other areas. We give here an application to databases.

Suppose a database DB contains the information P(a) and P(b) and neither P(c) nor P(d). Traditional database approaches would take this to mean that P(c) and P(d) are false. I.e., there is an assumption of complete data, often referred to as the 'closed world assumption' (Reiter [1978]). This is not to say that the closed world assumption is logically valid; rather, that in certain data sets, it happens to hold. This of course is a very limiting situation. For instance it does not allow for the possibility that some data simply has not yet been gathered, surely an extremely frequent occurence in real-world databases.

A more dramatic version of this is 'indefinite' data of the form P(c) or P(d). Here it is not simply that we do not know about c and d. We know that at least one of them has property P, but we do not know which. McCarthy's circumscription (among other approaches) provides a solution to this, in which from P(x) one can conclude, for instance x=a v x=b v x=c v x=d, given the database DB = {P(a), P(b), P(c) v P(d)}. Thus there is in force a kind of closed world assumption, but broadened so as to deal with indefinite data, what Minker [1982] calls 'the generalized closed world assumption'.

Indeed, we can regard the incomplete database as a special kind of indefinite database, in which the lack of information about P(c) is represented as an indefiniteness between P(c) and -P(c). Yet no assertion of the form P(c) v P(x) will do what is required. If x is different from c, then we are asserting more than is wanted, for now we are committing x also to be indefinite, not to mention that x and c are also being bound together in a special relation not part of our intention. If on the other hand, we let x be c itself, then P(c) v P(c) tells us too much, namely that c definitely has property P.

Other ideas in this vein include P(c) v -P(c) (a tautology which achieves nothing), and P(c) v P(ind) where ind is a new constant introduced for this purpose. The latter has some promise, but leaves us with the undesirable feature that now we can prove that something (either c or ind) has property P, this again not being the intended outcome.

With this background we then look at protected circumscription for a solution to this difficulty. Let S(x) be the predicate x=c. This will serve to protect c. Now if we use protected circumscription on P by the database DB = {P(a) P(b) Q(c) Q(d)} with S as stated, we find as expected that -P(c) cannot be concluded, although -P(d) can be concluded. In terms of minimal models, we first consider all objects that do not have property S (this is only c here) and that also must have property P in each model. These objects are only a and b, so these are the only ones we can conclude to have property P. On the other hand, we also examine all objects which do not have property S (again just c here) and which must fail to have property P in each minimal model. In this case the only such object is d, hence we conclude $-P(d)$.

In the case of (Horn) databases we have a generalization of the idea of Clark [1978] who, when discussing negation as failure, showed that an 'if and only if' condition was its analogue. For example, if $P(a)$ and $P(b)$ are known and we do not care about c or d, then we would write

$$(x = a) \vee (x = b) \longleftrightarrow P(x).$$

Now, if one wants to protect c while leaving d unprotected, our solution is simply to place $(x = c)$ on both the right and left hand sides of the above formula, to obtain

$$(x = a) \vee (x = b) \vee (x = c) \longleftrightarrow P(x) \vee (x = c).$$

Relating this to our protected circumscription shcema, we can re-write this as a conjunction of two formulas and then remove tautologies:

(1) $$(x=a) \vee (x=b) \vee (x=c) \longrightarrow P(x) \vee (x=c)$$

(2) $$(x=a) \vee (x=b) \vee (x=c) \longleftarrow P(x) \vee (x=c)$$

and then

(3) $$(x=a) \vee (x=b) \longrightarrow P(x)$$

(4) $$P(x) \longrightarrow (x=a) \vee (x=b) \vee (x=c)$$

(here we assume distinct constants stand for distinct entities). Let $Z(x)$ be $(x=a) \vee (x=b)$;

then

(5) $P(x) \,\&\, (x \neq c) \rightarrow Z(x)$ from (4)

and finally, letting $S(x)$ be $(x=c)$, we have

(6) $P/S(x) \rightarrow Z(x)$ from (5)

(7) $Z/S(x) \rightarrow P(x)$ from (3).

Hence, the generalization of Clark's is simply achieved for databases by the modified formula which is equivalent to our protected circumscription.

Acknowledgements

Our work obviously depends greatly on that of John McCarthy. We have also been influenced by work of and discussions with Ray Reiter. This paper was written with support from the following grants:

AFOSR-82-0303, for J. Minker and D. Perlis

NSFD MCS 79 19418, for J. Minker

U. of Md. Summer Research Award for D. Perlis

Bibliography

Clark, K. [1978] "Negation as Failure", In: Logic and Databases, (Gallaire, H. and Minker, J., Eds.) Plenum Press, NY 1978, 293-322.

Davis, M. [1980] "The Mathematics of Non-Monotonic Reasoning". Artificial Intelligence 13 (1980), 73-80.

McCarthy, J. [1980] "Circumscription--A Form of Non-Monotonic Reasoning". Artificial Intelligence 13 (1980), 27-39.

McDermott, D., and Doyle, J. [1980] "Non-Monotonic Logic I" Artificial Intelligence 13 (1980), 41-72.

Minker, J. [1982] "On Indefinite Databases and the Closed-World Assumption". Springer-Verlag Lecture Notes in Computer Science, v.138, 292-308. Sixth Conference on Automated Deduction. New York, NY. 1982.

Minker, J., and Perlis, D. [1984] "On the Semantics of Circumscription". Technical Report, Univ. of Maryland, 1984.

Reiter, R. [1980] "A Logic for Default Reasoning". Artificial Intelligence 13 (1980), 81-132.

Reiter, R. [1978] "On Closed World Databases". In: Logic and Data Bases, (Gallaire, H. and Minker, J., eds.) Plenum, 1978, 55-76.

Reiter, R. [1982] "Circumscription Implies Predicate Completion (Sometimes)". Proceedings of AAAI-82, 418-420.

IMPLEMENTATION STRATEGIES FOR PLAN-BASED DEDUCTION

Kenneth Forsythe and Stanislaw Matwin
Dept. of Computer Science
University of Ottawa
Ottawa, Ontario
K1N 6N5

ABSTRACT

This paper discusses some results of experimentation with a plan-based deduction system. The system incorporates an efficient intelligent backtracking strategy. During implementation, several important questions concerning different strategies to control the deduction process arose. These questions are answered in the paper, with special emphasis on the problem of generating redundant solutions.

1. INTRODUCTION

This paper presents different implementation strategies for a plan-based deduction method. The method, presented in [Pietrzykowski & Matwin 82] and further developed in [Matwin & Pietrzykowski 83], forms the basis of a logic programming system using intelligent backtracking.

Given an initial set of clauses with a goal statement, a mechanical theorem prover will attempt to refute the goal statement via resolution. There are many different algorithms upon which to base the resolution process (for example see [Chang and Lee 73]) but most of these incorporate a linear backtracking strategy or do not address the backtracking implementation at all. By linear

----

This work has been supported by National Sciences and Engineering Research Council of Canada grant No A2480.

backtracking we mean a strategy which backtracks sequentially through applicative goals in exactly the reverse order they were encountered, starting with the current goal. Unfortunately, this strategy will blindly explore every path until a solution is found. Since the number of paths grows exponentially with the number of clauses, it is advantageous to elimate paths which cannot lead to a solution before they are tried. This is the concept behind plan based deduction.

In plan based deduction, backtracking is not restricted to starting with the most current goal but can be applied to any goal. In practice, we limit this to those goals (called conflicts) whose removal from the plan will restore unifiability to it. The structure of the plan is such that backtracking terminates when the original goal statement is encountered. This property speeds up the worst case of linear backtracking by an exponential factor [Pietrzykowski & Matwin 82]. There are a number of other deduction algorithms in which graphs are used [Sickel 76], [Kowalski 75], [Chang, Slagle 79], [Bibel 83]. However, the approach presented here differs from all of them in terms of what a plan represents and how it is operated upon to obtain a refutation.

In [Chang, Slagle 79] connection graphs represent the search space and rewriting rules are obtained from it. Connection graphs determine sequences of substitutions, leading possibly to a refutation. Consistency of these substitutions is only checked after a whole sequence has been generated. If it turns out to be inconsistent, another sequence is tried, which is equivalent to backtracking. The problem of avoiding backtracking is not addressed, neither is the problem of redundancy as understood here.

In [Sickel 76], clause interconnection graphs are a representation of the total search space. Similarly to [Chang, Slagle 79], this representation is traversed in search of a consistent set of substitutions. This set of substitutions is generated incrementally, as opposed[to [Chang, Slagle 79]. The incremental approach does not, however, prevent the method from a backtracking behavior: the issue of the action, appropriate when inconsistency is detected, is not discussed.

Yet another, comprehensive approach is presented in [Kowalski 75]. Although the redundancy problem is discussed, it is presented differently from our approach, i.e. on the propositional calculus level. It is not obvious how the method suggested by [Kowalski 75] for deduction in prepositional calculus generalizes for predicate calculus. The approach, presented in this paper, follows suggestions in [Kowalski 75] and develops them into a full redundancy removal algorithm for predicate calculus.

Another method, presented in [Bibel 83], is different from all the ones mentioned above because of its non-clausal representation. It also uses graphs to represent the solution space, similarly to [Kowalski 75]. However, as in [Sickel 76], backtracking may occur, but this issue is not addressed at length.

In our plan based deduction system, the plan is a graph containing all the clauses currently involved in the resolution process where each clause is a node in the graph. A node consists of a key: the complementary literal selected for resolution, and its goals: all the remaining literals in the clause. The root of the graph is the original goal statement which consists of all goals. The deduction process consists of selecting clauses with complementary literals to resolve all the goals in the plan (i.e. the plan is closed). If this is accomplished and a most general unifier for the plan exists then a refutation for the goal statement has been found. If the plan is nonunifiable then conflicts are determined and removed from the plan so that the deduction process can be resumed and new clauses selected.

Unfortunately, the problem of generating redundant plans is inherent to this type of deduction system. In other words, unless some kind of restriction is placed on the conflict selection process, duplicate plans will be generated although the paths leading to these plans are unique. This overlapping of paths results from generating new plans that produce the same conflicts as those they were developed from. As there may be several different clauses from which to resolve a goal with, it is inevitable that different paths may derive the same plan.

This problem is further compounded when one wishes to obtain all the refutations possible for a given goal statement and set of clauses. This is important in many Logic Programming applications, particularly when the "generate and test" paradigm [Clocksin, Mellish 82] is applied. A solution to a problem, specified by a set of clauses, is obtained by first generating a superset of solutions. Each of them is then tested for satisfiability of conditions, which extract a true solution from the superset. To accomplish this, we introduce the concept of artificial conflicts. The deduction process in our system, as previously mentioned, consists of resolving goals introduced into the plan via nonunit clauses or by processing conflicts. To reactivate the deduction process on a closed unifiable plan we have to artificially induce conflicts into this plan in such a way that all solutions will eventually be generated. However, unless some restrictions are placed on the method of selecting artificial conflicts, redundant plans will also be generated.

The problem of generating duplicate plans is the main theme of this paper, but when these problems were realized a separate but related problem concerning the efficiency of developing plans was also encountered. It was found that sometimes, nodes are added to a plan, which later becomes nonunifiable, but do not influence the selection of conflicts and later get deleted through the backtracking mechanism.

All of these problems were encountered during the implementation of a plan based deduction system. This paper describes more completely the nature of these problems and the strategies used to solve them. In summary, we can paraphrase these problems as the following questions: 1) Removing redundancy - How to avoid generating redundant solutions while maintaining completeness? 2) Processing criteria - How to develop a plan efficiently so that only nodes relevent to the selection of conflicts are created? 3) Artificial conflicts - How to introduce artificial conflicts on a unifiable plan so that a complete solution set can be generated with minimal redundancy?

These three questions are discussed individually in sections 3, 4, and 5. In section 2, a preliminary review of terms and concepts is presented and section 6 contains our concluding remarks.

2. TERMS AND CONCEPTS

This section deals with introducing notation used in [Pietrzykowski & Matwin 82], for a plan-based deduction system. This system acts on a given goal statement and a set of clauses, collectively called the base, and attempts to find a refutation for the goal statement via resolution.

In a preprocessing phase, every literal in the base becomes associated with a list of all the other potentially unifiable complementary literals in the base. This list is referred to as the literal's potentials and it also contains the corresponding most general unifier for the two literals.

In the second phase of the resolution process, the dynamic processing phase, a refutation for the goal statement is attempted. This phase builds and maintains two structures: the plan which is a graph depicting which clauses have been selected for resolution, and the graph of dynamic constraints which records the most general unifier for the plan (i.e., records all the substitions which have occurred).

The plan is constructed of nodes, which corresponds to a clause in the base, each of which contains a key and zero or more goals. The key is the resolvent literal and the goals are the remaining literals in the clause. The only exception to this is the top node, which consists of all goals and is derived from the goal statement. The list of potentials associated with each literal is accessible by the goal representing that literal in a node.

To begin the dynamic processing phase, the original goal statement is inserted into the plan as the top node, which is considered to be the root of the graph and all other nodes become descendants of it. As nodes are inserted into the plan, every goal in them is classified as open. The resolution process consists of repeatedly selecting open goals to be processed. This is accomplished by choosing one of the goal's potentials as a resolvent, inserting the potential's clause into the plan, updating the most general unifier for the plan and changing the status of the goal from open to closed. This process continues until there are no more goals to resolve or all of the open goals have an empty list of potentials. In the first case, if the plan is unifiable then a refutation for the statement has been found otherwise a clash is said to have occured and the conflict checker phase is activated. In the second case there is no refutation for the goal statement.

The conflict checker phase removes the conflicting nodes from the plan, restores the graph of dynamic constraint and returns back to the dynamic processing phase. This is accomplished by determining which sets of nodes can be removed so that unifiability will be restored to the plan. These sets of nodes are represented via the goals they are resolvents of, where each set of goals is called a clash and each goal in a clash is called a conflict. Each of these clashes are processed individually so that completeness will be retained. To ensure that each of these clashes are processed on the correct plan and corresponding graph of dynamic constraints, the current state of the search space is copied to disk so that it can be retrieved later [Forsythe and Matwin 83]. Processing a clash consists of selecting each goal in turn and backtracking up the graph through successive father goals until a goal with a nonempty set of potentials is found. If no such goal is encountered then the process fails and another clash is tried. If the search is sucessful then all descendant nodes of that goal are pruned from the plan and the graph of dynamic constraints updated to reflect the most general unifer for the modified plan. The goals are then marked as open and dynamic processing resumed.

3. REMOVING REDUNDANCY

Consider the following set of clauses in example 1:

-P(x)-Q(x)
P(a)
P(b)
P(e)
Q(c)
Q(d)
Q(e)

Example 1.

Figure 1 shows a trace of the deduction process for the base in example 1 of which the first clause is the goal statement. In this figure each plan is represented by the name of the goals above the constants of the conmplementary literals to which the variables of each goal is bound. In braces alongside the constant is a list of constants belonging to the list of potential complementary literals which could have been chosen instead. Each line under a goal-complementary literal pair leads to a new plan generated by replacing the complementary literal with the first potential inside the braces. If the set of potentials is empty then this leads to a failure, indicated by a bar at the end of the line. If a line extends from more than one goal-complementary literal pair it means that both goals belong to the same clash and the two literals are replaced simultaneously. This shorthand notation is used as we are interested in the development of the total search space rather than the individual plan.

From this figure, we can see that the first plan is nonunifiable. The set of clashes associated with it consists of two elements, each containing one conflict. One element consists of the goal which introduced P(a) and the other consists of the goal which introduced Q(c). The left branch indicates that the conflict introducing P(a) was chosen and replaced with a new node containing P(b). The right branch indicates that Q(d) replaces

Q(c) as the new node. Both of the two new plans contain similar conflict sets which when resolved lead to further plans. All the possible plans which could be developed for this set of clauses are as shown. Notice that in figure 1, six branches lead to the same refutation. In fact the left subtree of every right branch duplicates the right subtree of the corresponding left branch. The inefficiency of this strategy (although it is complete) is unacceptable for any practical implementation and this section presents an algorithm to remove this inefficiency.

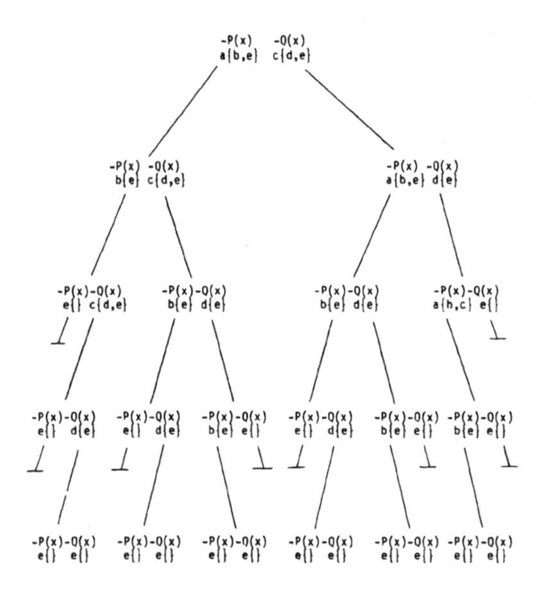

Figure 1. A trace of the deduction process for example 1.

A possible solution to this problem, suggested in [Bruynooghe 83], is to give each predicate an ordering which controls which alternatives to that predicate are permitted to be selected. This ordering restricts lower order predicates from generating solutions already obtained by a higher ordered one. We have employed a similiar strategy, but instead of giving each individual predicate an ordering we have given each clash in the set of clashes an ordering (it is possible that a conflict can occur in more than one clash though the clashes, themselves, are distinct).

Initially each clash in a set is given a unique order number which becomes the current order number as each particular clash is processed. If the resulting plan produces a new set of clashes then all the similar elements in the new set are given the same order number. By similar we mean any clash which contains the same conflicts as a clash in the generating plan. All the clashes in the new set which obtain an order number greater than the current order number are discarded. As shown in figure 1, it is common for most of the resulting clashes to be similar to the clashes of the plan it was derived from. In the special case where nonsimilar clashes are generated then each of these is given an order number (bounded by the current order number) which guarantees completeness of the search space.

The result of applying this strategy to the base of example 1 is shown in figure 2. The two clashes of one conflict each, are initially ordered as 1 and 2. The figure shows that the set of clashes of any plan derived from processing a clash with an order number of 1 is restricted to elements whose resulting order number is not greater than one. By censuring the paths leading from any given plan through this algorithm, we can eliminate the overlapping of the search space. Graphically, this can be interpreted as removal of the redundant left subtree of every right branch.

In order to measure the amount of improvement by this strategy we introduce the idea of counters for the the number of arcs in the graph we traverse, delete and insert. By arcs we mean every goal-key pair in the graph. For the search space using the original strategy as shown in figure 1 we determined that the number of traversals was 46, the number of insertions 20 and the number of deletions 18. Comparatively, for figure 1, the numbers of traversals, insertions and deletions are 22, 10 and 8, respectively. Also, the number of identical refutations found in figure 2 is zero as opposed to five in figure 1.

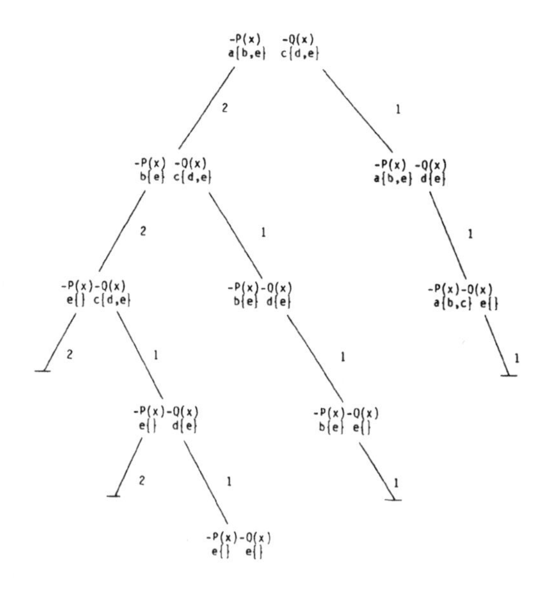

Figure 2. A trace for example 1 using the improved algorithm.

In order to measure the amount of improvement by this strategy we introduce the idea of counters for the the number of arcs in the graph we traverse, delete and insert. By arcs we mean every goal-key pair in the graph. For the search space using the original strategy as shown in figure 1 we determined that the number of traversals was 46, the number of insertions 20 and the number of deletions 18. Comparatively, for figure 1, the numbers of traversals, insertions and deletions are 22, 10 and 8, respectively. Also, the number of identical refutations found in figure 2 is zero as opposed to five in figure 1.

We now explain the ordering strategy for situations where a nonsimiliar set of clashes is generated. There are essentially two types of these situations: when a new open goal is developed thus introducing a totally new conflict set and when the same conflicts get rearranged into different clashes.

When a new open goal is developed clashes are given an order number dictated only in the way they are arranged. This method of ordering is obviously used on the original plan as the previous conflict set is empty and all goals are initially open. This situation can also occur when a clash has just been resolved which creates a unifiable plan causing the deduction process to choose a new open goal. This phenomena occurs because as soon as a conflict is determined the plan development process is interrupted and the conflict-checker phase initiated. As a result, the development of all the other open goals is suspended until the conflict is resolved (see section 4).

It is possible that a new set of clashes can be generated which cannot be paralleled to the previous set for an identical ordering. If this happens, any element in the new set for which there is no corresponding clash in the old set is given an order number high enough to maintain completeness. In practice this usually means giving these elements an order number equal to the current order number. This approach may lead to generating redundant solutions but we have not yet discovered a more efficient algorithm that will still generate a complete solution set. However, it may be possible to minimize the inefficiency by ordering the elements in the conflict set using heuristic strategies which would reduce the number of redundant solutions generated. For an example of how this strategy is implemented consider the clauses of example 2.

P(x)Q(x)R(x)
-P(a)
-P(e)
-Q(a)
-Q(b)
-Q(e)
-R(b)
-R(e)

Example 2.

An outline of the resolution process for the clauses in example 2 is given in figure 3. In the first plan of figure 3, predicates P and Q are in conflict with predicate R. Suppose the goals belonging to P and Q are given an order number of 2 and the goal belonging to R an order number of 1. New potentials are selected for P and Q and the second plan is developed. In this second plan the predicates Q and R are in conflict with predicate P, thus there is no parallel ordering between the goals in the two plans. Consequently, both elements in the new conflict set are given an order number of 2, the current order number. Resolving these two conflicts leads to a failure and a refutation. If we resolve the second conflict of the first plan we find that after generating a nonunifiable plan this path leads to a failure. It can be shown that the work done by this strategy in this example is less than half of what a linear backtracking algorithm would do.

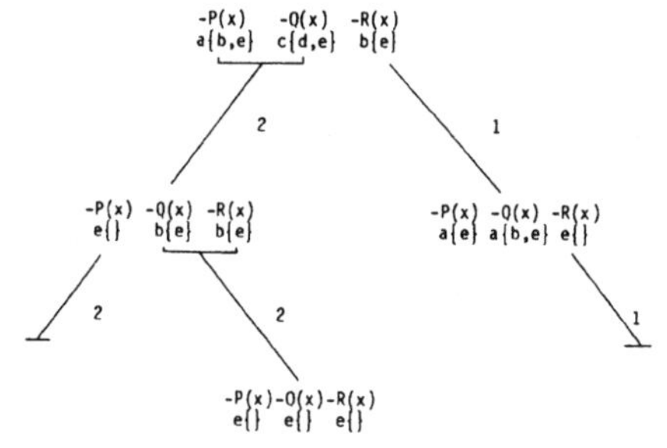

Figure 3. A trace for example 2 using the improved algorithm.

4. PROCESSING CRITERIA

In this section we would like to address the question of when to interrupt the processing of open goals and begin resolving clashes. Our initial strategy was to allow the resolution process to resolve all the open goals (i.e. generate a closed plan) before it began processing conflicts. However, analysis of this approach showed that there was much work that was wasted by developing goals that were later discarded. As an example consider the following base of clauses (this example shows only one of many situations where developing a closed plan is inefficient).

-P(x)-Q(x,y)
P(a)
Q(c,z)M(z)
Q(b,z)M(z)
Q(a,z)M(z)
-M(z)R(x,z)S(y,z)T(x,y)
-R(c,e)
-S(d,e)
-T(c,d)

Example 3.

If we consider the resolution process applied to this set of clauses we would find that the first attempt to refute the goal statement leads to a conflict between the term a in literal P and the term c in literal Q. If we develop the whole plan we would also resolve literals M, R, S, and T. However to resolve the conflict we need only to replace the clause containing c with another alternative (the clause containing literal Q(b,z)). In doing this we would remove the arc from -Q(x,z) to Q(c,z), effectively removing the arcs containing the complementary pairs of literals M, R, S and T, and replace Q(c,z) with Q(b,z). This in turn leads to a similiar situation where literals M, R, S and T are again resolved and deleted from the plan by replacing Q(b,z) with Q(a,z). Once more, literals M, R, S and T are resolved. This gives a closed plan without any conflicts (i.e. refutation) so the resolution process is finished.

If we determine the work done using this strategy we find that 13 insertions and 2 deletions were made before a refutation was found. However, if we interrupt the plan develoment process as soon as a conflict is encountered we would have avoided resolving literals M, R, S and T. In the third attempt, there are no conflicts generated so the literals M, R, S and T are resolved. Using this approach we find that only 7 insertions and 2 deletions are made, which is roughly half of the work that our original strategy does.

If in example 3, we replace the literal Q(a,z) with Q(d,z), we would find this new set of clauses has no solution. In this case the process would terminate after finding that Q(d,z) leads to a conflict in which there are no more alternatives. Consequently, the literals M, R, S and T would never be resolved and the total number of insertions would be 4 and the number of deletions 2.

5. ARTIFICIAL CONFLICTS

This section deals with the question of how to generate a complete solution set. The deduction algorithm is designed to process open goals, obtained by resolving literals contained in non-unit clauses or by removing conflicts, until a refutation is found. The idea behind artificial conflicts is that by designating specific goals in a closed unifiable plan as conflicts, the deduction process can be continually reactivated until all the solutions are found. The problem we would like to address is how to arrange the goals which are selected as artificial conflicts into clashes so that a complete solution set with minimal redundancy can be generated.

To obtain artificial conflicts from a closed unifiable plan we simply label each goal that introduces a unit clause as a conflict. This ensures that all goals with potentials will eventually be considered because of the nature of the bactracking strategy which checks each goal on the path from the selected goal to the root of the plan. The problem is how to arrange these conflicts into clashes so that every possible solution will be derived.

The first approach which comes to mind is to put each conflict into a separate clash. This strategy will obviously guarantee completenes as each path will eventually be backtracked. The effect of applying this strategy to the clauses of example 4, below, is shown in figure 4. (In this figure only the plans which represent soutions are shown.)

-P(x)-Q(x)-R(y)
P(a)
P(b)
Q(a)
Q(b)
R(c)
R(d)

Example 4.

In figure 4 we see that by selecting every goal introducing a unit clause as the only conflict in a clash, we obtain three clashes for the first solution space generated. Resolving each of these clashes gives three more solutions of which only two are unique. Applying this strategy to each of these solutions results in four more solutions of which only one is unique. So out of eight solutions generated four are redundant.

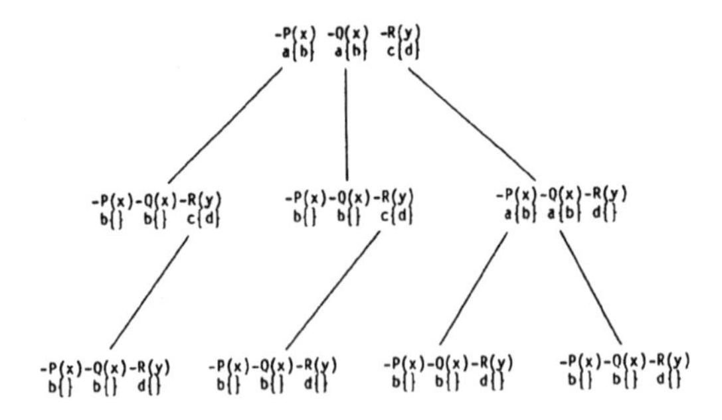

Figure 4. A complete solution set for example 4.

To derive a more efficient algorithm for this clash selection process we must first consider the sources of redundancy. The major cause of duplication is the lack of consideration for bound constants. If we examine figure 4 more closely we see that in the first solution the predicates P(a) and Q(a) are bound together through the variable x in the clause -P(x)-Q(x)-R(y). Resolving the clash containing P leads to the same solution as resolving the clash containing Q. This is because the binding between P and Q implies that to replace P one must also replace Q and vice-versa. Thus the first improvement we can make to the clause selection algorithm is to place all the goals bound through a variables together into one clash. This ensures that both predicates will be replaced at the same time. If a goal is bound to two different goals through two diffent bindings, which are not themselves bound together, then two separate clashes must be created with the common goal contained in both. This is neccessary to maintain completeness by allowing each binding to be processed individually.

The second cause of redundant solutions is the same as that described in section 3. Essentially, other duplications can occur because all the potentials associated with one clash are systematically processed with all the potentials of a second clash. Then the potentials for this second clash are processed with all the potentials of the first clash. (This is the duplication of subtrees phenomena described above). In section 3, we described a solution to this problem by introducing the idea of ordering the clashes in the set. This caused the deduction process to keep from choosing combinations of potentials and goals that had already been resolved. Applying this concept of ordering to clashes to artificial conflicts allows the deduction procedure to prevent duplicate solutions from being generated. In other words we wish to extend this concept of ordering clashes within a single solution search space to apply to the complete search space. That is, as new clashes of artificial conflicts are generated they are compared with the previous set of artificial conflicts and given the appropriate ordering. Any clash in the new set given an order number greater than the element which generated the set is discarded.

Adding the above two strategies to the clash selection process for artificial conflicts produces an algorithm which will generate a complete solution set with minimal redundancy. The result of applying this new strategy to the clauses of example 4 is shown in figure 5. One can see that a complete solution set is generated but without the redundancy of the original algorithm.

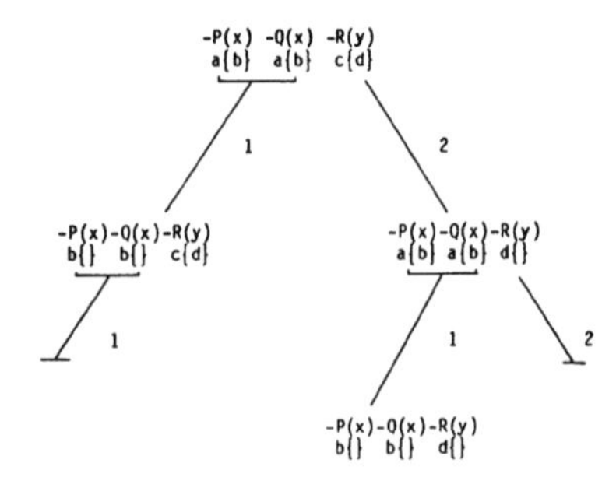

Figure 5. A complete nonredundant solution set for example 4.

CONCLUSION

An implementation of a plan based deduction system has now been completed. It involves some 6000 lines of PASCAL code and runs under CMS on an AMDAHL 470/V5.

Three important problems, encountered during early experimentation with this plan-based deduction system, have been presented. We have shown, using simplified examples, solutions to these problems. However, simplicity of the examples should not be misleading. They extract important experience gained during the usage of our system on larger logic programs, such as the graph-coloring problem used as illustration in [Pereira & Porto 80],]or the Huffman-Clewes theory of polyhedral scenes, suggested to us by M. van Emden.

Reviewing the results of experimentation and suggested implementation strategies, we feel that our research will lead to a practical and efficient deduction system. Our emphasis on attempting to control the system so that it avoids generating redundant solutions is particularly significant. Without some kind of constraint, the system tends to generate an unacceptably large number of identical solutions which makes it impractically slow and inflates its memory requirements. The problem lies with imposing a constraint which does not restrict the system from generating a complete solution set. Completeness is an important consideration when applying an automated deduction system to cope with the intensional clauses of a large data base. We believe that the strategy outlined in this paper provides such a constraint, where completeness (of [Matwin & Pietrzykowski 83]) is preserved and reduncancy is significantly decreased.

An open and interesting question, however remains. During deduction the system may proceed either depth-first or breadth-first when developing open goals. Is the choice of either strategy relevant for efficiency (if yes, in what way?) or is this dependant upon some topological properties of the plan being asserted?

REFERENCES

[Bibel 83]

Bibel, W. "Matings in Matrices", Communications of ACM, Vol 26, No 26, pp. 844-852, 1983.

[Bruynooghe 83]

Bruynooghe, M., "Deduction Revision by Intelligent Backtracking", Universidade Nova de Lisboa, Research Report, July 1983.

[Chang and Slagle 79]

Chang, C.L. and Slagle, J.R., "Using Rewriting Rules for Connection Graphs to Prove Theroms", Artificial Intelligence, Vol. 12, pp. 159-180, 1979.

[Clocksin and Mellish 82]
Clocksin, W.F. and Mellish, C.S., "Programming in Prolog", Springer Verlag, 1982.

[Forsythe & Matwin 83]
Forsythe, K and Matwin, S., "Copying of Multi-level Structures in a PASCAL Environment", submitted to Software - Practice and Experience, 1983.

[Kowalski 75]
Kowalski, R., "A Proof Procedure Using Connection Graphs", Journal of ACM, Vol 22, No 4, pp. 572-595, 1975.

[Matwin & Pietrzykowski 83]
Matwin, S and Pietrzykowski, T., "Intelligent Backtracking in Plan-Based Deduction", submitted to IEEE Trans. on Pattern Analysis and Machine Intelligence.

[Pereira & Porto 80]
Pereira, L.M., and Porto, A., "Selective Backtracking for Logic Programs", Procs. of CADE-5, pp. 306-317.

[Pietrzykowski & Matwin 82] Pietrzykowski, T. and Matwin, S., "Exponential Improvement of Exhaustive Backtracking: A Strategy for Plan-Based Deduction", Procs. of CADE-6, pp.223-239.

[Sickle 76]
Sickle, S. "A Search Technique for Clause Interconnectivity Graphs", IEEE Trans. on Computers, Vol 25, No 8, pp. 823-835, 1976.

A Programming Notation for Tactical Reasoning

David A. Schmidt

Computer Science Department
Edinburgh University
Edinburgh, Scotland*

Abstract: A notation for expressing the control algorithms (subgoaling strategies) of natural deduction theorem provers is presented. The language provides tools for building widely known, fundamental theorem proving strategies and is independent of the problem area and inference rule system chosen, facilitating formulation of high level algorithms that can be compared, analyzed, and even ported across theorem proving systems. The notation is a simplification and generalization of the tactic language of Edinburgh LCF. Examples using a natural deduction system for propositional logic are given.

0. Introduction

Logical systems of natural deduction (Pra) have demonstrated their usefulness in the development of traditional problem areas in formal logic and mathematics. Their application to computing related areas such as formal semantics (Hoa,Plo), data type specification (Gut), and program development (Cos,Nor) emphasizes the importance of understanding the notion of derivation and the strategies available for constructing proofs. Traditionally, these concerns have fallen in the realm of automated theorem proving (Ble,Boy,Coh,Gor), but the emphasis in this mechanized world often falls upon the number or difficulty of the theorems proved, rather than the style in which they are proved. Further, the boundaries between kind of logical system (natural deduction versus axiomatic), problem area of interest (first order logic, group theory, set theory, etc.), and the proof discovery strategy are often poorly delineated. If the theorem proving art is to be advanced and its most elegant ideas applied to new problem areas such as program development, the distinctions between these levels must be made clear, and the methodologies underlying proof discovery need to be expressed in a machine independent, understandable way.

This paper describes an initial version of a notation for expressing control algorithms for natural deduction theorem provers. The notation is independent of the specific problem area and rule system chosen, but it supplies the

---

*Present address: Computer Science Department, Kansas State University, Manhattan, Kansas 66506

fundamental tools for building useful subgoaling strategies from the inference rules supplied. The addition of control structures and a scoping mechanism allows definition of realistic algorithms. The language is a simplification and generalization of the tactic language of Edinburgh LCF (Gor).

After a brief review of natural deduction in section 1, section 2 outlines the basic features of Edinburgh LCF. The new notation is described in section 3, and section 4 presents an example algorithm for theorem proving in a subset of propositional logic.

1. Background

The form of natural deduction used is described in Prawitz (Pra); the examples in this paper use propositional logic (Lem), although any other problem area would do as well. Given a language L built up from propositions P,Q,R,..., and logical symbols $\wedge, \vee, \supset, \neg$, use the usual syntax rules to build the language of propositional logic. Arbitrary propositions (also called formulas) are denoted by A,B,C,..., and lists of propositions are represented by $\Gamma, \Delta, \Sigma, \ldots$. The rule schemes for inferring new facts from already established ones are

$$\wedge I: \frac{A \quad B}{A \wedge B} \qquad \wedge E\ell: \frac{A \wedge B}{B} \qquad \wedge Er: \frac{A \wedge B}{A}$$

$$\vee I\ell: \frac{A}{B \vee A} \qquad \vee Ir: \frac{A}{A \vee B} \qquad \vee E: \frac{A \vee B \quad \begin{matrix}(A)\\ C\end{matrix} \quad \begin{matrix}(B)\\ C\end{matrix}}{C}$$

$$\supset I: \frac{\begin{matrix}(A)\\ B\end{matrix}}{A \supset B} \qquad \supset E: \frac{A \supset B \quad A}{B}$$

$$\neg I: \frac{\begin{matrix}(A)\\ ff\end{matrix}}{\neg A} \qquad \neg E: \frac{\begin{matrix}(\neg A)\\ ff\end{matrix}}{A}$$

where ff abbreviates any formula $D \wedge \neg D$. A proof of a proposition C from propositions Δ is a tree whose leaves are the members of Δ and whose root is C. Each internal connection between parent and children nodes is justified by one of the inference rules. A rule with a parenthesized formula (Such as $\supset$I) causes the removal (discharge) of that parenthesized leaf node when applied to the tree. As an example,

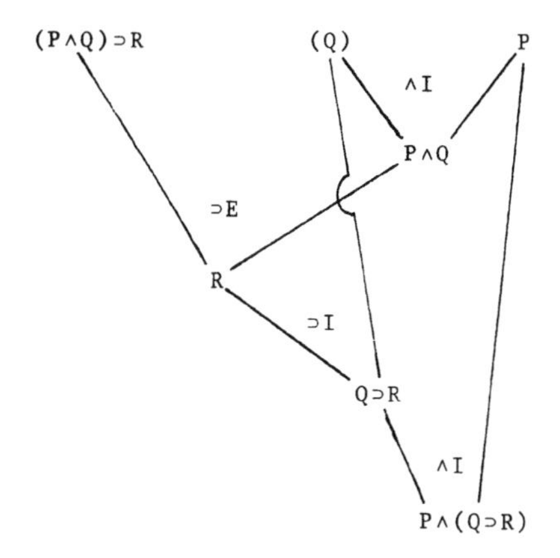

is a proof that $(P\wedge Q)\supset R$ and P infer $P\wedge(Q\supset R)$. Write $(P\wedge Q)\supset R,\ P \vdash P\wedge(Q\supset R)$ to abbreviate the tree; this expression is called a <u>theorem</u>. Note that the order in which the nodes of the tree were added does not affect the final result.

The following two results hold for all natural deduction systems:

i) if $\Gamma \vdash B$ and $B,\Delta \vdash C$, then $\Gamma,\Delta \vdash C$.

ii) if $\Gamma \vdash C$, then $\Gamma, A \vdash C$.

Call i) the <u>cut</u> principle and ii) the <u>pad</u> principle. Since the proofs of both are constructive, there exist associated functions <u>cut</u> and <u>pad</u> which build the deduction tree of the consequent theorem form the deduction tree(s) of the antecedent theorem(s). These functions will be useful for tree assembly.

2. LCF

<u>L</u>ogic for <u>C</u>omputable <u>F</u>unctions (Gor) is a software tool for developing natural deduction style proofs. A notable feature of the system is its functional depiction of deduction. Formulas are assigned the data type <u>form</u> and are built using formulas and logical connectives. Those formulas which are taken as axioms are given type <u>thm</u> (theorem). Inference rule schemes are functions which produce results of type thm from their arguments.

For example,

P	has type form
$P\wedge Q$	has type form
<u>assume</u>	has type form → thm
<u>assume</u>$(P\wedge Q)$	has type thm
$\supset E$	has type thm × thm → thm

$\supset E$ (<u>assume</u>$(P\supset(P\wedge Q))$, <u>assume</u>(P)) has type thm.

Expressions of type thm are written in their sequent form, e.g., $P\supset(P\wedge Q), P \vdash P\wedge Q$. The two expressions of type thm seen above are short examples of (forwards) LCF proofs, which are constructed by nested applications of the inference rule functions. This provides an element of security to the system, for only through the use of the axioms and the inference rules can new theorems be created.

The LCF system rises above its role as a mere proof checker due to its ability to perform goal directed (backwards) proofs, building a deduction tree from its root-- its goal-- to its leaves, i.e., assumptions. The basic approach is to take a goal, $\Delta \overset{?}{\vdash} C$, where Δ is a set of assumptions and C is the desired conclusion, and decompose C into a list of subgoals C1,...,Cn such that if proofs exist for $\Delta \overset{?}{\vdash} C1, \ldots, \Delta \overset{?}{\vdash} Cn$, then a proof of $\Delta \overset{?}{\vdash} C$ is also constructable. A function which decomposes a goal is called a tactic.

The functional formalization of goal directed proof is defined in LCF as

goal: form list × form

-- the assumption formula set Δ and the desired conclusion C;

tactic: goal ->(goal list × validation)

-- the decomposition step of goal $\Delta \overset{?}{\vdash} C$ into its subgoals plus a thm producing function;

validation: thm list ->thm

-- the thm producing function which produces $\Delta \vdash C$ from $\Delta \vdash C1, \ldots, \Delta \vdash Cn$, thus justifying the decomposition.

Using angle brackets to enclose lists and parentheses to bind pairs, here are the definitions of some LCF-style tactics:

IMPTAC: $\Delta \overset{?}{\vdash} A\supset B \mapsto (\langle\Delta, A \overset{?}{\vdash} B\rangle, \supset I)$

ANDTAC: $\Delta \overset{?}{\vdash} A\wedge B \mapsto (\langle\Delta \overset{?}{\vdash} A; \Delta \overset{?}{\vdash} B\rangle, \wedge I)$

TRIV: $\Delta, A \overset{?}{\vdash} A \mapsto (\langle\rangle, triv\langle\Delta;A\rangle)$

IDTAC: $\Delta \overset{?}{\vdash} A \mapsto (\langle\Delta \overset{?}{\vdash} A\rangle, \lambda\langle t\rangle.t)$

For example, ANDTAC accepts as an argument a goal $\Delta \overset{?}{\vdash} C$. If C has structure $A\wedge B$, the result is the list of subgoals $\langle\Delta \overset{?}{\vdash} A; \Delta \overset{?}{\vdash} B\rangle$. This decomposition is justified by the rule $\wedge I$. Note that if C is not a conjunction, the tactic fails (generates an exception). The validation $triv\langle\Delta;A\rangle$ maps an empty list of theorems to the axiom $\Delta, A \vdash A$. IDTAC is the identity map for tactics.

Systematic decomposition of goals is performed by composing tactical steps such as these until all subgoals reduce to empty subgoal lists. The forwards proof corresponding to the decompositions is obtained by applying the validation functions in the order inverse to that of the composition of the corresponding tactics. To aid in tactic (and validation) composition, tactic combinators known as <u>tacticals</u> are used. In the descriptions to follow, let g be a goal and @ denote the list append operator. The four tacticals used most frequently are:

i) THENL: tactic × tactic list -> tactic

THENL performs sequencing of tactics. An expression t THENL <ul;...;um> applies t to its input goal g and then applies tactic ui to the corresponding goal gi from the list of m subgoals that t produces. Formally,

```
(t THENL <ul;...;um>) g =
   let (<gl;...;gm>,v) = t g in
   let (G'i,v'i) = ui gi, for 1≤i≤m, in
      (G'1 @ ... @ G'm, v∘(v'1 × ... × v'm)).
```

The tactic fails if t g or any ui gi fails. The validation v∘(v'1 × ... × v'm) represents the function which accepts the list of theorems that satisfy all the subgoals in the list G'1 @ ... @ G'm. The function parcels out the theorems to each of the subvalidations v'i, so that the theorems satisfying goals <gl;...;gm> are produced. These theorems are in turn given to v for production of the original goal.

ii) THEN: tactic × tactic -> tactic

THEN is a form of THENL which is used when the number of subgoals produced from the tactic t may vary.

```
(t THEN u) g =
   let (<gl;...;gm>,v) = t g in
   let (G'i,v'i) = u gi, for 1≤i≤m, in
      (G'1 @ ... @ G'm, v∘(v'1 × ... × v'm)).
```

The tactic fails if t g or any u gi fails.

iii) ORELSE: tactic × tactic -> tactic

```
(t ORELSE u) g =
   t g, if t does not fail with argument g
   u g, otherwise.
```

ORELSE directs depth-first search.

iv) REPEAT: tactic -> tactic

The tactic REPEAT t applies t repeatedly to the subgoals produced by t until it cannot succeed on any more of the subgoals. It is defined recursively as

REPEAT t g = <u>let</u> <u>rec</u> f = (t THEN f) ORELSE IDTAC <u>in</u> f g.

As defined, REPEAT t can never fail.

Here are some simple tactics built using tacticals:

DECOMPOSE-TAC = ANDTAC ORELSE IMPTAC

attempts to simplify a goal $\Delta \overset{?}{\vdash} C$ by decomposing C into its subparts. A conjunction gets split via ANDTAC; if C is an implication, ANDTAC will fail, but IMPTAC will successfully remove the implication and assume the antecedent portion. DECOMPOSE-TAC fails if C is neither a conjunction nor an implication.

LOOP-TAC = (REPEAT DECOMPOSE-TAC) THEN TRIVTAC

applies DECOMPOSE-TAC repeatedly to remove all conjunction and implication symbols from C of $\Delta \overset{?}{\vdash} C$. Once all the connectives are removed, TRIVTAC is applied to each of the remaining subgoals to see if each is a trivial theorem. If so, the subgoal list empties and the strategy succeeds; if not, the tactic fails.

Applying LOOP-TAC to the goal given in section 1 yields the following trace of the subgoal lists:

LOOP-TAC $((A\wedge B)\supset C,\ A \overset{?}{\vdash} A\wedge(B\supset C)\ )$

=> $\langle (A\wedge B)\supset C, A \overset{?}{\vdash} A;\quad (A\wedge B)\supset C, A \overset{?}{\vdash} B\supset C\rangle$ by ANDTAC

=> $\langle (A\wedge B)\supset C, A \overset{?}{\vdash} A;\quad (A\wedge B)\supset C, A, B \overset{?}{\vdash} C\rangle$ by IMPTAC,

but note that nothing can be done to the first subgoal. The next iteration of REPEAT can cause no changes, so the loop is exited.

=> <u>failure</u>, as the first goal can be removed by TRIVTAC, but the second causes an exception. If TRIVTAC was replaced by (TRIVTAC ORELSE IDTAC), the subgoal list $\langle (A\wedge B)\supset C, A, B \overset{?}{\vdash} C\rangle$ would result. In either case, the problem is not completely solved.

The tacticals plus primitive tactics form a language for theorem proving. The discovery of useful proof algorithms for specific problem areas is a major topic of study in LCF-oriented research (Con,CoM,Les,Mon).

Two conceptual problems exist with LCF-style tactics. The first is that the system directly supports only backwards inference in the tactical mode. That is, a tactical step upon a goal $\Delta \vdash^{?} C$ is designed to simplify formula C with no effect on the formulas of Δ (aside from occasionally adding a new assumption). Often, a form of forwards chaining is desired while in tactical mode--the above example would certainly benefit. The existing system allows forwards chaining only through hand-coded simulation in which the user codes the subgoaling step and the validation function.

The second problem concerns the validity of tactical decomposition. An LCF tactic

$$\Delta \vdash^{?} C \;\mapsto\; (<\Gamma 1 \vdash^{?} B1; \;\ldots\; ; \Gamma n \vdash^{?} Bn>, \; v)$$

is <u>valid</u> if $v <\Gamma 1 \vdash B1; \;\ldots\; ; \Gamma n \vdash Bn> = \Delta \vdash C$. A user wishes to deal with valid tactics only, but allowing user-coded tactics to alleviate the first deficiency also allows the introduction of tactics which perform decompositions which are not justified by their corresponding validations. A sample invalid tactic is

$$\Delta \vdash^{?} A \wedge B \;\mapsto\; (<\Delta \vdash^{?} A>, \; (\lambda <t>.\wedge I(t, \; B \vdash B))).$$

"Tactics" such as these require run time checking in some form, even after an apparently successful decomposition. What's more, even a valid tactic can fall prey to a situation in which the wrong theorems are given to its validation. This situation certainly arises if an invalid tactic is composed with a valid one. In order to control these problems and develop a useful programming notation, a new view of tactical proof is needed.

3. Tactical proof

The task of tactical theorem proving is that of completing a partially constructed deduction tree. Both the leaves (assumptions) and root (conclusion) of the tree are known, and the proof is obtained by filling in the interior nodes of the tree in any order desired. The status of a partially completed proof of conclusion C from assumptions $A1,\ldots,Am$ is

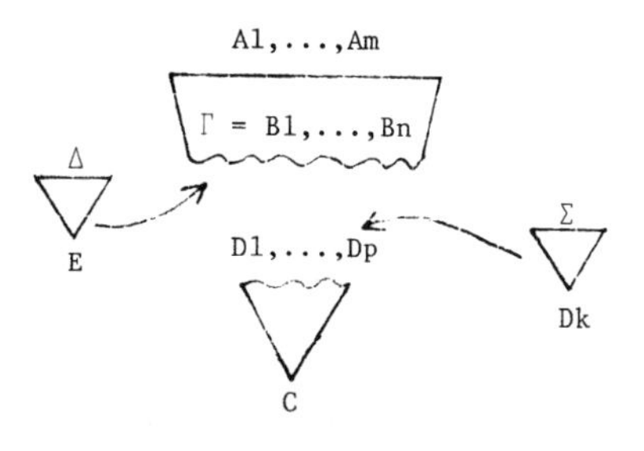

Some forwards inferences have been made from the assumptions to create the upper frontier of known facts Γ. Conclusion C has been simplified to create a lower frontier of subgoals $D1,\ldots,Dp$. A forwards step "pastes" a new fact E onto Γ whenever $\Delta \vdash E$ and $\Delta \subseteq \Gamma$. Similarly, a backwards decomposition of some Dk into formulas Σ occurs when $\Sigma \vdash Dk$. (A related concept of "proof window" appears in (And).)

These ideas can be formalized. Let a <u>goal</u> $\Delta \overset{?}{\vdash} C$ represent a partially completed tree with frontiers Δ and C. In the case that the lower frontier contains a list of formulas $D1,\ldots,Dp$, a list of goals $<\Delta 1 \overset{?}{\vdash} D1; \ldots ; \Delta p \overset{?}{\vdash} Dp>$ is used. Tactics are used to close frontiers, and they are naturally induced from the inference rules of the natural deduction system supplied. Given rule scheme

$$r = \frac{\begin{matrix}(A1) & \ldots & (Ak) & & \\ B1 & & Bk & B(k+1) & \ldots & Bm\end{matrix}}{C}$$

where propositions $B1,\ldots,Bm$ are used to deduce C, and dischargable assumptions $A1,\ldots,Ak$ are used to deduce $B1,\ldots,Bk$ respectively, there exist two fundamental tactic schemes based upon r:

i) Backwards:

$$(\vdash r) = \Delta \overset{?}{\vdash} C \mapsto <\Delta,A1 \overset{?}{\vdash} B1; \ldots ; \Delta,Ak \overset{?}{\vdash} Bk; \; \Delta \overset{?}{\vdash} B(k+1); \ldots ; \Delta \overset{?}{\vdash} Bm>.$$

The tactic is validated by rule r treated as a function of type thm list -> thm.

ii) Forwards:

$$(r\vdash) = B1,\ldots,Bm,\Delta \overset{?}{\vdash} D \mapsto <B1,\ldots,Bm,C,\Delta \overset{?}{\vdash} D>$$

and is validated by the function

$(\lambda <t>$:thm list. <u>cut</u>$(C)<r<B1 \vdash B1; \ldots ; Bm \vdash Bm>; t>$.

<u>Cut</u> is the function mentioned at the end of section 1:

<u>cut</u>$(D) <\Gamma \vdash D; \; \Delta,D \vdash E> = \Gamma,\Delta \vdash E$.

These tactic building operations may also be applied to a previously proved theorem $\Gamma \vdash D$, as the theorem can be viewed as an inference rule $\frac{\Gamma}{D}$.

Some useful examples of tactics built with the operators are

$$(\vdash\wedge I) = \Delta \overset{?}{\vdash} A\wedge B \mapsto <\Delta \overset{?}{\vdash} A; \; \Delta \overset{?}{\vdash} B>$$

$$(\vdash\supset I) = \Delta \overset{?}{\vdash} A\supset B \mapsto <\Delta,A \overset{?}{\vdash} B>$$

$$(\vdash(A\wedge\neg B \vdash \neg(A\supset B))) = \Delta \overset{?}{\vdash} \neg(A\supset B) \mapsto <\Delta \overset{?}{\vdash} A\wedge\neg B>.$$

All theorem provers make central use of tactics of the form ($\vdash$ opI), where op is a logical connective. Forwards examples include

$$(\wedge E\ell\vdash) = A\wedge B,\Delta \overset{?}{\vdash} C \mapsto \langle A\wedge B,B,\Delta \overset{?}{\vdash} C\rangle$$

$$(\supset E\vdash) = A\supset B,A,\Delta \overset{?}{\vdash} C \mapsto \langle A\supset B,A,B,\Delta \overset{?}{\vdash} C\rangle$$

$$((\neg(A\vee B) \vdash \neg A\wedge\neg B)\vdash) = \neg(A\vee B),\Delta \overset{?}{\vdash} C \mapsto \langle \neg(A\vee B),\neg A\wedge\neg B,\Delta \overset{?}{\vdash} C\rangle$$

(opE $\vdash$) tactics are used for forwards chaining strategies. The appendix presents a set of inference rules from which tactical skolemization is derived.

The following three tactics are not created using the above operators, but they are quite important and easily justified.

$$\text{IDTAC} = \Delta \overset{?}{\vdash} C \mapsto \langle\Delta \overset{?}{\vdash} C\rangle$$

$\text{TRIVTAC} = \Delta,A \overset{?}{\vdash} A \mapsto \langle\rangle$, as before, and

$\text{THIN}(A) = \Delta,A \overset{?}{\vdash} C \mapsto \langle\Delta \overset{?}{\vdash} C\rangle$, justified by <u>pad</u> (see section 1).

Tacticals are again used to build compound tactics. To keep this paper brief, the combinators of section 1 are reused.

The informal operational semantics of a tactic scheme t when applied to goal g is as follows: t attempts to perform a unification of g against its expected argument structure (its "domain"). If successful, the variables in t's domain scheme are instantiated to match g, and g is mapped to the goal list defined by the tactic's instantiated "range". If g can not be unified to the domain, failure results. The validation portion of a tactic is automatically generated from the operators used to build it. In this way, the validation accompanying a subgoal list is kept hidden from the user and safely maintained by the system. When a goal $g = \Delta \overset{?}{\vdash} C$ is decomposed by a tactic t to an empty subgoal list, i.e., $t\ g = \langle\rangle$, a system function <u>validate</u> can be invoked to activate the validation attached to $\langle\rangle$ and create the theorem (and its deduction tree): <u>validate</u>(t g) = $\Delta \vdash C$. The application of <u>validate</u> to a nonempty subgoal list fails.

More useful tactics can be built from partially or totally instantiated inference rule schemes. A substitution operation is used for this purpose. Given a rule or theorem scheme r, let $r[{}^{v1\ \ldots\ vn}_{e1\ \ldots\ en}]$ denote the instantiation of term and formula variables v1,...,vn in r by appropriately typed expressions e1,...,en, respectively. The expressions ei may also contain variables. For example,

$\supset\mathbf{E}[{}^{A}_{P\wedge C}]$ is the rule scheme $\dfrac{(P\wedge C)\supset B \quad (P\wedge C)}{B}$.

Domain restriction is also needed. Given a tactic tac and a goal expression $\Delta \vdash^? C$, possibly containing formula list and formula variables, let $t = tac\upharpoonright(\Delta \vdash^? C)$ denote tac restricted to operate only upon those goals whose structure matches $\Delta \vdash^? C$. If t is applied to an argument whose structure does not match, failure occurs. (The order of formulas in Γ in a goal $\Gamma \vdash^? P$ is <u>not</u> significant, and the unification of $\Gamma \vdash^? P$ with $\Delta \vdash^? C$ takes this into account.) The restriction operator is essentially a guard, and the reader may prefer a prefix notation; I use the postfix form to make use of the established operator $\upharpoonright$ and to point out the true nature of restriction.

Restriction and substitution work together with tacticals to establish a block structured language for theorem discovery. For example, the standard proof by cases tactic is defined from the $\vee E$ rule as

$$\begin{aligned} \text{CASES-TAC} = \;& (\vdash(\vee E[{}^{A}_{A'},{}^{B}_{B'}])\ \text{THENL} \\ & \langle \text{TRIVTAC};\ \text{THIN}(A'\vee B');\ \text{THIN}(A'\vee B')\rangle)\ \upharpoonright(A'\vee B',\Delta \vdash^? C) \\ = \;& A'\vee B',\Delta \vdash^? C \mapsto \langle \Delta,A' \vdash^? C;\ \Delta,B' \vdash^? C\rangle. \end{aligned}$$

Backwards chaining using $\supset E$ can be defined similarly, and, in general, such "match and thin" tactics using opE rules are a standard part of useful proof discovery algorithms. In a related fashion, forwards proof can be aided by pruning search space, as in

$$\begin{aligned} \text{REMOVE\_AND} = \;& ((\wedge E\ell[{}^{A}_{A'},{}^{B}_{B'}]\vdash)\ \text{THEN} \\ & (\wedge Er[{}^{A}_{A'},{}^{B}_{B'}]\vdash)\ \text{THEN}\ \ \text{THIN}(A'\wedge B'))\ \upharpoonright(A'\wedge B',\Delta \vdash^? C) \\ = \;& A'\wedge B',\Delta \vdash^? C \mapsto \langle A',B',\Delta \vdash^? C\rangle. \end{aligned}$$

Useful proof by contradiction strategies may also be expressed:

$$\begin{aligned} \text{REMOVE\_NOT} = \;& ((\vdash\neg I[{}^{A}_{A'}\ {}^{ff}_{C'\wedge\neg C'}])\ \text{THEN} \\ & (\vdash\wedge I)\ \text{THENL} \\ & \langle \text{THIN}(\neg C');\ \text{TRIVTAC}\rangle)\ \upharpoonright(\neg C',\Delta \vdash^? \neg A) \\ = \;& \neg C',\Delta \vdash^? \neg A' \mapsto \Delta,A' \vdash^? C'. \end{aligned}$$

4. An example

Consider the problem of constructing a theorem proving procedure for that subset of propositional logic utilizing only the conjunction and implication connectives. Given goal $\Delta \vdash^{?} C$, a standard theorem proving strategy first decomposes C by systematically removing its logical connectives (via backwards tactics) and then simplifies members of Δ (using forwards ones) until subgoals of the form $\Delta, A \vdash^{?} A$ appear. The procedure has the form

FIND = SIMPLIFY_RHS THEN SIMPLIFY_LHS.

The first step is straightforward:

SIMPLIFY_RHS = REPEAT (TRIVTAC ORELSE ($\vdash\supset$I) ORELSE ($\vdash\wedge$I)).

At each iteration, a check is made to see if an axiom form has been reached. If not, the primary logical connective of the goal is removed.

SIMPLIFY_LHS = REPEAT (TRIVTAC ORELSE
REMOVE_AND ORELSE REMOVE_IMPLIES).

The REMOVE_AND tactic is exactly that defined in section 3. Decomposition of an implication $A \supset B$ in a goal $A \supset B, \Gamma \vdash^{?} C$ is not trivial, as the $\supset$E rule requires the presence of A in Γ. More generally, if $\Gamma \vdash A$, then $\supset$E can be applied. Showing $\Gamma \vdash A$ requires a subroutining process: invoke FIND to solve for $\Gamma \vdash^{?} A$ and then validate the successful search, creating $\Gamma \vdash A$. The theorem can be used to build a forwards tactic in

REMOVE_IMPLIES = (((validate (FIND ($\Delta \vdash^{?}$ A'))) $\vdash$) THEN
($\supset$E$[{A \atop A'}, {B \atop B'}]$ $\vdash$) THEN
THIN(A'$\supset$B')) $\upharpoonright$(A'$\supset$B',$\Delta \vdash^{?}$ C).

This completes the formulation of FIND. Each tactic used in its construction effectively reduces the number of logical connectives in a subgoal by one, and so FIND always terminates. I conjecture that FIND is "complete" in the sense that any theorem provable using just the rules for conjunction and implication is also provable with FIND. The example of section 1 generates the following trace:

FIND($(P\wedge Q)\supset R, P \vdash^{?} P\wedge(Q\supset R)$)
=> < $(P\wedge Q)\supset R, P \vdash^{?} P$; $(P\wedge Q)\supset R, P \vdash^{?} Q\supset R$ >
=> < $(P\wedge Q)\supset R, P, Q \vdash^{?} R$ > since TRIVTAC handles the first subgoal, and ($\vdash\supset$I) does the second. FIND must apply SIMPLIFY_LHS, and the subroutine in REMOVE_IMPLIES generates a local subgoal:

$P, Q \vdash^{?} P\wedge Q$
=> <$P, Q \vdash^{?} P$; $P, Q \vdash^{?} Q$> => <>

so the search returns to the top level with $P, Q \vdash P\wedge Q$, yielding

$\Rightarrow \; \langle (P\wedge Q)\supset R, P, Q, P\wedge Q \vdash^{?} R \rangle$

$\Rightarrow \; \langle (P\wedge Q)\supset R, P, Q, P\wedge Q, R \vdash^{?} R \rangle$

$\Rightarrow \; \langle P, Q, P\wedge Q, R \vdash^{?} R \rangle$, all of which occur within REMOVE_IMPLIES.

$\Rightarrow \; \langle\rangle$ on the next iteration of SIMPLIFY_LHS.

The derivation tree built by validating the empty subgoal list is exactly that shown in section 1, as the cut and pad functions are tree stitching operations and not inference rules. The reader is encouraged to build the validation based upon the trace and generate the proof tree.

5. Conclusion

A notation for expressing proof discovery algorithms for natural deduction systems has been defined. It allows succinct specification of strategies widely used in all theorem provers and encourages the discovery of complementary ones as well. As the language is independent of the inference rule system used, it supports formulation of problem area-independent strategies and facilitates comparisons of control algorithms of different theorem proving systems. This separation of theorem discovery strategies fron the problem areas to be studied exposes the importance of properly formulating the underlying inference rule system of the theorem prover. Many technical difficulties disappear when a coherent, complementary set of rules is used as a foundation.

This paper's length prevented a closer examination of the control constructs for the language. The LCF tacticals served well for the simple examples, but an obvious need exists for more sophisticated combinators such as breadth-first search ("EITHER t1 OR t2") and conditional iteration ("REPEAT t UNLESS e"). This area merits study.

Acknowledgements: Brian Monahan produced many concrete examples of tactic generation in his work with the LCF system. He made many useful suggestions and carefully read an earlier version of this paper. Discussions with Colin Stirling and Robin Milner have also been helpful. Dorothy McKie's typing is gratefully acknowledged.

References

(And) Andrews, P.B. Transformaing matings into natural deduction proofs. 5th Conference on Automated Deduction, Les Arcs, France, 1980, LNCS 87, pp. 281-292.

(Ble) Bledsoe, W.W., and Tyson, M. The UT interactive theorem prover. Memo ATP-17, Mathematics Dept., University of Texas, Austin, 1975.

(Boy) Boyer, R.S., and Moore, J.S. A Computational Logic. Academic Press, New York, 1979.

(Cha) Chang, C., and Lee, R.E. Symbolic Logic and Mechanical Theorem Proving. Academic Press, New York, 1973.

(Coh) Cohen, P.R., and Feigenbaum, E.A., eds. The Handbook of Artificial Intelligence, Vol. 3. Pittman, New York, Ch. 12.

(Con) Cohn, A. The equivalence of two semantic definitions: a case study in LCF. Report CSR-76-81, Computer Science Dept., University of Edinburgh, Scotland, 1981.

(CoM) Cohn, A., and Milner, R. On using Edinburgh LCF to prove the correctness of a parsing algorithm. Report CSR-113-82, Computer Science Dept., University of Edinburgh, Scotland, 1982.

(Cos) Constable, R.L. Proofs as programs: a synopsis. Information Proc. Letters 16-3 (1983) 105-112.

(Gor) Gordon, M., Milner, R., and Wadsworth,C. Edinburgh LCF. LNCS 78, Springer-Verlag, Berlin, 1979.

(Gut) Guttag, J. Notes on type abstraction. IEEE Trans. on Software Engg. SE-6-1 (1980) 13-23.

(Hoa) Hoare, C.A.R. An axiomatic basis for computer programming. Comm. ACM 12 (1969) 576-580, 583.

(Lem) Lemmon, E.J. Beginning Logic. Nelson, London, 1965.

(Les) Leszczyłowski, J. An experiment with Edinburgh LCF. 5th Conference on Automated Deduction, Les Arcs, France, 1980, LNCS 87, pp. 170-181.

(Mon) Monahan, B. Ph.D. thesis, University of Edinburgh, forthcoming.

(Nor) Nordström, B. Programming in constructive set theory: some examples. ACM Conf. on Functional Programming Languages and Computer Architecture, Portsmouth, N.H., 1981, pp. 141-153.

(Plo) Plotkin, G. A structural approach to operational semantics. Report DAIMI FN-19, Computer Science Dept., University of Aarhus, Denmark, 1981.

(Pra) Prawitz, D. Natural Deduction. Almquist and Wiksel, Stockholm, 1965.

(Rob) Robinson, J.A. Logic:Form and Function. Edinburgh Univ. Press, Edinburgh, 1979.

(Sup) Suppes, P. Introduction to Logic. Van Nostrand, Princeton, 1957.

Appendix

Inference rule schemes for quantifiers have additional restrictions attached to control the use of free variables. Tactics generated from those rules must also follow these restrictions. The rules given here for the universal and existential quantifiers have restrictions which make their corresponding tactics into the skolemization procedures of Robinson (Rob); the rules are a slightly modified version of those in Suppes (Sup). The unification operation is treated as a function like cut or pad.

Given the usual syntax for first order logic, let x,y,z,... represent variables which may be quantified. A Skolem variable is a free, barred variable, e.g., $\bar{x},\bar{y},\bar{z},\ldots$, and a Skolem constant is a free variable, possibly subscripted by a list of Skolem variables (a "Skolem function"), e.g., $x(\bar{y})$. Let A^x_t denote the usual syntactic substitution of term t for all free occurrences of variable x in formula A. The rules are

$$\forall E : \frac{\forall xA}{A^x_{\bar{y}}}$$

$$\forall I : \frac{A^x_{y(\bar{z}1\ldots\bar{z}n)}}{\forall xA}$$

where none of $y,\bar{z}1,\ldots,\bar{z}n$ are free in any assumption upon which $A^x_{y(\bar{z}1\ldots\bar{z}n)}$ depends, and all Skolem variables in $A^x_{y(\bar{z}1\ldots\bar{z}n)}$ belong to $\{\bar{z}1,\ldots,\bar{z}n\}$.

The tactics $(\forall E\vdash)$ and $(\vdash\forall I)$ correspond to the usual skolemization routines for universal quantifiers. $\forall I$'s inverted restriction forces y to be newly generated; $\bar{z}1,\ldots,\bar{z}n$ are exactly the Skolem variables in A.

$$\exists I : \frac{A^x_t}{\exists xA}$$

$$\exists E : \frac{\exists xA}{A^x_{y(\bar{z}1\ldots\bar{z}n)}}$$

where y is not free in any assumption upon which $\exists xA$ depends, and all Skolem variables in A belong to $\{\bar{z}1,\ldots,\bar{z}n\}$.

The tactics $(\vdash\exists I)$ (take term t to be $\bar{y}$) and $(\exists E\vdash)$ correspond to the usual skolemization routines for existential quantifiers.

Unification is substitution:

$$\text{UNIFY\_BACKWARDS}\ (\bar{x},t) = \Delta \vdash^{?} C \mapsto \langle \Delta \vdash^{?} C^{\bar{x}}_{t} \rangle$$

$$\text{UNIFY\_FORWARDS}\ (\bar{x},t) = \Delta \vdash^{?} C \mapsto \langle \Delta^{\bar{x}}_{t} \vdash^{?} C \rangle .$$

Of course, a global state is needed to ensure unique generation of Skolem constants (and variables) and consistent unification across all subgoals. Here is a hypothetical tactical proof of an example using quantifiers:

$$\begin{array}{lllll}
 & \exists y \forall x F(y,x) & \vdash^{?} \forall x \exists y F(y,x) & & \\
\Rightarrow & \langle \exists y \forall x F(y,x) & \vdash^{?} \exists y F(y,x) & \rangle & \text{by } (\vdash \forall I) \\
\Rightarrow & \langle \exists y \forall x F(y,x) & \vdash^{?} F(\bar{y},x) & \rangle & \text{by } (\vdash \exists I) \\
\Rightarrow & \langle \forall x F(y,x) & \vdash^{?} F(\bar{y},x) & \rangle & \text{by } (\exists E \vdash) \\
\Rightarrow & \langle F(y,\bar{x}) & \vdash^{?} F(\bar{y},x) & \rangle & \text{by } (\forall E \vdash) \\
\Rightarrow & \langle F(y,x) & \vdash^{?} F(\bar{y},x) & \rangle & \text{by UNIFY_FORWARDS}(\bar{x},x) \\
\Rightarrow & \langle F(y,x) & \vdash^{?} F(y,x) & \rangle & \text{by UNIFY_BACKWARDS}(\bar{y},y) \\
\Rightarrow & \langle\rangle & & & \text{by TRIVTAC.}
\end{array}$$

This should be compared to the attempted proof of

$$\begin{array}{lllll}
 & \forall x \exists y F(y,x) & \vdash^{?} \exists y \forall x F(y,x) & & \\
\Rightarrow & \langle \forall x \exists y F(y,x) & \vdash^{?} \forall x F(\bar{y},x) & \rangle & \text{by } (\vdash \exists I) \\
\Rightarrow & \langle \forall x \exists y F(y,x) & \vdash^{?} F(\bar{y},x(\bar{y})) & \rangle & \text{by } (\vdash \forall I)\text{. Note the necessity of the argu-} \\
 & & & & \text{ment } \bar{y} \text{ in Skolem constant } x(\bar{y})\text{.} \\
\Rightarrow & \langle \exists y F(y,\bar{x}) & \vdash^{?} F(\bar{y},x(\bar{y})) & \rangle & \text{by } (\forall E \vdash) \\
\Rightarrow & \langle F(y(\bar{x}),\bar{x}) & \vdash^{?} F(\bar{y},x(\bar{y})) & \rangle & \text{by } (\exists E \vdash).
\end{array}$$

Attempts at unifying the structurally similar formulas can not succeed.

The Mechanization of Existence Proofs of Recursive Predicates

Ketan Mulmuley

Computer Science Department
Carnegie-Mellon University
Schenley Park
Pittsburgh, PA 15213, USA

1. Abstract

Proving the congruence of two semantics of a language is a well known problem. Milne[3] and Reynolds [5] gave techniques for proving such congruences. Both techniques hinge on proving the existence of certain recursively defined predicates. Milne's technique is more general than Reynolds', but the proofs based on that technique are known to be very complicated. In the last eight years many authors have expressed the need for a more systematic method and a mechanical aid to assist the proofs. In this paper we give a systematic method based on domain theory. The method works by building up appropriate cpos and continuous functions on them. Existence of a predicate then follows by using the Fixed Point Theorem. A mechanized tool has been developed on top of LCF to assist proofs based on this method. The paper refutes the fear expressed by many people that fixed-point theory could not be used to show existence of such predicates.

2. Introduction

The Scott-Strachey approach to giving semantics to a language is well known. In this approach each programming construct in a language is given a denotation or meaning in an appropriately constructed domain which is some kind of cpo (complete partial order). Of course there can more than one way of constructing such domains and denotations. The question which arises naturally is: how does one know that these ways are in some sense equivalent? Say the language is L and we have two semantics $\mathbb{L}_1$ and $\mathbb{L}_2$ which map L into the domains $\mathbf{D}_1$ and $\mathbf{D}_2$ respectively. One might start by constructing some predicate, say $P \in \mathbf{D}_1 \times \mathbf{D}_2$, which relates the equivalent values from two domains i.e $(d_1, d_2) \in P$ iff d_1 and

---

This research was supported in part by the Defense Advanced Research Projects Agency (DOD), ARPA Order No. 3597, monitored by the Air Force Avionics Laboratory under Contract F33615-81-K-1539, and in part by the U.S. Army Communications R&D Command under Contract DAAK80-81-K-0074. The views and conclusions contained in this document are those of the authors and should not be interpreted as representing the official policies, either expressed or implied, of the Defense Advanced Research Projects Agency or the U.S. Government.

d_2 are in some sense equivalent. The above question then reduces to asking: does $(\mathcal{L}_1(e), \mathcal{L}_2(e)) \in P$ for all $e \in L$? Sometimes one might succeed in constructing only a weaker predicate P such that $(d_1, d_2) \in P$ iff d_1 is *weaker* than d_2 in some sense. One then proves the equivalence of two semantics by proving the following two assertions independently:

(1) $\mathcal{L}_1(e)$ is *weaker* than $\mathcal{L}_2(e)$ for all $e \in L$.

(2) $\mathcal{L}_2(e)$ is *weaker* than $\mathcal{L}_1(e)$ for all $e \in L$.

The key problem then is the construction of an appropriate predicate and proving its existence, the issue we shall be addressing in this paper. To make the ideas concrete we shall consider one example from [8]. Consider the simple expression language, essentially the lambda calculus with atoms,

$$E ::= I \mid B \mid (\lambda I.E_1)E_2$$

Here I belongs to a syntactic domain **Ide** of identifiers and B belongs to a domain **B** of basic values. *Dynamic* scoping is intended. Let **D**, the domain of values and **C**, the domain of environments, be the least solutions satisfying $\mathbf{D} = \mathbf{C} \to \mathbf{B}$ and $\mathbf{C} = \mathbf{Ide} \to \mathbf{D}$. Let **E** be the syntactic domain of expressions. Let **F** denote the domain of contexts, $\mathbf{Ide} \to \mathbf{E}$. Then a denotational semantics, $\mathcal{E} : \mathbf{E} \to \mathbf{D}$ (or $\mathcal{E} : \mathbf{E} \to \mathbf{C} \to \mathbf{B}$) and an operational semantics, $\mathcal{O} : \mathbf{E} \to \mathbf{F} \to \mathbf{B}$ can be given easily:

$$\begin{array}{ll} \mathcal{E}[\![I]\!]c = (cI)c & \mathcal{O}[\![I]\!]f = \mathcal{O}(fI)f \\ \mathcal{E}[\![B]\!]c = B & \mathcal{O}[\![B]\!]f = B \\ \mathcal{E}[\![(\lambda I.E_1)E_2]\!]c = \mathcal{E}[\![E_1]\!](c[\mathcal{E}[\![E_2]\!]/I]) & \mathcal{O}[\![(\lambda I.E_1)E_2]\!]f = \mathcal{O}[\![E_1]\!](f[E_2/I]) \end{array}$$

The congruence problem is to show that $\mathcal{E}$ and $\mathcal{O}$ compute the same thing. More precisely let $convert : \mathbf{F} \to \mathbf{C}$(i.e $\mathbf{F} \to \mathbf{Ide} \to \mathbf{D}) = \lambda f \lambda I.(\mathcal{E}(fI))$. Then we have to show that for all $E : \mathbf{E}$ and $f : \mathbf{F}$, $\mathcal{E}[\![E]\!](convertf) = \mathcal{O}[\![E]\!]f$. It is easy to prove that $\mathcal{O}[\![E]\!]f \sqsubseteq \mathcal{E}[\![E]\!](convertf)$. The difficult part is to prove the other inequality. For that, one shows that the following recursively defined predicates, $\theta \subseteq \mathbf{D} \times \mathbf{E}$ and $\eta \subseteq \mathbf{C} \times \mathbf{F}$ exist, where

$$\begin{array}{l} \theta = \{(d,e) \mid \forall (c,f) \in \eta .\ dc \sqsubseteq \mathcal{O}ef\} \\ \eta = \{(c,f) \mid \forall I \in \mathbf{Ide}.\ (cI, fI) \in \theta\} \end{array} \tag{2.1}$$

Then it can be shown that for all $E : \mathbf{E}$ and $f : \mathbf{F}$, $(\mathcal{E}[\![E]\!], E) \in \theta$ and $(convertf, f) \in \eta$. Hence from (2.1) we get $\mathcal{E}[\![E]\!](convertf) \sqsubseteq \mathcal{O}[\![E]\!]f$. The above discussion shows the importance of recursively defined predicates in a congruence proof. Henceforth we shall confine our attention solely to the existence of such predicates.

The existence of the above predicates, θ and η, is shown in [8] using Milne's technique [3]. Later in this paper we will show how the proof based on our method can be mechanized. Both methods show that θ and η exist for *any* choice of $\mathcal{O}$

But if we replace $\sqsubseteq$ by $=$ or $\sqsupseteq$ in (2.1), the interesting question is: do the predicates still exist? This question was raised in [8] but was left open. We can answer that question partially now. For that $\mathcal{O}$ given above we still do not know the answer. However we shall show using diagonalization that there *exists* an $\mathcal{O}$ such that the predicates do not exist.

Consider the case of $\sqsupseteq$ first. Let **Ide** be a trivial one-point domain. Then **C** can be identified with **D** and **F** can be identified with **E**, so we have $\mathbf{D} = \mathbf{D} \rightarrow \mathbf{B}$, $\mathbf{F} = \mathbf{E}$ and $\theta = \eta$. And the question reduces to asking whether θ' exists where

$$\theta' = \{(d : \mathbf{D}, e : \mathbf{E}) \mid \forall (d', e') \in \theta' .\ dd' \sqsupseteq \mathcal{O}ee'\}$$

Let $f : \mathbf{D} = (\lambda x.xx)$. We are omitting the obvious injections and projections. It can be shown that $ff = \perp$(as in the well known Park's theorem.) Let $\mathcal{O} = \lambda e_1 \lambda e_2 .b$ where the constant $b : \mathbf{B} \sqsupset \perp$. We are assuming that **B** contains at least two elements. As $\mathcal{O}$ is constant, existence of θ' is equivalent to existence of some α where

$$\alpha = \{d : \mathbf{D} \mid \forall d' \in \alpha .\ dd' \sqsupseteq b\} \tag{2.2}$$

Assume for a contradiction that such an α does exist. Consider two cases. 1) $f \in \alpha$: in this case by (2.2), $ff \sqsupseteq b$ and hence $\perp \sqsupseteq b$. This is a contradiction. 2) $f \notin \alpha$: Consider any $d \in \alpha$. By (2.2) we get $dd \sqsupseteq b$. Hence $fd = (\lambda x.xx)d = dd \sqsupseteq b$. Thus for all $d \in \alpha$, $fd \sqsupseteq b$. Hence by (2.2) $f \in \alpha$. This is again a contradiction . We conclude that such an α does not exist.

If we replace $\sqsubseteq$ in (2.1) by $=$ exactly the same reasoning goes through.

The implications of this counterexample are many. It shows that the existence of recursively defined predicates is indeed a nontrivial problem. Secondly it weakens the hope that there exists a rich enough syntactic language such that any predicate expressed in that language exists. Reynolds [5] and Milne [3] gave general techniques to prove the existence of such predicates. Note that Reynolds' method can *not* be used to show the existence the above predicates, θ and η. There has been some confusion about this in the literature. For example, see lemma 2.2.1 in Peter Mosses' thesis [4]. There he says that the existence of a certain recursively defined predicate follows by Reynolds' method easily. But it is possible to give a somewhat similar looking predicate and show that it does *not* exist. Due to lack of space we shall not discuss that example here. But the idea is the same as in the above counterexample, namely self-application. We hope that one example makes our point clear. (Incidently I can not show that the particular predicate in [4] does not exist. So that remains an open question again). Milne's method can be used to show existence of θ and η (see [8]). The proof is undeniably complicated. Because the existence proofs of recursively defined predicates are so complicated, many authors have expressed the need for some mechanical aid in carrying out such proofs [2,8]. This problem has been open for many years. In this paper we shall show that such a mechanization is indeed possible. A mechanical aid has been actually implemented on top of LCF to assist in these existence proofs.

3. Basic Notations

It will be assumed that the reader is familiar with domain theory. All the results in this section can be found in [6]. Domains in this paper are assumed to be consistently complete algebraic cpos. We shall denote them by **C**,**D** etc. $\mathbf{C} \rightarrow \mathbf{D}$ will denote the usual domain of continuous functions from **C** to **D**. The universal domain will be denoted by **U**.

3.1. Definition.

(1) $r : \mathbf{D} \rightarrow \mathbf{D}$ is called a retract if $r \circ r = r$, where $\circ$ denotes composition. We shall denote the fixed point set of r by $|r|$, and by r if no ambiguity arises. Note that $|r|$ is a cpo. Its bottom element is $r(\perp)$; we shall denote it by $\perp_r$.

(2) A retract $r : \mathbf{D} \rightarrow \mathbf{D}$ is called a projection if $r \sqsubseteq I$, where $I : \mathbf{D} \rightarrow \mathbf{D}$ is the identity function.

(3) A projection $r : \mathbf{D} \rightarrow \mathbf{D}$ is called *finitary* if $|r|$ is isomorphic to a domain.

(4) Given retracts $r, s : \mathbf{D} \rightarrow \mathbf{D}$, we say that r is a subcpo of s (denoted by $r \preceq s$) if $s \circ r = r$. This is equivalent to saying $|r| \subseteq |s|$. ∎

3.2. Fact. If $f, g : \mathbf{D} \rightarrow \mathbf{D}$ are projections then $f \sqsubseteq g$ iff $|f| \subseteq |g|$ ∎

The domain of finitary projections of the universal domain **U** will be denoted by **V**. Each domain is isomorphic to the fixed point set of some finitary projection on **U** [6]; hence it can be regarded as an element of **V**. Therefore, we shall use the notation **C**, **D** etc to denote elements of **V** as well.

3.3. Fact. There exist $\rightarrow, \times, + : \mathbf{V} \times \mathbf{V} \rightarrow \mathbf{V}$ such that $|\mathbf{D} \rightarrow \mathbf{E}| \simeq |\mathbf{D}| \rightarrow |\mathbf{E}|$, $|\mathbf{D} \times \mathbf{E}| \simeq |\mathbf{D}| \times |\mathbf{E}|$, $|\mathbf{D} + \mathbf{E}| \simeq |\mathbf{D}| + |\mathbf{E}|$. ∎

In this paper a retract will perform a double duty. Sometimes we can think of it as denoting the set of its fixed points, sometimes we can think of it as a function. Thus if $r : \mathbf{D} \rightarrow \mathbf{D}$ is a retract, then $x : r$, $x \in r$, $x \in |r|$, $r\,x = x$ all make sense; in fact they mean the same thing. Similarly if $S \subseteq |\mathbf{D}|$ then $r(S)$ denotes the set $\{y \mid \exists x : \mathbf{D}.\ y = r\,x\}$. Of course the finitary projections are retracts, so all this applies to them too. *The duality between finitary projections of* **U**, *i.e. elements of* **V**, *and domains will be assumed throughout this paper.* Thus if $L \in \mathbf{V}^k$ then sometimes we shall think of it as a subdomain of $\mathbf{U}^k$, and sometimes as a function over $\mathbf{U}^k$.

3.4. Definition.

(1) The n-ary relation R on some domain $\mathbf{R}$ is said to be upward(downward) closed in its ith index iff $(x_1, \cdots, x_i, \cdots, x_n) \in R$ and $y_i \sqsupseteq (\sqsubseteq) x_i$ implies $(x_1, \cdots, y_i, \cdots, x_n) \in R$.

(2) The relation S is directed-complete if the l.u.b of any directed subset of S belongs to S.

(3) Cl(A,B), where A and B are sets of indices, is defined to be the set of all directed-complete relations R on domain $\mathbf{R}$ which are upward closed in ith index for all $i \in A$ and downward closed in jth index for all $j \in B$. The degree n of the relation and the domain $\mathbf{R}$ will usually be inferrable from the context. ∎

For any domain $\mathbf{D}$, with a partial order $\sqsubseteq$ on it, the partial order can be regarded a binary predicate and $\sqsubseteq \in \mathrm{Cl}(\{2\},\{1\})$. However, the equality predicate is not in $\mathrm{Cl}(\{2\},\{1\})$ and nor is $\sqsupseteq$. This fact might have been crucial in deciding the existence of a solution to (2.1). We shall use upward and downward closure properties of the predicates throughout this paper.

As $A \rightarrow B$ will denote the set of continuous functions, we shall denote the set of *all* functions from A to B by $A \mapsto\!\!\rightarrow B$.

4. Predicate Cpos

4.1. Predicate Cpo Definition.

(1) Given a retract $\mathbb{L} : \mathbf{V}^n \rightarrow \mathbf{V}^n$, the sets A and B, and an n-ary predicate $P \subseteq |\perp_{\mathbb{L}}|$ where $P \in \mathrm{Cl}(A,B)$ we define
$\| \mathbb{L}, A, B, P \| = \quad \{(\mathbf{L}, L) \mid \mathbf{L} \in |\mathbb{L}|,\ L \subseteq |\mathbf{L}|,\ L \in \mathrm{Cl}(A,B),\ (\perp_{\mathbb{L}})L \subseteq P,\ P \subseteq L\}$

(2) For $(\mathbf{L}, L), (\mathbf{R}, R) \in \| \mathbb{L}, A, B, P \|$ we say $(\mathbf{L}, L) \sqsubseteq (\mathbf{R}, R)$ iff
(a) $\mathbf{L} \sqsubseteq \mathbf{R}$. (This means $|\mathbf{L}| \subseteq |\mathbf{R}|$ by Fact 3.2).
(b) $\mathbf{L}(R) \subseteq L$
(c) $L \subseteq R$ ∎

4.2. Theorem. *$\| \mathbb{L}, A, B, P \|$ is a cpo w.r.t the ordering given in (def 4.1). Further $(\perp_{\mathbb{L}}, P)$ is the bottom element of this cpo.*

Proof. Suppose $X = \{(\mathbf{L}_j, L_j) \mid j \in J\}$ is an indexed directed subset of $\| \mathbb{L}, A, B, P \|$. Let $Y = \{\mathbf{L}_j \mid j \in J\}$. Then it can be shown that

$$lub(X) = (lub(Y), \{x : lub(Y) \mid \forall j \in J. \mathbf{L}_j\, x \in L_j\}) \tag{4.1}$$

The proof is similar to [prop 2.5.3 ,(3)]. ∎

4.3. Fact. If $(\mathbf{L}, L_1) \sqsubseteq (\mathbf{L}, L_2)$ then $L_1 = L_2$. ∎

Let $\| \mathfrak{L}, A, B, P \|$ and $\| \mathfrak{R}, C, D, Q \|$ be predicate cpos where $\mathfrak{L}$ is a retract over $\mathbf{V}^m$ and and $\mathfrak{R}$ is a retract over $\mathbf{V}^n$. We want to construct a map from $\| \mathfrak{L}, A, B, P \|$ to $\| \mathfrak{R}, C, D, Q \|$. As any element in $\| \mathfrak{L}, A, B, P \|$ is of the form $(\mathbf{L}, L)$ where $L \subseteq |\mathbf{L}|$, it is natural to consider a map consisting of two components (see [3] also). The first component will be map on domains, $T : \mathbf{V}^m \to \mathbf{V}^n$, which will map $\mathbf{L}$ to $T(\mathbf{L})$. The second component will be a map on predicates, which will map L to some predicate P on $T(\mathbf{L})$. We shall specify this map by some formula, $\mathbf{w}$, based on a domain variable $\mathbf{X} : \mathbf{V}^m$ and a predicate variable, X. Then P can be defined as the set of all elements in $T(\mathbf{L})$ which 'satisfy' $\mathbf{w}$ when $\mathbf{X}$ is 'interpreted' as $\mathbf{L}$ and X is 'interpreted' as L.

4.4. Definition. Suppose we are given a domain variable $\mathbf{X} : \mathbf{V}^m$ and a predicate variable X.

(1) The set of predicate symbols, $\Theta ::= \Theta(X)$, is defined as:

$$\Theta = X \mid \sqsubseteq \mid \sqsupseteq \mid = \mid E$$

where E is any constant predicate symbol.

(2) The set of domain terms, $\Gamma(\mathbf{X})$, is defined as the set of lambda calculus terms over $\mathbf{V}$ which can possibly contain $\mathbf{X} : \mathbf{V}^m$ as a free variable and some constant symbols.

(3) The set of simple terms, $\Delta(\mathbf{X})$, is defined as the set of all τ such that: τ is a term over $\mathbf{U}^k$, for some k, which can contain some free variables over $\mathbf{U}$ and some constant symbols. We shall also allow an occurrence of $\mathbf{X}$ in a term belonging to $\Delta(\mathbf{X})$.(Remember that $\mathbf{X} : \mathbf{V}^m$ can be considered as a function over $\mathbf{U}^m$)

(4) The set of well formed formulae (wffs), $\Phi = \Phi(\mathbf{X}, X)$, is then recursively defined as follows:

(1) "$\tau \in P$" $\in \phi$, where $\tau \in \Delta(\mathbf{X})$ and $P \in \Theta(X)$.

(2)"$\forall y \in S.\ \mathbf{g}$" and "$\exists y \in S.\ \mathbf{g}$" $\in \Phi$,
where y is a variable over $\mathbf{U}$, $S \in \Gamma(\mathbf{X})$ and $\mathbf{g} \in \Phi$.

(3) "$\mathbf{f} \wedge \mathbf{g}$" , "$\mathbf{f} \vee \mathbf{g}$" and "$\mathbf{f} \Rightarrow \mathbf{g}$" $\in \Phi$, where $\mathbf{f}, \mathbf{g} \in \Phi$. ∎

One can then routinely define freeness of a variable, $z : \mathbf{U}$, in a wff. As an example "$\forall x \in Fst(\mathbf{X}).\ ((\mathbf{X}(x, z), \bot) \in \sqsupseteq) \wedge ((x, x) \in X)$" is a wff in $\Phi(\mathbf{X} : \mathbf{V}^2, X)$, where $Fst : \mathbf{V}^2 \to \mathbf{V}$ is a constant symbol. It has one free variable, z, over $\mathbf{U}$. For the sake of simplicity we would write this formula as: "$\forall x \in Fst(\mathbf{X}).\ \mathbf{X}(x, z) \sqsupseteq \bot \wedge ((x, x) \in X)$". As syntactic sugar we define "$\forall x \in X.\ \mathbf{g}$" and "$\exists x \in X.\ \mathbf{g}$" as short forms for "$\forall x \in \mathbf{X}.\ x \in X \Rightarrow \mathbf{g}$" and "$\exists x \in \mathbf{X}.\ x \in X \wedge \mathbf{g}$". We shall also allow the obvious shortforms such as: "$\forall (y_1, \cdots, y_l) \in \cdots$".

Let $\mathbf{w} \in \Phi(\mathbf{X} : \mathbf{V}^m, X)$. Let an interpretation $\Im$ be a function such that:

(1) $\Im$ assigns to $\mathbf{X} : \mathbf{V}^m$ a domain $\mathbf{X}^\Im$ in $\mathbf{V}^m$ and to X a relation $X^\Im$ on $\mathbf{X}^\Im$; i.e, $X^\Im \subseteq |\mathbf{X}^\Im|$.

(2) $\Im$ assigns to every constant symbol $C : \mathbf{D}$ (which can occur in $\Delta(\mathbf{X})$ or $\Gamma(\mathbf{X})$) an element $C^\Im$ in $\mathbf{D}$.

(3) $\Im$ assigns to every constant predicate symbol E a predicate $E^\Im$ on some domain $\mathbf{E}^\Im$; i.e, $E^\Im \subseteq |\mathbf{E}^\Im|$.

(4) $\Im$ assigns to the predicate symbol '$\sqsubseteq$' the standard partial ordering predicate $\sqsubseteq$ on $\mathbf{U}^2$. Assignments to the predicate symbols '$\sqsupseteq$' and '$=$' are similar.

Let s be a function from the set of variables over $\mathbf{U}$ into $\mathbf{U}$. It is very easy to extend the interpretation to predicate symbols, domain terms and simple terms:

(1) Every $S : \mathbf{V} \in \Gamma(\mathbf{X})$ can be assigned an element $S^\Im$ in $\mathbf{V}$.

(2) Every $\tau : \mathbf{U}^k \in \Delta(\mathbf{X})$ can be assigned an element $\tau^\Im[s]$ in $\mathbf{U}^k$.

(3) Every $P \in \Theta(X)$ can be assigned a predicate $P^\Im$ on some domain $\mathbf{P}^\Im$; i.e, $P^\Im \subseteq |\mathbf{P}^\Im|$.

Now we inductively define what it means for $\Im$ to *satisfy* $\mathbf{w} \in \Phi(\mathbf{X}, X)$ *with* s, $\Im \models w[s]$:

(1) $\Im \models (\tau \in P)[s]$ iff $\mathbf{P}^\Im(\tau^\Im[s]) \in P^\Im$.

(2) $\Im \models (\forall y \in S.\ \mathbf{g})[s]$ iff $\forall a \in S^\Im.\ \Im \models \mathbf{g}[s(a/y)]$; $s(a/y)$ is exactly like s except that it maps y to a.The other case is similar.

(3) $\Im \models (\mathbf{f} \Rightarrow \mathbf{g})[s]$iff $\Im \models \mathbf{f}[s]$ **implies** $\Im \models \mathbf{g}[s]$. the other cases are similar.

Let $\mathbf{w}$ be a wff whose free variables over $\mathbf{U}$ are contained in the list $v_1, \ldots, v_k$. Let a be a tuple in $\mathbf{U}^k$. Then we say $\Im \models w[a]$ iff $\Im \models w[s]$ where s is a function which assigns the jth component of a, $\langle a \rangle_j$, to v_j. If τ is a term whose free variables over $\mathbf{U}$ are contained in the list $v_1, \ldots, v_k$, $\tau^\Im[a]$ is similarly defined. As the interpretations intended for the constant symbols and the constant predicate symbols should be clear from the context, we shall not mention them. In fact, if $\Im$ is the interpretation which assigns $\mathbf{L}$ to $\mathbf{X}$ and L to X, then we shall sometimes denote $\Im \models \mathbf{w}[s]$ by simply $(\mathbf{L}, L) \models \mathbf{w}[s]$.

Finally we are in a position to define a map on a predicate domain.

4.5. Definition. Let $\| \mathfrak{L}, A, B, P \|$ and $\| \mathfrak{R}, C, D, Q \|$ be predicate cpos, where $\mathfrak{L}$ is a retract over $\mathbf{V}^m$ and $\mathfrak{R}$ is a retract over $\mathbf{V}^n$. Then given $T : \mathbf{V}^m \to \mathbf{V}^n$, and $\mathbf{w} \in \Phi(\mathbf{X} : \mathbf{V}^m, X)$ whose free variables over U are contained in $v_1, \cdots, v_n$, the predicator, $||T, \mathbf{w}||$, is a map on $\| \mathcal{L}, A, B, P \|$, given by:

$$||T, \mathbf{w}||(\mathbf{L}, L) = (T(\mathbf{L}), \{a : T(\mathbf{L}) \mid (\mathbf{L}, L) \models \mathbf{w}[a]\}) \tag{4.2}$$

where $(\mathbf{L}, L) \in \| \mathcal{L}, A, B, P \|$. Clearly $||T, \mathbf{w}|| \in \| \mathcal{L}, A, B, P \| \mapsto\to \| \mathcal{R}, C, D, Q \|$ iff for all $(\mathbf{L}, L) \in \mathcal{L}$,

$$||T, \mathbf{w}||(\mathbf{L}, L) \in \| \mathcal{R}, C, D, Q \|$$

4.6. Theorem. *Suppose $||T, \mathbf{w}||$ is a map on $\| \mathcal{L}, A, B, P \|$. Then if $||T, \mathbf{w}||$ is monotonic it is continuous.*

Proof. Follows from (4.1), **Fact** 4.3 and the continuity of T. ∎

5. Goal Generator

Let the predicator $||T, \mathbf{w}||$ be a map on the predicate domain $\| \mathcal{L}, A, B, P \|$, where $\mathcal{L}$ is a retract over $\mathbf{V}^m$, $T \in \mathbf{V}^m \to \mathbf{V}^n$ and $\mathbf{w} \in \Phi(\mathbf{X} : \mathbf{V}^m, X)$. We describe a crucial algorithm which generates *sufficient* goals to guarantee monotonocity (and hence continuity by **Thm 4.6**) of $||T, \mathbf{w}||$. The algorithm is crucial because goals generated by it can be proved within the LCF formalism. Assume $(\mathbf{L}, L), (\underline{\mathbf{L}}, \underline{L}) \in \| \mathcal{L}, A, B, P \|$, where $(\mathbf{L}, L) \sqsubseteq (\underline{\mathbf{L}}, \underline{L})$. We want to prove $||T, \mathbf{w}||(\mathbf{L}, L) \sqsubseteq ||T, \mathbf{w}||(\underline{\mathbf{L}}, \underline{L})$. Using (4.2) and **Def** 4.1 (2), this follows if

(1) $T(\mathbf{L}) \sqsubseteq T(\underline{\mathbf{L}})$.

(2) For all $a \in T(\mathbf{L})$, $(\mathbf{L}, L) \models \mathbf{w}[a]$ implies $(\underline{\mathbf{L}}, \underline{L}) \models \mathbf{w}[a]$.

(3) For all $\underline{a} \in T(\underline{\mathbf{L}})$, $(\underline{\mathbf{L}}, \underline{L}) \models \mathbf{w}[\underline{a}]$ implies $(\mathbf{L}, L) \models \mathbf{w}[T(\mathbf{L})\underline{a}]$.

The first one is trivial. To prove (2) we generate the goal '$(\mathbf{L}, L) \models \mathbf{w}[a : T(\mathbf{L})]$ **implies** $(\underline{\mathbf{L}}, \underline{L}) \models \mathbf{w}[a]$', and simplify it using the following algorithm **Reduce1**. Note that the variable a in the goal is implicitly quantified over $T(\mathbf{L})$. To prove (3) we generate the goal '$(\underline{\mathbf{L}}, \underline{L}) \models \mathbf{w}[\underline{a} : T(\underline{\mathbf{L}})]$ **implies** $(\mathbf{L}, L) \models \mathbf{w}[T(\mathbf{L})\underline{a}]$' and simplify it using the following algorithm **Reduce2**. **Reduce1** and **Reduce2** are mutually recursive.

Reduce1 reduces a goal of the form '$(\mathbf{L}, L) \models \mathbf{w}[z]$ **implies** $(\underline{\mathbf{L}}, \underline{L}) \models \mathbf{w}[y]$' where $z \sqsubseteq y$.

Reduce2 reduces a goal of the form '$(\underline{\mathbf{L}}, \underline{L}) \models \mathbf{w}[v]$ **implies** $(\mathbf{L}, L) \models \mathbf{w}[u]$' where $u \sqsubseteq v$.

Reduce1: Let $\mathbf{w} \in \Phi(\mathbf{X} : \mathbf{V}^m, X)$ be a wff whose free variables over U are among $v_1, \ldots, v_n$. Let $\| \mathcal{L}, A, B, P \|$ be a predicate cpo where $\mathcal{L}$ is a retract over $\mathbf{V}^m$. Let $(\mathbf{L} : \mathbf{V}^m, L), (\underline{\mathbf{L}} : \mathbf{V}^m, \underline{L}) \in \| \mathcal{L}, A, B, P \|$ be such that $(\mathbf{L}, L) \sqsubseteq (\underline{\mathbf{L}}, \underline{L})$.

Let $\Im$ be the interpretation which assigns $\mathbf{L}$ to $\mathbf{X}$ and L to X. Let $\underline{\Im}$ be the interpretation which assigns $\underline{\mathbf{L}}$ to $\mathbf{X}$ and $\underline{L}$ to X. Let the goal be

$$\Im \models \mathbf{w}[z] \text{ implies } \underline{\Im} \models \mathbf{w}[y] \text{ i.e, } (\mathbf{L}, L) \models \mathbf{w}[z] \text{ implies } (\underline{\mathbf{L}}, \underline{L}) \models \mathbf{w}[y]$$

where $z : \mathbf{U}^n \sqsubseteq y : \mathbf{U}^n$. Then inductively we reduce this goal to the simpler goals. We shall justify only a few cases. The others are similar.

(1) $\mathbf{w} =$ "$\tau \in P$":
Then goal reduces to showing '$\mathbf{P}^{\Im}(\tau^{\Im}[z]) \in P^{\Im}$ implies $\mathbf{P}^{\underline{\Im}}(\tau^{\underline{\Im}}[y]) \in P^{\underline{\Im}}$'. Let $(r_1, \ldots, r_k) = \mathbf{P}^{\Im}(\tau^{\Im}[z])$ and $(\underline{r}_1, \ldots, \underline{r}_k) = \mathbf{P}^{\underline{\Im}}(\tau^{\underline{\Im}}[z])$, for some k. Then for all j, $r_j \sqsubseteq \underline{r}_j$. Assume that $\mathbf{P}^{\Im}(\tau^{\Im}[z] \in P^{\Im}$. As $(\mathbf{P}^{\Im}, P^{\Im}) \sqsubseteq (\mathbf{P}^{\underline{\Im}}, P^{\underline{\Im}})$, this is immediate because $(\mathbf{L}, L) \sqsubseteq (\underline{\mathbf{L}}, \underline{L})$, it follows that $(r_1, \cdots, r_m) = \mathbf{P}^{\Im}(\tau^{\Im}[z]) \in P^{\underline{\Im}}$. Suppose $P^{\underline{\Im}}$ is upward closed in the indices of some index set G. (For example, if P is the predicate symbol '$\sqsubseteq$', we know that $P^{\underline{\Im}} = \sqsubseteq$ is upward closed in the second index. If P is the predicate symbol X then, because $(\underline{\mathbf{L}}, \underline{L}) \in \| \mathsf{L}, A, B, P \|$, we know by **Def 4.1** that $P^{\underline{\Im}} = \underline{L}$ is upward closed in all indices of A). Then to show $(\underline{r}_1, \ldots, \underline{r}_k) \in P^{\underline{\Im}}$ it suffices to prove that $r_j = \underline{r}_j$ for all $j \notin G$. Hence for all $j \notin G$ we generate the goal

$$\langle \mathbf{P}^{\Im}(\tau^{\Im}[z]) \rangle_j = \langle \mathbf{P}^{\underline{\Im}}(\tau^{\underline{\Im}}[y]) \rangle_j$$

(2) (a) $\mathbf{w} =$ "$\forall q \in S.\ \mathbf{g}$":
The goal reduces to showing '$(\forall a \in S^{\Im}.\ \Im \models \mathbf{g}[a, z])$ implies $(\forall \underline{a} \in S^{\underline{\Im}}.\ \underline{\Im} \models \mathbf{g}[\underline{a}, y])$'. But for all $\underline{a} \in S^{\underline{\Im}}$, $S^{\Im}\underline{a} \in S^{\Im}$. Hence it suffices to show: $\forall \underline{a} \in S^{\underline{\Im}}.\ (\Im \models \mathbf{g}[S^{\Im}\underline{a}, z]$ implies $\underline{\Im} \models \mathbf{g}[\underline{a}, y])$. Therefore we generate the *sufficient* goal

$$\Im \models \mathbf{g}[S^{\Im}\underline{a}, z] \text{ implies } \underline{\Im} \models \mathbf{g}[\underline{a} : S^{\underline{\Im}}, y]$$

Note that $(S^{\Im}\underline{a}, z) \sqsubseteq (\underline{a}, y)$, for all $\underline{a} \in S^{\underline{\Im}}$. Hence the goal can be further reduced by **Reduce1** as its induction hypothesis is satisfied.

(b) $\mathbf{w} =$ "$\exists q \in S.\ \mathbf{g}$":
generate the goal

$$\Im \models \mathbf{g}[a : S^{\Im}, z] \text{ implies } \underline{\Im} \models \mathbf{g}[a, y]$$

Reduce it further by **Reduce1** recursively.

(3) (a) $\mathbf{w} =$ "$\mathbf{h} \Rightarrow \mathbf{g}$" :
We have to prove that:

($\Im \models$ h$[z]$ **implies** $\Im \models$ g$[z]$) **implies** ($\underline{\Im} \models$ h$[y]$ **implies** $\underline{\Im} \models$ g$[y]$) It is easy to see the sufficiency of the goals:

$$\underline{\Im} \models \mathrm{h}[y] \textbf{ implies } \Im \models \mathrm{h}[z] \quad \text{and}$$
$$\Im \models \mathrm{g}[z] \textbf{ implies } \underline{\Im} \models \mathrm{g}[y]$$

These can be reduced further by **Reduce2** and **Reduce1** respectively.
(b)w = "h $\wedge$ g"or "h $\vee$ g":
generate the goals:

$$\Im \models \mathrm{h}[z] \textbf{ implies } \underline{\Im} \models \mathrm{h}[y] \quad \text{and}$$
$$\Im \models \mathrm{g}[z] \textbf{ implies } \underline{\Im} \models \mathrm{g}[y]$$

These could be reduced further by **Reduce1**.

Reduce2: Let the goal be '$\underline{\Im} \models$ w$[v]$ **implies** $\Im \models$ w$[u]$' where w, $\Im$, $\underline{\Im}$ are as in **Reduce1** and $u : \mathrm{U}^n \sqsubseteq v : \mathrm{U}^n$. **Reduce2** simplifies the goal recursively. It is very similar to **Reduce1** and hence we shall not discuss it.

5.1. Example . Let us now prove the existence of a solution to (2.1). Remember that **D**, the domain of values, and **C**, the domain of environments, are the least solutions to the equations $\mathbf{D} = \mathbf{C} \to \mathbf{B}$ and $\mathbf{C} = Ide \to \mathbf{D}$. **E** is the domain of expressions and **F** is the domain of contexts, $Ide \to \mathbf{E}$. The general plan is as follows. We first construct the predicators $||T, \mathrm{w}||$ and $||S, \mathbf{f}||$ where

$$T : \mathbf{V}^2 \to \mathbf{V}^2 = \lambda(\mathrm{C}', \mathrm{F}').(\mathrm{C}' \to \mathbf{B}, \mathbf{E})$$
$$\mathrm{w}[d : \mathbf{U}, e : \mathbf{U}] \in \Phi(\mathrm{X} : \mathbf{V}^2, X) = \forall (c, f) \in X.apply\ d\ c \sqsubseteq \mathcal{O}\ e\ f$$
$$S : \mathbf{V}^2 \to \mathbf{V}^2 = \lambda(\mathrm{D}', \mathrm{E}').(Ide \to \mathrm{D}', Ide \to \mathrm{E}')$$
$$\mathbf{f}[c, f] \in \Phi(\mathrm{Y} : \mathbf{V}^2, Y) = \forall I \in Ide.(apply\ c\ I, apply\ f\ I) \in Y$$

Here $apply : \mathbf{U} \to \mathbf{U} \to \mathbf{U} = \lambda x \lambda y.(j_{\to}\ x)\ y$ where $j_{\to}$ is the projection from **U** to $\mathbf{U} \to \mathbf{U}$. Strictly speaking $\mathcal{O}$ here is the extension to $\mathbf{U} \to \mathbf{U} \to \mathbf{U}$ of the operational semantics, $\mathcal{O}$, given in **section** 2. Such an extension can be carried out uniquely.

Next we shall construct predicate cpos such that the above predicators will form *continuous* functions between them. That is, we choose certain retracts $\mathfrak{L}$, $\mathfrak{R}$ over $\mathbf{V}^2$ and predicates P and Q such that:

$$||T, \mathbf{w}|| \in \|\ \mathfrak{R}, \{2\}, \{1\}, Q\ \| \to \|\ \mathfrak{L}, \{2\}, \{1\}, P\ \|$$

$$||S, \mathbf{f}|| \in \|\ \mathfrak{L}, \{2\}, \{1\}, P\ \| \to \|\ \mathfrak{R}, \{2\}, \{1\}, Q\ \|$$

Note that we are considering only those relations which belong to $\mathrm{Cl}(\{2\}, \{1\})$. Continuity of the predicators will allow us to take the following fixed point.

$$(\mathbf{L}^*, L^*), (\mathbf{R}^*, R^*) = FIX(\lambda(\mathbf{L}, L), (\mathbf{R}, R).(||T, \mathrm{w}||\ (\mathbf{R}, R), ||S, \mathbf{f}||(\mathbf{L}, L))). \quad (5.1)$$

It will turn out that

$$\mathbf{L}^* = \mathbf{D} \times \mathbf{E}\ ,\ \mathbf{R}^* = \mathbf{C} \times \mathbf{F} \tag{5.2}$$

Then using (5.1), (5.2) and (4.2) it follows that (2.1) gets satisfied if we let $\theta = L$ and $\eta = R$. Of course several conditions ought to be satisfied before this can be done. Firstly to guarantee that the predicators are well defined we require that:

$$\text{for all } (\mathbf{R}, R) \in \|\ \mathfrak{R}, C, D, Q\ \|,\ \|T, \mathbf{w}\|(\mathbf{R}, R) \in \|\ \mathfrak{L}, A, B, P\ \| \tag{5.3}$$
$$\text{for all } (\mathbf{L}, L) \in \|\ \mathfrak{L}, A, B, P\ \|,\ \|S, \mathbf{f}\|(\mathbf{L}, L) \in \|\ \mathfrak{R}, C, D, Q\ \| \tag{5.4}$$

Secondly the algorithm given above generates some goals to ensure monotonicity (and hence continuity by thm 4.6) of the predicators. (The goals given below were the ones generated by the implemented version. It takes care of some trivial goals, carries out primitive simplification and generates the goals which are more readable.) The goals for $\|T, \mathbf{w}\|$ are:

$$\forall(\mathbf{C}', \mathbf{F}'), (\underline{\mathbf{C}}', \underline{\mathbf{F}}') \in \mathfrak{R}.\ (\mathbf{C}', \mathbf{F}') \sqsubseteq (\underline{\mathbf{C}}', \underline{\mathbf{F}}') \Rightarrow$$
$$\forall \underline{c} \in \underline{\mathbf{C}}', d \in \mathbf{C}' \rightarrow \mathbf{B}.\ \mathbf{B}(apply\ d\,(\mathbf{C}'\,\underline{c})) = \mathbf{B}(apply\ d\,\underline{c}) \tag{5.5}$$
$$\forall \underline{e} \in \mathbf{E},\ f \in \mathbf{F}'.\ \mathbf{B}(\mathcal{O}(\mathbf{E}\,\underline{e})\,f) = \mathbf{B}(\mathcal{O}\,\underline{e}\,f) \tag{5.6}$$

Similarly goals for $\|S, \mathbf{f}\|$ are:

$$\forall(\mathbf{D}', \mathbf{E}'), (\underline{\mathbf{D}}', \underline{\mathbf{E}}') \in \mathfrak{L}.\ (\mathbf{D}', \mathbf{E}') \sqsubseteq (\underline{\mathbf{D}}', \underline{\mathbf{E}}') \Rightarrow$$
$$\forall c \in \mathbf{Ide} \rightarrow \mathbf{D}',\ I \in \mathbf{Ide}.\ \mathbf{D}'(apply\ c\,(\mathbf{Ide}\ I)) = \mathbf{D}'(apply\ c\ I) \tag{5.7}$$
$$\forall \underline{f} \in \mathbf{Ide} \rightarrow \underline{\mathbf{E}}',\ I \in \mathbf{Ide}.\ \mathbf{E}'(apply\ ((\mathbf{Ide} \rightarrow \mathbf{E}')\,\underline{f})\ I) = \mathbf{E}'(apply\ \underline{f}\ I) \tag{5.8}$$

Note that (5.5) to (5.8) are goals which can be handled within the LCF formalism (remember that $x \in f$,where f is retract, is a shortform for $fx = x$.) It can be easily shown that the goals (5.5) to (5.7) are satisfied irrespective of the choice of $\mathfrak{L}$ and $\mathfrak{R}$. In fact in the implemented system there is a tactic STANDARDTAC which automatically proved them. In the deduction it made use of definition of $\rightarrow$: $\mathbf{V} \times \mathbf{V} \rightarrow \mathbf{V}$(we didn't give it here) and simple properties of finitary projections. That leaves us with (5.8) and some goals corresponding to (5.3) and (5.4) (we shall see them later). We shall construct $\mathfrak{R}$,$\mathfrak{L}$ as the fixed points of certain higher order functionals. *Then these goals can be proved in LCF using fixed point induction.* In the next section we shall give a technique for constructing these higher order functions.

6. Higher Order Functions

Suppose $T \in \mathbf{V}^m \rightarrow \mathbf{V}^n$. Can we always find a $\hat{T} : (\mathbf{V}^m \rightarrow \mathbf{V}^m) \rightarrow (\mathbf{V}^n \rightarrow \mathbf{V}^n)$ such that for given a retract $\mathfrak{L} : \mathbf{V}^m \rightarrow \mathbf{V}^m$, $\hat{T}(\mathfrak{L})$ is also a retract and $|\hat{T}(\mathfrak{L})| = T(|\mathfrak{L}|)$? The answer is no. For example,let $proj_i^n : \mathbf{V}^n \rightarrow \mathbf{V} =$

$\lambda(x_1, \cdots, x_n).x_i$. It can be shown that there exists a retract $\mathsf{L} : \mathbf{V}^n \to \mathbf{V}^n$ such that $proj_i^n(|\mathsf{L}|)$ is not even a cpo. Obviously some restrictions are needed.

Let $\overrightarrow{proj_i^n} : \mathbf{V} \to \mathbf{V}^n = \lambda x_i.(\perp, \cdots, x_i, \cdots, \perp)$. A retract $\mathsf{L} : \mathbf{V}^n \to \mathbf{V}^n$ will be called *well behaved (wb)* if for all i, $(proj_i^n \circ \mathsf{L} \circ \overrightarrow{proj_i^n}) : \mathbf{V} \to \mathbf{V}$ is a retract and $|proj_i^n \circ \mathsf{L} \circ \overrightarrow{proj_i^n}| = proj_i^n(|\mathsf{L}|)$. Then it can be easily shown that the set of *well behaved* retracts on $\mathbf{V}^n$ forms a cpo; call it $\mathcal{W}b(\mathbf{V}^n)$.

6.1. Definition. A pair $\langle T : \mathbf{V}^m \to \mathbf{V}^n, \overline{T} : \mathbf{V}^n \to \mathbf{V}^m \rangle$ is called a dual if given $\mathsf{L} : \mathcal{W}b(\mathbf{V}^m)$, $(T \circ \mathsf{L} \circ \overline{T}) \in \mathcal{W}b(\mathbf{V}^n)$ and $|T \circ \mathsf{L} \circ \overline{T}| = T(|\mathsf{L}|)$. We shall call $\overline{T}$ a right inverse of T. ∎

It is obvious that $< proj_i^n, \overrightarrow{proj_i^n} >$ is a dual.

6.2. Definition. $T : \mathbf{V}^m \to \mathbf{V}^n$ is said to be *nice* if it can be constructed from $\to, \times, + : \mathbf{V} \times \mathbf{V} \to \mathbf{V}$, $I : \mathbf{V} \to \mathbf{V}$, constants $\mathbf{C} : \mathcal{W}b(\mathbf{V}^k)$ (k is arbitary), $\circ$, fnpair, $proj_i^k$ (k is arbitary). Here $\circ$ is the composition, $fnpair\, f\, g = \lambda x.(f\,x, g\,x)$, and I is the identity function. ∎

6.3. Theorem. *If T is nice then it has a right inverse* $\overline{T}$.

Proof. Though the proof is nontrivial it is technical so we shall omit it ∎

The result we are interested in is:

6.4. corollary. If $T : \mathbf{V}^m \to \mathbf{V}^n$ is nice, there exists $\hat{T} : \mathcal{W}b(\mathbf{V}^m) \to \mathcal{W}b(\mathbf{V}^n)$ such that $|\hat{T}(\mathsf{L})| = T(|\mathsf{L}|)$.

Proof. Let $\hat{T} = \lambda \mathsf{L}.\ T \circ \mathsf{L} \circ \overline{T}$. ∎

Nice functions form a fairly rich class. Most of the functions on finitary projections used to build reflexive domains are nice. For example T and S in example 5.1 are nice. In addition we have at our disposal the following simple combinators:

(1) *strict* : $\mathcal{W}b(\mathbf{V}^n) \to \mathcal{W}b(\mathbf{V}^n)$, where strict is defined by : $strict\, f = \lambda x.$ if $x = \perp$ then $\perp$ else $f(x)$.

(2) $\otimes : \mathcal{W}b(\mathbf{V}^n) \times \mathcal{W}b(\mathbf{V}^m) \to \mathcal{W}b(\mathbf{V}^{n+m})$ where $\otimes$ be defined by $(f \otimes g)(x, y) = (f\,x, g\,y)$. ∎

6.5. (**Continuation of example 5.1**). We shall choose now L, R, P, Q such that the conditions stated in **Example 5.1** are satisfied. Conditions (5.5) to (5.8) guarantee monotonocity of $||T, \mathbf{w}||$ and $||S, \mathbf{f}||$. We have already seen how (5.5) to (5.7) could be proved. Hence only (5.8) remains to be proved. We state it again.

$$\begin{aligned}&\forall (\mathbf{D}', \mathbf{E}'), (\underline{\mathbf{D}}', \underline{\mathbf{E}}') \in \mathsf{L}.\ (\mathbf{D}', \mathbf{E}') \sqsubseteq (\underline{\mathbf{D}}', \underline{\mathbf{E}}') \Rightarrow \\ &\forall \underline{f} \in \mathbf{Ide} \to \underline{\mathbf{E}}', I \in \mathbf{Ide}.\ \mathbf{E}'(apply\,((\mathbf{Ide} \to \mathbf{E}')\, \underline{f})\, I) = \mathbf{E}'(apply\, \underline{f}\, I)\end{aligned} \tag{6.1}$$

By the monotonocity of $||T, \mathbf{w}||$, $||S, \mathbf{f}||$ (above conditions will guarantee that) and **corollary** 6.4, (5.3) and (5.4) can be reduced to:

$$
\begin{array}{lll}
(a) & \hat{T}(\mathcal{R}) \preceq \mathcal{L} \text{ and } \hat{S}(\mathcal{L}) \preceq \mathcal{R} & (6.2) \\
(b) & \text{given } G \subseteq |\mathbf{G}|, \text{ if } G \in \mathrm{Cl}(\{2\},\{1\}) & \\
 & \quad \text{then } |\mathbf{w}|(G) \text{ and } |\mathbf{f}|(G) \in \mathrm{Cl}(\{2\},\{1\}) & (6.3) \\
(c) & ||T,\mathbf{w}||(\bot_{\mathcal{R}}, Q) \sqsupseteq (\bot_{\mathcal{L}}, P),\ ||S,\mathbf{f}||(\bot_{\mathcal{L}}, P) \sqsupseteq (\bot_{\mathcal{R}}, Q) & (6.4)
\end{array}
$$

where $|\mathbf{w}|(G) = \{x : T(\mathbf{G}) \mid (\mathbf{G}, G) \models w[x]\}$, and $|\mathbf{f}|(G)$ is similarly defined. By the preceding theory we are justified in choosing $\mathcal{L}$ and $\mathcal{R}$ as the fixed points:

$$
\begin{array}{l}
\mathcal{L} : \mathcal{W}b(\mathbf{V}^2), \mathcal{R} : \mathcal{W}b(\mathbf{V}^2) = \\
\quad FIX(\lambda(\mathcal{L}', \mathcal{R}').((strict(\hat{T}\,\mathcal{R}')), \hat{S}(\mathcal{L}'))
\end{array} \qquad (6.5)
$$

Then it can be shown in LCF that $\bot_{\mathcal{L}} : \mathbf{V} \times \mathbf{V} = (\bot, \bot)$, $\bot_{\mathcal{R}} = (\bot, \bot)$. The fixed point sets of $\mathcal{L}$ and $\mathcal{R}$ look simply as shown in fig 6.1.

(D, E) • • (C, Ide → E) = (C, F)

(⊥ → B, E) • • (Ide → (⊥ → B), Ide → E)

(⊥, ⊥) • • (⊥, ⊥)

|𝓛| |𝓡|

fig 6.1

Let $P(\subseteq | \bot_{\mathcal{L}} |) = \{(\bot, \bot)\}$ and $Q(\subseteq \bot_{\mathcal{R}}) = \{(\bot, \bot)\}$. Then (6.3) and (6.4) follow trivially. *But they can not be proved within the LCF formalism, hence have to be proved separately.* (6.2) can be proved almost automatically by using the standard set of tactics provided by the implemented system. These tactics use straightforward properties of *wb* retracts , right inverses etc which we have not stated in this paper. (6.1) can be shown using *fixed point induction* (refer to (6.5)). Thus all the conditions stated in **Example 5.1** can be proved. The fixed point operation in (5.1) is then justified. Finally (5.2) follows by one more fixed point induction. Now the recursive predicates are obtained as was indicated in **Example** 5.1. Thus except for (6.3) and (6.4) everything else has been mechanized.

As a side remark it is interesting to note what happens if $\sqsubseteq$ in (2.1) is replaced by $=$ or $\sqsupseteq$. In that case the goals which are generated by the LCF interface can no longer be proved for *all* $\mathcal{O}$. But we have already proved that in this case there exists $\mathcal{O}$ for which the predicates do not exist. Thus though the goals generated by the LCF interface are by no means *necessary,* they seem to be necessary in some *weak* sense.

7. Implementation

Now the basic design of the system should be clear.It consists of two parts; an LCF interface and a working environment.

7.1. LCF interface. The input to this stage is a representation of the predicate domains and predicators. A function on predicates is specified by a wff. The output of this stage is the set of goals which will guarantee the continuity of the predicators. We have already discussed how this is done. These goals are proved in the working environments.

7.2. Working Environment. The working environment provides a hierarchy of LCF theories and a standard set of tactics. The hierarchy consists of theories of universal domain, finitary projection, duals, well behaved retracts etc. Tactics are provided which take care of goals which frequently arise. They are programmed in ML using well known programming principles such as tacticals and simple resolution. Though the implementation is nontrivial, these aspects of ML programming and LCF are well understood, hence we shall not discuss them here (for example, see [1]).

8. Example

We shall consider one bigger example. This is taken from [2]. It is was used to show correctness of a LISP implementation. Let $\mathbf{D}$, a domain of values, and $\mathbf{C}$, a domain of environments, be the least solutions of the equations $\mathbf{D} = \mathbf{C} \rightarrow \mathbf{B}$ and $\mathbf{C} = \mathbf{Ide} \rightarrow \mathbf{D}$ as in **section 2**. We want to show the existence of recursively defined predicates:

$$\begin{aligned} &\theta \subseteq \mathbf{D} \times \{Z \mid Z \subseteq \mathbf{Ide}\} \\ &=^Z \subseteq \mathbf{C} \times \mathbf{C} \qquad\qquad \text{(one for each } Z \subseteq \mathbf{Ide}) \end{aligned}$$

Where intuitively $(v, Z) \in \theta$ means the free variables of v are included in Z and $(\rho, \rho') \in =^Z$ means ρ and ρ' "strongly" agree for all $z \in Z$. Formally we require that:

$$\begin{aligned} \theta &= \{(v, Z) \mid \forall Y, \rho, \rho'.(Z \subseteq Y \Rightarrow (\rho =^Y \rho' \Rightarrow v\,\rho = v\,\rho'))\} \\ =^Z &= \{(\rho, \rho') \mid \forall z \in Z.\rho\, z = \rho' z \text{ and } (\rho\, z, Z) \in \theta\} \end{aligned}$$

By rearrangement we get :

$$\begin{aligned} \theta = \{(v, Z) \mid \forall Y, \rho, \rho'.Z \subseteq Y \Rightarrow \\ (\forall y \in Y.\rho\, y = \rho' y \text{ and } (\rho\, y, Y) \in \theta) \Rightarrow \\ v\,\rho = v\,\rho'\} \end{aligned}$$

We shall show the existence of θ, then that of $=^Z$ easily follows.

Let Pide be the flat domain of subsets of Ide. We show that $\|T, \mathbf{w}\| \in \| \mathcal{L}, \{\}, \{\}, P \| \rightarrow \| \mathcal{L}, \{\}, \{\}, P \|$. where:

$$
\begin{aligned}
&T = (\lambda(\mathbf{D}', \mathbf{P}').((\mathbf{I}de \rightarrow \mathbf{D}') \rightarrow \mathbf{B}, \mathbf{P}ide) \\
&\mathbf{w}[v : \mathbf{U}, Z : \mathbf{U}] \in \Phi(\mathbf{X} : \mathbf{V}^2, X) = \\
&\quad \forall Y \in \mathbf{P}ide. \forall \rho \in \mathbf{I}de \rightarrow (proj_1^2(\mathbf{X})). \forall \rho' \in \mathbf{I}de \rightarrow (proj_1^2(\mathbf{X})).\ Z \subseteq Y \Rightarrow \\
&\quad\quad ((\forall y \in \mathbf{I}de.y \text{ memberof } Y \Rightarrow \\
&\quad\quad\quad (\rho\, y = \rho'\, y \wedge (\rho\, y, Y) \in X)) \Rightarrow v\, \rho = v\, \rho') \\
&\mathcal{L} = FIX(\lambda \mathcal{L}'.(strict(\hat{proj}_1^2(\hat{T}(\mathcal{L}'))) \otimes (\lambda \mathbf{P}'.\mathbf{P}ide))) \\
&P = \{(\perp, Q) \mid Q \in \mathbf{P}ide\}
\end{aligned}
$$

The wff **w** contains two constant predicate symbols 'memberof' and '$\subseteq$'.The symbol '$\subseteq$'is interpreted over $\mathbf{P}ide \times \mathbf{P}ide$ and 'memberof' is interpreted over $\mathbf{I}de \times \mathbf{P}ide$ in the obvious manner. Sixteen goals were generated by the LCF interface to guarantee continuity of the above predicator. Fifteen were generated by the goal generation algorithm of section 5. All of them were proved by standard tactics automatically except the following:

$$
\forall (\mathbf{D}', \mathbf{P}'), (\underline{\mathbf{D}}', \underline{\mathbf{P}}') \in \mathcal{L}.\ (\mathbf{D}', \mathbf{P}') \sqsubseteq (\underline{\mathbf{D}}', \underline{\mathbf{P}}') \Rightarrow (\forall Y \in \mathbf{P}ide.\ \mathbf{P}'\, Y = \underline{\mathbf{P}}'\, Y)
$$

This can be proved very easily using fixed point induction (refer to defn of $\mathcal{L}$ above) by the user. The remaining goal was '$\hat{T}(\mathcal{L}) \preceq \mathcal{L}$'. This could be proved *almost* automatically. Once the continuity of the predicator is proved, existence of the predicate follows immediately as in **Example** 5.1.

9. Scope And Conclusion

It is hoped that this work will meet a long felt need for a mechanical aid in proving the existence of recursively defined predicates. We would like to make few comments. Everything we have said so far easily generalizes to the case when predictors have more than one argument. Secondly, complicated predicators can always be constructed from simpler predicators. Hopefully continuity of these simpler ones would have been proved already. In the general case retracts underlying predicate cpos do not look linear as in fig 6.1. Due to lack of space we could not discuss more examples, but it should suffice to say that existence of all the predicates in [2,9,7] can be proved in the present system. Also corresponding to every relational functor in [5], it is easy to construct a predicator and prove its continuity in the present system. In the next paper we hope to report on these and bigger predicates. Right now the major bottleneck of the system seems to be the speed.

10. Acknowledgement

This work was done under the guidance of Prof. Dana Scott. I would like to thank him for his insights and advice. Special thanks to Steve Brookes and Glynn

Winskel for many helpful discussions and a great help in improving readability of the paper. Thanks also to Roberto Minio and Bill Scherlis for helping me with TEX.

11. Bibliography

(1) Cohn,Avra: *The Equivalence of Two Semantic Definitions: A Case Study In LCF*;
Internal Report, University Of Edinburgh Report, (1981).

(2) Gordon, Michael: *Towards a Semantic Theory of Dynamic Binding*; Memo AIM-265, Computer Science Department, Stanford University.

(3) Milne,Robert and Strachey,Christopher: *A Theory of Programming Language Semantics*; Chapman and Hall, London, and John Wiley, New York (1976).

(4) Mosses, Peter: *Mathematical Semantics and Compiler Generation* ; Ph.D thesis, Oxford University Computing Laboratory, Programming Research Group (1975).

(5) Reynolds, J.C.: *On the Relation Between Direct and Continuation Semantics* ; pp. 141-156 of proceedings of the Second Colloquium on Automata, Languages and Programming, Saarbrücken, Springer-verlag, Berlin (1974).

(6) Scott,Dana: *Lectures On a Mathematical Theory of Computation*; Technical Monograph PRG-19 (May 1981). Oxford University Computing Laboratory, Programming Research Group.

(7) Sethi, Ravi and Tang Adrian: *Constructing Call-by-value Continuation Semantics*; Journal of the Association for Computing Machinery,vol 27. No.3. July 1980.pp.580-597.

(8) Stoy,Joseph: *Denotational Semantics: The Scott-Strachey Approach to Programming Language Theory*; (MIT Press, Cambridge, MA, 1977).

(9) Stoy,Joseph: *The Congruence of Two Programming Language Definitions*; Theoretical Computer Science 13 (1981) 151-174. North-Holland Publishing company.

Solving Word Problems in Free Algebras Using Complexity Functions

Alex Pelin

Department of Computer and Information Science
Temple University
Philadelphia, Pa 19122

and

Jean H. Gallier

Department of Computer and Information Science
University of Pennsylvania
Moore School of Electrical Engineering D2
Philadelphia, Pa 19104

Abstract: We present a new method for solving word problems using *complexity functions*. Complexity functions are used to compute normal forms. Given a set of (conditional) equations E, complexity functions are used to convert these equations into reductions (rewrite rules decreasing the complexity of terms). Using the top-down reduction extension Rep induced by a set of equations E and a complexity function, we investigate properties which guarantee that any two (ground) terms t_1 and t_2 are congruent modulo the congruence $\cong_E$ if and only if $Rep(t_1)=Rep(t_2)$. Our method actually consists in computing Rep incrementally, as the composition of a sequence of top-down reduction extensions induced by possibly different complexity functions. This method relaxes some of the restrictions imposed by the Church-Rosser property.

1. Introduction

We present a method for computing the normal form of a term with respect to a set of (conditional) equations. Given a *signature* Σ and a countable set of *variables* V, the initial algebra over Σ is denoted as T_Σ and the free algebra generated by V is denoted as $T_\Sigma(V)$ ($T_\Sigma(V)$ is the algebra of Σ-terms with variables from V). Let E be a set of *equations* (or conditional equations) over $T_\Sigma(V)$. We are interested in solving the *word problem* for T_Σ, that is, the problem of determining for any two terms t_1 and t_2 in T_Σ whether t_1 and t_2 are congruent modulo the congruence $\cong_E$ induced by the set of equations E.

Our method is to compute normal forms for the terms in T_{Σ}. Since the word problem is undecidable in general, we are interested in classes of sets of equations for which normal forms are effectively computable and can be characterized by (deterministic) context-free languages.

Given a set of equations E and a set of ground terms L such that all ground substitution instances of terms occurring in E are in L, a function f : L -> L is a <u>representation function</u> for (L,E), if for all terms t_1, t_2 in L, $t_1 \cong_E t_2$ if and only if $f(t_1) = f(t_2)$. In words, t_1 and t_2 are equivalent modulo $\cong_E$ if and only if their representatives are identical. Our goal is to find a representation function Rep : $T_{\Sigma} \rightarrow T_{\Sigma}$ for (T_{Σ}, E).

Actually, the representation function Rep is computed as the composition $R_n . R_{n-1} \ldots R_1$ of representation functions.

With each function R_i is associated an input set of ground terms L_{i-1} and a set of equations A_{i-1} over L_{i-1}. The equations (conditional equations) in A_{i-1} are called <u>Axioms</u>. Let $Th(A_{i-1})$ be the set of theorems derivable from A_{i-1}. The representation function R_i selects a subset E_i of $Th(A_{i-1})$ which it attempts to "eliminate". The elimination is accomplished by transforming the equations into <u>reduction rules</u>. This is done by defining a <u>complexity function</u> over L_{i-1} which <u>suits</u> E_i (this concept is related to that of a norm-decreasing set of rules in Gallier and Book [4]).

A complexity function f over L_{i-1} assigns to each term t in L_{i-1} an element f(t) of a well-ordered set $(N^k, >)$, where N denotes the set of nonnegative integers, and k is a positive integer. A function $f : L_{i-1} \rightarrow (N^k, >)$ is a complexity function if it is recursive, monotone, and has the subterm property, that is, whenever t_1 is a subterm of t_2 then $f(t_2) > f(t_1)$.

We say that a complexity function f *strongly suits* an equation $l \doteq r$ if for all ground substitutions s, $f(s(l)) > f(s(r))$, or for all ground substitutions s, $f(s(r)) > f(s(l))$. If f suits $l \doteq r$ and for all ground substitutions s, $f(s(l))>f(s(r))$, we say that $l \doteq r$ is a reduction (with respect to f). A complexity function f strongly suits a conditional equation $e_1,\ldots,e_n \Rightarrow e$ if f strongly suits e and for all equations e_i, $1 \le i \le n$, for all ground substitutions s, $f(s(e)) > f(s(e_i))$. The complexity of an instance of an equation $s(l) \doteq s(r)$ is defined as $\max\{f(s(l)),f(s(r))\}$.

A weaker version of the suitability concept which is useful in treating conditional axioms is the concept of *weak suitability* defined below.

A complexity function f *weakly suits* an equation $l \doteq r$ if for all ground substitutions s, $f(s(l)) = f(s(r))$ implies that $s(l)=l(r)$, that is, s(l) and s(r) are identical.

Both strong and weak suitability are used to generate *meta-reductions*. A meta-reduction has the form $(C) \Rightarrow l \rightarrow r$, where C is a recursive predicate in a meta-language which contains names for the well-order $(N^k,>)$ the complexity functions, and l and r are terms in T_Σ. For all such meta-reductions, $f(s(l)) > f(s(r))$ if the condition C evaluates to true. Note that our approach is somewhat similar to that taken in Brand, Darringer and Joyner [2].

Once we have selected the set of theorems E_i and we have found a complexity function f which suits $\langle L_{i-1},E_i \rangle$ (that is, suits all equations or conditional equations in E_i), we define R_i to be the *top-down reduction extension* of the set $\mathbf{R}_i$ of reductions generated by E_i and f.

The <u>top-down reduction extension</u> α of a set of meta-rules $\mathbf{R}_i$ is defined as follows:

For every term t in L_{i-1}, we have the following cases:

(1) If for a meta-rule (C) l -> r and a ground substitution s, s(l)=t and C evaluates to true, then $\alpha(t)=\alpha(s(r))$.

(2) If case 1 does not apply and t has the form $g(t_1,...,t_n)$, then compute recursively $\alpha(t_1),...,\alpha(t_n)$. If for any i, $1\leq i\leq n$, $\alpha(t_i)\neq t_i$, then $\alpha(t)=\alpha(g(\alpha(t_1),...,\alpha(t_n)))$.

(3) If neither of the above cases applies, then $\alpha(t)=t$.

In case 3, t is called an atom.

Let L_i be the range of R_i and let A_i be the system of axioms obtained by applying R_i to the axioms in A_{i-1}. L_i consists of the set of atoms of the top-down reduction extension R_i. If the set of theorems E_i is well chosen, some axioms from A_{i-1} become string identities in A_i and can be eliminated. If it is not possible to eliminate axioms, a proper choice for E_i may produce a simpler syntactic form for the set of terms L_i or for the axioms A_i.

We say that a top-down reduction extension $R_i : \langle L_{i-1},A_{i-1}\rangle \rightarrow \langle L_i,A_i\rangle$ has the <u>α-property for</u> L_{i-1} if for every operator f of rank n and for every n terms $t_1,...,t_n$ in L_{i-1}, for every j, $1\leq j\leq n$, if $f(t_1,...,t_n) \in L_{i-1}$ then $R_i(f(t_1,...,t_j,...,t_n))=R_i(f(t_1,...,R_i(t_j),...,t_n))$.

The representation function Rep can be seen as the composition of the functions given by the sequence

(4) $\langle L_0,A_0\rangle \xrightarrow{R_1} \langle L_1,A_1\rangle \longrightarrow \ldots \xrightarrow{R_n} \langle L_n,\emptyset\rangle$.

It can be shown that if the composition $R_n.R_{n-1}...R_1$ has the α-property, then it has the representation property for $\langle L_0,A_0\rangle$. L_n is then the set of normal forms with respect to Rep.

We will now give criteria for obtaining "useful" theorems in E_i. There are essentially two methods.

The first method is to force local confluence in the set of rules associated with $\langle L_{i-1}, A_{i-1} \rangle$ using the method of critical pairs of Knuth and Bendix [7]. However, the resulting equation is not transformed into a rewrite rule but added to E_i.

The second method is to use conditional meta-rules. The complexity functions guarantee that the reductions in $\mathbf{R}_i$ terminate, that is, that there is no infinite sequence of rewrite steps.

Using conditional meta-rules allows us to deal with systems of rules in which rules which increase the complexity are allowed, provided that the number of applications of such rules is finite.

Our approach allows us to deal with sets of rules which are not necessarily confluent, and we list below some of its advantages.

First, the method is a stepwise refinement technique in which one chooses to eliminate axioms one by one. At each step, the complexity functions may be different. We also do not require that there is a single complexity function which transforms all the equations into rewrite rules as in Knuth and Bendix [7]. For example, if one wants to compute the prenex disjunctive normal form of a first-order sentence, one can proceed as follows:

(1) Relabel variables which appear more than once

(2) Push negations all the way inward

(3) Push quantifiers in front

(4) Distribute $\wedge$ over $\vee$

(5) Eliminate duplicates

If one attempted to perform these five steps in a single transformation, it would be very complicated. With our approach, this breakdown seems quite natural. Also, our approach structures the process of computing normal forms. The complexity functions serve a double role. They can be used as the "control part" of the algorithm for computing normal forms, and they can be used to prove termination. Since there is a certain flexibility in choosing the complexity functions, desired normal forms can be computed.

For example, if we consider a semigroup with a finite number of generators $a_1,\ldots,a_n$ and have the single associativity axiom

(5) $+x+yz \doteq ++xyz$, we can define complexity functions $f,g : L \to N$, where L is the language given by the context-free grammar $\langle\{S\},\{a_1,\ldots,a_n\},P,S\rangle$ whose productions are shown in (6):

(6) $S \to +SS$

$S \to a_1 \mid a_2 \mid \ldots \mid a_n$

(7) $f(a_i) = 1$

$f(+xy) = 1 + 2f(x) + f(y)$

(8) $g(a_i) = 1$

$g(+xy) = 1 + g(x) + 2g(y)$

It is easily verified that both f and g strongly suit (5). For example, $f(+x+yz)=1+2f(x)+1+2f(y)+f(z)= 2+2f(x)+2f(y)+f(z) < 3+4f(x)+2f(y)+f(z)=f(++xyz)$.

The complexity function f transforms (5) into the reduction $R_1=\{++xyz \to +x+yz\}$ and the complexity function g yields the reduction $R_2=\{+x+yz \to ++xyz\}$. The range of β, the top-down reduction extension of R_1, is the language L_1 generated by the grammar whose set of productions is shown in (9):

(9) $S \to a_1 \mid a_2 \mid \ldots \mid a_n \mid +a_1S \mid \ldots \quad +a_nS$.

It can also be shown that β has the α-property, and similarly for the top-down reduction extension of R_2.

2. The Formalism

We will follow the notation and definitions found in Huet and Oppen [6]. Given a finite signature (S,Σ,τ), the initial algebra T_Σ is defined in the usual way ([6]). Terms in T_Σ will be represented in prefix norm. Then, the set $T_\Sigma{}^s$ of terms of sort s is a deterministic context-free language.

Given an S-sorted set of variables V, the free algebra over V is denoted as $T_\Sigma(V)$ and consists of terms with variables.

A *Σ-equation of sort s* is a pair $\langle M,N\rangle$ of terms in $T_\Sigma(V)^s$, and will also be denoted as $M\doteq N$. A *conditional equation* is an expression of the form $e_1,\ldots,e_n\Rightarrow e$, where $e_1,\ldots,e_n,e$ are equations.

A *presentation* is a triple $P=\langle\Sigma,V,E\rangle$, where Σ is a finite signature, V an S-indexed set of variables and E a finite set of (conditional) equations. Substitutions and E-unification are defined as in Huet and Oppen [6].

Let k be a positive integer, and $>$ a recursive well-ordering on N^k. A *complexity function* for Σ is a function $h : T_\Sigma \to N^k$ such that the following conditions hold:

(1) For every operator f such that $\tau(f)=s_1 x\ldots x s_n \to s$, for any n terms $t_1\in T_\Sigma{}^{s_1},\ldots,t_n\in T_\Sigma{}^{s_n}$,

$h(f(t_1,\ldots,t_n))>h(t_i)$, for all i, $1\leq i\leq n$.

Condition (1) is called the *subterm property*

(2) h is a recursive function.

Given a complexity function h, for an operator f such that $\tau(f)=s_1 x \ldots x s_n \rightarrow s$, we say that h is monotone in f if for every i, $1 \leq i \leq n$, and all terms $t_1 \in T_{\Sigma}^{s_1}, \ldots, t_n \in T_{\Sigma}^{s_n}$, the following condition holds:

(3) $h(t_i') > h(t_i)$ implies
$h(f(t_1, \ldots, t_i', \ldots, t_n)) > h(f(t_1, \ldots, t_i', \ldots, t_n))$.

We say that a complexity function is monotone if it is monotone in every operator in Σ.

The concepts of strong and weak suitability were introduced in the Introduction and are not repeated here. A complexity function h strongly (weakly) suits a set of (conditional) equations if it strongly (weakly) suits every (conditional) equation in the set.

Given a set of (conditional) equations E and a complexity function h which suits E (weakly or strongly), we define the set of rules associated with E under h, denoted as R(E,h) or for short R, as follows:

(4) If $l \doteq r \in E$ and h strongly suits $l \doteq r$ then

(i) If for all s, $h(s(l)) > h(s(r))$, then $l \rightarrow r$ is in R(E,h)

(ii) otherwise $r \rightarrow l$ is in R(E,h)

(5) If $l \doteq r \in E$ and h weakly suits $l \doteq r$ then both meta-reductions $(h(l) > h(r)) \Rightarrow l \rightarrow r$ and $(h(r) > h(l)) \Rightarrow r \rightarrow l$ are in R(E,h)

(6) If $e : e_1, \ldots, e_n \Rightarrow l \doteq r \in E$ and h strongly suits e then $(e_1 \wedge \ldots \wedge e_n) \Rightarrow l \rightarrow r$ is in R(E,h) if $h(s(l)) > h(s(r))$ for all ground substitutions s, or $(e_1 \wedge \ldots \wedge e_n) \Rightarrow r \rightarrow l \in R(E,h)$ otherwise.

(7) If $e : e_1, \ldots, e_n \Rightarrow l \doteq r \in E$ and h weakly suits e then both meta-rules $((h(l) > h(r)) \wedge e_1 \wedge \ldots \wedge e_n) \Rightarrow l \rightarrow r$ and $((h(r) > h(l)) \wedge e_1 \wedge \ldots \wedge e_n) \Rightarrow r \rightarrow l$ are in R(E,h).

We say that R(E,h) is functional if for all terms $t \in T_{\Sigma}$ and pairs of meta-rules $(C_1) \Rightarrow l_1 \to r_1$ and $(C_2) \Rightarrow l_2 \to r_2$ in R(E,h), if there exists substitutions s_1 and s_2 such that $s_1(l_1)=s_2(l_2)=t$ and both $C_1(t)$ and $C_2(t)$ are true, then $s_1(r_1)=s_2(r_2)$.

The set R(E,h) is linear if for any meta-rule $(C) \Rightarrow l \to r$, every variable occurs at most once in l. The top-down reduction extension α of R(E,h) has been defined in the Introduction.

3. Technical Results

In Lemmas 1-11, E is assumed to be a finite set of Σ-equations, h is a complexity function which weakly suits E, R is the set of meta-rules associated with E under h, and β is the top-down reduction extension of R to T_{Σ}.

Lemma 1

If h is monotone then β is a recursive function from T_{Σ} to the power-set of T_{Σ}.

Lemma 2

If h is monotone and R is functional then β is a recursive function from T_{Σ} to T_{Σ}.

Lemmas 3-8 characterize the range of β.

Lemma 3

If h is monotone then for all terms t in T_{Σ}, $\beta(\beta(t))=\beta(t)$.

Lemma 4

If h is monotone then for all terms t in T_{Σ}, $h(t)=h(\beta(t))$ if and only if t is an atom (that is, $\beta(t)=t$).

As a corollary we have:

<u>Lemma</u> 5

If h is monotone, the range of β is a recursive set.

<u>Lemma</u> 6

If h is monotone, f is an operator of type $\tau(f)=s_1 x \ldots x s_n \to s$, if $f(v_1,\ldots,v_n)$ and l are not unifiable for any variables $v_1,\ldots,v_n$ (with each v_i of sort s_i) and meta-rule (C) => l->r, then $\beta(f(t_1,\ldots,t_n))=f(\beta(t_1),\ldots,\beta(t_n))$.

If the number of variables of each sort is unbounded, the unification condition can be replaced by:
$f(v_1,\ldots,v_n)$ and l are not unifiable for an n-tuple of distinct variables $v_1,\ldots,v_n$, each v_i of sort s_i.

Lemmas 7 and 8 characterize the range of β using concepts from formal language theory.

A set R of meta-reductions said to be <u>pure</u> if it does not contain conditional meta-rules and h strongly suits E.

<u>Lemma</u> 7

Let R be pure and linear, h be monotone and L a subset of T_Σ such that $\beta(L)$ is a subset of L. If L is accepted by a (deterministic) bottom-up finite tree automaton then $\beta(L)$ is also accepted by a (deterministic) bottom-up finite tree automaton.

As a corollary we obtain:

<u>Lemma</u> 8

If the terms of the language L in Lemma 7 are represented in prefix notation then $\beta(L)$ is accepted by a deterministic pushdown automaton.

The proofs of Lemmas 1-8 can be found in Pelin and Gallier [9]. Lemmas 7 and 8 have a constructive proof, that is, automata accepting the range of β can be effectively constructed.

If R is not linear or contains conditional equations, $\beta(L)$ is not necessarily deterministic or even context-free.

Next, we show some results which relate the α-property presented in the Introduction to the concept of local confluence (Huet [5]).

<u>Lemma</u> 9

Let R be pure and functional and h be monotone. The top-down reduction extension β has the α-property if and only if R is locally confluent.

<u>Lemma</u> 10

Let R be pure and functional and h be monotone. If R is locally confluent, then β has the representation property for E.

By extending the concept of confluence to conditional meta-rules, the purity condition can be dropped. In this case, Lemma 9 becomes:

<u>Lemma</u> 11

Let h be monotone. The top-down reduction extension β is a function with the α-property if and only if R is locally confluent.

<u>Lemma</u> 12

Assume that h is monotone, that β has the α-property, and let $L'=\beta(L)$ be the range of β. Given an equation $l \doteq r$ with terms in L, $\beta(s(l))=\beta(s(r))$ for every substitution s with range L if and only if $\beta(s(l))=\beta(s(r))$ for every substitution s with range L'.

Let us consider a sequence $\langle L_0,A_0\rangle \dashrightarrow \langle L_1,A_1\rangle \dashrightarrow \ldots \dashrightarrow \langle L_n,A_n\rangle$, where for each i, $1 \leq i \leq n$, we have a set of equations E_i transformed by a suitable complexity function h_i into a system of reductions R_i. Let β_i be the top-down reduction extension of R_i to L_{i-1}, $L_i=\beta_i(L_{i-1})$, $A_i=\beta_i(A_{i-1})$ and $\beta=\beta_n \cdot \beta_{n-1} \cdots \beta_1$.

<u>Lemma</u> 13

If conditions (i),(ii),(iii) below are satisfied then β has the representation property for $\langle A_0,L_0\rangle$.

(i) Each R_i is confluent

(ii) $A_n=\emptyset$

(iii) For all i,j, $1 \leq i < j \leq n$, term $t \in L_{i-1}$, meta-reduction (C) => l->r in R_j and tree-address u in dom(t):

If the subtree t/u at u is unifiable with l and C(t/u) holds then $\beta_i(t)=\beta_j(t[u \leftarrow \beta_j(t/u)])$ (where $t[u \leftarrow \beta_j(t/u)]$ denotes the tree obtained by replacing the subtree rooted at u in t with $\beta_j(t/u)$).

Given the initial pair $\langle L_0,A_0\rangle$, we must decide which axioms will be eliminated first. One criteria is that the sets A_i $(i \geq 1)$ should be simple. Sometimes, it is advantageous to split a step into substeps as shown in Lemma 14.

<u>Lemma</u> 14

Let R_1 and R_2 be two sets of reductions over a language L and let β_1 and β_2 be their top-down reduction extensions. Let L_1 be the range of β_1 and L_2 the range of β_2. If β_1 and β_2 have the α-property for L, the union of R_1 and R_2 is locally confluent, and $\beta_1(L_1)$ is a subset of L_1 then $\beta_2 \cdot \beta_1$ has the representation property for the system of equations generated by $R_1 \cup R_2$.

Next, we present examples illustating the above techniques and results.

Example 1 (Stacks of natural numbers with errors)

The set of sorts is S={nat,stack}, Σ={pop,push,top,0,succ,Λ,e,E}, and the typing function is:

$\tau(0)=\tau(e)=$ -> nat

$\tau(\Lambda)=\tau(E)=$ -> stack

τ(push)= stack x nat -> stack

τ(pop) = stack -> stack

τ(top) = stack -> nat

τ(succ) = nat -> nat

Variables of sort nat will be denoted as n_k and variables of sort stack as s_k. The axioms for the data type stack of natural numbers are:

(A) Axioms for stacks

1. push E n $\doteq$ E
2. push s e $\doteq$ E
3. pop E $\doteq$ E
4. pop Λ $\doteq$ E
5. top Λ $\doteq$ e
6. top E $\doteq$ e
7. pop push s 0 $\doteq$ s
8. pop push s n $\doteq$ s => pop push s Succ n $\doteq$ s
9. top push Λ n $\doteq$ n
10. top push s n $\doteq$ n, top push s m $\doteq$ m => top push push s n m $\doteq$ m
11. Succ e $\doteq$ e

The symbol Λ stands for the empty stack, E for the error stack, 0 is the natural number zero, Succ is the successor function, e is the error natural number, push is the push function, pop pops the top of the stack and top returns the top of the stack (without altering the stack). Axioms 1,2,3,6 and 11 state that once an error occurred, any subsequent operation will yield an error as result. Axioms 4 and 5 state that poping or retrieving the top of the empty stack yields an error. Axioms 7 and 8 are weaker versions of pop push s n $\doteq$ s which is not valid for stacks with errors. Axioms 9 and 10 are weaker versions of top push s n $\doteq$ n which holds for s≠E. However, s≠E => top push s n $\doteq$ n is not a Horn formula, and this is the reason why they have been introduced.

It can be verified that the complexity function length (length of a string) strongly suits (A). It can also be verified that the set of reductions obtained from (A) is functional. Hence, by Lemma 2, the top-down reduction extension β is recursive. In order to determine the range of β, Lemma 15 can be proved.

<u>Lemma</u> 15

For all stacks s and natural numbers n:
β(pop push s n) ≠ β(s) => β(n)=e $\wedge$
β(top push s n) ≠ β(n) => β(s)=E $\wedge$
length(β(pop s)) < length(pop s) $\wedge$
length(β(top s)) < length(top s).

The proof proceeds by induction on the length of push s n.

<u>Lemma</u> 16

The range of β is a context-free language.

<u>Lemma</u> 17

β has the representation property for <L_0,A> where L_0 is the language generated by the context-free grammar given below:
S -> Λ | E | push S N | pop S

N -> e | 0 | top S | Succ N

Example 2 (Finite sets of natural numbers)

The set of sorts is S={nat,set}, Σ={add,succ,$\emptyset$,0}, and the typing function is:

τ(0) = -> nat

τ($\emptyset$) = -> set

τ(Succ) = nat -> nat

τ(add) = set x nat -> set

Variables of sort set will be denoted as s_k and variables of sort nat as n_k. The set of terms in prefix notation of the initial algebra T_Σ is denoted as L, and is generated by the following context-free grammar:

S' -> S | N

S -> $\emptyset$ | add S N

N -> 0 | Succ N

The set of axioms for finite sets of natural numbers is given below:

(B) Axioms for finite sets of natural numbers

add add s_0 n_0 n_1 $\doteq$ add add s_0 n_1 n_0

add add s_0 n_0 n_0 $\doteq$ add s_0 n_0

In order to transform (B) into a set of meta-rewrite rules, we define the complexity function h as follows:

h(0)=h($\emptyset$)=1

h(Succ n)=1+h(n)

h(add s_0 n_0)=1+2h(s_0)+h(n)

It can be shown that h weakly suits <L,B>. The system of meta-rules generated from B by h is the following:

$(h(n_0)>h(n_1)) \Rightarrow (\text{add add } s_0\ n_0\ n_1 \Rightarrow \text{add } s_0\ n_1\ n_0)$

$\text{add add } s_0\ n_0\ n_0 \rightarrow \text{add } s_0\ n_0$

There is only one conditional meta-rule because the two conditional meta-rules generated from the first equation in (B) are isomorphic. It can be shown that the system of meta-rules is locally confluent, and therefore, the top-down reduction extension β has the representation property. However, the range of β is not context-free. This happens because the second rule is not linear. In general, it is difficult to relax the conditions of Lemma 7.

Example 3 (Closure system).

The set of sorts is $S=\{N\}$, $\Sigma=\{g,I,+,-\}$, and the typing function is:

$\tau(g)= \quad \rightarrow N$

$\tau(I)=\tau(-)= N \rightarrow N$

$\tau(+) = N \times N \rightarrow N$

The set L_0 of terms in prefix form in the initial algebra T_Σ is given by the following context-free grammar:

S' -> G | S

G -> g | IG

S -> +SS | -S

The set of axioms (C) is given below:

(C) Axioms for a closure system

1. $--x \doteq -x$
2. $-+xy \doteq +-y-x$
3. $+xy \doteq +yx$
4. $+x+yz \doteq ++xyz$

First, we pick $E_1=\{--x \doteq -x,\ -+xy \doteq +-y-x\}$. It can be verified that the complexity function h_1 given below strongly suits E_1:

$h_1(-x)=2h_1(x)$

$h_1(+xy)=1+h_1(x)+h_1(y)$

$h_1(Ix)=1$

$h_1(g)=1$

The reduction system $R_1=\{--x \rightarrow -x,\ -+xy \rightarrow +-y-x\}$ is obtained. Let β_1 be the top-down reduction extension of R_1. It can be shown that R_1 is locally confluent and thus, β_1 has the α-property. By Lemma 8, the range of β_1 is a deterministic context-free language.

It can also be shown that if we take the reductions $R_1'=\{--x \rightarrow -x\}$ and $R_1''=\{-+xy \rightarrow +-y-x\}$ separately, they each have the local confluence property. Let β_1' be the top-down extension of R_1' and β_1'' the top-down extension of R_1''. It can be shown that $\beta_1'(\beta_1''(L_0))$ is a subset of $\beta_1''(L_0)$, but that $\beta_1''(\beta_1'(L_0))$ is not a subset of $\beta_1'(L_0)$. Applying Lemma 14, we have $\beta_1=\beta_1''\cdot\beta_1'$.

We have chosen to eliminate equations (1) and (2) in (C) because, by Lemma 6, $\beta_1(+xy)=+\beta_1(x)\beta_1(y)$. $A_1=\beta_1(A_0)$ is the set:

$+xy \doteq +yx$

$+x+yz \doteq ++xyz$

At this point, we try to eliminate associativity. For that, we define the function h_2 as follows:

$h_2(g)=1$

$h_2(Ix)=1$

$h_2(-x)=1$

$h_2(+xy)=1+h_2(x)+2h_2(y)$

It can be shown that h_2 strongly suits $E_2=\{+x+yz \doteq ++xyz\}$, and we let β_2 be the top-down reduction extension of $R_2=\{+x+yz \to ++xyz\}$. $A_2=\beta_2(A_1)$ has the form $\{\beta_2(+xy)\doteq\beta_2(+yx) \mid$ for all ground terms x and y in $L_1\}$. In order to eliminate the set of axioms A_2, we use reductions obtained by forcing local confluence on the set of rules obtained by orienting the axioms of A_1 from left to right.

Since $+x+yz \Longrightarrow ++xyz$, $+x+yz \Longrightarrow +x+zy \Longrightarrow ++xzy$, we use the equation $++xyz\doteq++xzy$. Let us call the terms of the form g, -g, $I^n g$ or $-I^n g$ ($n\geq 1$) literals. We choose $E_3=\{++xyz\doteq++xzy\} \cup \{ +yz\doteq+zy \mid$ for all literals y,z in $L_2\}$. If we define the complexity function h_3 so that h_3 assigns a natural number to each literal and this assignment is injective, and if $h_3(+xy)=1+2h_3(x)+h_3(y)$, we can show that h_3 weakly suits E_3. The system of meta-rules obtained from h_3 is $R_3=\{(h_3(y)>h_3(z)) \Rightarrow ++xyz\to++xzy,\ (h_3(y)>h_3(z)) \Rightarrow +yz\to+zy\}$. Let β_3 be the top-down reduction extension of R_3 to L_2. Using Lemma 12, it can be shown that $\beta_3(\beta_2(+xy))=\beta_3(\beta_2(+yx))$. Hence, $A_3=\beta_3(A_2)=\emptyset$. It can also be shown that $\beta_3\cdot\beta_2\cdot\beta_1$ has the α-property. Thus, $\beta_3\cdot\beta_2\cdot\beta_1$ has the representation property for $\langle L_0,C\rangle$.

4. Conclusions

Complexity functions play an important role in computing normal forms. Various authors such as Book [1], Gallier and Book [4], Huet [5], Lankford and Ballantyne [8], Knuth and Bendix [7] and others have used complexity functions for generating reductions. Dershowitz [3] and Plaisted [10] use particular complexity functions to prove termination of rewriting systems. In our approach, complexity functions serve both to prove termination and generate reductions, in an incremental fashion.

Of course, our definition of a complexity function is very general and it would be useful to identify which classes of complexity functions can be associated with particular types of axioms. For instance, we have shown that linear functions (functions of the form $h(ft_1,\ldots,t_n))=c_0+c_1h(t_1)+\ldots+c_nh(t_n))$ can be used for associativity and commutativity axioms. However, the distributivity axiom $*x+yz \doteq +*xy*xz$ requires a quadratic function to obtain the reduction rule $*x+yz \rightarrow +*xy*xz$. The following function does the job:

$h(*xy)=2h(x)h(y)$

$h(+xy)=1+h(x)+h(y)$.

Also, some axioms, such as the commutativity axiom for a groupoid with an infinite number of generators, require complexity functions over N^k, for $k>1$.

Complexity functions give an upper bound on the number of steps needed to carry out a computation. For instance, a linear complexity function yields an exponential upper bound.

Complexity functions are also useful to carry out proofs by induction. Finally, connections with the work of Siekman and Szabo [11] should be explored.

References

[1] Book,R., Confluent and other Types of Thue Systems, JACM 29 (1982), 171-182.

[2] Brand,D.,Darringer,J., and Joyner,W., Completeness of Conditional Reductions, IBM Technical Report RC-7404 (1978), T.J. Watson Research Center, Yorktown Heights, N.Y.

[3] Dershowitz,N., Orderings for Term-Rewriting Systems, Theoretical Computer Science 17 (1982), 279-301.

[4] Gallier,J.H. and Book,R.V., Reductions in Tree-Rewriting Systems, to appear in Theoretical Computer Science (1984).

[5] Huet,G., Confluent Reductions: Abstract Properties and Applications to Term-Rewriting Systems, JACM 27(4) (1980), 797-821.

[6] Huet,G. and Oppen,D., Equations and Rewrite Rules, in Formal Languages: Perspectives and Open Problems, R.V. Book, Ed., Academic Press (1980), 349-405.

[7] Knuth,D. and Bendix,P., Simple Word Problems in Universal Algebras, in Computational Problems in Abstract Algebra, Leach J., Ed., Pergamon Press (1970), 263-297.

[8] Lankford,D.S. and Ballantyne,A.M., Decision Procedures for Simple Equational Theories with Permutative Axioms: Complete Sets of Permutative Reductions, Report ATP-37, Department of Mathematics and Computer Science, University of Texas, Austin, Texas (1977).

[9] Pelin A. and Gallier,J.H., Computing Normal Forms Using Complexity Functions over N^k, in preparation.

[10] Plaisted,D., Well-Founded Orderings for Proving Termination of Systems of Rewrite Rules, Report R-78-932, Department of Computer Science, University of Illinois, Urbana, Ill. (1978).

[11] Siekman,J. and Szabo,P., Universal Unification and Regular Equational ACFM Theories, Technical Report, University of Karlsruhe (1981).

Solving a Problem in Relevance Logic with an Automated Theorem Prover

Hans-Jürgen Ohlbach
Institut für Informatik I
University of Kaiserslautern
Postfach 3049
D-675 Kaiserslautern

Graham Wrightson
Victoria University
Dep. of Computer Science
Private Bag
Wellington, New Zealand

Abstract

A new challenging problem for automated theorem provers (ATP) is presented. It is from the field of relevance logic and is known as "converse of contraction". We firstly give some background information about relevance logic and the problem itself and then discuss a proof for the theorem which has been found by the Markgraf Karl Refutation Procedure (MKRP), a resolution based theorem prover, under development at the Universities of Karlsruhe and Kaiserslautern.

For solving this problem we implemented a new method to control the application of "generator clauses" like Exy => Ef(x)f(y). An uncontrolled application of such clauses produces arbitrarily deeply nested and often useless terms f(f(f... which in general cannot be avoided with a global term depth limit because deeply nested terms of more heterogenous structure may occur in the proof.

I. 0-Order Relevance Logic

I.1 Introduction

Relevance logic was apparently first treated by Ackermann and Church and has been intensively refined and developed mainly by Anderson and Belnap [AB75]. The main motivation was to avoid certain paradoxes of implication which are present in classical formal logic. For example, ex falso quod libet, which has been shown to lead to all sorts of unpleasant results, is a valid formula in classical logic. The cause of the matter seems to lie in the definition of implication, which leads to other counterintuitive properties as well. For example the two true sentences "Grass is green" and "Two plus two equals four" lead to the true sentence "Grass is green implies two plus two equals four". Yet another feature of classical logic which does seem to be adequate for many applications lies in negation: by not denying a statement it cannot be concluded that the statement is affirmed.

I.2 Syntax of 0-order Relevance Logic RL

We closely follow [RM72] in presenting the syntax and semantics of RL. RL is built up syntactically in the usual way from a denumerably infinite set S of sentential parameters, the unary connective - (called relevant negation), the usual truthfunctional connectives & and v, the binary connective → (called relevant implication).

The axiom schemata of RL are:

A1. $A \to A$
A2. $A \to ((A \to B) \to B)$

A3. $(A \to B) \to ((B \to C) \to (A \to C))$
A4. $(A \to (A \to B)) \to (A \to B)$
A5. $A \& B \to A$
A6. $A \& B \to B$
A7. $(A \to B) \& (A \to C) \to (A \to B \& C)$
A8. $A \to A \vee B$
A9. $B \to A \vee B$
A10. $(A \to C) \& (B \to C) \to (A \vee B \to C)$
A11. $A \& (B \vee C) \to A \& B \vee A \& C$
A12. $(A \to -B) \to (B \to -A)$
A13. $--A \to A$

and the deduction rule schemata are:

R1. From $A \to B$ and A, infer B
R2. From A and B infer A & B

I.3 Semantics of RL

A relevant model structure (rms) is a quadruple (0,K,R,*), where K is a set of objects called set-ups, $0 \in K$, R is a ternary relation on K, and * is a unary operation on K, satisfying the postulates p1-p6 below.

A binary relation $<$ and a quaternary relation R^2 on K are defined as abbreviations:

For all a,b,c,d in K

d1. $a < b$:= R0ab
d2. R^2abcd := $\exists$ x (Rabx and Rxcd and x in K)

The postulates for all a,b,c,d in K are

p1. R0aa
p2. Raaa
p3. $R^2abcd \Rightarrow R^2acbd$
p4. $R^20abc \Rightarrow Rabc$
p5. $Rabc \Rightarrow Rca^*b^*$
p6. $a^{**} = a$

The definition d2 is generalized in the following way:

d3. $\underbrace{R^nabc...def}_{n+2} := \exists\, x\underbrace{R^rabc...x}_{r+2}$ and $\underbrace{R^{n-r}x...def}_{n-r+2}$

and x in K and $1 \leq r \leq n-1$

From the postulates the following lemmas can easily be derived:

L1. $R^nabc...def \Leftrightarrow R^{n+1}0abc...def$
L2. All R^n-relations, which can be formed by a permutation of the first n+1 set-ups in the relation $R^nabc...def$, can be shown to follow from $R^nabc...def$

L3. Each relation R^n implies a relation R^{n+1} by replacement of any set-up a in R^n by aa.

A valuation **v** of RL in an rms (0,K,R,* is a function from S x K to the set of classical truth-values {T,F}, which satisfies the following condition for all p in S and a,b in K:

(1) a < b and **v**(p,a) = T => **v** p,b) = T

The interpretation I associated with **v** is a function from FORMULAS x K to {T,F} which satisfies the following conditions for all p in S, A,B in FORMULAS, and a in K:

(i)	I(p,a) = **v**(p,a)	
(ii)	I(A & B,a) = T	iff I(A,a) = T and I(B,a) = T
(iii)	I(A v B,a) =	= T or I(B,a) = T
(iv)	I(A → B,a) = T	iff, for all a,b,c in K, Rabc and I(A,b) = T => I(B,c) = T
(v)	I(-A,a) = T	iff I(A,a*) = F

A formula A is true on a valuation **v**, or on the associated interpretation I at a set-up a in K, if I(A,a) = T, otherwise A is false at a.

A is verified on **v** or on the associated I if I(A,0) = T and otherwise A is falsified on **v**.

A is valid in an rms iff A is verified on all valuations therein.

A is R-valid iff A is valid in all rms.

A entails B on a valuation **v** or in the associated Interpretation I provided that, for all a in K, I(A,a) = T => I(B,a) = T

A entails B in a rms iff A entails B on all valuations therein.

A R-entails B iff A entails B in all rms.

The following 3 lemmas and their proofs can be found in [RM72]

L4. a < b and I(A,a) = T => I(A,b) = T

L5. A entails B on **v** iff A → B is verified on **v**

L6. A entails B in **v** iff A → B is verified on **v**. So A entails B in an rms iff A → B is valid therein, and A R-entails B iff A → B is R-valid.

Fundamental Lemmas

From the definition of an interpretation the following lemmas result:

(a)	I(A → B,x) = T and ∀y,z in K Rxyz => (I(A,y) = F or I(B,z) = T)
(b)	I(A → B,x) = F => (∃ y,z in K Rxyz and I(A,y) = T and I(B,z) = F)
(c)	I(A & B,x) = T => I(A,x) = T and I(B,x) = T
(d)	I(A & B,x) = F => I(A,x) = F or I(B,x) = F
(e)	I(A v B,x) = T => I(A,x) = T or I(B,x) = T
(f)	I(A v B,x) = F => I(A,x) = F and I(B,x) = F
(g)	I(-A,x) = T => I'A,x*) = F
(h)	I(-A,x) = F => I(A,x*) = T

I.4 An Example

To illustrate the non-triviality in applying the fundamental lemmas in order to show the R-validity or R-invalidity of a formula of RL we give an example "done by intuition". We use the notation (line number, applied "rule", line no or nos to which the rule is applied.) and e.g. "A,F in a" for I(A,a) = F. We take the contraction axiom A4. The tree must close in the classical (or extensional) sense, i.e. in each path in the tree must be an atom and its extensionally negated atom both holding in the same set-up. In order to achieve this it is necessary to manipulate the relations determined by the use of theorem (b) in the example R0ab and Rbcd, so that suitable relations are created in order to close the tree extensionally, i.e. we need to find in this case Racx and Rxcd. In other words, the tree needs to be closed intensionally as well as extensionally.

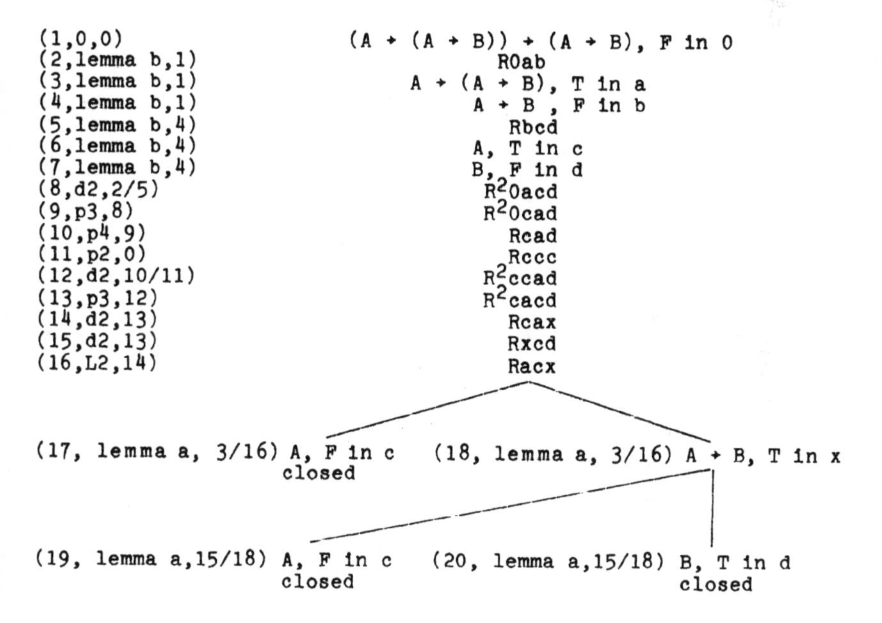

II. The Problem

The problem we have dealt with is known as "converse of contraction":

(1) $(A \to (B \to B)) \to (A \to (A \to (B \to B)))$

and is given in [AB 75] on page 96 as being a theorem of $T_{\to}$, a subsystem of RL.

Now notice that

(I) $\not\vdash_{RL} (A \to B) \to A \to (A \to B)$

Whereas

(II) $\vdash_{RL}$ (A → (A → B)) → (A → B) (Contraction Axiom)

(II) concurs with one of the intuitions of relevance implication → namely that in X → Y, Y cannot contain more information than is being put into it by X. (I) on the other hand contradicts this intuition. However it is not a theorem of RL. The oddity about RL though is that (1) is a theorem <u>and</u> contradicts the intuition.

It was Prof. N.D. Belnap, one of the fathers of relevance logic, who gave us this problem. Therefore we call it sometimes Belnap's theorem. The proof is so tricky that even Belnap himself had forgotten how to do it. Dr. Michael Mc.Robbie's theorem prover at La Trobe University, Melbourne, Australia solved it in less than a second, but his system works proof-theoretically and is especially designed to handle RL and the related systems, i.e. it is a special purpose ATP. Dr. Bob Meyer at ANU is one of the very few specialists in relevance logic and he was able to solve the problem by hand using semantic tableaux. An ordinary resolution proof, which is much more complicated than a proof-theoretical solution, however, was not known so far.

The real difficulty and hence the significance for resolution based theorem provers (and for human logicians) is due to the enormous search space which is generated by the postulate p3:

∀abcdx Rabx & Rxcd => ∃y Racy & Rybd

Using the symmetry of R (L2) and p2: ∀x Rxxx with this axiom one can deduce from each clause of the form Rabc <u>eight</u> new clauses to which d2 can immediately be applied again. A rough estimate yields a total of several million clauses which have to be generated before solving (1) using a straightforward search strategy.

III. <u>Solution of the Problem by the MKR-Procedure</u> [KMR 83]

We need the following axioms and lemmas for the axiomization of relevance implication:
(In the sequel "-" is the predicate logic negation sign)

p1:	∀a	R0aa
p2:	∀a	Raaa
p3:	∀a,b,c,d,x	Rabx AND Rxcd => ∃y Racy AND Rybd
p4:	∀a,b,c,x	R0ax AND Rxbc => Rabc
L2:	∀a,b,c	Rabc => Rbac

Lemma a: ∀C,D,x,y,z T(→(C D) x) AND Rxyz => -T(C y) OR T(D z)
Lemma b: ∀C,D,x -T(→(C D) x) => ∃y,z Rxyz AND T(C y) AND -T(D z)

where T(→(C D) x) means I(C → D, x) = T
and - T(→(C D) x) means I(C → D, x) = F

In this notation the theorem is
T(→(→(A →(B B)) → (A →(A →(B B)))) 0)

The following list is an original protocol of the proof found by the MKR-Procedure with some minor manual modifications to make it more readable and to fit it into the format of this paper.

```
*****************************************************************
*                                                               *
*     MARKGRAF KARL REFUTATION PROCEDURE, VERSION 14-MAR-83     *
*                                                               *
*     DATE:     28-MAR-83       19:47:07                        *
*                                                               *
*****************************************************************
```

FORMULAE GIVEN TO THE THEOREM PROVER:

AXIOMS:

```
∀ X              R(0 X X)
∀ X              R(X X X)
∀ V,W,X,Y,Z      R(V W X) AND R(X Y Z) IMPL ∃ U R(V Y U) AND
                                                    R(U  W  Z)
∀ W,X,Y,Z        R(0 W X) AND R(X Y Z) IMPL R(W Y Z)
∀ C,D,W,X,Y,Z    T(→(C D) W) AND R(W X Y) IMPL -T(C X) OR T(D Y)
∀ C,D,W,         -T(→(C D) W) IMPL ∃ X,Y R(W X Y) AND
                                            T(C X) AND -T(D Y)
                  SYMMETRIC (R)
```

```
THEOREM:          T(→(→(A →(B B)) → (A →(A →(B B)))) 0)
```

```
*********************
*   INITIAL GRAPH   *
*********************
```

AXIOMS:

```
A1:  ∀ X              R(0 X X)
A2:  ∀ X              R(X X X)
A3:  ∀ V,W,X,Y,Z      -R(V W X) OR -R(X Y Z) OR R(V Y F_1(Z W Y V))
A4:  ∀ V,W,X,Y,Z      -R(V W X) OR -R(X Y Z) OR R(F_1(Z W Y V) W Z)
A5:  ∀ W,X,Y,Z        -R(0 W X) OR -R(X Y Z) OR R(W Y Z)
A6: ∀ V,W,X,Y,Z     -T(→(V  W)  X)  OR  -R(X  Y  Z)  OR  -T(V  Y)  OR
                                                           T(W Z)
A7:  ∀ X,Y,Z        T(→(X Y) Z) OR R(Z F_3(Z Y X) F_2(Z Y X)))
A8:  ∀ X,Y,Z        T(→(X Y) Z) OR T(X F_3(Z Y X))
A9:  ∀ X,Y,Z        T(→(X Y) Z) OR -T(Y F_2(Z Y X))
```

THEOREM:

```
T10:       -T(→(→(A →(B B)) → (A →(A →(B B)))) 0)
```

```
ABBREVIATIONS: a = F_3(0  →(A →(A →(B B))) →(A →(B B)))
               a'= F_2(0  →(A →(A →(B B))) →(A →(B B)))
               b = F_3(a' →(A →(B B)) A)
               c = F_2(a' →(A →(B B)) A)
               d = F_3(c  →(B B) A)
               e = F_2(c  →(B B) A)
               f = F_3(e  B B)
               g = F_2(e  B B)
               h = F_1(c a b a)
               i = F_1(e h d a)
               j = F_1(g i f h)
```

```
T10 + A9   => R1 : -T(→(A →(A →(B B))) a´)
R1  + A9   => R2 : -T(→(A →(B B)) c)
R2  + A9   => R3 : -T(→(B B) e)
R3  + A9   => R4 : -T(B g)
R3  + A8   => R5 : T(B f)
R1  + A8   => R6 : T(B b)
T10 + A8   => R7 : T(→(A →(B B)) a)
T10 + A7   => R8 : R(0 a a´)
R1  + A7   => R9 : R(a´ b c)
A5  + R9   => R10: R(a b c) OR -R(0 a a´)
R10 + R8   => R11: R(a b c)
A3  + A2   => R12: R(a b h) OR -R(a b c)
R12 + R11  => R13: R(a b h)
A6  + R13  => R14: T(→(B B) h) OR -T(A b) OR -T(→(A →(B B)) a)
R14 + R7   => R15: T(→(B B) h) OR -T(A b)
R15 + R6   => R16: T(→(B B) h)
R3  + A7   => R17: R(e f g)
R2  + A7   => R18: R(c d e)
A4  + A2   => R19: R(h a c) OR -R(a b c)
R19 + R11  => R20: R(h a c)
A4  + R20  => R21: R(i h e) OR -R(c d e)
R21 + R18  => R22: R(i h e)
A3  + R22  => R23: R(h f j) OR -R(e f g)
R23 + R17  => R24: R(h f j)
A6  + R24  => R25: T(B j) OR -T(B f) OR -T(→(B B) h)
R25 + R16  => R26: T(B j) OR -T(B f)
R26 + R5   => R27: T(B j)
R2  + A8   => R28: T(A d)
A3  + R20  => R29: R(a d i) OR -R(c d e)
R29 + R18  => R30: R(a d i)
A6  + R30  => R31: T(→(B B) i) OR -T(A d) OR -T(→(A →(B B)) a)
R31 + R7   => R32: T(→(B B) i) OR -T(A d)
R32 + R28  => R33: T(→(B B) i)
A4  + R22  => R34: R(j i g) OR -R(e f g)
R34 + R17  => R35: R(j i g)
A6  + R35  => R36: T(B g) OR -T(B j) OR -T(→(B B)) i)
R36 + R33  => R37: T(B g) OR -T(B j)
R37 + R27  => R38: T(B g)
R4  + R38  => R39: EMPTY

GRAPH SUCCESSFULLY REFUTED.

CPU-TIME USED:                      516.33 SECONDS
NUMBER OF LINKS GENERATED:          311
NUMBER OF CLAUSES GENERATED:        49
NUMBER OF CLAUSES DELETED:          15
LEVEL OF PROOF:                     14
G-PENETRANCE:                       0.97
D-PENETRANCE:                       1.00

THE FOLLOWING CLAUSES WERE USED IN THE PROOF:

A8 A9 T10 R1 R2 R3 R5 R6 R7 A7 R8 R9 A5 R10 R11 A2 A3 R12 R13 A6
R14 R15 R16 R17 R18 A4 R19 R20 R21 R22 R23 R24 R25 R26 R27 R28
R29 R30 R31 R32 R33 R34 R35 R36 R37 R38 R4 R39.

END OF PROOF:        28-MAR-83       20:13:37
```

To demonstrate one of the difficulties which is hidden by the use of abbreviations in the protocol, we list an unabbreviated version of only one of the resolvents:

```
R35:
R(F_1(F_2(F_2(F_2(F_2(0 →(A →(A →(B B))) →(A →(B B)))
                  →(A →(B B))
                  A)
              →(B B)
              A)
          B
          B)
      F_1(F_2(F_2(F_2(0 →(A →(A →(B B))) →(A →(B B)))
                  →(A →(B B))
                  A)
              →(B B)
              A)
          F_1(F_2(F_2(0 →(A →(A →(B B))) →(A →(B B)))
                  →(A →(B B))
                  A)
              F_3(0 →(A →(A →(B B))) →(A →(B B)))
              F_3(F_2(0 →(A →(A →(B B))) →(A →(B B)))
                  →(A →(B B))
                  A)
              F_3(0 →(A →(A →(B B))) →(A →(B B)))
          F_3(F_2(F_2(0 →(A →(A →(B B))) →(A →(B B)))
                  →(A →(B B))
                  A)
              →(B B))
              A)
          F_3(0 →(A →(A →(B B))) →(A →(B B))))
      F_3(F_2(F_2(F_2(0 →(A →(A →(B B))) →(A →(B B)))
                      →(A →(B B))
                      A)
                  →(B B)
                  A)
              B
              B)
          F_1(F_2(F_2(0 →(A →(A →(B B))) →(A →(B B)))
                  →(A →(B B))
                  A)
              F_3(0 →(A →(A →(B B))) →(A →(B B)))
              F_3(F_2(0 →(A →(A →(B B))) →(A →(B B)))
                  →(A →(B B))
                  A)
              F_3(0 →(A →(A →(B B))) →(A →(B B)))))
      F_1(F_2(F_2(F_2(0 →(A →(A →(B B))) →(A →(B B)))
                  →(A →(B B))
                  A)
              →(B B))
              A)
          F_1(F_2(F_2(0 →(A →(A →(B B))) →(A →(B B)))
                  →(A →(B B))
                  A)
              F_3(0 →(A →(A →(B B))) →(A →(B B)))
              F_3(F_2(0 →(A →(A →(B B))) →(A →(B B)))
                  →(A →(B B))
                  A)
              F_3(0 →(A →(A →(B B))) →(A →(B B))))
```

```
     F_3(F_2(F_2(0 →(A →(A →(B B))) →(A →(B B)))
                   →(A →(B B))
                   A)
           →(B B))
           A)
     F_3(0 →(A →(A →(B B))) →(A →(B B))))
F_2(F_2(F_2(F_2(0 →(A →(A →(B B))) →(A →(B B)))
                   →(A →(B B))
                   A)
           →(B B)
           A)
     B
     B))
```

It is most interesting to note that the proof is almost identical to the one found by Bob Meyer, only the order of the deductions is different. A more natural order is used below where we translated the proof back into a relevance logic notation.

```
T10:          (A → (B → B)) → (A → (A → (B → B))) F in 0

T10,A8          => R7:    A → (B → B)          T in a
T10,A9          => R1:    A → (A → (B → B))    T in a´

T10,A7          => R8:    R0aa´
R1,A7           => R9:    Ra´bc
A5,R8,R9        => R11:   Rabc

R1,A8           => R6:    A                    T in b
R1,A9           => R2:    A → (B → B)          F in c
R2,A8           => R28:   A                    T in d
R2,A9           => R3:    B → B                F in e
R3,A8           => R5:    B                    T in f
R3,A9           => R4:    B                    F in g

R2,A7           => R18:   Rcde
R3,A7           => R17:   Refg
A3,A2,R11       => R13:   Rabh
A4,A2,R11       => R20:   Rhac
A3,R18,R20      => R30:   Radi
A4,R18,R20      => R22:   Rihe
A3,R17,R22      => R24:   Rhfj
A4,R17,R22      => R35:   Rjig

A6,R7,R28,R30   => R33:   B → B                T in i
A6,R7,R13,R14   => R16:   B → B                T in h
A6,R5,R16,R24   => R27:   B                    T in j
A6,R4,R27,R33,R33     =>           EMPTY
                                   *****
```

IV. How did the MKR-Procedure Manage this Problem?

The deduction machine of the MKR-Procedure is primarily based on an extension of Kowalski's "Connection Graph Calculus" [KO 75]. In this calculus the clauses are represented as nodes of a graph, where the edges (links) represent the deduction possibilities. Presently there are two main components responsible for the strength of the system.

1. A powerful reduction module which is able to detect redundant clauses and links of various types (subsumption, tautologies, purity etc.) [E 81], [WA 81]. Link reduction is as important as clause reduction, because each deleted link facilitates considerably the work of the second main component of the system:

2. The Terminator Module [AO 83]:
This module is designed to extract refutation trees from unit refutable clause graphs [HR 78]. The extraction mechanism itself can be compared to UR-Resolution, where one resolves between a non-unit clause and as many unit clauses as necessary to produce a new unit clause. The difference lies in the exploitation of the information of the clause graph by the Terminator module. However all the strategies and heuristics for guiding the search for a refutation tree can be used in UR-Resolution too, and vice versa.

The proof of Belnap's theorem was found by the Terminator module, but not without an additional mechanism, which we had to implement for solving this problem. The difficulties which could not be overcome even by the AURA system at Argonne National Laboratory [OW 83] come from the axioms A3 and A4 together with A2 and the symmetry of R.

```
A3:  -Rvwx    -Rxyz    RvyF_1(zwyv)
A4:  -Rvwx    -Rxyz    RF_1(zwyv)wz
A2:   Rxxx
```

As soon as a unit clause Rabc (which is equivalent to Rbac) has been generated during the search for a proof, four UR-Resolution steps with A2 and A3 are possible, producing four new unit clauses:

```
U1:  RabF_1(caba)       U2:  RbaF_1(cbab)
U3:  RacF_1(cbca)       U4:  RbcF_1(cacb)
```

and analogously four new unit clauses can be deduced from A4. These eight formulas can be immediately used to produce another 64 new unit clauses etc. All these atoms are in some sense variants of each other, that means they contain the same terms, but with a deeper nested F_1 function.

In this example it is absolutely necessary to restrict the exponential growth of such unit clauses. A limitation of the term depth, however, is not possible, because the limit has to be so high, as one can see from R35, that it is useless.

Therefore a more sophisticated control of the application of A3 and A4 is necessary. We solved this problem with a mechanism which can be applied to arbitrary "generator clauses", i.e. clauses which tend to produce deeply nested terms with the same subterm structure like f(f(f(f ... or f(g(f(g(f(g ... etc.

The Generator Control Mechanism

Definitions:

Let L and K be two literals of a clause with the same predicate symbol and opposite sign. We call L a generator literal if there exists a substitution u with u|K| == |L| (== means equal after variable renaming), and u contains a component v ← f(t1) and t1 is a termlist with variables occurring in it. (The third literals of A3 and A4 and the first literal of A6 are of this type.)

A unit clause is called a descendant of a literal L, if it is an instance of L, produced by an UR-Resolution step.

Example 1

```
A3:        <-Rvwx            -Rxyz            RvyF_1(zwyv)>
              |                 |
              |                 |
A2:         <Rxxx>    U1:    <Rabc>
```

U2 = <RabF_1(caba)> is a descendant of the third literal of A3.

Unit clauses which are used in the same UR step are called partner units. In the example above U1 is the partner of A2 and vice versa.

Furthermore A2 and U1 are called parents of U2.

The Mechanism:

We attach to each unit clause a mark which may be NIL or an integer > 0:

Every descendant of a non generator literal is marked with NIL.

Descendants of generator literals are marked as follows:

1. If at least one parent unit PU is marked with a number n ≠ NIL and <u>every</u> partner unit of PU is a parent unit of PU as well, then U is marked with n-1.

2. Otherwise, if at least one parent unit PU is marked with a number n ≠ NIL then U is marked with n-m, where m is the number of partners of PU which are also marked with a number ≠ NIL.

3. In any other case U is marked with a number LIMIT > 0, given by the user.

In the "converse of contraction" -example we have set LIMIT to 1. In this case the mechanism works as follows:

- There is no influence to deductions from A6.
- UR-Resolutions with A3 and A4 like that one in example 1 are allowed, if U1 is marked with NIL. U2 however will be marked with n=1. Therefore a further UR-Resolution with A3, A2 and U2 or A4, A2 and U2 is forbidden, because A2 is a partner of U2 as well as one of its parents.
- UR-Resolutions with A3 or A4 like

Example 2

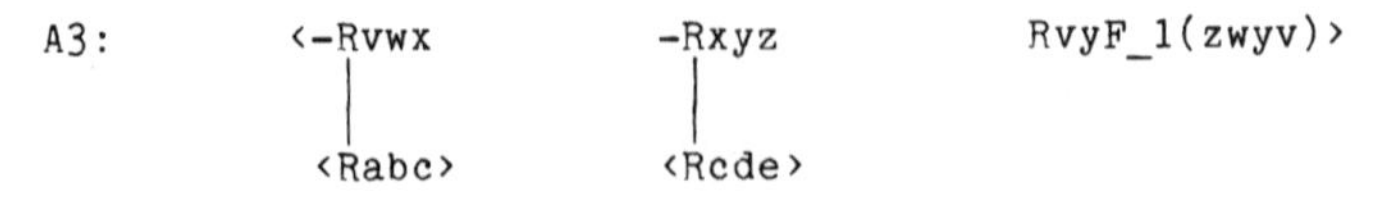

where both unit clauses are marked with 1 are forbidden, because rule 2 yields n-m = 0 in this case.

The user can control the search space restriction with the parameter LIMIT. Mainly this parameter defines how deeply a function may be nested in a direct sequence.

If for example we have a clause

<-Px Pf(x)>

and LIMIT is set to 2, then a term f(f(f(... derived from this clause is not possible, but a term f(f(g(f(f ... is allowed because all intermediate steps with other clauses restore the mark of the descendant units to NIL.

General Utility of the Mechanism

A general valuation of the utility of this method is very difficult, because there exists no representative set of examples for statistical investigations:

The following assertions however can be made:

- Belnap's problem is almost unsolvable without the mechanism.
- We have other examples of similar complexity which have not been solved so far by our theorem prover; however the form of the axioms give rise to the assumption, that the mechanism can help solving them, too.
- We inspected the comparative study of Minker and Wilson [MW 76]. 45 of altogether 98 examples in this study contain generator clauses; mainly substitution axioms for the equality predicate. Tests with these examples yielded reductions of the search space of up to 70%. The reductions are more significant for problems with a large prove depth and with a large percentage of generator clauses among the other axioms.

Finally we can say, that the generator control mechanism is useless for textbook examples. Its application can however be very helpful in case of really difficult problems.

References

[AB 75] A.R. Anderson, N.D. Belnap
Entailment: The Logic of Relevance and Necessity
Vol. 1, Princeton University Press, 1975

[AO 83] G. Antoniou, H.J. Ohlbach
Terminator
Proc. of Eights IJCAI, Karlsruhe 1983

[E 81] N. Eisinger
Subsumption and Connection Graphs
Proc. of GWAI-81, Bad Honnef 1981

[HR 78] M.C. Harrison, N. Rubin
Another Generalization of Resolution
JACM 25:3 341-351 1978

[KMR83] Karl Mark G. Raph
The Markgraf Karl Refutation Procedure: Spring 1983
University of Karlsruhe, Interner Bericht

[KO 75] R. Kowalski
A Proof Procedure Using Connection Graphs
JACM 22:4 1975

[MW 76] G.A. Wilson, J. Minker
Resolution Refinement and Search Strategies: A Comparative Study. Trans. on Comp., vol. C-25, no.8, 1976

[OW 83] R. Overbeck, L. Wos
Private Communication (IJCAI 83)

[WA 81] C. Walther
Elimination of Redundant Links in Extended Connection Graphs.
University of Karlsruhe, Interner Bericht 10/81

[RM 72] R. Routley, R. Meyer
The Semantics of Entailment in Truth, Syntax and Modality.
Leblanc(ed) North-Holland 1972

l. 117: Fundamentals of Computation Theory. Proceedings, 1981. lited by F. Gécseg. XI, 471 pages. 1981.

l. 118: Mathematical Foundations of Computer Science 1981. oceedings, 1981. Edited by J. Gruska and M. Chytil. XI, 589 pages. 81.

l. 119: G. Hirst, Anaphora in Natural Language Understanding: Survey. XIII, 128 pages. 1981.

l. 120: L. B. Rall, Automatic Differentiation: Techniques and Applitions. VIII, 165 pages. 1981.

. 121: Z. Zlatev, J. Wasniewski, and K. Schaumburg, Y12M Soon of Large and Sparse Systems of Linear Algebraic Equations. 128 pages. 1981.

. 122: Algorithms in Modern Mathematics and Computer Science. ceedings, 1979. Edited by A. P. Ershov and D. E. Knuth. XI, 487 ges. 1981.

. 123: Trends in Information Processing Systems. Proceedings, 1. Edited by A. J. W. Duijvestijn and P. C. Lockemann. XI, 349 ges. 1981.

. 124: W. Polak, Compiler Specification and Verification. XIII, 9 pages. 1981.

. 125: Logic of Programs. Proceedings, 1979. Edited by E. geler. V, 245 pages. 1981.

. 126: Microcomputer System Design. Proceedings, 1981. Edited M. J. Flynn, N. R. Harris, and D. P. McCarthy. VII, 397 pages. 1982.

l. 127: Y.Wallach, Alternating Sequential/Parallel Processing. 329 pages. 1982.

128: P. Branquart, G. Louis, P. Wodon, An Analytical Descrip- of CHILL, the CCITT High Level Language. VI, 277 pages. 1982.

. 129: B. T. Hailpern, Verifying Concurrent Processes Using nporal Logic. VIII, 208 pages. 1982.

. 130: R. Goldblatt, Axiomatising the Logic of Computer Programg. XI, 304 pages. 1982.

. 131: Logics of Programs. Proceedings, 1981. Edited by D. Kozen. 429 pages. 1982.

132: Data Base Design Techniques I: Requirements and Logical uctures. Proceedings, 1978. Edited by S.B. Yao, S.B. Navathe, Weldon, and T.L. Kunii. V, 227 pages. 1982.

133: Data Base Design Techniques II: Proceedings, 1979. ted by S.B. Yao and T.L. Kunii. V, 229–399 pages. 1982.

134: Program Specification. Proceedings, 1981. Edited by J. unstrup. IV, 426 pages. 1982.

135: R.L. Constable, S.D. Johnson, and C.D. Eichenlaub, An oduction to the PL/CV2 Programming Logic. X, 292 pages. 1982.

136: Ch. M. Hoffmann, Group-Theoretic Algorithms and Graph norphism. VIII, 311 pages. 1982.

137: International Symposium on Programming. Proceedings, 2. Edited by M. Dezani-Ciancaglini and M. Montanari. VI, 406 es. 1982.

138: 6th Conference on Automated Deduction. Proceedings, 2. Edited by D.W. Loveland. VII, 389 pages. 1982.

139: J. Uhl, S. Drossopoulou, G. Persch, G. Goos, M. Dausmann, Vinterstein, W. Kirchgässner, An Attribute Grammar for the antic Analysis of Ada. IX, 511 pages. 1982.

40: Automata, Languages and programming. Edited by M. Nielnd E.M. Schmidt. VII, 614 pages. 1982.

41: U. Kastens, B. Hutt, E. Zimmermann, GAG: A Practical piler Generator. IV, 156 pages. 1982.

42: Problems and Methodologies in Mathematical Software uction. Proceedings, 1980. Edited by P.C. Messina and A. Murli. 71 pages. 1982.

Vol. 143: Operating Systems Engineering. Proceedings, 1980. Edited by M. Maekawa and L.A. Belady. VII, 465 pages. 1982.

Vol. 144: Computer Algebra. Proceedings, 1982. Edited by J. Calmet. XIV, 301 pages. 1982.

Vol. 145: Theoretical Computer Science. Proceedings, 1983. Edited by A.B. Cremers and H.P. Kriegel. X, 367 pages. 1982.

Vol. 146: Research and Development in Information Retrieval. Proceedings, 1982. Edited by G. Salton and H.-J. Schneider. IX, 311 pages. 1983.

Vol. 147: RIMS Symposia on Software Science and Engineering. Proceedings, 1982. Edited by E. Goto, I. Nakata, K. Furukawa, R. Nakajima, and A. Yonezawa. V. 232 pages. 1983.

Vol. 148: Logics of Programs and Their Applications. Proceedings, 1980. Edited by A. Salwicki. VI, 324 pages. 1983.

Vol. 149: Cryptography. Proceedings, 1982. Edited by T. Beth. VIII, 402 pages. 1983.

Vol. 150: Enduser Systems and Their Human Factors. Proceedings, 1983. Edited by A. Blaser and M. Zoeppritz. III, 138 pages. 1983.

Vol. 151: R. Piloty, M. Barbacci, D. Borrione, D. Dietmeyer, F. Hill, and P. Skelly, CONLAN Report. XII, 174 pages. 1983.

Vol. 152: Specification and Design of Software Systems. Proceedings, 1982. Edited by E. Knuth and E. J. Neuhold. V, 152 pages. 1983.

Vol. 153: Graph-Grammars and Their Application to Computer Science. Proceedings, 1982. Edited by H. Ehrig, M. Nagl, and G. Rozenberg. VII, 452 pages. 1983.

Vol. 154: Automata, Languages and Programming. Proceedings, 1983. Edited by J. Diaz. VIII, 734 pages. 1983.

Vol. 155: The Programming Language Ada. Reference Manual. Approved 17 February 1983. American National Standards Institute, Inc. ANSI/MIL-STD-1815A-1983. IX, 331 pages. 1983.

Vol. 156: M.H. Overmars, The Design of Dynamic Data Structures. VII, 181 pages. 1983.

Vol. 157: O. Østerby, Z. Zlatev, Direct Methods for Sparse Matrices. VIII, 127 pages. 1983.

Vol. 158: Foundations of Computation Theory. Proceedings, 1983. Edited by M. Karpinski, XI, 517 pages. 1983.

Vol. 159: CAAP'83. Proceedings, 1983. Edited by G. Ausiello and M. Protasi. VI, 416 pages. 1983.

Vol. 160: The IOTA Programming System. Edited by R. Nakajima and T. Yuasa. VII, 217 pages. 1983.

Vol. 161: DIANA, An Intermediate Language for Ada. Edited by G. Goos, W.A. Wulf, A. Evans, Jr. and K.J. Butler. VII, 201 pages. 1983.

Vol. 162: Computer Algebra. Proceedings, 1983. Edited by J.A. van Hulzen. XIII, 305 pages. 1983.

Vol. 163: VLSI Engineering. Proceedings. Edited by T.L. Kunii. VIII, 308 pages. 1984.